T0222679

PRINCIPLES and TECHNIQUES in COMBINATORICS
Solutions Manual

PRINCIPLES and TECHNIQUES
in COMBINATORICS
Solutions Manual

FOO Kean Pew

LIN Mingyan Simon

University of Illinois at Urbana-Champaign, USA

World Scientific

NEW JERSEY · LONDON · SINGAPORE · BEIJING · SHANGHAI · HONG KONG · TAIPEI · CHENNAI · TOKYO

Published by

World Scientific Publishing Co. Pte. Ltd.

5 Toh Tuck Link, Singapore 596224

USA office: 27 Warren Street, Suite 401-402, Hackensack, NJ 07601

UK office: 57 Shelton Street, Covent Garden, London WC2H 9HE

Library of Congress Cataloging-in-Publication Data

Names: Lin, Mingyan Simon, author. | Foo, Kean Pew, author. | supplement to (work):
    Chen, Chuan Chong. Principles and techniques in combinatorics.
Title: Principles and techniques in combinatorics : solutions manual / by
    Simon Lin Mingyan, Kean Pew Foo.
Description: New Jersey : World Scientific, [2018] | Solution manual to a textbook published in
    1992 under title Principles and techniques in combinatorics by Chen Chuan-Chong.
Identifiers: LCCN 2018032351 | ISBN 9789813238848 (paperback : alk. paper)
Subjects: LCSH: Combinatorial analysis--Problems, exercises, etc. |
    Combinatorial analysis--Textbooks. | LCGFT: Problems and exercises. | Textbooks.
Classification: LCC QA164.5 .L56 2018 | DDC 511/.6--dc23
LC record available at https://lccn.loc.gov/2018032351

British Library Cataloguing-in-Publication Data

A catalogue record for this book is available from the British Library.

For any available supplementary material, please visit
https://www.worldscientific.com/worldscibooks/10.1142/10955#t=suppl

Printed in Singapore

Foreword

This solutions manual to the textbook has taken a long and meandering path to fruition. Over the years since the book was first published in 1992, there were some attempts to produce a solutions manual. It took the determined effort of a scholar, Foo Kean Pew, to make the first serious crack at it.

Kean Pew graduated with First Class Honours in Mathematics from the National University of Singapore (NUS) in 1991. Given his keen interest in mathematics, he worked on the solutions to the exercises during his spare time while under the employment of Singapore Airlines. Around the end of 1995, he came to us with a draft of the solutions to a number of exercises and a list of the outstanding problems yet to be solved. We had intermittent discussions since then and the last appointment to meet was on 31 July 1996. When he did not turn up, we thought he was busy with work commitments. So we were shocked and saddened when his father Mr Alan Foo visited us at the end of 1997 to tell us that Kean Pew had passed away during surgery for intestinal lymphangiectasia. Little did we know that he had been diagnosed with the condition since his high school days. In his memory, his father donated his pension money to the Singapore Mathematical Society to establish the Foo Kean Pew Memorial Prize.

With Kean Pew passing on, the project languished. Fortunately, along came another brilliant scholar, Simon Lin, who upon hearing about the story behind the project very gallantly agreed to complete it. At that time, he was an undergraduate student in mathematics at NUS. He served as the President of the NUS Mathematics Society during the academic year

2011/12. His enthusiasm and hard work finally saw the completion of this solutions manual. The efforts of these two scholars will undoubtedly be helpful to students of combinatorics.

Chen Chuan-Chong and Koh Khee-Meng

Preface

I first came across the book *Principles and Techniques in Combinatorics* when I was a secondary school student participating in the Mathematical Olympiad training sessions organised by the Singapore Mathematical Society. It was from this blue book (which my peers affectionately referred to due to its blue cover) that I learnt enumerative combinatorics, but it wasn't till my undergraduate days when, with the help of Kean Pew's insightful solutions to the problems, that I had a chance to work through the forest of problems in this book.

This solutions manual contains detailed solutions to each of the 489 problems that appeared in the textbook, and it is my hope that this solutions manual will be a valuable supplement to the textbook proper, and that students, teachers and professors will benefit from the manual itself. The manual itself is primarily written from a self-contained approach, and to that end, I have sought to make the solutions as elementary as possible while invoking as little theorems outside the scope of the textbook as possible. In addition, I have also included the statement of the theorems that we have used in the solutions whenever they appear outside of the scope of the textbook, as well as the sources of the solutions that were taken at the end of the solution. The advanced reader may have realised that it is possible to obtain neater and more elegant solutions to some of the problems in the book using more sophisticated machinery. However, I have eschewed that approach in favour of a more first-principle approach in the hope that the solutions will help the reader further augment their understanding of the recurring concepts in this book.

Over the course of typesetting the solutions, I have also realised that there were minor errors that appeared in some of the problems, as well as the answers to the problems. I have addressed the errors by pointing them out in the solutions.

This solutions manual grew out of a typesetting project while I was an undergraduate student at the National University of Singapore. I would like to thank Professors Wong Yan Loi and Tay Tiong Seng for providing me with the opportunity to typeset the solutions manual, as well as their guidance during my Mathematical Olympiad training days. I would also like to thank Chang Hai Bin and Theo Fanuela for their support in the typesetting of the solutions manual, as well as Dr Chan Onn and Dr Lee Seng Hwang for his tireless help in vetting the solutions manual. Thanks also go to Dr Leong Yu Kiang for his many judicious suggestions that improved the foreword and preface.

Any alternative solutions and corrections to the existing solutions are very welcome.

Lin Mingyan, Simon

Contents

Foreword	v
Preface	vii
Exercise 1	1
Exercise 2	19
Exercise 3	33
Exercise 4	39
Exercise 5	49
Exercise 6	61
Solutions to Exercise 1	77
Solutions to Exercise 2	133
Solutions to Exercise 3	203
Solutions to Exercise 4	229
Solutions to Exercise 5	301
Solutions to Exercise 6	361

Exercise 1

1. Find the number of ways to choose a pair $\{a, b\}$ of distinct numbers from the set $\{1, 2, \ldots, 50\}$ such that

   (i) $|a - b| = 5$;   (ii) $|a - b| \leq 5$.

2. There are 12 students in a party. Five of them are girls. In how many ways can these 12 students be arranged in a row if

   (i) there are no restrictions?
   (ii) the 5 girls must be together (forming a block)?
   (iii) no 2 girls are adjacent?
   (iv) between two particular boys A and B, there are no boys but exactly 3 girls?

3. m boys and n girls are to be arranged in a row, where $m, n \in \mathbf{N}$. Find the number of ways this can be done in each of the following cases:

   (i) There are no restrictions;
   (ii) No boys are adjacent $(m \leq n + 1)$;
   (iii) The n girls form a single block;
   (iv) A particular boy and a particular girl must be adjacent.

4. How many 5-letter words can be formed using $A, B, C, D, E, F, G, H,$ $I, J,$

   (i) if the letters in each word must be distinct?
   (ii) if, in addition, A, B, C, D, E, F can only occur as the first, third or fifth letters while the rest as the second or fourth letters?

5. Find the number of ways of arranging the 26 letters in the English alphabet in a row such that there are exactly 5 letters between x and y.

6. Find the number of *odd* integers between 3000 and 8000 in which no digit is repeated.

7. Evaluate

$$1 \cdot 1! + 2 \cdot 2! + 3 \cdot 3! + \cdots + n \cdot n!,$$

where $n \in \mathbf{N}$.

8. Evaluate

$$\frac{1}{(1+1)!} + \frac{2}{(2+1)!} + \cdots + \frac{n}{(n+1)!},$$

where $n \in \mathbf{N}$.

9. Prove that for each $n \in \mathbf{N}$,

$$(n+1)(n+2) \cdots (2n)$$

is divisible by 2^n. (Spanish Olympiad, 1985)

10. Find the number of common positive divisors of 10^{40} and 20^{30}.

11. In each of the following, find the number of positive divisors of n (inclusive of n) which are multiples of 3:

    (i) $n = 210$;   (ii) $n = 630$;   (iii) $n = 151200$.

12. Show that for any $n \in \mathbf{N}$, the number of positive divisors of n^2 is always odd.

13. Show that the number of positive divisors of "$\underbrace{111 \ldots 1}_{1992}$" is even.

14. Let $n, r \in \mathbf{N}$ with $r \le n$. Prove each of the following identities:

    (i) $P_r^n = n P_{r-1}^{n-1}$,
    (ii) $P_r^n = (n - r + 1) P_{r-1}^n$,
    (iii) $P_r^n = \frac{n}{n-r} P_r^{n-1}$, where $r < n$,
    (iv) $P_r^{n+1} = P_r^n + r P_{r-1}^n$,
    (v) $P_r^{n+1} = r! + r(P_{r-1}^n + P_{r-1}^{n-1} + \cdots + P_{r-1}^r)$.

15. In a group of 15 students, 5 of them are female. If exactly 3 female students are to be selected, in how many ways can 9 students be chosen from the group

    (i) to form a committee?
    (ii) to take up 9 different posts in a committee?

16. Ten chairs have been arranged in a row. Seven students are to be seated in seven of them so that no two students share a common chair. Find the number of ways this can be done if no two empty chairs are adjacent.

17. Eight boxes are arranged in a row. In how many ways can five distinct balls be put into the boxes if each box can hold at most one ball and no two boxes without balls are adjacent?

18. A group of 20 students, including 3 particular girls and 4 particular boys, are to be lined up in two rows with 10 students each. In how many ways can this be done if the 3 particular girls must be in the front row while the 4 particular boys be in the back?

19. In how many ways can 7 boys and 2 girls be lined up in a row such that the girls must be separated by exactly 3 boys?

20. In a group of 15 students, 3 of them are female. If at least one female student is to be selected, in how many ways can 7 students be chosen from the group

   (i) to form a committee?
   (ii) to take up 7 different posts in a committee?

21. Find the number of $(m + n)$-digit binary sequences with m 0's and n 1's such that no two 1's are adjacent, where $n \leq m + 1$.

22. Two sets of parallel lines with p and q lines each are shown in the following diagram:

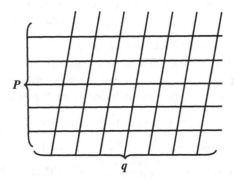

   Find the number of parallelograms formed by the lines.

23. There are 10 girls and 15 boys in a junior class, and 4 girls and 10 boys in a senior class. A committee of 7 members is to be formed from these 2 classes. Find the number of ways this can be done if the committee must have exactly 4 senior students and exactly 5 boys.

24. A box contains 7 identical white balls and 5 identical black balls. They are to be drawn randomly, one at a time without replacement, until the box is empty. Find the probability that the 6th ball drawn is white, while before that exactly 3 black balls are drawn.

25. In each of the following cases, find the number of shortest routes from O to P in the street network shown below:

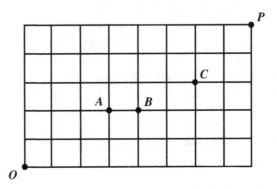

    (i) The routes must pass through the junction A;
   (ii) The routes must pass through the street AB;
  (iii) The routes must pass through junctions A and C;
  (iv) The street AB is closed.

26. Find the number of ways of forming a group of $2k$ people from n couples, where $k, n \in \mathbf{N}$ with $2k \leq n$, in each of the following cases:

    (i) There are k couples in such a group;
   (ii) No couples are included in such a group;
  (iii) At least one couple is included in such a group;
  (iv) Exactly two couples are included in such a group.

27. Let $S = \{1, 2, \ldots, n+1\}$ where $n \geq 2$, and let

$$T = \{(x, y, z) \in S^3 \mid x < z \quad \text{and} \quad y < z\}.$$

Show by counting $|T|$ in two different ways that

$$\sum_{k=1}^{n} k^2 = |T| = \binom{n+1}{2} + 2\binom{n+1}{3}.$$

28. Consider the following set of points in the $x - y$ plane:

$$A = \{(a, b) \mid a, b \in \mathbf{Z}, \ 0 \leq a \leq 9 \quad \text{and} \quad 0 \leq b \leq 5\}.$$

Find

    (i) the number of rectangles whose vertices are points in A;
   (ii) the number of squares whose vertices are points in A.

29. Fifteen points $P_1, P_2, \ldots, P_{15}$ are drawn in the plane in such a way that besides P_1, P_2, P_3, P_4, P_5 which are collinear, no other 3 points are collinear. Find

    (i) the number of straight lines which pass through at least 2 of the 15 points;

    (ii) the number of triangles whose vertices are 3 of the 15 points.

30. In each of the following 6-digit natural numbers:

$$333333, \ 225522, \ 118818, \ 707099,$$

    every digit in the number appears at least twice. Find the number of such 6-digit natural numbers.

31. In each of the following 7-digit natural numbers:

$$1001011, \ 5550000, \ 3838383, \ 7777777,$$

    every digit in the number appears at least 3 times. Find the number of such 7-digit natural numbers.

32. Let $X = \{1, 2, 3, \ldots, 1000\}$. Find the number of 2-element subsets $\{a, b\}$ of X such that the product $a \cdot b$ is divisible by 5.

33. Consider the following set of points in the $x - y$ plane:

$$A = \{(a, b) \mid a, b \in \mathbf{Z} \quad \text{and} \quad |a| + |b| \leq 2\}.$$

    Find

    (i) $|A|$;

    (ii) the number of straight lines which pass through at least 2 points in A;   and

    (iii) the number of triangles whose vertices are points in A.

34. Let P be a convex n-gon, where $n \geq 6$. Find the number of triangles formed by any 3 vertices of P that are pairwise nonadjacent in P.

35. 6 boys and 5 girls are to be seated around a table. Find the number of ways that this can be done in each of the following cases:

    (i) There are no restrictions;

    (ii) No 2 girls are adjacent;

    (iii) All girls form a single block;

    (iv) A particular girl G is adjacent to two particular boys B_1 and B_2.

36. Show that the number of r-circular permutations of n distinct objects, where $1 \leq r \leq n$, is given by $\frac{n!}{(n-r)! \cdot r}$.

37. Let $k, n \in \mathbb{N}$. Show that the number of ways to seat kn people around k distinct tables such that there are n people in each table is given by $\frac{(kn)!}{n^k}$.

38. Let $r \in \mathbb{N}$ such that

$$\frac{1}{\binom{9}{r}} - \frac{1}{\binom{10}{r}} = \frac{11}{6\binom{11}{r}}.$$

Find the value of r.

39. Prove each of the following identities:

(a) $\binom{n}{r} = \frac{n}{r}\binom{n-1}{r-1}$, where $n \geq r \geq 1$;

(b) $\binom{n}{r} = \frac{n-r+1}{r}\binom{n}{r-1}$, where $n \geq r \geq 1$;

(c) $\binom{n}{r} = \frac{n}{n-r}\binom{n-1}{r}$, where $n > r \geq 0$;

(d) $\binom{n}{m}\binom{m}{r} = \binom{n}{r}\binom{n-r}{m-r}$, where $n \geq m \geq r \geq 0$.

40. Prove the identity $\binom{n}{r} = \binom{n}{n-r}$ by (BP).

41. Let $X = \{1, 2, \ldots, n\}$, $\mathcal{A} = \{A \subseteq X \mid n \notin A\}$, and $\mathcal{B} = \{A \subseteq X \mid n \in A\}$. Show that $|\mathcal{A}| = |\mathcal{B}|$ by (BP).

42. Let $r, n \in \mathbb{N}$. Show that the product

$$(n+1)(n+2)\cdots(n+r)$$

of r consecutive positive integers is divisible by $r!$.

43. Let A be a set of kn elements, where $k, n \in \mathbb{N}$. A k-grouping of A is a partition of A into k-element subsets. Find the number of different k-groupings of A.

44. Twenty five of King Arthur's knights are seated at their customary round table. Three of them are chosen – all choices of three being equally likely – and are sent off to slay a troublesome dragon. Let P be the probability that at least two of the three had been sitting next to each other. If P is written as a fraction in lowest terms, what is the sum of the numerator and denominator? (AIME, 1983/7) (Readers who wish to get more information about the AIME may write to Professor Walter E. Mientka, AMC Executive Director, Department of Mathematics & Statistics, University of Nebraska, Lincohn, NE 68588-0322, USA.)

45. One commercially available ten-button lock may be opened by depressing – in any order – the correct five buttons. The sample shown

below has $\{1, 2, 3, 6, 9\}$ as its combination. Suppose that these locks are re-designed so that sets of as many as nine buttons or as few as one button could serve as combinations. How many additional combinations would this allow? (AIME, 1988/1)

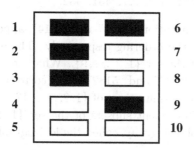

46. In a shooting match, eight clay targets are arranged in two hanging columns of three each and one column of two, as pictured. A marksman is to break all eight targets according to the following rules: (1) The marksman first chooses a column from which a target is to be broken. (2) The marksman must then break the lowest remaining unbroken target in the chosen column. If these rules are followed, in how many different orders can the eight targets be broken? (AIME, 1990/8)

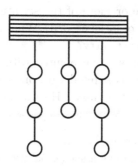

47. Using the numbers $1, 2, 3, 4, 5$, we can form $5! (= 120)$ 5-digit numbers in which the 5 digits are all distinct. If these numbers are listed in increasing order:

$$\underset{\text{1st}}{12345}, \quad \underset{\text{2nd}}{12354}, \quad \underset{\text{3rd}}{12435}, \quad \ldots, \quad \underset{\text{120th}}{54321},$$

find (i) the position of the number 35421; (ii) the 100th number in the list.

48. The $P_3^4(= 24)$ 3-permutations of the set $\{1, 2, 3, 4\}$ can be arranged in the following way, called the lexicographic ordering:

$$123, \ 124, \ 132, \ 134, \ 142, \ 143, \ 213, \ 214, \ 231, \ 234,$$
$$241, \ 243, \ 312, \ \ldots, \ 431, \ 432.$$

Thus the 3-permutations "132" and "214" appear at the 3rd and 8th positions of the ordering respectively. There are $P_4^9(= 3024)$ 4-permutations of the set $\{1, 2, \ldots, 9\}$. What are the positions of the 4-permutations "4567" and "5182" in the corresponding lexicographic ordering of the 4-permutations of $\{1, 2, \ldots, 9\}$?

49. The $\binom{5}{3}(= 10)$ 3-element subsets of the set $\{1, 2, 3, 4, 5\}$ can be arranged in the following way, called the lexicographic ordering:

$$\{1, 2, 3\}, \ \{1, 2, 4\}, \ \{1, 2, 5\}, \ \{1, 3, 4\}, \ \{1, 3, 5\}, \ \{1, 4, 5\},$$
$$\{2, 3, 4\}, \ \{2, 3, 5\}, \ \{2, 4, 5\}, \ \{3, 4, 5\}.$$

Thus the subset $\{1, 3, 5\}$ appears at the 5th position of the ordering. There are $\binom{10}{4}$ 4-element subsets of the set $\{1, 2, \ldots, 10\}$. What are the positions of the subsets $\{3, 4, 5, 6\}$ and $\{3, 5, 7, 9\}$ in the corresponding lexicographic ordering of the 4-element subsets of $\{1, 2, \ldots, 10\}$?

50. Six scientists are working on a secret project. They wish to lock up the documents in a cabinet so that the cabinet can be opened when and only when three or more of the scientists are present. What is the smallest number of locks needed? What is the smallest number of keys each scientist must carry?

51. A 10-storey building is to be painted with some 4 different colours such that each storey is painted with one colour. It is not necessary that all 4 colours must be used. How many ways are there to paint the building if

   (i) there are no other restrictions?
   (ii) any 2 adjacent storeys must be painted with different colours?

52. Find the number of all multi-subsets of $M = \{r_1 \cdot a_1, r_2 \cdot a_2, \ldots, r_n \cdot a_n\}$.

53. Let $r, b \in \mathbf{N}$ with $r \leq n$. A permutation $x_1 x_2 \cdots x_{2n}$ of the set $\{1, 2, \ldots, 2n\}$ is said to have property $P(r)$ if $|x_i - x_{i+1}| = r$ for at least one i in $\{1, 2, \ldots, 2n - 1\}$. Show that, for each n and r, there are more permutations with property $P(r)$ than without.

54. Prove by a combinatorial argument that each of the following numbers is always an integer for each $n \in \mathbf{N}$:

(i) $\dfrac{(3n)!}{2^n \cdot 3^n}$,

(ii) $\dfrac{(6n)!}{5^n \cdot 3^{2n} \cdot 2^{4n}}$,

(iii) $\dfrac{(n^2)!}{(n!)^n}$,

(iv) $\dfrac{(n!)!}{(n!)^{(n-1)!}}$.

55. Find the number of r-element multi-subsets of the multi-set

$$M = \{1 \cdot a_1, \infty \cdot a_2, \infty \cdot a_3, \ldots, \infty \cdot a_n\}.$$

56. Six distinct symbols are transmitted through a communication channel. A total of 18 blanks are to be inserted between the symbols with at least 2 blanks between every pair of symbols. In how many ways can the symbols and blanks be arranged?

57. In how many ways can the following 11 letters: $A, B, C, D, E, F, X,$ X, X, Y, Y be arranged in a row so that every Y lies between two X's (not necessarily adjacent)?

58. Two n-digit integers (leading zero allowed) are said to be *equivalent* if one is a permutation of the other. For instance, 10075, 01057 and 00751 are equivalent 5-digit integers.

    (i) Find the number of 5-digit integers such that no two are equivalent.

    (ii) If the digits 5, 7, 9 can appear at most once, how many nonequivalent 5-digit integers are there?

59. How many 10-letter words are there using the letters a, b, c, d, e, f if

    (i) there are no restrictions?

    (ii) each vowel (a and e) appears 3 times and each consonant appears once?

    (iii) the letters in the word appear in alphabetical order?

    (iv) each letter occurs at least once and the letters in the word appear in alphabetical order?

60. Let $r, n, k \in \mathbf{N}$ such that $r \geq nk$. Find the number of ways of distributing r identical objects into n distinct boxes so that each box holds at least k objects.

61. Find the number of ways of arranging the 9 letters $r, s, t, u, v, w, x, y, z$ in a row so that y always lies between x and z (x and y, or y and z need not be adjacent in the row).

62. Three girls A, B and C, and nine boys are to be lined up in a row. In how many ways can this be done if B must lie between A and C, and A, B must be separated by exactly 4 boys?

63. Five girls and eleven boys are to be lined up in a row such that from left to right, the girls are in the order: G_1, G_2, G_3, G_4, G_5. In how many ways can this be done if G_1 and G_2 must be separated by at least 3 boys, and there is at most one boy between G_4 and G_5?

64. Given $r, n \in \mathbf{N}$ with $r \geq n$, let $L(r, n)$ denote the number of ways of distributing r distinct objects into n identical boxes so that no box is empty and the objects in each box are arranged in a row. Find $L(r, n)$ in terms of r and n.

65. Find the number of integer solutions to the equation:

$$x_1 + x_2 + x_3 + x_4 + x_5 + x_6 = 60$$

in each of the following cases:

(i) $x_i \geq i - 1$ for each $i = 1, 2, \ldots, 6$;
(ii) $x_1 \geq 2$, $x_2 \geq 5$, $2 \leq x_3 \leq 7$, $x_4 \geq 1$, $x_5 \geq 3$ and $x_6 \geq 2$.

66. Find the number of integer solutions to the equation:

$$x_1 + x_2 + x_3 + x_4 = 30$$

in each of the following cases:

(i) $x_i \geq 0$ for each $i = 1, 2, 3, 4$;
(ii) $2 \leq x_1 \leq 7$ and $x_i \geq 0$ for each $i = 2, 3, 4$;
(iii) $x_1 \geq -5$, $x_2 \geq -1$, $x_3 \geq 1$ and $x_4 \geq 2$.

67. Find the number of quadruples (w, x, y, z) of nonnegative integers which satisfy the inequality

$$w + x + y + z \leq 1992.$$

68. Find the number of nonnegative integer solutions to the equation:

$$5x_1 + x_2 + x_3 + x_4 = 14.$$

69. Find the number of nonnegative integer solutions to the equation:

$$rx_1 + x_2 + \cdots + x_n = kr,$$

where $k, r, n \in \mathbf{N}$.

70. Find the number of nonnegative integer solutions to the equation:

$$3x_1 + 5x_2 + x_3 + x_4 = 10.$$

71. Find the number of positive integer solutions to the equation:

$$(x_1 + x_2 + x_3)(y_1 + y_2 + y_3 + y_4) = 77.$$

72. Find the number of nonnegative integer solutions to the equation:

$$(x_1 + x_2 + \cdots + x_n)(y_1 + y_2 + \cdots + y_n) = p,$$

where $n \in \mathbb{N}$ and p is a prime.

73. There are 5 ways to express "4" as a sum of 2 nonnegative integers in which the order counts:

$$4 = 4 + 0 = 3 + 1 = 2 + 2 = 1 + 3 = 0 + 4.$$

Given $r, n \in \mathbb{N}$, what is the number of ways to express r as a sum of n nonnegative integers in which the order counts?

74. There are 6 ways to express "5" as a sum of 3 positive integers in which the order counts:

$$5 = 3 + 1 + 1 = 2 + 2 + 1 = 2 + 1 + 2 = 1 + 3 + 1 = 1 + 2 + 2 = 1 + 1 + 3.$$

Given $r, n \in \mathbb{N}$ with $r \geq n$, what is the number of ways to express r as a sum of n positive integers in which the order counts?

75. A positive integer d is said to be *ascending* if in its decimal representation: $d = d_m d_{m-1} \cdots d_2 d_1$ we have

$$0 < d_m \leq d_{m-1} \leq \cdots \leq d_2 \leq d_1.$$

For instance, 1337 and 2455566799 are ascending integers. Find the number of ascending integers which are less than 10^9.

76. A positive integer d is said to be *strictly ascending* if in its decimal representation: $d = d_m d_{m-1} \cdots d_2 d_1$ we have

$$0 < d_m < d_{m-1} < \cdots < d_2 < d_1.$$

For instance, 145 and 23689 are strictly ascending integers. Find the number of strictly ascending integers which are less than (i) 10^9, (ii) 10^5.

77. Let $A = \{1, 2, \ldots, n\}$, where $n \in \mathbb{N}$.

   (i) Given $k \in A$, show that the number of subsets of A in which k is the maximum number is given by 2^{k-1}.
   (ii) Apply (i) to show that

$$\sum_{i=0}^{n-1} 2^i = 2^n - 1.$$

78. In a given circle, $n \geq 2$ arbitrary chords are drawn such that no three are concurrent within the interior of the circle. Suppose m is the number of points of intersection of the chords within the interior. Find, in terms of n and m, the number r of line segments obtained through dividing the chords by their points of intersection. (In the following example, $n = 5, m = 3$ and $r = 11$.)

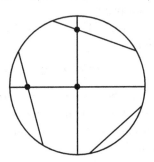

79. There are $p \geq 6$ points given on the circumference of a circle, and every two of the points are joined by a chord.

    (i) Find the number of such chords.

    Assume that no 3 chords are concurrent within the interior of the circle.

    (ii) Find the number of points of intersection of these chords within the interior of the circle.

    (iii) Find the number of line segments obtained through dividing the chords by their points of intersection.

    (iv) Find the number of triangles whose vertices are the points of intersection of the chords within the interior of the circle.

80. In how many ways can $n + 1$ different prizes be awarded to n students in such a way that each student has at least one prize?

81. (a) Let $n, m, k \in \mathbf{N}$, and let $\mathbf{N}_k = \{1, 2, \ldots, k\}$. Find

    (i) the number of mappings from $\mathbf{N}_n$ to $\mathbf{N}_m$.
    (ii) the number of 1-1 mappings from $\mathbf{N}_n$ to $\mathbf{N}_m$, where $n \leq m$.

    (b) A mapping $f: \mathbf{N}_n \to \mathbf{N}_m$ is *strictly increasing* if $f(a) < f(b)$ whenever $a < b$ in $\mathbf{N}_n$. Find the number of strictly increasing mappings from $\mathbf{N}_n$ to $\mathbf{N}_m$, where $n \leq m$.

    (c) Express the number of mappings from $\mathbf{N}_n$ *onto* $\mathbf{N}_m$ in terms of $S(n, m)$ (the Stirling number of the second kind).

82. Given $r, n \in \mathbf{Z}$ with $0 \leq n \leq r$, the Stirling number $s(r, n)$ of the first kind is defined as the number of ways to arrange r distinct objects around n identical circles such that each circle has at least one object. Show that

   (i) $s(r, 1) = (r - 1)!$ for $r \geq 1$;
   (ii) $s(r, 2) = (r - 1)!(1 + \frac{1}{2} + \frac{1}{3} + \cdots + \frac{1}{r-1})$ for $r \geq 2$;
   (iii) $s(r, r - 1) = \binom{r}{2}$ for $r \geq 2$;
   (iv) $s(r, r - 2) = \frac{1}{24}r(r - 1)(r - 2)(3r - 1)$ for $r \geq 2$;
   (v) $\sum_{n=0}^{r} s(r, n) = r!$.

83. The Stirling numbers of the first kind occur as the coefficients of x^n in the expansion of

$$x(x + 1)(x + 2) \cdots (x + r - 1).$$

For instance, when $r = 3$,

$$x(x + 1)(x + 2) = 2x + 3x^2 + x^3$$
$$= s(3, 1)x + s(3, 2)x^2 + s(3, 3)x^3;$$

and when $r = 5$,

$$x(x + 1)(x + 2)(x + 3)(x + 4)$$
$$= 24x + 50x^2 + 35x^3 + 10x^4 + x^5$$
$$= s(5, 1)x + s(5, 2)x^2 + s(5, 3)x^3 + s(5, 4)x^4 + s(5, 5)x^5.$$

Show that

$$x(x + 1)(x + 2) \cdots (x + r - 1) = \sum_{n=0}^{r} s(r, n)x^n,$$

where $r \in \mathbf{N}$.

84. Given $r, n \in \mathbf{Z}$ with $0 \leq n \leq r$, the Stirling number $S(r, n)$ of the second kind is defined as the number of ways of distributing r distinct objects into n identical boxes such that no box is empty. Show that

   (i) $S(r, 2) = 2^{r-1} - 1$;
   (ii) $S(r, 3) = \frac{1}{2}(3^{r-1} + 1) - 2^{r-1}$;
   (iii) $S(r, r - 1) = \binom{r}{2}$;
   (iv) $S(r, r - 2) = \binom{r}{3} + 3\binom{r}{4}$.

85. Let $(x)_0 = 1$ and for $n \in \mathbf{N}$, let

$$(x)_n = x(x - 1)(x - 2) \cdots (x - n + 1).$$

The Stirling numbers of the second kind occur as the coefficients of $(x)_n$ when x^r is expressed in terms of $(x)_n$'s. For instance, when $r = 2, 3$ and 4, we have, respectively,

$$x^2 = x + x(x-1) = (x)_1 + (x)_2$$
$$= S(2,1)(x)_1 + S(2,2)(x)_2,$$
$$x^3 = x + 3x(x-1) + x(x-1)(x-2)$$
$$= S(3,1)(x)_1 + S(3,2)(x)_2 + S(3,3)(x)_3,$$
$$x^4 = x + 7x(x-1) + 6x(x-1)(x-2) + x(x-1)(x-2)(x-3)$$
$$= S(4,1)(x)_1 + S(4,2)(x)_2 + S(4,3)(x)_3 + S(4,4)(x)_4.$$

Show that for $r = 0, 1, 2, \ldots,$

$$x^r = \sum_{n=0}^{r} S(r,n)(x)_n.$$

86. Suppose that m chords of a given circle are drawn in such a way that no three are concurrent in the interior of the circle. If n denotes the number of points of intersection of the chords within the circle, show that the number of regions divided by the chords in the circle is $m + n + 1$.

87. For $n \geq 4$, let $r(n)$ denote the number of interior regions of a convex n-gon divided by all its diagonals if no three diagonals are concurrent within the n-gon. For instance, as shown in the following diagrams, $r(4) = 4$ and $r(5) = 11$. Prove that $r(n) = \binom{n}{4} + \binom{n-1}{2}$.

88. Let $n \in \mathbf{N}$. How many solutions are there in ordered positive integer pairs (x, y) to the equation

$$\frac{xy}{x+y} = n?$$

(Putnam, 1960)

89. Let $S = \{1, 2, 3, \ldots, 1992\}$. In each of the following cases, find the number of 3-element subsets $\{a, b, c\}$ of S satisfying the given condition:

    (i) $3|(a+b+c)$;

    (ii) $4|(a+b+c)$.

90. A sequence of 15 random draws, one at a time with replacement, is made from the set

$$\{A, B, C, \ldots, X, Y, Z\}$$

of the English alphabet. What is the probability that the string:

$$UNIVERSITY$$

occurs as a block in the sequence?

91. A set $S = \{a_1, a_2, \ldots, a_r\}$ of positive integers, where $r \in \mathbf{N}$ and $a_1 < a_2 < \cdots < a_r$, is said to be *m-separated* ($m \in \mathbf{N}$) if $a_i - a_{i-1} \geq m$, for each $i = 2, 3, \ldots, r$. Let $X = \{1, 2, \ldots, n\}$. Find the number of r-element subsets of X which are m-separated, where $0 \leq r \leq n - (m-1)(r-1)$.

92. Let $a_1, a_2, \ldots, a_n$ be positive real numbers, and let S_k be the sum of products of $a_1, a_2, \ldots, a_n$ taken k at a time. Show that

$$S_k S_{n-k} \geq \binom{n}{k}^2 a_1 a_2 \cdots a_n,$$

for $k = 1, 2, \ldots, n-1$. (APMO, 1990)

93. For $\{1, 2, 3, \ldots, n\}$ and each of its nonempty subsets, a unique *alternating sum* is defined as follows: Arrange the numbers in the subset in decreasing order and then, beginning with the largest, alternately add and subtract successive numbers. (For example, the alternating sum for $\{1, 2, 4, 6, 9\}$ is $9 - 6 + 4 - 2 + 1 = 6$ and for $\{5\}$ it is simply 5.) Find the sum of all such alternating sums for $n = 7$. (AIME, 1983/13)

94. A gardener plants three maple trees, four oak trees and five birch trees in a row. He plants them in random order, each arrangement being equally likely. Let $\frac{m}{n}$ in lowest terms be the probability that no two birch trees are next to one another. Find $m + n$. (AIME, 1984/11)

95. In a tournament each player played exactly one game against each of the other players. In each game the winner was awarded 1 point, the loser got 0 points, and each of the two players earned 1/2 point if the game was a tie. After the completion of the tournament, it was found that exactly half of the points earned by each player were earned in games against the ten players with the least number of points. (In particular, each of the ten lowest scoring players earned half of her/his points against the other nine of the ten.) What was the total number of players in the tournament? (AIME, 1985/14)

96. Let S be the sum of the base 10 logarithms of all of the proper divisors of $1,000,000$. (By a proper divisor of a natural number we mean a positive integral divisor other than 1 and the number itself.) What is the integer nearest to S? (AIME, 1986/8)

97. In a sequence of coin tosses one can keep a record of the number of instances when a tail is immediately followed by a head, a head is immediately followed by a head, etc. We denote these by *TH,HH*, etc. For example, in the sequence *HHTTHHHHTHHTTTT* of 15 coin tosses we observe that there are five *HH*, three *HT*, two *TH* and four *TT* subsequences. How many different sequences of 15 coin tosses will contain exactly two *HH*, three *HT*, four *TH* and five *TT* subsequences? (AIME, 1986/13)

98. An ordered pair (m, n) of nonnegative integers is called "simple" if the addition $m + n$ in base 10 requires no carrying. Find the number of simple ordered pairs of nonnegative integers that sum to
(i) 1492;   (AIME, 1987/1)   (ii) 1992.

99. Let m/n, in lowest terms, be the probability that a randomly chosen positive divisor of 10^{99} is an integer multiple of 10^{88}. Find $m + n$. (AIME, 1988/5)

100. A convex polyhedron has for its faces 12 squares, 8 regular hexagons, and 6 regular octagons. At each vertex of the polyhedron one square, one hexagon, and one octagon meet. How many segments joining vertices of the polyhedron lie in the interior of the polyhedron rather than along an edge or a face? (AIME, 1988/10)

101. Someone observed that $6! = 8 \cdot 9 \cdot 10$. Find the largest positive integer n for which $n!$ can be expressed as the product of $n - 3$ consecutive positive integers. (AIME, 1990/11)

102. Let $S = \{1, 2, \ldots, n\}$. Find the number of subsets A of S satisfying the following conditions:
$A = \{a, a + d, \ldots, a + kd\}$ for some positive integers a, d and k, and $A \cup \{x\}$ is no longer an A.P. with common difference d for each $x \in S \setminus A$.
(Note that $|A| \geq 2$ and any sequence of two terms is considered as an A.P.) (Chinese Math. Competition, 1991)

103. Find all natural numbers $n > 1$ and $m > 1$ such that

$$1!3!5! \cdots (2n - 1)! = m!.$$

(Proposed by I. Cucurezeanu, see *Amer. Math. Monthly*, **94** (1987), 190.)

104. Show that for $n \in \mathbf{N}$,

$$\sum_{r=0}^{n} P_r^n = \lfloor n!e \rfloor,$$

where $\lfloor x \rfloor$ denotes the greatest integer $\leq x$ and $e = 2.718\cdots$. (Proposed by D. Ohlsen, see *The College Math. J.*, **20** (1989), 260.)

105. Let $S = \{1, 2, \ldots, 1990\}$. A 31-element subset A of S is said to be *good* if the sum $\sum_{a \in A} a$ is divisible by 5. Find the number of 31-element subsets of S which are good. (Proposed by the Indian Team at the 31st IMO.)

106. Let S be a 1990-element set and let $\mathcal{P}$ be a set of 100-ary sequences $(a_1, a_2, \ldots, a_{100})$, where a_i's are distinct elements of S. An ordered pair (x, y) of elements of S is said to *appear* in $(a_1, a_2, \ldots, a_{100})$ if $x = a_i$ and $y = a_j$ for some i, j with $1 \leq i < j \leq 100$. Assume that every ordered pair (x, y) of elements of S appears in at most one member in $\mathcal{P}$. Show that

$$|\mathcal{P}| \leq 800.$$

(Proposed by the Iranian Team at the 31st IMO.)

107. Let $M = \{r_1 \cdot a_1, r_2 \cdot a_2, \ldots, r_n \cdot a_n\}$ be a multi-set with $r_1 + r_2 + \cdots + r_n = r$. Show that the number of r-permutations of M is equal to the number of $(r-1)$-permutations of M.

108. Prove that it is impossible for seven distinct straight lines to be situated in the Euclidean plane so as to have at least six points where exactly three of these lines intersect and at least four points where exactly two of these lines intersect. (Putnam, 1973)

109. For what $n \in \mathbf{N}$ does there exist a permutation $(x_1, x_2, \ldots, x_n)$ of $(1, 2, \ldots, n)$ such that the differences $|x_k - k|, 1 \leq k \leq n$, are all distinct? (Proposed by M. J. Pelling, see *Amer. Math. Monthly*, **96** (1989), 843–844.)

110. Numbers $d(n, m)$, where n, m are integers and $0 \leq m \leq n$, are defined by

$$d(n, 0) = d(n, n) = 1 \quad \text{for all } n \geq 0$$

and

$$m \cdot d(n, m) = m \cdot d(n-1, m) + (2n - m) \cdot d(n-1, m-1)$$

for $0 < m < n$. Prove that all the $d(n, m)$ are integers. (Great Britain, 1987)

111. A difficult mathematical competition consisted of a Part I and a Part II with a combined total of 28 problems. Each contestant solved 7 problems altogether. For each pair of problems, there were exactly two contestants who solved both of them. Prove that there was a contestant who, in Part I, solved either no problems or at least four problems. (USA MO, 1984/4)

112. Suppose that five points in a plane are situated so that no two of the straight lines joining them are parallel, perpendicular, or coincident. From each point perpendiculars are drawn to all the lines joining the other four points. Determine the maximum number of intersections that these perpendiculars can have. (IMO, 1964/5)

113. Let n distinct points in the plane be given. Prove that fewer than $2n^{\frac{3}{2}}$ pairs of them are at unit distance apart. (Putnam, 1978)

114. If c and m are positive integers each greater than 1, find the number $n(c, m)$ of ordered c-tuples $(n_1, n_2, \ldots, n_c)$ with entries from the initial segment $\{1, 2, \ldots, m\}$ of the positive integers such that $n_2 < n_1$ and $n_2 \le n_3 \le \cdots \le n_c$. (Proposed by D. Spellman, see *Amer. Math. Monthly*, **94** (1987), 383–384.)

115. Let $X = \{x_1, x_2, \ldots, x_m\}, Y = \{y_1, y_2, \ldots, y_n\}$ $(m, n \in \mathbf{N})$ and $A \subseteq X \times Y$. For $x_i \in X$, let

$$A(x_i, \cdot) = (\{x_i\} \times Y) \cap A$$

and for $y_j \in Y$, let

$$A(\cdot, y_j) = (X \times \{y_j\}) \cap A.$$

(i) Prove the following *Fubini Principle*:

$$\sum_{i=1}^{m} |A(x_i, \cdot)| = |A| = \sum_{j=1}^{n} |A(\cdot, y_j)|.$$

(ii) Using (i), or otherwise, solve the following problem: There are $n \ge 3$ given points in the plane such that any three of them form a right-angled triangle. Find the largest possible value of n.

(23rd Moscow MO)

Exercise 2

1. The number 4 can be expressed as a sum of one or more positive integers, taking order into account, in 8 ways:

$$4 = 1 + 3 = 3 + 1 = 2 + 2 = 1 + 1 + 2$$
$$= 1 + 2 + 1 = 2 + 1 + 1 = 1 + 1 + 1 + 1.$$

   In general, given $n \in \mathbf{N}$, in how many ways can n be so expressed?

2. Find the number of $2n$-digit binary sequences in which the number of 0's in the first n digits is equal to the number of 1's in the last n digits.

3. Let $m, n, r \in \mathbf{N}$. Find the number of r-element multi-subsets of the multi-set

$$M = \{a_1, a_2, \ldots, a_n, m \cdot b\}$$

   in each of the following cases:

   (i) $r \leq m, r \leq n$;
   (ii) $n \leq r \leq m$;
   (iii) $m \leq r \leq n$.

4. Ten points are marked on a circle. How many distinct convex polygons of three or more sides can be drawn using some (or all) of the ten points as vertices? (Polygons are distinct unless they have exactly the same vertices.) (AIME, 1989/2)

5. Find the coefficient of x^5 in the expansion of $(1 + x + x^2)^8$.

6. Find the coefficient of x^6 in the expansion of $(1 + x + x^2)^9$.

7. Find the coefficient of x^{18} in the expansion of

$$(1 + x^3 + x^5 + x^7)^{100}.$$

8. Find the coefficient of x^{29} in the expansion of

$$(1 + x^5 + x^7 + x^9)^{1000}.$$

19

9. In the expansion of

$$(1 + x + x^2 + \cdots + x^{10})^3,$$

what is the coefficient of

(i) x^5?   (ii) x^8?

10. Given an n-element set X, where $n \in \mathbf{N}$, let $\mathcal{O} = \{A \subseteq X \mid |A| \text{ is odd}\}$ and $\mathcal{E} = \{A \subseteq X \mid |A| \text{ is even}\}$. Show that $|\mathcal{O}| = |\mathcal{E}|$ by establishing a bijection between $\mathcal{O}$ and $\mathcal{E}$.

11. Find the number of permutations of the multi-set $\{m \cdot 1, n \cdot 2\}$, where $m, n \in \mathbf{N}$, which must contain the m 1's.

12. Let $1 \leq r \leq n$ and consider all r-element subsets of the set $\{1, 2, \ldots, n\}$. Each of these subsets has a *largest* member. Let $H(n, r)$ denote the arithmetic mean of these largest members. Find $H(n, r)$ and simplify your result (see Example 2.5.2).

13. For $n \in \mathbf{N}$, let $\Delta(n)$ denote the number of triangles XYZ in the nth subdivision of an equilateral triangle ABC (see Figure 2.5.2) such that $YZ//BC$, and X and A are on the same side of YZ. Evaluate $\Delta(n)$. (For other enumeration problems relating to this, see M. E. Larsen, The eternal triangle – A history of a counting problem, *The College Math. J.*, **20** (1989), 370–384.)

14. Find the coefficients of x^n and x^{n+r} ($1 \leq r \leq n$) in the expansion of

$$(1 + x)^{2n} + x(1 + x)^{2n-1} + x^2(1 + x)^{2n-2} + \cdots + x^n(1 + x)^n.$$

15. A polynomial in x is defined by

$$a_0 + a_1 x + a_2 x^2 + \cdots + a_{2n} x^{2n} = (x + 2x^2 + \cdots + nx^n)^2.$$

Show that

$$\sum_{i=n+1}^{2n} a_i = \frac{n(n + 1)(5n^2 + 5n + 2)}{24}.$$

16. Show that

$$P_r^r + P_r^{r+1} + \cdots + P_r^{2r} = P_r^{2r+1},$$

where r is a nonnegative integer.

17. Given $r, n, m \in \mathbf{N}^*$ with $r \leq n$, show that

$$P_r^n + P_r^{n+1} + \cdots + P_r^{n+m} = \frac{1}{r + 1} \left(P_{r+1}^{n+m+1} - P_{r+1}^n \right).$$

(See Problem 2.35.)

18. Show that

(i) for even $n \in \mathbf{N}$,

$$\binom{n}{i} < \binom{n}{j} \quad \text{if } 0 \le i < j \le \frac{n}{2};$$

and

$$\binom{n}{i} > \binom{n}{j} \quad \text{if } \frac{n}{2} \le i < j \le n.$$

(ii) for odd $n \in \mathbf{N}$,

$$\binom{n}{i} < \binom{n}{j} \quad \text{if } 0 \le i < j \le \frac{1}{2}(n-1);$$

and

$$\binom{n}{i} > \binom{n}{j} \quad \text{if } \frac{1}{2}(n+1) \le i < j \le n.$$

19. Give three different proofs for each of the following identities:

   (i) $\binom{2(n+1)}{n+1} = \binom{2n}{n+1} + 2\binom{2n}{n} + \binom{2n}{n-1}$;

   (ii) $\binom{n+1}{m} = \binom{n}{m-1} + \binom{n-1}{m} + \binom{n-1}{m-1}$.

20. Give a combinatorial proof for the identity

$$\binom{n}{m}\binom{m}{r} = \binom{n}{r}\binom{n-r}{m-r}.$$

21. Show that for $n \in \mathbf{N}^*$,

$$\sum_{r=0}^{n} \frac{(2n)!}{(r!)^2((n-r)!)^2} = \binom{2n}{n}^2.$$

22. By using the identity $(1-x^2)^n = (1+x)^n(1-x)^n$, show that for each $m \in \mathbf{N}^*$ with $m \le n$,

$$\sum_{i=0}^{2m}(-1)^i\binom{n}{i}\binom{n}{2m-i} = (-1)^m\binom{n}{m},$$

and

$$\sum_{i=0}^{2m+1}(-1)^i\binom{n}{i}\binom{n}{2m+1-i} = 0.$$

Deduce that

$$\sum_{i=0}^{n}(-1)^i\binom{n}{i}^2 = \begin{cases} (-1)^{\frac{n}{2}}\binom{n}{\frac{n}{2}} & \text{if } n \text{ is even} \\ 0 & \text{if } n \text{ is odd}. \end{cases}$$

23. What is the value of the sum

$$S = m! + \frac{(m+1)!}{1!} + \frac{(m+2)!}{2!} + \cdots + \frac{(m+n)!}{n!} \quad ?$$

(Beijing Math. Contest (1962))

Prove each of the following identities in Problems 24–43, where $m, n \in \mathbf{N}^*$:

24. $\displaystyle\sum_{r=0}^{n} 3^r \binom{n}{r} = 4^n$,

25. $\displaystyle\sum_{r=0}^{n} (r+1)\binom{n}{r} = (n+2)2^{n-1}$,

26. $\displaystyle\sum_{r=0}^{n} \frac{1}{r+1}\binom{n}{r} = \frac{1}{n+1}(2^{n+1} - 1)$,

27. $\displaystyle\sum_{r=0}^{n} \frac{(-1)^r}{r+1}\binom{n}{r} = \frac{1}{n+1}$,

28. $\displaystyle\sum_{r=m}^{n} \binom{n}{r}\binom{r}{m} = 2^{n-m}\binom{n}{m}$ for $0 < m \leq n$,

29. $\displaystyle\sum_{r=0}^{m} (-1)^r \binom{n}{r} = \begin{cases} (-1)^m \binom{n-1}{m} & \text{if } m < n \\ 0 & \text{if } m = n > 0, \end{cases}$

30. $\displaystyle\sum_{r=0}^{m} (-1)^{m-r}\binom{n}{r} = \binom{n-1}{m}$ for $m \leq n - 1$,

31. $\displaystyle\sum_{r=0}^{n} (-1)^r r\binom{n}{r} = 0$ for $n > 1$,

32. $\displaystyle\sum_{r=0}^{n-1} \binom{2n-1}{r} = 2^{2n-2}$,

33. $\displaystyle\sum_{r=0}^{n} \binom{2n}{r} = 2^{2n-1} + \frac{1}{2}\binom{2n}{n}$,

34. $\displaystyle\sum_{r=0}^{n} r\binom{2n}{r} = n2^{2n-1}$,

35. $\displaystyle\sum_{r=k}^{m} \binom{n+r}{n} = \binom{n+m+1}{n+1} - \binom{n+k}{n+1}$ for $k \in \mathbf{N}^*$ and $k \leq m$,

36. $\displaystyle\sum_{r=1}^{n} \binom{n}{r}\binom{n-1}{r-1} = \binom{2n-1}{n-1}$,

37. $\sum_{r=m}^{n} (-1)^r \binom{n}{r}\binom{r}{m} = \begin{cases} (-1)^m & \text{if } m = n \\ 0 & \text{if } m < n, \end{cases}$

38. $\sum_{r=1}^{n-1} (n-r)^2 \binom{n-1}{n-r} = n(n-1)2^{n-3}$,

    (See Problem 2.47(i).)

39. $\sum_{r=0}^{n} \binom{2n}{r}^2 = \frac{1}{2}\left\{\binom{4n}{2n} + \binom{2n}{n}^2\right\}$,

40. $\sum_{r=0}^{n} \binom{n}{r}^2 \binom{r}{n-k} = \binom{n}{k}\binom{n+k}{k}$ for $k \in \mathbf{N}^*$, $0 \le k \le n$,

41. $\sum_{r=0}^{n} \binom{n}{r}\binom{m+r}{n} = \sum_{r=0}^{n} \binom{n}{r}\binom{m}{r}2^r$,

42. $\sum_{r=0}^{m} \binom{m}{r}\binom{n}{r}\binom{p+r}{m+n} = \binom{p}{m}\binom{p}{n}$, for $p \in \mathbf{N}$, $p \ge m, n$;

    (Li Shanlan, 1811–1882)

43. $\sum_{r=0}^{m} \binom{m}{r}\binom{n}{r}\binom{p+m+n-r}{m+n} = \binom{p+m}{m}\binom{p+n}{n}$ for $p \in \mathbf{N}$.

    (Li Shanlan)

---

44. Prove the following identities using the technique of counting shortest routes in a grid:

    (i) $\sum_{r=0}^{n} \binom{n}{r} = 2^n$,

    (ii) $\sum_{k=0}^{n} \binom{r+k}{r} = \binom{r+n+1}{r+1}$.

45. Use the technique of finding the number of shortest routes in rectangular grid to prove the following identity:

    $$\binom{p}{q}\binom{r}{0} + \binom{p-1}{q-1}\binom{r+1}{1} + \cdots + \binom{p-q}{0}\binom{r+q}{q} = \binom{p+r+1}{q}.$$

46. Give a combinatorial proof for the following identity:

    $$\sum_{r=1}^{n} r\binom{n}{r} = n \cdot 2^{n-1}.$$

47. Given $n \in \mathbf{N}$, show that

    (i) $\sum_{r=1}^{n} r^2 \binom{n}{r} = n(n+1)2^{n-2}$;

(Putnam, 1962)

(ii) $\displaystyle\sum_{r=1}^{n} r^3 \binom{n}{r} = n^2(n+3)2^{n-3}$;

(iii) $\displaystyle\sum_{r=1}^{n} r^4 \binom{n}{r} = n(n+1)(n^2+5n-2)2^{n-4}$.

48. (i) Prove that for $r, k \in \mathbf{N}$,

$$r^k = \sum_{i=0}^{k} \binom{k}{i}(r-1)^{k-i}.$$

(ii) For $n, k \in \mathbf{N}$, let

$$R(n,k) = \sum_{r=1}^{n} r^k \binom{n}{r} \quad \text{and} \quad R(n,0) = \sum_{r=0}^{n} \binom{n}{r}.$$

Show that

$$R(n,k) = n \cdot \sum_{j=0}^{k-1} \binom{k-1}{j} R(n-1,j).$$

Remark. Two Chinese teachers, Wei Guozhen and Wang Kai (1988) showed that

$$\sum_{r=1}^{n} r^k \binom{n}{r} = \sum_{i=1}^{k} S(k,i) \cdot P_i^n \cdot 2^{n-i},$$

where $k \le n$ and $S(k,i)$'s are the Stirling numbers of the second kind.

49. Prove that

$$\sum_{r=1}^{n} \frac{1}{r} \binom{n}{r} = \sum_{r=1}^{n} \frac{1}{r}(2^r - 1).$$

50. Give two different proofs for the following identity:

$$\sum_{r=1}^{n} r \binom{n}{r}^2 = n \binom{2n-1}{n-1},$$

where $n \in \mathbf{N}$.

51. Let p be a prime. Show that

$$\binom{p}{r} \equiv 0 \pmod{p}$$

for all r such that $1 \le r \le p - 1$. Deduce that $(1+x)^p \equiv (1+x^p) \pmod{p}$.

52. Let p be an odd prime. Show that

$$\binom{2p}{p} \equiv 2 \pmod{p}.$$

53. Let n, m, p be integers such that $1 \leq p \leq m \leq n$.

   (i) Express, in terms of $S(n,p)$, the number of mappings $f: \mathbf{N}_n \to \mathbf{N}_m$ such that $|f(\mathbf{N}_n)| = p$.
   (ii) Express, in terms of $S(n,k)$'s, where $p \leq k \leq m$, the number of mappings $f: \mathbf{N}_n \to \mathbf{N}_m$ such that $\mathbf{N}_p \subseteq f(\mathbf{N}_n)$.

54. Recall that for nonnegative integers $n, r, H_r^n = \binom{r+n-1}{r}$. Prove each of the following identities:

   (a) $H_r^n = \frac{n}{r} H_{r-1}^{n+1}$;
   (b) $H_r^n = \frac{n+r-1}{r} H_{r-1}^n$;
   (c) $H_r^n = H_{r-1}^n + H_r^{n-1}$;
   (d) $\displaystyle\sum_{k=0}^{r} H_k^n = H_r^{n+1}$;
   (e) $\displaystyle\sum_{k=1}^{r} k H_k^n = n H_{r-1}^{n+2}$;
   (f) $\displaystyle\sum_{k=0}^{r} H_k^m H_{r-k}^n = H_r^{m+n}$.

55. For $n, k \in \mathbf{N}$ with $n \geq 2$ and $1 \leq k \leq n$, let

$$d_k(n) = \left| \binom{n}{k} - \binom{n}{k-1} \right|,$$

$$d_{\min}(n) = \min\{d_k(n) \mid 1 \leq k \leq n\}.$$

Show that

   (i) $d_{\min}(n) = 0$ iff n is odd;
   (ii) For odd n, $d_k(n) = 0$ iff $k = \frac{1}{2}(n+1)$.

Let $d_{\min}^*(n) = \min\{d_k(n) \mid 1 \leq k \leq n, \ k \neq \frac{1}{2}(n+1)\}$. Show that

   (iii) For $n \neq 4$, $d_{\min}^*(n) = n - 1$;
   (iv) For $n \neq 4$ and $n \neq 6$,

$$d_k(n) = n - 1 \text{ iff } k = 1 \text{ or } k = n;$$

   (v) For $n = 6$, $d_k(6) = 5$ iff $k = 1, 3, 4$ or 6.

(See Z. Shan and E. T. H. Wang, The gaps between consecutive binomial coefficients, *Math. Magazine*, **63** (1990), 122–124.)

56. Prove that

  (i) $\binom{\binom{n}{2}}{2} = 3\binom{n+1}{4}$ for $n \in \mathbf{N}$;

  (ii) $\binom{\binom{n}{2}}{3} > \binom{\binom{n}{3}}{2}$ for $n \in \mathbf{N}, n \geq 3$;

  (iii) $\binom{\binom{n}{r}}{2} = \sum_{j=1}^{r} \binom{\binom{r}{j}+\epsilon_j}{2} \binom{n+r-j}{2r}$,

  where $\epsilon_j = \begin{cases} 1 & \text{if } j \text{ is odd} \\ 0 & \text{if } j \text{ is even} \end{cases}$ and $n, r \in \mathbf{N}$ with $r \leq n$;

  (iv) $\binom{\binom{n}{r}}{2} = \sum_{j=1}^{r} \binom{2j-1}{j} \binom{r+j}{2j} \binom{n}{r+j}$,

  for $n, r \in \mathbf{N}$ with $r \leq n$.

(For more results on these iterated binomial coefficients, see S. W. Golomb, Iterated binomial coefficients, *Amer. Math. Monthly,* **87** (1980), 719–727.)

57. Let $a_n = 6^n + 8^n$. Determine the remainder on dividing a_{83} by 49. (AIME, 1983/6)

58. The increasing sequence $1, 3, 4, 9, 10, 12, 13, \ldots$ consists of all those positive integers which are powers of 3 or sums of distinct powers of 3. Find the 100th term of this sequence (where 1 is the 1st term, 3 is the 2nd term, and so on). (AIME, 1986/7)

59. The polynomial $1 - x + x^2 - x^3 + \cdots + x^{16} - x^{17}$ may be written in the form $a_0 + a_1 y + a_2 y^2 + a_3 y^3 + \cdots + a_{16} y^{16} + a_{17} y^{17}$, where $y = x + 1$ and the a_i's are constants. Find the value of a_2. (AIME, 1986/11)

60. In an office, at various times during the day, the boss gives the secretary a letter to type, each time putting the letter on top of the pile in the secretary's in-box. When there is time, the secretary takes the top letter off the pile and types it. There are nine letters to be typed during the day, and the boss delivers them in the order $1, 2, 3, 4, 5, 6, 7, 8, 9$.

    While leaving for lunch, the secretary tells a colleague that letter 8 has already been typed, but says nothing else about the morning's typing. The colleague wonders which of the nine letters remain to be typed after lunch and in what order they will be typed. Based upon the above information, how many such *after-lunch typing orders* are possible? (That there are no letters left to be typed is one of the possibilities.) (AIME, 1988/15)

61. Expanding $(1 + 0.2)^{1000}$ by the binomial theorem and doing no further

manipulation gives

$$\binom{1000}{0}(0.2)^0 + \binom{1000}{1}(0.2)^1 + \binom{1000}{2}(0.2)^2 + \cdots + \binom{1000}{1000}(0.2)^{1000}$$
$$= A_0 + A_1 + A_2 + \cdots + A_{1000},$$

where $A_k = \binom{1000}{k}(0.2)^k$ for $k = 0, 1, 2, \ldots, 1000$. For which k is A_k the largest? (AIME, 1991/3)

62. Prove that the number of distinct terms in the expansion of

$$(x_1 + x_2 + \cdots x_m)^n$$

is given by $H_n^m = \binom{n+m-1}{n}$.

63. Show by two different methods that

$$\binom{n}{n_1, n_2, \ldots, n_m} = \sum_{i=1}^{m} \binom{n-1}{n_1, \ldots, n_i - 1, n_{i+1}, \ldots, n_m}.$$

64. For $n, m \in \mathbf{N}$, show that

$$\sum \binom{n}{n_1, n_2, \ldots, n_m} = m! S(n, m),$$

where the sum is taken over all m-ary sequences $(n_1, n_2, \ldots, n_m)$ such that $n_i \neq 0$ for all i, and $S(n, m)$ is a Stirling number of the second kind.

65. Prove that

$$\sum \binom{n}{n_1, n_2, \ldots, n_m}(-1)^{n_2+n_4+n_6+\cdots} = \begin{cases} 1 & \text{if } m \text{ is odd} \\ 0 & \text{if } m \text{ is even,} \end{cases}$$

where the sum is taken over all m-ary sequences $(n_1, n_2, \ldots, n_m)$ of nonnegative integers with $\sum_{i=1}^{m} n_i = n$.

66. Prove the following generalized Vandermonde's identity for multinomial coefficients: for $p, q \in \mathbf{N}$,

$$\binom{p+q}{k_1, k_2, \ldots, k_m}$$
$$= \sum \binom{p}{j_1, j_2, \ldots, j_m}\binom{q}{k_1 - j_1, k_2 - j_2, \ldots, k_m - j_m},$$

where the sum is taken over all m-ary sequences $(j_1, j_2, \ldots, j_m)$ of nonnegative integers with $j_1 + j_2 + \cdots + j_m = p$.

67. Given any prime p and $m \in \mathbf{N}$, show that

$$\binom{p}{n_1, n_2, \ldots, n_m} \equiv 0 \pmod{p},$$

if $p \neq n_i$, for any $i = 1, 2, \ldots, m$.

Deduce that

$$\left(\sum_{i=1}^{m} x_i\right)^p \equiv \sum_{i=1}^{m} x_i^p \pmod{p}.$$

68. Let p be a prime, and $n \in \mathbf{N}$. Write n in base p as follows:

$$n = n_0 + n_1 p + n_2 p^2 + \cdots + n_k p^k,$$

where $n_i \in \{0, 1, \ldots, p-1\}$ for each $i = 1, 2, \ldots, k$.

Given $m \in \mathbf{N}$, show that the number of terms in the expansion of $(x_1 + x_2 + \cdots + x_m)^n$ whose coefficients are not divisible by p is

$$\prod_{i=0}^{k} \binom{n_i + m - 1}{m - 1}.$$

(See F. T. Howard, The number of multinomial coefficients divisible by a fixed power of a prime, *Pacific J. Math.*, **50** (1974), 99–108.)

69. Show that

$$\sum_{r=0}^{n} \binom{n}{r}^2 \binom{2n + m - r}{2n} = \binom{m + n}{n}^2.$$

(Li Jen Shu)

70. Show that

$$\sum_{r=1}^{n} \frac{(-1)^{r-1}}{r} \binom{n}{r} = \sum_{k=1}^{n} \frac{1}{k},$$

where $n \in \mathbf{N}$.

71. Given $r \in \mathbf{N}$ with $r \geq 2$, show that

$$\sum_{n=r}^{\infty} \frac{1}{\binom{n}{r}} = \frac{r}{r-1}.$$

(See H. W. Gould, *Combinatorial Identities*, Morgantown, W. V. (1972), 18–19.)

72. Let $S = \{1, 2, \ldots, n\}$. For each $A \subseteq S$ with $A \neq \emptyset$, let $M(A) = \max\{x \mid x \in A\}$, $m(A) = \min\{x \mid x \in A\}$ and $\alpha(A) = M(A) + m(A)$. Evaluate the arithmetic mean of all the $\alpha(A)$'s when A runs through all nonempty subsets of S.

73. Given $a_n = \displaystyle\sum_{k=0}^{n} \binom{n}{k}^{-1}$, $n \in \mathbf{N}$,

show that

$$\lim_{n \to \infty} a_n = 2.$$

(Putnam, November 1958)

74. Let $(z)_0 = 1$ and for $n \in \mathbb{N}$, let

$$(z)_n = z(z-1)(z-2)\cdots(z-n+1).$$

Show that

$$(x+y)_n = \sum_{i=0}^{n} \binom{n}{i} (x)_i (y)_{n-i},$$

for all $n \in \mathbb{N}^*$. (Putnam, 1962)

75. In how many ways can the integers from 1 to n be ordered subject to the condition that, except for the first integer on the left, every integer differs by 1 from some integer to the left of it? (Putnam, 1965)

76. Show that, for any positive integer n,

$$\sum_{r=0}^{\lfloor \frac{n-1}{2} \rfloor} \left\{ \frac{n-2r}{n} \binom{n}{r} \right\}^2 = \frac{1}{n} \binom{2n-2}{n-1}.$$

(Putnam, 1965)

77. Show that for $n \in \mathbb{N}$ with $n \geq 2$,

$$\sum_{r=1}^{n} r \sqrt{\binom{n}{r}} < \sqrt{2^{n-1} n^3}.$$

(Spanish MO, 1988)

78. Let $n, r \in \mathbb{N}$ with $r \leq n$ and let k be the HCF of the following numbers:

$$\binom{n}{r}, \binom{n+1}{r}, \ldots, \binom{n+r}{r}.$$

Show that $k = 1$.

79. Show that there are no four consecutive binomial coefficients $\binom{n}{r}$, $\binom{n}{r+1}$, $\binom{n}{r+2}$, $\binom{n}{r+3}$ ($n, r \in \mathbb{N}$ with $r+3 \leq n$) which are in arithmetic progression. (Putnam, 1972)

80. Find the greatest common divisor (i.e., HCF) of

$$\binom{2n}{1}, \binom{2n}{3}, \ldots, \binom{2n}{2n-1}.$$

(Proposed by N. S. Mendelsohn, see *Amer. Math. Monthly*, **78** (1971), 201.)

81. Let $n \in \mathbb{N}$. Show that $\binom{n}{r}$ is odd for each $r \in \{0, 1, 2, \ldots, n\}$ iff $n = 2^k - 1$ for some $k \in \mathbb{N}$.

82. An unbiased coin is tossed n times. What is the expected value of $|H - T|$, where H is the number of heads and T is the number of tails? In other words, evaluate in *closed form*:

$$\frac{1}{2^{n-1}} \sum_{k < \frac{n}{2}} (n - 2k) \binom{n}{k}.$$

("closed form" means a form not involving a series.) (Putnam, 1974)

83. Prove that

$$\binom{pa}{pb} \equiv \binom{a}{b} \pmod{p}$$

for all integers p, a and b with p a prime, and $a \geq b \geq 0$. (Putnam, 1977)

84. The geometric mean (G.M.) of k positive numbers $a_1, a_2, \ldots, a_k$ is defined to be the (positive) kth root of their product. For example, the G.M. of $3, 4, 18$ is 6. Show that the G.M. of a set S of n positive numbers is equal to the G.M. of the G.M.'s of all nonempty subsets of S. (Canadian MO, 1983)

85. For $n, k \in \mathbf{N}$, let $S_k(n) = 1^k + 2^k + \cdots + n^k$. Show that

(i)

$$\sum_{k=0}^{m-1} \binom{m}{k} S_k(n) = (n + 1)^m - 1,$$

(ii)

$$S_m(n) - \sum_{k=0}^{m} (-1)^{m-k} \binom{m}{k} S_k(n) = n^m,$$

where $m \in \mathbf{N}$.

86. Let $P(x)$ be a polynomial of degree n, $n \in \mathbf{N}$, such that $P(k) = 2^k$ for each $k = 1, 2, \ldots, n + 1$. Determine $P(n + 2)$. (Proposed by M. Klamkin, see *Pi Mu Epsilon*, **4** (1964), 77, Problem 158.)

87. Let $X = \{1, 2, \ldots, 10\}$, $\mathcal{A} = \{A \subset X \mid |A| = 4\}$, and $f : \mathcal{A} \to X$ be an arbitrary mapping. Show that there exists $S \subset X$, $|S| = 5$ such that

$$f(S - \{r\}) \neq r$$

for each $r \in S$.

88. (i) Applying the arithmetic-geometric mean inequality on

$$\binom{n+1}{1}, \binom{n+1}{2}, \dots, \binom{n+1}{n},$$

show that

$$(2^{n+1} - 2)^n \geq n^n \prod_{r=1}^{n} \binom{n+1}{r},$$

where $n \in \mathbf{N}$.

(ii) Show that

$$(n!)^{n+1} = \left(\prod_{r=1}^{n} r^r \right) \left(\prod_{r=1}^{n} (r!) \right).$$

(iii) Deduce from (i) and (ii) or otherwise, that

$$\left(\frac{n(n+1)!}{2^{n+1} - 2} \right)^{\frac{n}{2}} \leq \frac{(n!)^{n+1}}{\prod_{r=1}^{n} r^r}.$$

(iv) Show that the equality in (iii) holds iff $n = 1$ or $n = 2$. (See *The College Math. J.*, **20** (1989), 344.)

89. Find, with proof, the number of positive integers whose base-n representation consists of distinct digits with the property that, except for the leftmost digit, every digit differs by ± 1 from some digit further to the left. (Your answer should be an explicit function of n in simplest form.) (USA MO, 1990/4)

90. Let $S_n = \sum_{k=0}^{n} \binom{3n}{3k}$. Prove that

$$\lim_{n \to \infty} (S_n)^{\frac{1}{3n}} = 2.$$

(Bulgarian Spring Competition, 1985)

91. (i) If $f(n)$ denotes the number of 0's in the decimal representation of the positive integer n, what is the value of the sum

$$S = 2^{f(1)} + 2^{f(2)} + \dots + 2^{f(\underbrace{9999999999}_{10})} ?$$

(ii) Let a be a nonzero real number, and $b, k, m \in \mathbf{N}$. Denote by $f(k)$ the number of zeros in the base $b+1$ representation of k. Compute

$$S_n = \sum_{k=1}^{n} a^{f(k)},$$

where $n = (b+1)^m - 1$.

Remark. Part (i) was a 1981 Hungarian Mathematical Competition problem. Part (ii) is a generalization of part (i), and was formulated by M. S. Klamkin (see *Crux Mathematicorum*, **9** (1983), 17–18).

92. Prove that the number of binary sequences of length n which contain exactly m occurrences of "01" is $\binom{n+1}{2m+1}$. (Great Britain MO, 1982/6)

93. There are n people in a gathering, some being acquaintances, some strangers. It is given that every 2 strangers have exactly 2 common friends, and every 2 acquaintances have no common friends. Show that everyone has the same number of friends in the gathering. (23rd Moscow MO)

94. Let $n \in \mathbf{N}^*$. For $p = 1, 2, \ldots$, define

$$A_p(n) = \sum_{0 \le k \le \frac{n}{2}} (-1)^k \left\{ \binom{n}{k} - \binom{n}{k-1} \right\}^p.$$

   Prove that, whenever n is odd, $A_2(n) = nA_1(n)$. (Proposed by H. W. Gould, see *Amer. Math. Monthly*, **80** (1973), 1146.)

95. Let $n \in \mathbf{N}^*$. For $p = 1, 2, \ldots$, define

$$B_p(n) = \sum_{k=0}^{\lfloor \frac{n}{2} \rfloor} \left\{ \binom{n}{k} - \binom{n}{k-1} \right\}^p.$$

   Evaluate $B_2(n)$. (Proposed by E. T. Ordman, see *Amer. Math. Monthly*, **80** (1973), 1066.)

96. Show that

   (i) $\binom{2n-1}{r}^{-1} = \frac{2n}{2n+1} \left\{ \binom{2n}{r}^{-1} + \binom{2n}{r+1}^{-1} \right\}$, where $r, n \in \mathbf{N}$;

   (ii) $\displaystyle\sum_{r=1}^{2n-1} (-1)^{r-1} \binom{2n-1}{r}^{-1} \sum_{j=1}^{r} \frac{1}{j} = \frac{2n}{2n+1} \sum_{r=1}^{2n} \frac{1}{r}$.

   (Proposed by I. Kaucký, see *Amer. Math. Monthly*, **78** (1971), 908.)

97. Given $\ell, m, n \in \mathbf{N}^*$ with $\ell, n \le m$, evaluate the double sum

$$\sum_{i=0}^{\ell} \sum_{j=0}^{i} (-1)^j \binom{m-i}{m-\ell} \binom{n}{j} \binom{m-n}{i-j}.$$

   (Proposed by D. B. West, see *Amer. Math. Monthly*, **97** (1990), 428–429.)

98. Show that

$$\sum_{r=0}^{n} \binom{n}{r} \binom{p}{r+s} \binom{q+r}{m+n} = \sum_{r=0}^{n} \binom{n}{r} \binom{q}{m+r} \binom{p+r}{n+s},$$

   where $m, n, p, q, s \in \mathbf{N}^*$. (See R. C. Lyness, The mystery of the double sevens, *Crux Mathematicorum*, **9** (1983), 194–198.)

Exercise 3

1. Show that among any 5 points in an equilateral triangle of unit side length, there are 2 whose distance is at most $\frac{1}{2}$ units apart.

2. Given any set C of $n+1$ distinct points ($n \in \mathbf{N}$) on the circumference of a unit circle, show that there exist $a, b \in C$, $a \neq b$, such that the distance between them does not exceed $2\sin\frac{\pi}{n}$.

3. Given any set S of 9 points within a unit square, show that there always exist 3 distinct points in S such that the area of the triangle formed by these 3 points is less than or equal to $\frac{1}{8}$. (Beijing Math. Competition, 1963)

4. Show that given any set of 5 numbers, there are 3 numbers in the set whose sum is divisible by 3.

5. Let A be a set of $n+1$ elements, where $n \in \mathbf{N}$. Show that there exist $a, b \in A$ with $a \neq b$ such that $n \mid (a - b)$.

6. Let $A = \{a_1, a_2, \ldots, a_{2k+1}\}$, where $k \geq 1$, be a set of $2k+1$ positive integers. Show that for any permutation $a_{i_1}, a_{i_2}, \ldots, a_{i_{2k+1}}$ of A, the product

$$\prod_{j=1}^{2k+1} (a_{i_j} - a_j)$$

is always even.

7. Let $A \subseteq \{1, 2, \ldots, 2n\}$ such that $|A| = n+1$, where $n \in \mathbf{N}$. Show that there exist $a, b \in A$, with $a \neq b$ such that $a \mid b$.

8. Let A be a subset of $\{1, 2, \ldots, 2n\}$ such that $|A| = n+1$. Show that there exist $a, b \in A$ such that a and b are coprime.

9. Show that among any group of n people, where $n \geq 2$, there are at least two people who know exactly the same number of people in the group (assuming that "knowing" is a symmetry relation).

10. Let $C = \{r_1, r_2, \ldots, r_{n+1}\}$ be a set of $n + 1$ real numbers, where $0 \le r_i < 1$ for each $i = 1, 2, \ldots, n + 1$. Show that there exist r_p, r_q in C, where $p \ne q$, such that $|r_p - r_q| < \frac{1}{n}$.

11. Show that given any set A of 13 distinct real numbers, there exist $x, y \in A$ such that
$$0 < \frac{x - y}{1 + xy} \le 2 - \sqrt{3}.$$

12. Consider a set of $2n$ points in space, $n > 1$. Suppose they are joined by at least $n^2 + 1$ segments. Show that at least one triangle is formed. Show that for each n it is possible to have $2n$ points joined by n^2 segments without any triangles being formed. (Putnam, 1956)

13. Let there be given nine lattice points (points with integral coordinates) in the three dimensional Euclidean space. Show that there is a lattice point on the interior of one of the line segments joining two of these points. (Putnam, 1971)

14. (i) A point (a_1, a_2) in the $x - y$ plane is called a *lattice point* if both a_1 and a_2 are integers. Given any set L_2 of 5 lattice points in the $x - y$ plane, show that there exist 2 distinct members in L_2 whose midpoint is also a lattice point (not necessarily in L_2).

    More generally, we have:

    (ii) A point $(a_1, a_2, \ldots, a_n)$ in the space $\mathbf{R}^n$, where $n \ge 2$ is an integer, is called a *lattice point* if all the a_i's are integers. Show that given any set L_n of $2^n + 1$ lattice points in $\mathbf{R}^n$, there exist 2 distinct members in L_n whose midpoint is also a lattice point (but not necessarily in L_n).

15. Let A be any set of 20 distinct integers chosen from the arithmetic progression $1, 4, 7, \ldots, 100$. Prove that there must be two distinct integers in A whose sum is 104. (Putnam, 1978)

16. Let A be a set of 6 points in a plane such that no 3 are collinear. Show that there exist 3 points in A which form a triangle having an interior angle not exceeding $30°$. (26th Moscow MO)

17. Let $n \ge 3$ be an odd number. Show that there is a number in the set
$$\{2^1 - 1, 2^2 - 1, \ldots, 2^{n-1} - 1\}$$
which is divisible by n. (USSR MO, 1980)

18. There are n people at a party. Prove that there are two people such that, of the remaining $n - 2$ people, there are at least $\lfloor n/2 \rfloor - 1$ of them, each of whom either knows both or else knows neither of the two. Assume that "knowing" is a symmetric relation, and that $\lfloor x \rfloor$ denotes the greatest integer less than or equal to x. (USA MO, 1985/4)

19. For a finite set A of integers, denote by $s(A)$ the sum of numbers in A. Let S be a subset of $\{1, 2, 3, \ldots, 14, 15\}$ such that $s(B) \neq s(C)$ for any 2 disjoint subsets B, C of S. Show that $|S| \leq 5$. (USA MO, 1986)

20. In the rectangular array

$$
\begin{matrix}
a_{11} & a_{12} & \cdots & a_{1n} \\
a_{21} & a_{22} & \cdots & a_{2n} \\
\vdots & \vdots & & \vdots \\
a_{m1} & a_{m2} & \cdots & a_{mn}
\end{matrix}
$$

of $m \times n$ real numbers, the difference between the maximum and the minimum element in each row is at most d, where $d > 0$. Each column is then rearranged in decreasing order so that the maximum element of the column occurs in the first row, and the minimum element occurs in the last row. Show that in the rearranged array the difference between the maximum and the minimum elements in each row is still at most d. (Swedish Math. Competition, 1986)

21. We are given a regular decagon with all diagonals drawn. The number "+1" is attached to each vertex and to each point where diagonals intersect (we consider only internal points of intersection). We can decide at any time to simultaneously change the sign of all such numbers along a given side or a given diagonal. Is it possible after a certain number of such operations to have changed all the signs to negative? (International Mathematics Tournament of the Towns, Senior, 1984)

22. In a football tournament of one round (each team plays each other once, 2 points for win, 1 point for draw and 0 points for loss), 28 teams compete. During the tournament more than 75% of the matches finished in a draw. Prove that there were two teams who finished with the same number of points. (International Mathematics Tournament of the Towns, Junior, 1986)

23. Fifteen problems, numbered 1 through 15, are posed on a certain examination. No student answers two consecutive problems correctly. If 1600 candidates sit the test, must at least two of them have the identical answer patterns? (Assume each question has only 2 possible answers, right or wrong, and assume that no student leaves any question unanswered.) (24th Spanish MO, 1989)

24. Suppose that $a_1 < a_2 \leq \cdots \leq a_n$ are natural numbers such that $a_1 + \cdots + a_n = 2n$ and such that $a_n \neq n + 1$. Show that if n is even, then for some subset K of $\{1, 2, \ldots, n\}$ it is true that $\sum_{i \in K} a_i = n$. Show that this is true also if n is odd when we make the additional

assumption that $a_n \neq 2$. (Proposed by J. Q. Longyear, see *Amer. Math. Monthly*, **80** (1973), 946–947.)

25. Let X be a nonempty set having n elements and C be a colour set with $p \geq 1$ elements. Find the greatest number p satisfying the following property: If we colour in an arbitrary way each subset of X with colours from C such that each subset receives only one colour, then there exist two distinct subsets A, B of X such that the sets $A, B, A \cup B, A \cap B$ have the same colour. (Proposed by I. Tomescu, see *Amer. Math. Monthly*, **95** (1988), 876–877.)

26. Consider the system of p equations in $q = 2p$ unknowns $x_1, x_2, \ldots, x_q$:

$$a_{11}x_1 + a_{12}x_2 + \cdots + a_{1_q}x_q = 0$$
$$a_{21}x_1 + a_{22}x_2 + \cdots + a_{2_q}x_q = 0$$
$$\cdots\cdots\cdots\cdots\cdots\cdots\cdots\cdots\cdots$$
$$a_{p1}x_1 + a_{p2}x_2 + \cdots + a_{pq}x_q = 0$$

with every coefficient a_{ij} a member of the set $\{-1, 0, 1\}$. Prove that the system has a solution $(x_1, x_2, \ldots, x_q)$ such that

(a) all x_i $(j = 1, 2, \ldots, q)$ are integers,
(b) there is at least one value of j for which $x_j \neq 0$,
(c) $|x_j| \leq q$ $(j = 1, 2, \ldots, q)$.

(IMO, 1976/5)

27. An international society has its members from six different countries. The list of members contains 1978 names, numbered $1, 2, \ldots, 1978$. Prove that there is at least one member whose number is the sum of the numbers of two members from his own country, or twice as large as the number of one member from his own country. (IMO, 1978/6)

28. Let $p, q \in \mathbf{N}$. Show that in any given sequence of $R(p, q)$ distinct integers, there is either an increasing subsequence of p terms or a decreasing subsequence of q terms.

29. Show that given any sequence of $pq + 1$ distinct real numbers, where p and q are nonnegative integers, there is either an increasing subsequence of $p + 1$ terms or a decreasing subsequence of $q + 1$ terms. (P. Erdös and G. Szekeres (1935))

30. Show that

(a) $R(p, q) = R(q, p)$, for all $p, q \in \mathbf{N}$;
(b) $R(2, q) = q$, for all $q \in \mathbf{N}$.

31. Let $p, p', q, q' \in \mathbf{N}$ with $p' \leq p$ and $q' \leq q$. Show that

(i) $R(p', q') \le R(p, q)$;

(ii) $R(p - 1, q) \le R(p, q) - 1$ for $p \ge 2$;

(iii) $R(p', q') = R(p, q)$ iff $p' = p$ and $q' = q$.

32. For $p, q \in \mathbf{N}$, show that

$$R(p, q) \le \binom{p + q - 2}{p - 1}.$$

33. Show that

$$R(3, q) \le \frac{1}{2}(q^2 + 3)$$

for $q \ge 1$.

34. Show that $R(3, 5) = 14$.

35. Show that

(a) $R(4, 4) \le 18$,

(b) $R(3, 6) \le 19$.

36. Show that

(a) $R(p_1, p_2, \ldots, p_k) = 1$ if $p_i = 1$ for some $i \in \{1, 2, \ldots, k\}$;

(b) $R(p, 2, 2, \ldots, 2) = p$ for $p \ge 2$.

37. Let $k, p_1, p_2, \ldots, p_k \in \mathbf{N}$ with $k \ge 2$. Show that

$$R(p_1, p_2, \ldots, p_k) = R(p_1, p_2, \ldots, p_k, 2).$$

38. Given any k integers $p_i \ge 2$, $i = 1, 2, \ldots, k$, where $k \ge 2$, show that

$$R(p_1, p_2, \ldots, p_k) \le \sum_{i=1}^{k} R(p_1, \ldots, p_{i-1}, p_i - 1, p_{i+1}, \ldots p_k) - (k - 2).$$

39. Let $k \in \mathbf{N}$, $p_1, p_2, \ldots, p_k \in \mathbf{N}^*$ and $p = \sum_{i=1}^{k} p_i$. Show by induction on p that

$$R(p_1 + 1, p_2 + 1, \ldots, p_k + 1) \le \frac{p!}{p_1! p_2! \cdots p_k!}.$$

40. For $k \in \mathbf{N}$ with $k \ge 2$, let R_k denote $R(\underbrace{3, 3, \ldots, 3}_{k})$. Show that

(a) (i) $R_k \le k(R_{k-1} - 1) + 2$ for $k \ge 3$;

(ii) $R_k \le \lfloor k!e \rfloor + 1$;

(iii) $R_4 \le 66$.

(R. E. Greenwood and A. M. Gleason, *Canad. J. Math.*, **7** (1955), 1–7.)

(b) $R_k \ge 2^k + 1$.

41. Let $k \in \mathbf{N}$ and let $\{S_1, S_2, \ldots, S_k\}$ be any partition of the set $N_n = \{1, 2, \ldots, n\}$, where $n = R(\underbrace{3, 3, \ldots, 3}_{k})$. Show that there exist $i \in \{1, 2, \ldots, k\}$, and some integers a, b, c (not necessarily distinct) in S_i such that $a + b = c$.

42. Show that

   (i) $R(3, 3, 2) = 6$,
   (ii) $R(3, 3, 3) \leq 17$. (See also Example 3.4.1.)

43. A p-clique is *monochromatic* if all its edges are coloured by the same colour.

   (a) Show that for any colouring of the edges of the 6-clique K_6 by 2 colours: blue or red, there are at least two monochromatic 3-cliques (not necessarily disjoint).

   (b) Give a colouring of the edges of K_6 by 2 colours such that there are no three monochromatic 3-cliques.

44. The edges of the 7-clique K_7 are coloured by 2 colours: blue or red. Show that there are at least four monochromatic 3-cliques in the resulting configuration.

45. Given any colouring of the edges of an n-clique $K_n (n \in \mathbf{N}, n \geq 3)$ by 2 colours, let $T(n)$ denote the number of monochromatic 3-cliques in the resulting configuration. Show that

$$T(n) \geq \begin{cases} \frac{1}{3}k(k-1)(k-2) & \text{if } n = 2k, \\ \frac{2}{3}k(k-1)(4k+1) & \text{if } n = 4k+1, \\ \frac{2}{3}k(k+1)(4k-1) & \text{if } n = 4k+3. \end{cases}$$

   (A. W. Goodman, *Amer. Math. Monthly*, **66** (1959), 778–783.)

46. Each of the 36 line segments joining 9 distinct points on a circle is coloured either red or blue. Suppose that each triangle determined by 3 of the 9 points contains at least one red side. Prove that there are four points such that the 6 segments connecting them are red. (Canadian MO, 1976)

Exercise 4

1. A group of 102 students took examinations in Chinese, English and Mathematics. Among them, 92 passed Chinese, 75 English and 63 Mathematics; at most 65 passed Chinese and English, at most 54 Chinese and Mathematics, and at most 48 English and Mathematics. Find the largest possible number of the students that could have passed all the three subjects.

2. (a) Let A, B and C be finite sets. Show that

   (i) $|\bar{A} \cap B| = |B| - |A \cap B|$;
   (ii) $|\bar{A} \cap \bar{B} \cap C| = |C| - |A \cap C| - |B \cap C| + |A \cap B \cap C|$.

   (b) Find the number of integers in the set $\{1, 2, \ldots, 10^3\}$ which are not divisible by 5 nor by 7 but are divisible by 3.

3. Find the number of integers in the set $\{1, 2, \ldots, 120\}$ which are divisible by exactly 'm' of the integers: $2, 3, 5, 7$, where $m = 0, 1, 2, 3, 4$. Find also the number of primes which do not exceed 120.

4. How many positive integers n are there such that n is a divisor of at least one of the numbers $10^{40}, 20^{30}$? (Putnam 1983)

5. Find the number of positive divisors of at least one of the numbers: $10^{60}, 20^{50}, 30^{40}$.

6. Find the number of integers in each of the following sets which are not of the form n^2 or n^3, where $n \in \mathbf{N}$:

   (i) $\{1, 2, \ldots, 10^4\}$,
   (ii) $\{10^3, 10^3 + 1, \ldots, 10^4\}$.

7. Prove Theorem 4.2.1 by

   (a) induction on q;
   (b) Corollary 2 to Theorem 4.3.1.

8. A year is a *leap* year if it is either (i) a multiple of 4 but not a multiple of 100, or (ii) a multiple of 400. For example, 1600 and 1924 were leap years while 2200 will not be. Find the number of leap years between 1000 and 3000 inclusive.

9. Each of n boys attends a school gathering with both of his parents. In how many ways can the $3n$ people be divided into groups of three such that each group contains a boy, a male parent and a female parent, and no boy is with both of his parents in his group?

10. A man has 6 friends. At dinner in a certain restaurant, he has met each of them 12 times, every two of them 6 times, every three of them 4 times, every four of them 3 times, every five twice and all six only once. He has dined out 8 times without meeting any of them. How many times has he dined out altogether?

11. Three identical black balls, four identical red balls and five identical white balls are to be arranged in a row. Find the number of ways that this can be done if all the balls with the same colour do not form a single block.

12. How many arrangements of $a, a, a, b, b, b, c, c, c$ are there such that

   (i) no three consecutive letters are the same?
   (ii) no two consecutive letters are the same?

13. Find the number of shortest routes from corner X to corner Y in the following rectangular grid if the segments AB, BC and BD are all deleted.

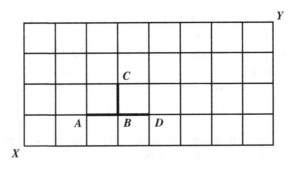

14. Find the number of integer solutions to the equation
$$x_1 + x_2 + x_3 = 28$$
where $3 \leq x_1 \leq 9, 0 \leq x_2 \leq 8$ and $7 \leq x_3 \leq 17$.

15. Find the number of integer solutions to the equation
$$x_1 + x_2 + x_3 = 40$$

where $6 \leq x_1 \leq 15, 5 \leq x_2 \leq 20$ and $10 \leq x_3 \leq 25$.

16. Find the number of integer solutions to the equation

$$x_1 + x_2 + x_3 + x_4 = 20$$

where $1 \leq x_1 \leq 5, 0 \leq x_2 \leq 7, 4 \leq x_3 \leq 8$ and $2 \leq x_4 \leq 6$.

17. Let $k, n, r \in \mathbf{N}$. Show that the number of integer solutions to the equation

$$x_1 + x_2 + \cdots + x_n = r$$

such that $0 \leq x_i \leq k$ for each $i = 1, 2, \ldots, n$ is given by

$$\sum_{i=0}^{n} (-1)^i \binom{n}{i} \binom{r - (k+1)i + n - 1}{n - 1}.$$

18. Let $k, n, r \in \mathbf{N}$. Show that the number of integer solutions to the equation

$$x_1 + x_2 + \cdots + x_n = r$$

such that $1 \leq x_i \leq k$ for each $i = 1, 2, \ldots, n$ is given by

$$\sum_{i=0}^{n} (-1)^i \binom{n}{i} \binom{r - ki - 1}{n - 1}.$$

19. Find the number of ways of arranging n couples $\{H_i, W_i\}$, $i = 1, 2, \ldots, n$, in a row such that H_i is not adjacent to W_i for each $i = 1, 2, \ldots, n$.

20. Let $p, q \in \mathbf{N}$ with p odd and $q > 1$. There are pq beads of q different colours: $1, 2, \ldots, q$ with exactly p beads in each colour. Assuming that beads of the same colour are identical, in how many ways can these beads be put in a string in such a way that

   (i) beads of the same colour must be in a single block?
   (ii) beads of the same colour must be in two separated blocks?
   (iii) beads of the same colour must be in at most two blocks?
   (iv) beads of the same colour must be in at most two blocks and the size of each block must be at least 2?

21. (a) Find the number of ways of distributing r identical objects into n distinct boxes such that no box is empty, where $r \geq n$.
   (b) Show that

$$\sum_{i=0}^{n-1} (-1)^i \binom{n}{i} \binom{r + n - i - 1}{r} = \binom{r - 1}{n - 1},$$

   where $r, n \in \mathbf{N}$ with $r \geq n$.

22. (a) Let B be a subset of A with $|A| = n$ and $|B| = m$. Find the number of r-element subsets of A which contain B as a subset, where $m \le r \le n$.

   (b) Show that for $m, r, n \in \mathbf{N}$ with $m \le r \le n$,

$$\binom{n-m}{n-r} = \sum_{i=0}^{m} (-1)^i \binom{m}{i} \binom{n-i}{r}.$$

23. (a) For $n \in \mathbf{N}$, find the number of binary sequences of length n which do not contain "01" as a block.

   (b) Show that

$$n+1 = \sum_{i=0}^{\lfloor \frac{n}{2} \rfloor} (-1)^i \binom{n-i}{i} 2^{n-2i}.$$

24. In each of the following configurations, each vertex is to be coloured by one of the λ different colours. It how many ways can this be done if any two vertices which are joined by a line segment must be coloured by different colours?

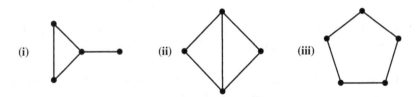

(i)   (ii)   (iii)

25. n persons are to be allocated to q distinct rooms. Find the number of ways that this can be done if only m of the q rooms have exactly k persons each, where $1 \le m \le q$ and $qk \le n$.

26. Suppose that $A = \{k \cdot x_1, k \cdot x_2, \ldots, k \cdot x_n\}$ is a multiset, where $k, n \in \mathbf{N}$. For $m \in \mathbf{N}^*$ with $m \le n$, let $\alpha(m)$ denote the number of ways to arrange the members of A in a row such that the number of blocks containing all the k elements of the same type in the arrangement is exactly m. Show that

$$\alpha(m) = \frac{(-1)^m}{(k!)^n} \binom{n}{m} \sum_{i=m}^{n} (-1)^i \binom{n-m}{i-m} (k!)^i \cdot \{kn - i(k-1)\}!$$

27. Prove identities (4.6.2)–(4.6.7).

   ([HSW]; for (4.6.7), see E. T. H. Wang, E2947, *Amer. Math. Monthly*, **89** (1982), 334.)

28. For $n \in \mathbf{N}$, let C_n denote the number of permutations of the set $\{1, 2, \ldots, n\}$ in which k is never followed immediately by $k + 1$ for each $k = 1, 2, \ldots, n - 1$.

    (i) Find C_n;
    (ii) Show that $C_n = D_n + D_{n-1}$ for each $n \in \mathbf{N}$.

29. Let $m, n \in \mathbf{N}$ with $m < n$. Find, in terms of D_k's, the number of derangements $a_1, a_2, \ldots, a_n$ of $\mathbf{N}_n$ such that

$$\{a_1, a_2, \ldots, a_m\} = \{1, 2, \ldots, m\}.$$

30. Let $m, n \in \mathbf{N}$ with $n \geq 2m$. Find the number of derangements $a_1, a_2, \ldots, a_n$ of $\mathbf{N}_n$ such that

$$\{a_1, a_2, \ldots, a_m\} = \{m + 1, m + 2, \ldots, 2m\}$$

in each of the following cases:

    (i) $n = 2m$;
    (ii) $n = 2m + 1$;
    (iii) $n = 2m + r, r \geq 2$.

31. Apply identity (4.6.8) to prove identities (4.6.10) and (4.6.11).
32. Given $n \in \mathbf{N}$, show that D_n is even iff n is odd.
33. Let $D_n(k) = D(n, n, k)$. Show that

    (i) $D_n(k) = \binom{n}{k} D_{n-k}$;
    (ii) $\binom{n}{0} D_0 + \binom{n}{1} D_1 + \cdots + \binom{n}{n} D_n = n!$, where $D_0 = 1$;
    (iii) $(k + 1) D_{n+1}(k + 1) = (n + 1) D_n(k)$.

34. Let $D_n(k)$ be the number of permutations of the set $\{1, 2, \ldots, n\}, n \geq 1$, which have exactly k fixed points (i.e., $D_n(k) = D(n, n, k)$). Prove that

$$\sum_{k=0}^{n} k \cdot D_n(k) = n!.$$

(IMO, 1987/1)

35. Let $D_n(k)$ denote $D(n, n, k)$. Show that

$$D_n(0) - D_n(1) = (-1)^n$$

for each $n \in \mathbf{N}$.

36. Let $D_n(k)$ denote $D(n, n, k)$. Prove that

$$\sum_{k=0}^{n} (k - 1)^2 D_n(k) = n!.$$

(West Germany MO, 1987)

37. Let $D_n(k)$ denote $D(n, n, k)$. Prove that

$$\sum_{k=r}^{n} k(k-1)\cdots(k-r+1)D_n(k) = n!,$$

where $r, n \in \mathbf{N}^*$ with $r \leq n$. (D. Hanson, *Crux Mathematicorum*, **15**(5) (1989), 139.)

38. (a) Without using equality (4.7.1), show that

   (i) the Euler φ-function is a multiplicative function; that is, $\varphi(mn) = \varphi(m)\varphi(n)$ whenever $m, n \in \mathbf{N}$ with $(m, n) = 1$.

   (ii) for a prime p and an integer $i \geq 1$,

   $$\varphi(p^i) = p^i - p^{i-1}.$$

   (b) Derive equality (4.7.1) from (i) and (ii).

39. (i) Compute $\varphi(100)$ and $\varphi(300)$.

   (ii) Show that $\varphi(m)|\varphi(n)$ whenever $m|n$.

40. Show that for $n \in \mathbf{N}$,

$$\sum(\varphi(d) \mid d \in \mathbf{N}, d|n) = n.$$

41. Let $m, n \in \mathbf{N}$ with $(m, n) = h$. Show by using equality (4.7.1) that

$$\varphi(mn) \cdot \varphi(h) = \varphi(m) \cdot \varphi(n) \cdot h.$$

42. Show that for $n \in \mathbf{N}$ with $n \geq 3, \varphi(n)$ is always even.

43. Let $n \in \mathbf{N}$ with $n \geq 2$. Show that if n has exactly k distinct prime factors, then

$$\varphi(n) \geq n \cdot 2^{-k}.$$

44. Let $n \in \mathbf{N}$ with $n \geq 2$. Show that if n has exactly k distinct odd prime factors, then

$$2^k|\varphi(n).$$

45. Does there exist an $n \in \mathbf{N}$ such that $\varphi(n) = 14$? Justify your answer.

46. For $n \in \mathbf{N}$, show that

$$\varphi(2n) = \begin{cases} \varphi(n) & \text{if } n \text{ is odd} \\ 2\varphi(n) & \text{if } n \text{ is even.} \end{cases}$$

47. For $m, r, q \in \mathbf{N}$ with $m \leq r \leq q$, let

$$A(m, r) = \sum_{k=m}^{r} (-1)^{k-m} \binom{k}{m} \omega(k).$$

Thus Theorem 4.3.1 says that $E(m) = A(m, q)$. Prove that

(i) if m and r have the same parity (i.e., $m \equiv r \pmod 2$), then

$$E(m) \leq A(m, r);$$

(ii) if m and r have different parities, then

$$E(m) \geq A(m, r);$$

(iii) strict inequality in (i) (resp., (ii)) holds iff $\omega(t) > 0$ for some t with $r < t \leq q$.

(See K. M. Koh, Inequalities associated with the principle of inclusion and exclusion, *Mathematical Medley*, Singapore Math. Soc. **19** (1991), 43–52.)

48. Prove the following Bonferroni inequality:

$$\sum_{k=j}^{q} (-1)^{k-j} \omega(k) \geq 0$$

for each $j = 0, 1, \ldots, q$.

49. (i) Let $A_1, A_2, \ldots, A_n$ be n finite sets. Show that

$$\left| \bigcup_{k=1}^{n} A_k \right| \geq \sum_{k=1}^{n} |A_k| - \sum_{1 \leq i < j \leq n} |A_i \cap A_j|.$$

(ii) Apply (i) to prove the following (see Example 1.5.4): A permutation of n couples $\{H_1, W_1, H_2, W_2, \ldots, H_n, W_n\}$ ($n \geq 1$) in a row is said to have property P if at least one couple H_i and W_i ($i = 1, 2, \ldots, n$) are adjacent in the row. Show that for each n there are more permutations with property P than without.

50. Let $B_0 = 1$ and for $r \in \mathbf{N}$, let $B_r = \sum_{k=1}^{r} S(r, k)$. The number B_r is called the rth *Bell* number (see Section 1.7). Show that

(i) Corollary 1 to Theorem 4.5.1 can be written as

$$S(r, k) = \frac{1}{k!} \sum_{j=0}^{k} (-1)^{k-j} \binom{k}{j} j^r,$$

where $r, k \in \mathbf{N}$;

(ii) $B_r = e^{-1} \sum_{j=0}^{\infty} \frac{j^r}{j!}$.

51. For $n \in \mathbf{N}^*$ and $r \in \mathbf{N}$, let

$$a_n = \sum_{i=0}^{n} (-1)^i \frac{r}{i+r} \binom{n}{i}.$$

Show that

$$a_n = \frac{n}{n+r} a_{n-1}.$$

Deduce that

$$a_n = \frac{1}{\binom{n+r}{r}}.$$

52. We follow the terminology given in Theorem 4.3.1. For $1 \le m \le q$, let $L(m)$ denote the number of elements of S that possess *at least* m of the q properties. Show that

$$L(m) = \sum_{k=m}^{q} (-1)^{k-m} \binom{k-1}{m-1} \omega(k).$$

Note. One possible proof is to follow the argument given in the proof of Theorem 4.3.1 and to apply the identity given in the preceding problem.

53. For $k = 1, 2, \ldots, 1992$, let A_k be a set such that $|A_k| = 44$. Assume that $|A_i \cap A_j| = 1$ for all $i, j \in \{1, 2, \ldots, 1992\}$ with $i \ne j$. Evaluate

$$\left| \bigcup_{k=1}^{1992} A_k \right|.$$

54. Twenty-eight random draws are made from the set

$$\{1, 2, 3, 4, 5, 6, 7, 8, 9, A, B, C, D, J, K, L, U, X, Y, Z\}$$

containing 20 elements. What is the probability that the sequence

$$CUBAJULY1987$$

occurs in that order in the chosen sequence? (Belgium, 1987)

55. A sequence of 35 random draws, one at a time with replacement, is made from the set of the English alphabet:

$$\{A, B, C, \ldots, X, Y, Z\}.$$

What is the probability that the string

$$MERRYCHRISTMAS$$

occurs as a block in the sequence?

56. In a group of 1990 people, each person has at least 1327 friends. Show that there are 4 people in the group such that every two of them are friends (assuming that friendship is a mutual relationship). (Proposed by France at the 31st IMO.)

57. Let $\mathbf{C}$ be the set of complex numbers, and let $S = \{z \in \mathbf{C} \mid |z| = 1\}$. For each mapping $f : S \to S$ and $k \in \mathbf{N}$, define the mapping $f^k : S \to S$ by $f^k(z) = \underbrace{f(f(\cdots(f(z))\cdots))}_{k}$. An element $w \in S$ is called an *n-periodic point* $(n \in \mathbf{N})$ *of* f if

$$f^i(\omega) \neq \omega \quad \text{for all } i = 1, 2, \ldots, n-1, \quad \text{but } f^n(w) = w.$$

Suppose $f : S \to S$ is a mapping defined by

$$f(z) = z^m \quad (m \in \mathbf{N}).$$

Find the number of 1989-periodic points of f. (Chinese Math. Competition, 1989)

58. For $m, n \in \mathbf{N}$, let $\mathcal{M}$ be the set of all $m \times n$ $(0, 1)$-matrices. Let

$$\mathcal{M}_r = \{M \in \mathcal{M} \mid M \text{ has at least one zero row}\}$$

and

$$\mathcal{M}_c = \{M \in \mathcal{M} \mid M \text{ has at least one zero column}\}.$$

Show that the number of matrices in $(\mathcal{M} \setminus \mathcal{M}_r) \cap \mathcal{M}_c$ is given by

$$\sum_{i=1}^{n} (-1)^{i-1} \binom{n}{i} (2^{n-i} - 1)^m.$$

(C. J. Everett and P. R. Stein, *Discrete Math.*, **6** (1973), 29.)

59. For $n, m \in \mathbf{N}$ with $m \leq n$, let $P_n(m)$ denote the number of permutations of $\{1, 2, \ldots, n\}$ for which m is the first number whose position is left unchanged. Thus $P_n(1) = (n-1)!$ and $P_n(2) = (n-1)! - (n-2)!$. Show that

(i) $P_n(m) = \sum_{i=0}^{m-1} (-1)^i \binom{m-1}{i} (n-1-i)!$;

(ii) $P_n(m+1) = P_n(m) - P_{n-1}(m)$ for each $m = 1, 2, \ldots, n-1$.

(See Problem 979, *Math. Magazine*, **50** (1977), 269–270.)

60. Let P be a nonempty, finite set with p members, and Q be a finite set with q members. Let $N_k(p, q)$ be the number of binary relations of cardinality k with domain P and range Q. (Equivalently, $N_k(p, q)$ is the number of $p \times q$ matrices of 0's and 1's with exactly k entries equal to 1 and no row or column identically 0.) Compute $\sum_{k=1}^{pq} (-1)^{k-1} N_k(p, q)$. (Proposed by S. Leader, see *Amer. Math. Monthly*, **80** (1973), 84.)

61. Let D_n and M_n denote the derangement number and the Ménage number, respectively. Prove or disprove that the sequence $\{M_n/D_n\}, n = 4, 5, 6, \ldots$ is monotonically increasing and $\lim_{n \to \infty}(M_n/D_n) = 1/e$. (Proposed by E. T. H. Wang, see *Amer. Math. Monthly*, **87** (1980), 829–830.)

62. Show that for $n \in \mathbf{N}$ and $r \in \mathbf{N}^*$,

$$\sum_{k=0}^{n} k^r \binom{n}{k} D_{n-k} = n! \sum_{m=0}^{\min\{r,n\}} S(r,m).$$

Deduce that for $n \geq r$,

$$\sum_{k=0}^{n} k^r \binom{n}{k} D_{n-k} = B_r \cdot n!,$$

where B_r is the rth Bell number. (See *Amer. Math. Monthly*, **94** (1987), 187–189.)

63. Let $S = \{1, 2, 3, \ldots, 280\}$. Find the smallest integer n such that each n-element subset of S contains at least 5 numbers which are pairwise relatively prime. (IMO, 1991/3)

Exercise 5

1. Find the coefficient of x^{20} in the expansion of $(x^3 + x^4 + x^5 + \cdots)^3$.
2. Find the coefficients of x^9 and x^{14} in the expansion of $(1 + x + x^2 + \cdots + x^5)^4$.
3. Prove Theorem 5.1.1 (iv), (vi), (viii), (ix) and (x).
4. Find the generating function for the sequence (c_r), where $c_0 = 0$ and $c_r = \sum_{i=1}^{r} i^2$ for $r \in \mathbf{N}$. Hence show that
$$\sum_{i=1}^{r} i^2 = \binom{r+1}{3} + \binom{r+2}{3}.$$

5. Find the generating function for the sequence (c_r), where $c_r = \sum_{i=0}^{r} i2^i$ with $r \in \mathbf{N}^*$. Hence show that
$$\sum_{i=0}^{r} i2^i = 2 + (r-1)2^{r+1}.$$

6. (i) For $r \in \mathbf{N}^*$, let $a_r = \frac{1}{4^r}\binom{2r}{r}$. Show that the generating function for the sequence (a_r) is given by $(1-x)^{-\frac{1}{2}}$.
   (ii) Using the identity
$$(1-x)^{-1} = (1-x)^{-\frac{1}{2}}(1-x)^{-\frac{1}{2}},$$
   show that
$$\sum_{k=0}^{n} \binom{2k}{k}\binom{2(n-k)}{n-k} = 4^n$$
   for each $n \in \mathbf{N}^*$.

7. Show that
$$\sum_{r=1}^{n} r\binom{n}{r}\binom{m}{r} = n\binom{n+m-1}{n}.$$

49

8. Find the number of ways to distribute 10 identical pieces of candy to 3 children so that no child gets more than 4 pieces.

9. Find the number of ways to distribute 40 identical balls to 7 distinct boxes if box 1 must hold at least 3, and at most 10, of the balls.

10. Find the number of ways to select $2n$ balls from n identical blue balls, n identical red balls and n identical white balls, where $n \in \mathbf{N}$.

11. In how many ways can 100 identical chairs be divided among 4 different rooms so that each room will have $10, 20, 30, 40$ or 50 chairs?

12. Let a_r be the number of ways of distributing r identical objects into 5 distinct boxes so that boxes $1, 3$ and 5 are not empty. Let b_r be the number of ways of distributing r identical objects into 5 distinct boxes so that each of the boxes 2 and 4 contains at least two objects.

   (i) Find the generating function for the sequence (a_r).
   (ii) Find the generating function for the sequence (b_r).
   (iii) Show that $a_r = b_{r+1}$ for each $r = 1, 2, \ldots$.

13. For $r \in \mathbf{N}^*$, let a_r denote the number of integer solutions to the equation

$$x_1 + x_2 + x_3 = r$$

   where $3 \le x_1 \le 9, 0 \le x_2 \le 8$ and $7 \le x_3 \le 17$. Find the generating function for (a_r), and determine the value of a_{28}.

14. In how many ways can 3000 identical pencils be divided up, in packages of 25, among four student groups so that each group gets at least 150, but not more than 1000, of the pencils?

15. Find the number of selections of 10 letters from "$F, U, N, C, \ T, I, O$" that contain at most three U's and at least one O.

16. Find the generating function for the sequence (a_r) in each of the following cases: a_r is

   (i) the number of selections of r letters (not necessarily distinct) from the set $\{D, R, A, S, T, I, C\}$ that contain at most 3 D's and at least 2 T's;
   (ii) the number of partitions of r into parts of sizes $1, 2, 3, 5$, and 8;
   (iii) the number of partitions of r into distinct parts of sizes 5, 10, and 15;
   (iv) the number of partitions of r into distinct odd parts;
   (v) the number of partitions of r into distinct even parts;
   (vi) the number of integer solutions to the inequality

$$x_1 + x_2 + x_3 + x_4 + x_5 \le r$$

with $1 \leq x_i \leq 6$ for each $i = 1, 2, \ldots, 5$.

17. Find the number of $4n$-element multi-subsets of the multi-set

$$\{(3n) \cdot x, (3n) \cdot y, (3n) \cdot z\},$$

where $n \in \mathbf{N}$.

18. Find the number of $3n$-element multi-subsets of the multi-set

$$M = \{n \cdot z_1, n \cdot z_2, \ldots, n \cdot z_m\},$$

where $n, m \in \mathbf{N}$ and $n, m \geq 3$.

19. What is the probability that a roll of 5 distinct dice yields a sum of 17?

20. Find the generating function for the sequence (a_r), where a_r is the number of ways to obtain a sum of r by a roll of *any* number of distinct dice.

21. For $k, m \in \mathbf{N}$ and $r \in \mathbf{N}^*$, let a_r denote the number of ways of distributing r identical objects into $2k + 1$ distinct boxes such that the first $k + 1$ boxes are non-empty, and b_r denote the number of ways of distributing r identical objects into $2k+1$ distinct boxes such that each of the last k boxes contains at least m objects.

   (i) Find the generating function for the sequence (a_r);
   (ii) Find the generating function for the sequence (b_r);
   (iii) Show that $a_r = b_{r+(m-1)k-1}$.

22. Find the generating function for the sequence (a_r), where a_r is the number of integer solutions to the equation

$$x_i + 2x_2 + 3x_3 + 4x_4 = r$$

with $x_i \geq 0$ for each $i = 1, 2, 3, 4$.

23. For $r \in \mathbf{N}^*$, let a_r denote the number of ways of selecting 4 distinct integers from $\{1, 2, \ldots, r\}$ such that no two are consecutive. Find the generating function for (a_r) and deduce that $a_r = \binom{r-3}{4}$.

24. For $r \in \mathbf{N}^*$, and $m, t \in \mathbf{N}$, let a_r denote the number of m-element subsets $\{n_1, n_2, \ldots, n_m\}$ of the set $\{1, 2, \ldots, r\}$, where $n_1 < n_2 < \cdots < n_m$ and $n_{i+1} - n_i \geq t$ for each $i = 1, 2, \ldots, m - 1$. Find the generating function for (a_r) and deduce that

$$a_r = \binom{r - (m - 1)(t - 1)}{m}.$$

(See Problem 1.91.)

25. For $r \in \mathbf{N}^*$, let a_r be the number of integer solutions to the inequality

$$x_1 + x_2 + x_3 + x_4 \leq r,$$

where $3 \leq x_1 \leq 9, 1 \leq x_2 \leq 10, x_3 \geq 2$ and $x_4 \geq 0$. Find the generating function for the sequence (a_r) and the value of a_{20}.

26. Prove that if $-1 < \alpha < 0$ and $n \in \mathbf{N}^*$, then

$$\binom{2\alpha}{2n} \geq (2n+1)\binom{\alpha}{n}^2;$$

while if $\alpha < -1$, the inequality is reversed. (Proposed by S. I. Rosencrans, see *Amer. Math. Monthly*, **79** (1972), 1136.)

27. For $n \in \mathbf{N}$, let

$$a_{n-1} = \sum_{k=0}^{n-1} \left\{ \binom{n}{0} + \binom{n}{1} + \cdots + \binom{n}{k} \right\} \cdot$$

$$\left\{ \binom{n}{k+1} + \binom{n}{k+2} + \cdots + \binom{n}{n} \right\}.$$

Let $B(x)$ be the generating function for the sequence (b_k), where $b_k = \binom{n}{0} + \binom{n}{1} + \cdots + \binom{n}{k}$.

   (i) Show that

$$B(x) = \frac{(1+x)^n}{1-x}.$$

   (ii) Show that

$$a_{n-1} = \sum_{r=0}^{n-1} \binom{2n}{r}(n-r).$$

   (iii) Show that

$$a_{n-1} = \frac{n}{2}\binom{2n}{n}.$$

(G. Chang and Z. Shan, 1984.)

28. For $m, n \in \mathbf{N}$ and $r \in \mathbf{N}^*$, a generalized quantity $\binom{n}{r}_m$ of binomial coefficients is defined as follows:

$$\binom{1}{r}_m = \begin{cases} 1 & \text{if } 0 \leq r \leq m-1 \\ 0 & \text{otherwise,} \end{cases}$$

and

$$\binom{n}{r}_m = \sum_{i=0}^{m-1} \binom{n-1}{r-i}_m \quad \text{for } n \geq 2.$$

Note that $\binom{n}{r}_2 = \binom{n}{r}$. Show that

(i) $\binom{n}{r}_m$ is the number of integer solutions to the equation

$$x_1 + x_2 + \cdots + x_n = r$$

with $0 \le x_i \le m - 1$ for each $i = 1, 2, \ldots, n$;

(ii) $\binom{n}{0}_m = 1$;

(iii) $\binom{n}{1}_m = n$, where $m \ge 2$;

(iv) $\binom{n}{r}_m = \binom{n}{s}_m$, where $r + s = n(m - 1)$;

(v) $\sum_{r=0}^{n(m-1)} \binom{n}{r}_m = m^n$;

(vi) the generating function for $\left(\binom{n}{r}_m\right)_{r=0,1,2,\ldots}$ is $(1+x+\cdots+x^{m-1})^n$;

(vii) $\sum_{r=0}^{n(m-1)} (-1)^r \binom{n}{r}_m = \begin{cases} 0 & \text{if } m \text{ is even} \\ 1 & \text{if } m \text{ is odd} \end{cases}$

(viii) $\sum_{r=1}^{n(m-1)} r\binom{n}{r}_m = \dfrac{n(m-1)m^n}{2}$;

(ix) $\sum_{r=1}^{n(m-1)} (-1)^{r-1} r\binom{n}{r}_m = \begin{cases} 0 & \text{if } m \text{ is even} \\ \dfrac{n(1-m)}{2} & \text{if } m \text{ is odd} \end{cases}$

(x) $\sum_{i=0}^{r} \binom{p}{i}_m \binom{q}{r-i}_m = \binom{p+q}{r}_m$, where $p, q \in \mathbf{N}$;

(xi) $\binom{n}{r}_m = \sum_{i=0}^{n} (-1)^i \binom{n}{i}\binom{n-1+r-mi}{n-1}$.

(See C. Cooper and R. E. Kennedy, A dice-tossing problem, *Crux Mathematicorum*, **10** (1984), 134–138.)

29. Given $n \in \mathbf{N}$, evaluate the sum

$$S_n = \sum_{r=0}^{n} 2^{r-2n} \binom{2n - r}{n}.$$

(Proposed by the Israeli Team at the 31st IMO.)

30. For each $r \in \mathbf{N}^*$, let

$$a_r = 1 \cdot 4 \cdot 7 \cdots (3r + 1).$$

Show that the exponential generating function for the sequence (a_r) is given by $(1 - 3x)^{-\frac{1}{3}}$.

31. For $n \in \mathbf{N}$, find the number of ways to colour the n squares of a $1 \times n$ chessboard using the colours: blue, red and white, if each square is coloured by a colour and an even number of squares are to be coloured red.

32. Find the number of n-digit quaternary sequences that contain an odd number of 0's, an even number of 1's and at least one 3.

33. For $n \in \mathbf{N}$, find the number of words of length n formed by the symbols: $\alpha, \beta, \gamma, \delta, \epsilon, \lambda$ in which the total number of α's and β's is (i) even, (ii) odd.

34. For $r \in \mathbf{N}^*$, find the number of ways of distributing r distinct objects into 5 distinct boxes such that each of the boxes $1, 3$, and 5 must hold an odd number of objects while each of the remaining boxes must hold an even number of objects.

35. Prove the following summations for all real z:

    (i) $\sum_{k=0}^{n} \binom{z}{2k}\binom{z-2k}{n-k}2^{2k} = \binom{2z}{2n}$,

    (ii) $\sum_{k=0}^{n} \binom{z+1}{2k+1}\binom{z-2k}{n-k}2^{2k+1} = \binom{2z+2}{2n+1}$.

    (Proposed by M. Machover and H. W. Gould, see *Amer. Math. Monthly*, **75** (1968), 682.)

36. Prove that

$$\sum_{r=1}^{n}\sum_{k=0}^{r}(-1)^{k+1}\frac{k}{r}\binom{r}{k}k^{n-1} = 0,$$

    where $n = 2, 3, 4, \ldots$. (Proposed by G. M. Lee, see *Amer. Math. Monthly*, **77** (1970), 308–309.)

37. Prove that

$$\sum \frac{1}{k_1!k_2!\cdots k_n!} = \frac{1}{r!}\binom{n-1}{r-1},$$

    where the sum is taken over all $k_1, k_2, \ldots, k_n \in \mathbf{N}^*$ with $\sum_{i=1}^{n} k_i = r$ and $\sum_{i=1}^{n} ik_i = n$.

    (Proposed by D. Ž. Djoković, see *Amer. Math. Monthly*, **77** (1970), 659.)

38. Ten female workers and eight male workers are to be assigned to work in one of four different departments of a company. In how many ways can this be done if

    (i) each department gets at least one worker?

    (ii) each department gets at least one female worker?

    (iii) each department gets at least one female worker and at least one male worker?

39. For $r \in \mathbf{N}^*$, find the number of r-permutations of the multi-set

$$\{\infty \cdot \alpha, \infty \cdot \beta, \infty \cdot \gamma, \infty \cdot \lambda\}$$

    in which the number of α's is odd while the number of λ's is even.

40. For $r \in \mathbf{N}^*$ and $n \in \mathbf{N}$, let $a_r = F(r, n)$, which is the number of ways to distribute r distinct objects into n distinct boxes so that no box is empty (see Theorem 4.5.1). Thus $a_r = n!S(r, n)$, where $S(r, n)$ is a

Stirling number of the second kind. Find the exponential generating function for the sequence (a_r), and show that for $r \geq 2$,

$$\sum_{m=0}^{\infty} (-1)^m m! S(r, m+1) = 0.$$

41. For $n \in \mathbf{N}$, let $A_n(x)$ be the exponential generating function for the sequence $(S(0, n), S(1, n), \ldots, S(r, n), \ldots)$. Find $A_n(x)$ and show that

$$\frac{d}{dx} A_n(x) = n A_n(x) + A_{n-1}(x),$$

where $n \geq 2$.

42. Let $B_0 = 1$ and for $r \in \mathbf{N}$, let $B_r = \sum_{k=1}^{r} S(r, k)$. The numbers B_r's are called the Bell numbers (see Section 1.7). Show that the exponential generating function for the sequence (B_r) is given by $e^{e^x - 1}$.

43. Let $n \in \mathbf{N}$ and $r \in \mathbf{N}^*$.

   (a) Find the number of ways of distributing r distinct objects into n distinct boxes such that the objects in each box are ordered.

   (b) Let a_r denote the number of ways to select at most r objects from r distinct objects and to distribute them into n distinct boxes such that the objects in each box are ordered. Show that

      (i) $a_r = \sum_{i=0}^{r} \binom{r}{i} n^{(i)}$, where $n^{(i)} = n(n+1) \cdots (n+i-1)$ with $n^{(0)} = 1$;

      (ii) the exponential generating function for the sequence (a_r) is given by

      $$e^x (1-x)^{-n}.$$

44. Find the generating function for the sequence (a_r) in each of the following cases: a_r is the number of ways of distributing r identical objects into

   (i) 4 distinct boxes;
   (ii) 4 distinct boxes so that no box is empty;
   (iii) 4 identical boxes so that no box is empty;
   (iv) 4 identical boxes.

45. For $n \in \mathbf{N}$, show that the number of partitions of n into parts where no even part occurs more than once is equal to the number of partitions of n in which parts of each size occur at most three times.

46. For $r \in \mathbf{N}^*$ and $n \in \mathbf{N}$, let a_r be the number of integer solutions to the equation

$$x_1 + x_2 + \cdots + x_n = r,$$

where $x_1 \geq x_2 \geq \cdots \geq x_n \geq 1$. Find the generating function for the sequence (a_r).

47. For $r \in \mathbf{N}^*$ and $n \in \mathbf{N}$, let b_r be the number of integer solutions to the equation

$$x_1 + x_2 + \cdots + x_n = r,$$

where $x_1 \geq x_2 \geq \cdots \geq x_n \geq 0$. Find the generating function for the sequence (b_r).

48. For $r \in \mathbf{N}^*$ and $n \in \mathbf{N}$, let a_r denote the number of ways to distribute r identical objects into n identical boxes, and b_r denote the number of integer solutions to the equation

$$\sum_{k=1}^{n} k x_k = r$$

with $x_k \geq 0$ for each $k = 1, 2, \ldots, n$. Show that $a_r = b_r$ for each $r \in \mathbf{N}^*$.

49. For $r \in \mathbf{N}^*$, let a_r denote the number of partitions of r into distinct powers of 2.

   (i) Find the generating function for (a_r);
   (ii) Show that $a_r = 1$ for all $r \geq 1$;
   (iii) Give an interpretation of the result in (ii).

50. Show that for $n \in \mathbf{N}$, the number of partitions of $2n$ into *distinct even* parts is equal to the number of partitions of n into *odd* parts.

51. Let $k, n \in \mathbf{N}$. Show that the number of partitions of n into odd parts is equal to the number of partitions of kn into distinct parts whose sizes are multiples of k.

52. Let $p(n)$ be the number of partitions of n. Show that

$$p(n) \leq \frac{1}{2}(p(n+1) + p(n-1)),$$

where $n \in \mathbf{N}$ with $n \geq 2$.

53. For $n, k \in \mathbf{N}$ with $k \leq n$, let $p(n, k)$ denote the number of partitions of n into exactly k parts.

   (i) Determine the values of $p(5, 1), p(5, 2), p(5, 3)$ and $p(8, 3)$.

(ii) Show that

$$\sum_{k=1}^{m} p(n,k) = p(n+m,m),$$

where $m \in \mathbf{N}$ and $m \le n$.

54. (i) With $p(n,k)$ as defined in the preceding problem, determine the values of $p(5,3), p(7,2)$ and $p(8,3)$.
    (ii) Show that

$$p(n-1,k-1) + p(n-k,k) = p(n,k).$$

55. Given $n, k \in \mathbf{N}$ with $n \le k$, show that

$$p(n+k,k) = p(2n,n) = p(n).$$

56. For $n, k \in \mathbf{N}$ with $k \le n$, show that

$$p(n,k) \ge \frac{1}{k!}\binom{n-1}{k-1}.$$

57. Given $n, k \in \mathbf{N}$, show that the number of partitions of n into k *distinct* parts is equal to $p\left(n - \binom{k}{2}, k\right)$.

58. Prove the corollary to Theorem 5.3.5.

59. (i) Prove Theorem 5.3.3.
    (ii) Prove Theorem 5.3.4.

60. For positive integers n, let $C(n)$ be the number of representations of n as a sum of nonincreasing powers of 2, where no power can be used more than three times. For example, $C(8) = 5$ since the representations for 8 are:

$$8, 4+4, 4+2+2, 4+2+1+1, \text{ and } 2+2+2+1+1.$$

Prove or disprove that there is a polynomial $Q(x)$ such that $C(n) = \lfloor Q(n) \rfloor$ for all positive integers n.
(Putnam, 1983.)

61. For $n \in \mathbf{N}$, let $C(n)$ be the number defined in the preceding problem. Show that the generating function for the sequence $(C(n))$ is given by

$$\frac{1}{(1+x)(1-x)^2}.$$

Deduce that $C(n) = \lfloor \frac{n+2}{2} \rfloor$ for each $n \in \mathbf{N}$.

62. (a) (i) List all partitions of 8 into 3 parts.
        (ii) List all noncongruent triangles whose sides are of integer length a, b, c such that $a + b + c = 16$.

(iii) Is the number of partitions obtained in (i) equal to the number of noncongruent triangles obtained in (ii)?

(b) For $r \in \mathbf{N}^*$, let a_r denote the number of noncongruent triangles whose sides are of integer length a, b, c such that $a + b + c = 2r$, and let b_r denote the number of partitions of r into 3 parts.

(i) Show by (BP) that $a_r = b_r$ for each $r \in \mathbf{N}^*$.
(ii) Find the generating function for (a_r).

63. A partition P of a positive integer n is said to be *self-conjugate* if P and its conjugate have the same Ferrers diagram.

(i) Find all the self-conjugate partitions of 15.
(ii) Find all the partitions of 15 into distinct odd parts.
(iii) Show that the number of the self-conjugate partitions of n is equal to the number of partitions of n into distinct odd parts.

64. Show that the number of self-conjugate partitions of n with largest size equal to m is equal to the number of self-conjugate partitions of $n - 2m + 1$ with largest size not exceeding $m - 1$.

65. (i) The largest square of asterisks in the upper left-hand corner of the Ferrers diagram is called the *Durfee* square of the diagram. Find the generating function for the number of self-conjugate partitions of r whose Durfee square is an $m \times m$ square, where $m \in \mathbf{N}$.

(ii) Deduce that

$$\prod_{k=0}^{\infty}(1 + x^{2k+1}) = 1 + \sum_{m=1}^{\infty} \frac{x^{m^2}}{\prod_{k=1}^{m}(1 - x^{2k})}.$$

66. Let $A(x)$ be the generating function for the sequence $(p(r))$ where $p(r)$ is the number of partitions of r.

(i) Find $A(x)$;
(ii) Use the notion of Durfee square to prove that

$$\left[\prod_{k=1}^{\infty}(1 - x^k)\right]^{-1} = 1 + \sum_{m=1}^{\infty} \frac{x^{m^2}}{\prod_{k=1}^{m}(1 - x^k)^2}.$$

67. By considering isosceles right triangles of asterisks in the upper left-hand corner of a Ferrers diagram, show that

$$\prod_{k=1}^{\infty}(1 + x^{2k}) = 1 + \sum_{m=1}^{\infty} \frac{x^{m(m+1)}}{\prod_{k=1}^{m}(1 - x^{2k})}.$$

68. Let $p, q, r \in \mathbf{N}$ with $p < r$ and $q < r$. Show that the number of partitions of $r - p$ into $q - 1$ parts with sizes not exceeding p, is equal to the number of partitions of $r - q$ into $p - 1$ parts with sizes not exceeding q.

69. For $n \in \mathbf{N}$, let $p_e(n)$ (resp., $p_o(n)$) denote the number of partitions of n into an even (resp., odd) number of *distinct* parts. Show that

$$p_e(n) - p_o(n) = \begin{cases} (-1)^k & \text{if } n = \frac{k(3k\pm1)}{2} \\ 0 & \text{otherwise.} \end{cases}$$

70. Prove the following Euler's pentagonal number theorem:

$$\prod_{k=1}^{\infty}(1 - x^k) = \sum_{m=-\infty}^{\infty} (-1)^m x^{\frac{1}{2}m(3m-1)}.$$

71. For $n \in \mathbf{N}$, show that

$$p(n) - p(n - 1) - p(n - 2) + p(n - 5) + p(n - 7)$$
$$+ \cdots + (-1)^m p\left(n - \frac{1}{2}m(3m - 1)\right) + \cdots$$
$$+ (-1)^m p\left(n - \frac{1}{2}m(3m + 1)\right) + \cdots = 0.$$

72. For $j \in \mathbf{N}^*$, let $\beta(j) = \frac{3j^2+j}{2}$. Prove the following Euler identity:

$$\sum_{j \text{ even}} p(n - \beta(j)) = \sum_{j \text{ odd}} p(n - \beta(j))$$

by (BP), where $n \in \mathbf{N}$.
(See D. M. Bressoud and D. Zeilberger, Bijecting Euler's partitions-recurrence, *Amer. Math. Monthly*, **92** (1985), 54–55.)

73. For $r, n \in \mathbf{N}$, let $f(r, n)$ denote the number of partitions of n of the form

$$n = n_1 + n_2 + \cdots + n_s,$$

where, for $i = 1, 2, \ldots, s - 1, n_i \geq rn_{i+1}$, and let $g(r, n)$ denote the number of partitions of n, where each part is of the form $1 + r + r^2 + \cdots + r^k$ for some $k \in \mathbf{N}^*$. Show that

$$f(r, n) = g(r, n).$$

(See D. R. Hickerson, A partition identity of the Euler type, *Amer. Math. Monthly*, **81** (1974), 627–629.)

Exercise 6

1. Solve

$$a_n = 3a_{n-1} - 2a_{n-2},$$

   given that $a_0 = 2$ and $a_1 = 3$.

2. Solve

$$a_n - 6a_{n-1} + 9a_{n-2} = 0,$$

   given that $a_0 = 2$ and $a_1 = 3$.

3. Solve

$$a_n = \frac{1}{2}(a_{n-1} + a_{n-2}),$$

   given that $a_0 = 0$ and $a_1 = 1$.

4. Solve

$$a_n - 4a_{n-1} + 4a_{n-2} = 0,$$

   given that $a_0 = -\frac{1}{4}$ and $a_1 = 1$.

5. Solve

$$2a_n = a_{n-1} + 2a_{n-2} - a_{n-3},$$

   given that $a_0 = 0$, $a_1 = 1$ and $a_2 = 2$.

6. Solve

$$a_n - 6a_{n-1} + 11a_{n-2} - 6a_{n-3} = 0,$$

   given that $a_0 = \frac{1}{3}$, $a_1 = 1$ and $a_2 = 2$.

7. Solve

$$a_n = -a_{n-1} + 16a_{n-2} - 20a_{n-3},$$

   given that $a_0 = 0$, $a_1 = 1$ and $a_2 = -1$.

8. Find the general solution of the recurrence relation

$$a_n + a_{n-1} - 3a_{n-2} - 5a_{n-3} - 2a_{n-4} = 0.$$

9. Solve

$$a_n = \frac{1}{2}a_{n-1} - 3,$$

given that $a_0 = 2(3 + \sqrt{3})$.

10. Solve

$$a_n - 3a_{n-1} = 3 \cdot 2^n - 4n,$$

given that $a_1 = 2$.

11. Solve

$$a_n - a_{n-1} = 4n - 1,$$

given that $a_0 = 1$.

12. Solve

$$a_n = pa_{n-1} + q,$$

given that $a_0 = r$, where p, q and r are constants.

13. Let (a_n) be a sequence of numbers such that

   (i) $a_0 = 1, a_1 = \frac{3}{5}$ and
   (ii) the sequence $(a_n - \frac{1}{10}a_{n-1})$ is a geometric progression with common ratio $\frac{1}{2}$.
   Find a general formula for a_n, $n \geq 0$.

14. Solve

$$a_n^4 a_{n-1} = 10^{10},$$

given that $a_0 = 1$ and $a_n > 0$ for all n.

15. A sequence (a_n) of positive numbers satisfies

$$a_n = 2\sqrt{a_{n-1}}$$

with the initial condition $a_0 = 25$. Show that $\lim_{n \to \infty} a_n = 4$.

16. A sequence (a_n) of numbers satisfies

$$\left(\frac{a_n}{a_{n-1}}\right)^2 = \frac{a_{n-1}}{a_{n-2}}$$

with the initial conditions $a_0 = \frac{1}{4}$ and $a_1 = 1$. Solve the recurrence relation.

17. Solve

$$a_n + 3a_{n-1} = 4n^2 - 2n + 2^n,$$

given that $a_0 = 1$.

18. Solve

$$a_n - 2a_{n-1} + 2a_{n-2} = 0,$$

given that $a_0 = 1$ and $a_1 = 2$.

19. Solve

$$a_n - 4a_{n-1} + 4a_{n-2} = 2^n,$$

given that $a_0 = 0$ and $a_1 = 3$.

20. Solve

$$a_n - a_{n-1} - 2a_{n-2} = 4n,$$

given that $a_0 = -4$ and $a_1 = -5$.

21. Solve

$$a_n + a_{n-1} - 2a_{n-2} = 2^{n-2},$$

given that $a_0 = a_1 = 0$.

22. Solve

$$a_n - 3a_{n-1} + 2a_{n-2} = 2^n,$$

given that $a_0 = 0$ and $a_1 = 5$.

23. Solve

$$a_n + 5a_{n-1} + 6a_{n-2} = 3n^2,$$

given that $a_0 = 0$ and $a_1 = 1$.

24. Let (a_n) be a sequence of numbers satisfying the recurrence relation

$$pa_n + qa_{n-1} + ra_{n-2} = 0$$

with the initial conditions $a_0 = s$ and $a_1 = t$, where p, q, r, s, t are constants such that $p + q + r = 0$, $p \neq 0$ and $s \neq t$. Solve the recurrence relation.

25. Let (a_n) be a sequence of numbers satisfying the recurrence relation

$$a_n = \frac{pa_{n-1} + q}{ra_{n-1} + s}$$

where p, q, r and s are constants with $r \neq 0$.

(i) Show that

$$ra_n + s = p + s + \frac{qr - ps}{ra_{n-1} + s}. \tag{1}$$

(ii) By the substitution $ra_n + s = \frac{b_{n+1}}{b_n}$, show that (1) can be reduced to the second order linear homogeneous recurrence relation for (b_n):

$$b_{n+1} - (p + s)b_n + (ps - qr)b_{n-1} = 0.$$

26. Solve

$$a_n = \frac{3a_{n-1}}{2a_{n-1} + 1},$$

given that $a_0 = \frac{1}{4}$.

27. Solve

$$a_n = \frac{3a_{n-1} + 1}{a_{n-1} + 3},$$

given that $a_0 = 5$.

28. A sequence (a_n) of numbers satisfies the condition

$$(2 - a_n)a_{n+1} = 1, \quad n \geq 1.$$

Find $\lim_{n \to \infty} a_n$.

29. For $n \in \mathbf{N}$, recall that D_n is the number of derangements of the set $\mathbf{N}_n$. Prove by a combinatorial argument that

$$D_n = (n - 1)(D_{n-1} + D_{n-2}).$$

30. For $n \in \mathbf{N}$, let a_n denote the number of ternary sequences of length n in which no two 0's are adjacent. Find a recurrence relation for (a_n) and solve the recurrence relation.

31. Let $C_0, C_1, C_2, \ldots$ be the sequence of circles in the Cartesian plane defined as follows:

(1) C_0 is the circle $x^2 + y^2 = 1$,

(2) for $n = 0, 1, 2, \ldots$, the circle C_{n+1} lies in the upper half-plane and is tangent to C_n as well as to both branches of the hyperbola $x^2 - y^2 = 1$.

Let a_n be the radius of C_n.

(i) Show that $a_n = 6a_{n-1} - a_{n-2}$, $n \geq 2$.

(ii) Deduce from (i) that a_n is an integer and
$$a_n = \frac{1}{2}[(3 + 2\sqrt{2})^n + (3 - 2\sqrt{2})^n].$$

(Proposed by B. A. Reznick, see *Amer. Math. Monthly*, **96** (1989), 262.)

32. The $n \times n$ determinant a_n is defined by
$$a_n = \begin{vmatrix} p+q & pq & 0 & 0 & \cdots & 0 & 0 \\ 1 & p+q & pq & 0 & \cdots & 0 & 0 \\ 0 & 1 & p+q & pq & \cdots & 0 & 0 \\ 0 & 0 & 1 & p+q & \cdots & 0 & 0 \\ \vdots & \vdots & \vdots & \vdots & \ddots & \vdots & \vdots \\ 0 & 0 & 0 & 0 & \cdots & p+q & pq \\ 0 & 0 & 0 & 0 & \cdots & 1 & p+q \end{vmatrix}$$
where p and q are nonzero constants. Find a recurrence relation for (a_n), and solve the recurrence relation.

33. Consider the following $n \times n$ determinant:
$$a_n = \begin{vmatrix} pq+1 & q & 0 & 0 & \cdots & 0 \\ p & pq+1 & q & 0 & \cdots & 0 \\ 0 & p & pq+1 & q & \cdots & 0 \\ 0 & 0 & p & pq+1 & \cdots & 0 \\ \vdots & \vdots & \vdots & \vdots & \ddots & \vdots \\ 0 & 0 & 0 & 0 & \cdots & q \\ 0 & 0 & 0 & 0 & \cdots & pq+1 \end{vmatrix}$$
where p and q are nonzero constants. Find a recurrence relation for (a_n), and solve the recurrence relation.

34. Given $n \in \mathbf{N}$, find the number of n-digit positive integers which can be formed from $1, 2, 3, 4$ such that 1 and 2 are not adjacent.

35. A $2 \times n$ rectangle ($n \in \mathbf{N}$) is to be paved with 1×2 identical blocks and 2×2 identical blocks. Let a_n denote the number of ways that can be done. Find a recurrence relation for (a_n), and solve the recurrence relation.

36. For $n \in \mathbf{N}$, let a_n denote the number of ways to pave a $3 \times n$ rectangle $ABCD$ with 1×2 identical dominoes. Clearly, $a_n = 0$ if n is odd. Show that
$$a_{2r} = \frac{1}{2\sqrt{3}}\left\{(\sqrt{3}+1)(2+\sqrt{3})^r + (\sqrt{3}-1)(2-\sqrt{3})^r\right\},$$
where $r \in \mathbf{N}$. (Proposed by I. Tomescu, see *Amer. Math. Monthly*, **81** (1974), 522–523.)

37. Solve the system of recurrence relations:

$$\begin{cases} a_{n+1} = a_n - b_n \\ b_{n+1} = a_n + 3b_n, \end{cases}$$

given that $a_0 = -1$ and $b_0 = 5$.

38. Solve the system of recurrence relations:

$$\begin{cases} a_n + a_{n-1} + 2b_{n-1} = 0 \\ b_n - 2a_{n-1} - 3b_{n-1} = 0, \end{cases}$$

given that $a_0 = 1$ and $b_0 = 0$.

39. Solve the system of recurrence relations:

$$\begin{cases} 10a_n = 9a_{n-1} - 2b_{n-1} \\ 5b_n = -a_{n-1} + 3b_{n-1}, \end{cases}$$

given that $a_0 = 4$ and $b_0 = 3$.

40. Solve the system of recurrence relations:

$$\begin{cases} 3a_n - 2a_{n-1} - b_{n-1} = 0 \\ 3b_n - a_{n-1} - 2b_{n-1} = 0, \end{cases}$$

given that $a_0 = 2$ and $b_0 = -1$.

41. Let (a_n) and (b_n) be two sequences of positive numbers satisfying the recurrence relations:

$$\begin{cases} a_n^2 = a_{n-1}b_n \\ b_n^2 = a_nb_{n-1} \end{cases}$$

with the initial conditions $a_0 = \frac{1}{8}$ and $b_0 = 64$. Show that

$$\lim_{n\to\infty} a_n = \lim_{n\to\infty} b_n,$$

and find the common limit.

42. For $n \in \mathbf{N}^*$, let a_n, b_n, c_n and d_n denote the numbers of binary sequences of length n satisfying the respective conditions:

	Number of 0's	Number of 1's
a_n	even	even
b_n	even	odd
c_n	odd	even
d_n	odd	odd

(i) Show that

$$a_n = b_{n-1} + c_{n-1},$$
$$b_n = a_{n-1} + d_{n-1} = c_n,$$
$$d_n = b_{n-1} + c_{n-1}.$$

(ii) Let $A(x), B(x), C(x)$ and $D(x)$ be, respectively, the generating functions of the sequences (a_n), (b_n), (c_n) and (d_n). Show that

$$A(x) = \frac{1 - 2x^2}{1 - 4x^2},$$

$$B(x) = C(x) = \frac{x}{1 - 4x^2},$$

$$D(x) = \frac{2x^2}{1 - 4x^2}.$$

(iii) Deduce from (ii) that

$$a_n = (-2)^{n-2} + 2^{n-2} \ (n \geq 1),$$
$$b_n = c_n = -(-2)^{n-2} + 2^{n-2} \ (n \geq 0),$$
$$d_n = (-2)^{n-2} + 2^{n-2} \ (n \geq 1).$$

43. Three given sequences (a_n), (b_n) and (c_n) satisfy the following recurrence relations:

$$a_{n+1} = \frac{1}{2}(b_n + c_n - a_n),$$

$$b_{n+1} = \frac{1}{2}(c_n + a_n - b_n),$$

and

$$c_{n+1} = \frac{1}{2}(a_n + b_n - c_n),$$

with the initial conditions $a_0 = p$, $b_0 = q$ and $c_0 = r$, where p, q, r are positive constants.

(i) Show that $a_n = \frac{1}{3}(p+q+r)(\frac{1}{2})^n + (-1)^n \frac{1}{3}(2p-q-r)$ for all $n \geq 0$.

(ii) Deduce that if $a_n > 0$, $b_n > 0$ and $c_n > 0$ for all $n \geq 0$, then $p = q = r$.

44. For $n \in \mathbf{N}$, let F_n denote the nth Fibonacci number. Thus

$$F_1 = 1, F_2 = 1, F_3 = 2, F_4 = 3, F_5 = 5, F_6 = 8, \ldots$$

and by Example 6.3.1,

$$F_{n+2} = F_n + F_{n+1}$$

and

$$F_n = \frac{1}{\sqrt{5}} \left\{ \left(\frac{1 + \sqrt{5}}{2} \right)^n - \left(\frac{1 - \sqrt{5}}{2} \right)^n \right\}.$$

Show that

(i) $\displaystyle\sum_{r=1}^{n} F_r = F_{n+2} - 1;$

(ii) $\displaystyle\sum_{r=1}^{n} F_{2r} = F_{2n+1} - 1;$

(iii) $\displaystyle\sum_{r=1}^{n} F_{2r-1} = F_{2n};$

(iv) $\displaystyle\sum_{r=1}^{n} (-1)^{r+1} F_r = (-1)^{n+1} F_{n-1} + 1.$

45. Show that for $m, n \in \mathbf{N}$ with $n \geq 2$,

(i) $F_{m+n} = F_m F_{n-1} + F_{m+1} F_n;$

(ii) $\begin{pmatrix} 1 & 1 \\ 1 & 0 \end{pmatrix}^n = \begin{pmatrix} F_{n+1} & F_n \\ F_n & F_{n-1} \end{pmatrix};$

(iii) $F_{n+1}F_{n-1} - F_n^2 = (-1)^n;$

(iv) $F_{n+1}^2 = 4F_n F_{n-1} + F_{n-2}^2, \ n \geq 3;$

(v) $(F_n, F_{n+1}) = 1$, where (a, b) denotes the HCF of a and b.

Remark. In general, $(F_m, F_n) = F_{(m,n)}$. Also, $F_m | F_n$ iff $m|n$.

46. Show that for $n \geq 2$

(i) $F_n^2 + F_{n-1}^2 = F_{2n-1};$

(ii) $F_{n+1}^2 - F_{n-1}^2 = F_{2n};$

(iii) $F_{n+1}^3 + F_n^3 - F_{n-1}^3 = F_{3n}.$

47. Show that

(i) $\displaystyle\sum_{r=1}^{n} F_r^2 = F_n F_{n+1},$

(ii) $\displaystyle\sum_{r=1}^{2n-1} F_r F_{r+1} = F_{2n}^2,$

(iii) $\displaystyle\sum_{r=1}^{2n} F_r F_{r+1} = F_{2n+1}^2 - 1.$

48. Show that for $n \in \mathbf{N}^*$,

$$F_{n+1} = \sum_{r=0}^{\lfloor \frac{n}{2} \rfloor} \binom{n-r}{r}.$$

49. Show that for $m, n \in \mathbf{N}$,

$$\sum_{r=0}^{n} \binom{n}{r} F_{m+r} = F_{m+2n}.$$

50. Show that

$$\lim_{n \to \infty} \frac{F_n}{F_{n+1}} = \frac{\sqrt{5} - 1}{2} \approx 0.618.$$

Note. The constant $\frac{\sqrt{5}-1}{2}$ is called the *golden number*.

51. Beginning with a pair of baby rabbits, and assuming that each pair gives birth to a new pair each month starting from the 2nd month of its life, find the number a_n of pairs of rabbits at the end of the nth month. (Fibonacci, Liber Abaci, 1202.)

52. Show that for $n \in \mathbf{N}^*$,

$$F_{n+1} = \begin{vmatrix} 1 & -1 & 0 & 0 & \ldots & 0 & 0 \\ 1 & 1 & -1 & 0 & \ldots & 0 & 0 \\ 0 & 1 & 1 & -1 & \ldots & 0 & 0 \\ \vdots & \vdots & \vdots & \vdots & \ddots & \vdots & \vdots \\ 0 & 0 & 0 & 0 & \ldots & 1 & -1 \\ 0 & 0 & 0 & 0 & \ldots & 1 & 1 \end{vmatrix}.$$

53. A man wishes to climb an n-step staircase. Let a_n denote the number of ways that this can be done if in each step he can cover either one step or two steps. Find a recurrence relation for (a_n).

54. Given $n \in \mathbf{N}$, find the number of binary sequences of length n in which no two 0's are adjacent.

55. For $n \in \mathbf{N}$ with $n \geq 2$, let a_n denote the number of ways to express n as a sum of positive integers greater than 1, taking order into account. Find a recurrence relation for (a_n) and determine the value of a_n.

56. Find the number of subsets of $\{1, 2, \ldots, n\}$, where $n \in \mathbf{N}$, that contain no consecutive integers. Express your answer in terms of a Fibonacci number.

57. Prove that

$$\sum_{j=0}^{n} \frac{\binom{n}{2j-n-1}}{5^j} = \frac{1}{2}(0.4)^n F_n.$$

(Proposed by S. Rabinowitz, see *Crux Mathematicorum*, **10** (1984), 269.)

58. Call an ordered pair (S, T) of subsets of $\{1, 2, \ldots, n\}$ *admissible* if $s > |T|$ for each $s \in S$, and $t > |S|$ for each $t \in T$. How many admissible ordered pairs of subsets of $\{1, 2, \ldots, 10\}$ are there? Prove your answer. (Putnam, 1990)

59. For each $n \in \mathbf{N}$, let a_n denote the number of natural numbers N satisfying the following conditions: the sum of the digits of N is n and each digit of N is taken from $\{1, 3, 4\}$. Show that a_{2n} is a perfect square for each $n = 1, 2, \ldots$. (Chinese Math. Competition, 1991)

60. Find a recurrence relation for a_n, the number of ways to place parentheses to indicate the order of multiplication of the n numbers $x_1 x_2 x_3 \cdots x_n$, where $n \in \mathbf{N}$.

61. For $n \in \mathbf{N}$, let b_n denote the number of sequences of $2n$ terms:

$$z_1, z_2, \ldots, z_{2n},$$

where each z_i is either 1 or -1 such that

(1) $\sum_{i=1}^{2n} z_i = 0$ and

(2) $\sum_{i=1}^{k} z_i \geq 0$ for each $k = 1, 2, \ldots, 2n - 1$.

(i) Find b_n for $n = 1, 2, 3$.

(ii) Establish a bijection between the set of all sequences of $2n$ terms as defined above and the set of all parenthesized expressions of the $n + 1$ numbers $x_1 x_2 \cdots x_n x_{n+1}$.

62. For $n \in \mathbf{N}$, let a_n denote the number of ways to pair off $2n$ distinct points on the circumference of a circle by n nonintersecting chords. Find a recurrence relation for (a_n).

63. Let $p(x_1, x_2, \ldots, x_n)$ be a polynomial in n variables with constant term 0, and let $\#(p)$ denote the number of distinct terms in p after terms with like exponents have been collected. Thus for example $\#((x_1+x_2)^5) = 6$. Find a formula for $\#(q_n)$ where

$$q_n = x_1(x_1 + x_2)(x_1 + x_2 + x_3) \cdots (x_1 + \cdots + x_n).$$

(Proposed by J. O. Shallit, see *Amer. Math. Monthly*, **93** (1986), 217–218.)

64. Find the total number of ways of arranging in a row the $2n$ integers $a_1, a_2, \ldots, a_n, b_1, b_2, \ldots, b_n$ with the restriction that for each i, a_i precede b_i, a_i precede a_{i+1} and b_i precede b_{i+1}. (Proposed by E. Just, see *Amer. Math. Monthly*, **76** (1969), 419–420.)

65. Mr. Chen and Mr. Lim are the two candidates taking part in an election. Assume that Mr. Chen receives m votes and Mr. Lim receives n votes, where $m, n \in \mathbf{N}$ with $m > n$. Find the number of ways that the ballots can be arranged in such a way that when they are counted, one at a time, the number of votes for Mr. Chen is always more than that for Mr. Lim.

66. For $n \in \mathbf{N}$, let a_n denote the number of mappings $f : \mathbf{N}_n \to \mathbf{N}_n$ such that if $j \in f(\mathbf{N}_n)$, then $i \in f(\mathbf{N}_n)$ for all i with $1 \le i \le j$.

    (i) Find the values of a_1, a_2 and a_3 by listing all such mappings f.
    (ii) Show that

$$a_n = \sum_{k=1}^{n} \binom{n}{k} a_{n-k}.$$

    (iii) Let $A(x)$ be the exponential generating function for (a_n), where $a_0 = 1$. Show that

$$A(x) = \frac{1}{2 - e^x}.$$

    (iv) Deduce that

$$a_n = \sum_{r=0}^{\infty} \frac{r^n}{2^{r+1}}.$$

67. Define S_0 to be 1. For $n \ge 1$, let S_n be the number of $n \times n$ matrices whose elements are nonnegative integers with the property that $a_{ij} = a_{ji}$ $(i, j = 1, 2, \ldots, n)$ and where $\sum_{i=1}^{n} a_{ij} = 1$, $(j = 1, 2, \ldots, n)$. Prove

    (a) $S_{n+1} = S_n + nS_{n-1}$,

    (b) $\displaystyle\sum_{n=0}^{\infty} S_n \frac{x^n}{n!} = e^{x + \frac{x^2}{2}}.$

    (Putnam, 1967)

68. A sequence (a_n) of numbers satisfies the following conditions:

    (1) $a_1 = \frac{1}{2}$ and
    (2) $a_1 + a_2 + \cdots + a_n = n^2 a_n$, $n \ge 1$.

    Determine the value of a_n.

69. What is the sum of the greatest odd divisors of the integers $1, 2, 3, \ldots, 2^n$, where $n \in \mathbf{N}$? (West German Olympiad, 1982)
    (*Hint*: Let a_n be the sum of the greatest odd divisors of $1, 2, 3, \ldots, 2^n$. Show that $a_n = a_{n-1} + 4^{n-1}$.)

70. Let d_n be the determinant of the $n \times n$ matrix in which the element in the ith row and the jth column is the absolute value of the difference of i and j. Show that

$$d_n = (-1)^{n-1}(n-1)2^{n-2}.$$

    (Putnam, 1969)

71. A sequence (a_n) of natural numbers is defined by $a_1 = 1$, $a_2 = 3$ and
$$a_n = (n+1)a_{n-1} - na_{n-2} \quad (n \geq 2).$$
Find all values of n such that $11|a_n$.

72. A sequence (a_n) of positive numbers is defined by
$$a_n = \frac{1}{16} \left(1 + 4a_{n-1} + \sqrt{1 + 24a_{n-1}} \right)$$
with $a_0 = 1$. Find a general formula for a_n.

73. A sequence (a_n) of numbers is defined by
$$2a_n = 3a_{n-1} + \sqrt{5a_{n-1}^2 + 4} \quad (n \geq 1)$$
with $a_0 = 0$. Show that for all $m \geq 1$, $1992 \not| a_{2m+1}$.

74. Solve the recurrence relation
$$na_n = (n-2)a_{n-1} + (n+1),$$
given that $a_0 = 0$.

75. Solve the recurrence relation
$$n(n-1)a_n - (n-2)^2 a_{n-2} = 0,$$
given that $a_0 = 0$ and $a_1 = 1$.

76. A sequence (a_n) of numbers satisfies the recurrence relation
$$(a_n - a_{n-1})f(a_{n-1}) + g(a_{n-1}) = 0$$
with the initial condition $a_0 = 2$, where
$$f(x) = 3(x-1)^2 \quad \text{and} \quad g(x) = (x-1)^3.$$
Solve the recurrence relation.

77. A sequence (a_n) of numbers satisfies the recurrence relation
$$n(n-1)a_n = (n-1)(n-2)a_{n-1} - (n-3)a_{n-2}$$
with the initial conditions $a_0 = 1$ and $a_1 = 2$.
Find the value of
$$\sum_{k=0}^{1992} \frac{a_k}{a_{k+1}}.$$

78. Let $a(n)$ be the number of representations of the positive integer n as the sums of 1's and 2's taking order into account. For example, since
$$4 = 1 + 1 + 2 = 1 + 2 + 1 = 2 + 1 + 1$$
$$= 2 + 2 = 1 + 1 + 1 + 1,$$
then $a(4) = 5$. Let $b(n)$ be the number of representations of n as the sum of integers greater than 1, again taking order into account and counting the summand n. For example, since $6 = 4+2 = 2+4 = 3+3 = 2 + 2 + 2$, we have $b(6) = 5$. Show that for each n, $a(n) = b(n+2)$. (Putnam, 1957)

79. Show that the sum of the first n terms in the binomial expansion of $(2 - 1)^{-n}$ is $\frac{1}{2}$, where $n \in \mathbf{N}$. (Putnam, 1967)

80. Prove that there exists a unique function f from the set $\mathbf{R}^+$ of positive real numbers to $\mathbf{R}^+$ such that

$$f(f(x)) = 6x - f(x)$$

and $f(x) > 0$ for all $x > 0$. (Putnam, 1988)

81. Let $T_0 = 2$, $T_1 = 3$, $T_2 = 6$, and for $n \geq 3$,

$$T_n = (n + 4)T_{n-1} - 4nT_{n-2} + (4n - 8)T_{n-3}.$$

The first few terms are

$$2, 3, 6, 14, 40, 152, 784, 5168, 40576.$$

Find, with proof, a formula for T_n of the form $T_n = A_n + B_n$, where (A_n) and (B_n) are well-known sequences. (Putnam, 1990)

82. Let $\{a_n\}$ and $\{b_n\}$ denote two sequences of integers defined as follows:

$$a_0 = 1, \quad a_1 = 1, \quad a_n = a_{n-1} + 2a_{n-2} \quad (n \geq 2),$$
$$b_0 = 1, \quad b_1 = 7, \quad b_n = 2b_{n-1} + 3b_{n-2} \quad (n \geq 2).$$

Thus, the first few terms of the sequences are:

$$a : 1, 1, 3, 5, 11, 21, \ldots$$
$$b : \ 1, 7, 17, 55, 161, 487, \ldots$$

Prove that, except for the "1", there is no term which occurs in both sequences. (USA MO, 1973)

83. The sequence $\{x_n\}$ is defined as follows: $x_1 = 2$, $x_2 = 3$, and

$$x_{2m+1} = x_{2m} + x_{2m-1}, \quad m \geq 1$$
$$x_{2m} = x_{2m-1} + 2x_{2m-2}, \quad m \geq 2.$$

Determine x_n (as a function of n). (Austrian MO, 1983)

84. Determine the number of all sequences $(x_1, x_2, \ldots, x_n)$, with $x_i \in \{a, b, c\}$ for $i = 1, 2, \ldots, n$ that satisfy $x_1 = x_n = a$ and $x_i \neq x_{i+1}$ for $i = 1, 2, \ldots, n - 1$. (18th Austrian MO)

85. The sequence $x_1, x_2, \ldots$ is defined by the equalities $x_1 = x_2 = 1$ and

$$x_{n+2} = 14x_{n+1} - x_n - 4, \quad n \geq 1.$$

Prove that each number of the given sequence is a perfect square. (Bulgarian MO, 1987)

86. How many words with n digits can be formed from the alphabet $\{0, 1, 2, 3, 4\}$, if adjacent digits must differ by exactly one? (West Germany, 1987)

87. The sequence (a_n) of integers is defined by

$$-\frac{1}{2} < a_{n+1} - \frac{a_n^2}{a_{n-1}} \leq \frac{1}{2}$$

with $a_1 = 2$ and $a_2 = 7$. Show that a_n is odd for all values of $n \geq 2$. (British MO, 1988)

88. In the network illustrated by the figure below, where there are n adjacent squares, what is the number of paths (not necessarily shortest) from A to B which do not pass through any intersection twice?

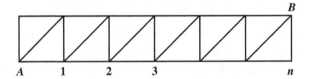

(Proposed by P. Andrews and E. T. H. Wang, see *CRUX Mathematicorum*, **14** (1988), 62–64.)

89. Let $a_1 = 1$ and $a_{n+1} = a_n + \lfloor \sqrt{a_n} \rfloor$ for $n \in \mathbf{N}$. Show that a_n is a square iff $n = 2^k + k - 2$ for some $k \in \mathbf{N}$. (Proposed by T. C. Brown, see *Amer. Math. Monthly*, **85** (1978), 52–53.)

90. Determine all pairs (h, s) of positive integers with the following property: If one draws h horizontal lines and another s lines which satisfy

   (i) they are not horizontal,
   (ii) no two of them are parallel,
   (iii) no three of the $h + s$ lines are concurrent,

   then the number of regions formed by these $h + s$ lines is 1992. (APMO, 1992)

91. Show that

$$S(r, n) = \sum_{k=n-1}^{r-1} \binom{r-1}{k} S(k, n-1),$$

where $r \geq n \geq 2$.

92. Let $B_0 = 1$ and for $r \in \mathbf{N}$, let $B_r = \sum_{n=1}^{r} S(r, n)$ denote the rth Bell number (see Section 1.7). Show that

$$B_r = \sum_{k=0}^{r-1} \binom{r-1}{k} B_k,$$

where $r \geq 1$.

93. Two sequences $P(m, n)$ and $Q(m, n)$ are defined as follows (m, n are integers). $P(m, 0) = 1$ for $m \geq 0$, $P(0, n) = 0$ for $n \geq 1$, $P(m, n) = 0$ for $m, n < 0$. $P(m, n) = \sum_{j=0}^{n} P(m - 1, j)$ for $m \geq 1$. $Q(m, n) = P(m - 1, n) + P(m - 1, n - 1) + P(m - 1, n - 2)$ for $m \geq 1$. Express $Q(m, n)$ in terms of m and n for $m \geq 1$. (Proposed by L. Kuipers, see *Amer. Math. Monthly*, **76** (1969), 97–98.)

94. For $n, k \in \mathbf{N}$, let $S_k(n) = \sum_{j=1}^{n} j^k$ (see Problem 2.85). Show that

(i) $S_k(n) = n^{k+1} - \sum_{r=0}^{k-1} \binom{k}{r} S_{r+1}(n - 1)$ for $n \geq 2$,

(ii) $(k + 1)S_k(n) = (n + 1)^{k+1} - (n + 1)^k - \sum_{r=0}^{k-2} \binom{k}{r} S_{r+1}(n)$.

Solutions to Exercise 1

1. (i) We note that a pair $\{a, b\} \subseteq \{1, 2, \ldots, 50\}$ satisfies $|a - b| = 5$ if and only if $\{a, b\} = \{t, t + 5\}$ where $t \in \{1, 2, \ldots, 45\}$. So the number of ways to choose a pair $\{a, b\}$ of distinct numbers such that $|a - b| = 5$ is 45.

   (ii) By a similar argument as in part (i), we see that the number of ways to choose a pair $\{a, b\}$ of distinct numbers such that $|a - b| \leq 5$ is equal to $45 + 46 + 47 + 48 + 49 = 235$.

2. (i) The number of ways is 12!.

   (ii) There are 8! ways to permute the 7 boys and the block of girls, and there are 5! ways to permute the 5 girls inside the block. By (MP), the number of such ways is equal to $8!5!$.

   (iii) We can first arrange the boys in 7! ways, and then arrange the 5 girls such that no 2 of them are adjacent in P_5^8 ways. By (MP), the number of such ways is equal to $7! \cdot P_5^8$.

   (iv) First, there are 2 ways to arrange the boys A and B. Then there are P_3^5 ways to arrange exactly 3 girls between A and B. Finally, there are 8! ways to arrange the remaining 7 people and the block containing A, B and the 3 girls. By (MP), the number of such ways is equal to $2 \cdot P_3^5 \cdot 8! = 120 \cdot 8!$.

3. (i) The number of ways is $(m + n)!$.

   (ii) We can first arrange the girls in $n!$ ways, and then arrange the m boys such that no 2 of them are adjacent in P_m^{n+1} ways. By (MP), the number of such ways is equal to $n! \cdot P_m^{n+1}$.

   (iii) There are $(m + 1)!$ ways to permute the m boys and the block of girls, and there are $n!$ ways to permute the n girls inside the block. By (MP), the number of such ways is equal to $(m + 1)!n!$.

   (iv) First, there are 2 ways to arrange the particular boy and the

particular girl. Then there are $(m + n - 1)!$ ways to arrange the
remaining $m+n-2$ people and the block containing the particular
boy and the particular girl. By (MP), the number of such ways
is equal to $2 \cdot (m + n - 1)!$.

4. (i) The number of such words is P_5^{10}.

   (ii) There are P_3^6 ways to choose the first, third and fifth letters from
   the letters A, B, C, D, E, F, and P_2^4 ways to choose the second
   and fourth letters from G, H, I, J. By (MP), the number of such
   words is equal to $P_3^6 \cdot P_2^4$.

5. First, there are 2 ways to arrange the letters x and y. Then there
are P_5^{24} ways to arrange exactly 5 letters between x and y. Finally,
there are 20! ways to arrange the remaining 19 letters and the block
containing x, y and the 5 letters. By (MP), the number of such ways
is equal to $2 \cdot P_5^{24} \cdot 20!$.

6. If the thousands digit is even, then there are 2 ways to choose the
thousands digit, 5 ways to choose the ones digit, and P_2^8 ways to
choose the remaining digits. Else, if the thousands digit is odd, then
there are 3 ways to choose the thousands digit, 4 ways to choose the
ones digit, and P_2^8 ways to choose the remaining digits. So the number
of odd integers between 3000 and 8000 such that no digit is repeated
is $2 \cdot 5 \cdot P_2^8 + 3 \cdot 4 \cdot P_2^8 = 1232$.

7. First Solution: We have

$$
\begin{aligned}
&1 \cdot 1! + 2 \cdot 2! + \cdots + n \cdot n! \\
&= (2 - 1) \cdot 1! + (3 - 1) \cdot 2! + \cdots + (n + 1 - 1) \cdot n! \\
&= 2! - 1! + 3! - 2! + \cdots + (n + 1)! - n! \\
&= (n + 1)! - 1.
\end{aligned}
$$

Second Solution: Let us count the number of permutations $a_1 \cdots a_{n+1}$
of the set $\{1, \ldots, n + 1\}$ that satisfy $(a_1, \ldots, a_{n+1}) \neq (1, \ldots, n + 1)$ in
two different ways. On one hand, the total number of permutations
$a_1 a_2 \cdots a_{n+1}$ of the set $\{1, 2, \ldots, n + 1\}$ is equal to $(n + 1)!$, so the
number of permutations $a_1 a_2 \cdots a_{n+1}$ of the set $\{1, 2, \ldots, n + 1\}$ that
satisfy $(a_1, a_2, \ldots, a_{n+1}) \neq (1, 2, \ldots, n + 1)$ is $(n + 1)! - 1$.
For the alternative way of counting, let us count the number of permutations $b_1 b_2 \cdots b_{n+1}$ of the set $\{1, 2, \ldots, n + 1\}$ that satisfy $b_i = i$
for all $i < k$ and $b_k \neq k$ for some $k \in \{1, 2, \ldots, n\}$. There are
$n + 1 - (k - 1) - 1 = n - k + 1$ choices for b_k, and subsequently,
there are $(n - k + 1)!$ choices to obtain a permutation $b_{k+1} b_{k+2} \cdots b_{n+1}$

of the set $\{1, 2, \ldots, n+1\} \setminus \{1, 2, \ldots, k-1, b_k\}$. By (AP), the number permutations $a_1 a_2 \cdots a_{n+1}$ of the set $\{1, 2, \ldots, n+1\}$ that satisfy $(a_1, a_2, \ldots, a_{n+1}) \neq (1, 2, \ldots, n+1)$ is equal to

$$\sum_{k=1}^{n} (n-k+1)(n-k+1)! = 1 \cdot 1! + 2 \cdot 2! + \cdots + n \cdot n!.$$

The desired result follows.

8. First Solution: We have

$$\frac{1}{(1+1)!} + \frac{2}{(2+1)!} + \cdots + \frac{n}{(n+1)!}$$
$$= \frac{2-1}{2!} + \frac{3-1}{3!} + \cdots + \frac{n+1-1}{(n+1)!}$$
$$= \frac{1}{1!} - \frac{1}{2!} + \frac{1}{2!} - \frac{1}{3!} + \cdots + \frac{1}{n!} - \frac{1}{(n+1)!}$$
$$= 1 - \frac{1}{(n+1)!}.$$

Second Solution: Let us calculate the probability that a permutation $a_1 a_2 \cdots a_{n+1}$ of the set $\{1, 2, \ldots, n+1\}$ satisfies the property that there exists some $i \in \{1, 2, \ldots, n\}$ such that $a_i > a_{i+1}$ in two different ways. On one hand, the only permutation $a_1 a_2 \cdots a_{n+1}$ of the set $\{1, 2, \ldots, n+1\}$ that do not satisfy the above property is the permutation $123 \cdots n+1$, so the number of permutations $a_1 a_2 \cdots a_{n+1}$ of the set $\{1, 2, \ldots, n+1\}$ that satisfy the above property is equal to $(n+1)! - 1$. Consequently, the desired probability is equal to $1 - \frac{1}{(n+1)!}$. For the alternative way of calculating the desired probability, let us count the number of permutations $b_1 \cdots b_{n+1}$ of the set $\{1, \ldots, n+1\}$ that satisfy $b_i < b_{i+1}$ for all $i < k$ and $b_k > b_{k+1}$ for some $k \in \{1, 2, \ldots, n\}$. There are $\binom{n+1}{k+1}$ choices to choose $k+1$ numbers out of the set $\{1, 2, \ldots, n+1\}$ for the numbers $b_1, b_2, \ldots, b_{k+1}$. Next, there are k choices for b_k to make it such that the numbers $b_1, b_2, \ldots, b_{k+1}$ satisfy $b_i < b_{i+1}$ for all $i < k$ and $b_k > b_{k+1}$. Finally, there are $(n-k)!$ choices to obtain a permutation $b_{k+2} b_{k+3} \cdots b_{n+1}$ of the set $\{1, 2, \ldots, n+1\} \setminus \{b_1, b_2, \ldots, b_{k+1}\}$. By (AP), the number permutations $a_1 a_2 \cdots a_{n+1}$ of the set $\{1, 2, \ldots, n+1\}$ that satisfy the above property is equal to

$$\sum_{k=1}^{n} \binom{n+1}{k+1} \cdot k \cdot (n-k)!$$

$$= \sum_{k=1}^{n} \frac{(n+1)!}{(k+1)!(n-k)!} \cdot k \cdot (n-k)!$$

$$= (n+1)! \sum_{k=1}^{n} \frac{k}{(k+1)!}$$

$$= (n+1)! \left[\frac{1}{(1+1)!} + \frac{2}{(2+1)!} + \cdots + \frac{n}{(n+1)!} \right].$$

The probability that a permutation $a_1 a_2 \cdots a_{n+1}$ of the set $\{1, 2, \ldots, n+1\}$ satisfies the required property is equal to

$$\frac{1}{(n+1)!} \sum_{k=1}^{n} \binom{n+1}{k+1} \cdot k \cdot (n-k)!$$

$$= \frac{1}{(1+1)!} + \frac{2}{(2+1)!} + \cdots + \frac{n}{(n+1)!}$$

$$= 1 - \frac{1}{(n+1)!}.$$

The desired result follows.

9. By Example 1.4.4, the number of ways to form n teams from a group of $2n$ people is given by $\frac{(2n)!}{n! \cdot 2^n}$, which is an integer. We note that

$$\frac{(2n)!}{n! \cdot 2^n} = \frac{1 \cdot 2 \cdots \cdots n \cdot (n+1)(n+2) \cdots (2n)}{1 \cdot 2 \cdots \cdots n \cdot 2^n}$$

$$= \frac{(n+1)(n+2) \cdots (2n)}{2^n}.$$

This implies that $\frac{(n+1)(n+2) \cdots (2n)}{2^n}$ is an integer. Hence

$$(n+1)(n+2) \cdots (2n)$$

is divisible by 2^n as desired.

10. We have $10^{40} = 2^{40} \cdot 5^{40}$ and $20^{30} = 2^{60} \cdot 5^{30}$. So the greatest common divisor of 10^{40} and 20^{30} is equal to $2^{40} \cdot 5^{30}$. Hence the number of common positive divisors of 10^{40} and 20^{30} is equal to $(40+1)(30+1) = 1271$.

11. (i) We note that $210 = 2 \cdot 3 \cdot 5 \cdot 7$. So the number of positive divisors of 210 which are multiples of 3 is equal to $2 \cdot 1 \cdot 2 \cdot 2 = 8$.

(ii) We note that $630 = 2 \cdot 3^2 \cdot 5 \cdot 7$. So the number of positive divisors of 630 which are multiples of 3 is equal to $2 \cdot 2 \cdot 2 \cdot 2 = 16$.

(iii) We note that $151200 = 2^5 \cdot 3^3 \cdot 5^2 \cdot 7$. So the number of positive divisors of 630 which are multiples of 3 is equal to $6 \cdot 3 \cdot 3 \cdot 2 = 108$.

12. Let the prime factorisation of n be $p_1^{m_1} \cdot p_2^{m_2} \cdots p_k^{m_k}$. Then the prime factorisation of n^2 is $p_1^{2m_1} \cdot p_2^{2m_2} \cdots p_k^{2m_k}$. So the number of positive divisors of n is equal to $(2m_1+1)(2m_2+1)\cdots(2m_k+1)$, which is odd as required.

13. We note that the sum of digits of $\underbrace{111\cdots11}_{1992}$ is 1992, which is divisible by 3 but not by 9. Hence, we see that the prime factorisation of $\underbrace{111\cdots11}_{1992}$ is $3 \cdot p_1^{m_1} \cdot p_2^{m_2} \cdots p_k^{m_k}$, where $p_1, p_2, \ldots p_k$ are primes larger than 3. Hence the number of positive divisors of $\underbrace{111\cdots11}_{1992}$ is equal to $2(m_1+1)(m_2+1)\cdots(m_k+1)$, which is even as required.

14. (i) Let us count the number of ways to arrange r boys in a row from a set of n boys in 2 different ways. On one hand the number of such ways is equal to P_r^n by definition. On the other hand, there are exactly n ways to choose a boy from the set for the leftmost position in the row, and P_{r-1}^{n-1} ways to arrange the remaining $r-1$ boys in a row. By (MP), the number of such ways is equal to $n \cdot P_{r-1}^{n-1}$ so we have

$$P_r^n = n \cdot P_{r-1}^{n-1}$$

as desired.

(ii) Let us count the number of ways to arrange r boys in a row from a set of n boys in 2 different ways. On one hand the number of such ways is equal to P_r^n by definition. On the other hand, there are exactly P_{r-1}^n ways to arrange $r-1$ boys in a row from a set of n boys, and there are exactly $n-r+1$ ways to choose a boy for the last position in the row. By (MP), the number of such ways is equal to $(n-r+1) \cdot P_{r-1}^n$ so we have

$$P_r^n = (n-r+1) \cdot P_{r-1}^n$$

as desired.

(iii) By using a similar argument as in parts (i) and (ii), we see that the number of ways to arrange $r+1$ boys in a row from a set of n boys is both equal to $n \cdot P_r^{n-1}$ and $(n-r) \cdot P_r^n$. Hence we have $n \cdot P_r^{n-1} = (n-r) \cdot P_r^n$, or equivalently,

$$P_r^n = \frac{n}{n-r} P_r^{n-1}$$

as desired.

(iv) Let us count the number of ways to arrange r boys in a row from a set of $n+1$ boys in 2 different ways. On one hand the number of such ways is equal to P_r^{n+1} by definition. On the other hand, let us first fix a boy A from the set of $n+1$ boys. If A is in the row, then there are exactly r ways to place A in any of the r positions in the row, and subsequently there are P_{r-1}^n ways to arrange the remaining $r-1$ boys from the remaining set of n boys. Else, if A is not in the row, then the number of ways is precisely P_r^n by definition. So the total number of ways is equal to $P_r^n + r \cdot P_{r-1}^n$, and hence we have

$$P_r^{n+1} = P_r^n + r \cdot P_{r-1}^n$$

as desired.

(v) First Solution: By part (iv), we have

$$
\begin{aligned}
P_r^{n+1} &= r \cdot P_{r-1}^n + P_r^n \\
&= r \cdot P_{r-1}^n + r \cdot P_{r-1}^{n-1} + P_r^{n-1} \\
&= r \cdot P_{r-1}^n + r \cdot P_{r-1}^{n-1} + r \cdot P_{r-1}^{n-2} + P_r^{n-2} \\
&= \cdots \\
&= r \cdot P_{r-1}^n + r \cdot P_{r-1}^{n-1} + r \cdot P_{r-1}^{n-2} + \cdots + r \cdot P_{r-1}^r + P_r^r \\
&= r! + r(P_{r-1}^n + P_{r-1}^{n-1} + P_{r-1}^{n-2} + \cdots + P_{r-1}^r).
\end{aligned}
$$

Second Solution: Let us count the number of r-permutations of the set $S = \{1, 2, \ldots, n+1\}$ in two different ways. On one hand, the number of such permutations is equal to P_r^{n+1} by definition. For the other method of counting, let us first denote the set of r-permutations of the set S by T, and define a function $f: T \to S$ by $f(a_1 a_2 \cdots a_r) = \min\limits_{i=1}^{r} a_i$. Then it is easy to verify that $f(T) = \{1, 2, \ldots, n-r+2\}$. For each $k \in \{1, 2, \ldots, n-r+2\}$, let us count the number of r-permutations $a_1 a_2 \cdots a_r$ of the set S for which $f(a_1 a_2 \cdots a_r) = k$. There are r ways to first place the number k, and subsequently, there are P_{r-1}^{n+1-k} ways to choose $r-1$ numbers out of the set $\{k+1, k+2, \ldots, n+1\}$, and arrange them in the remaining $r-1$ places. By (AP) and (MP), the number of r-permutations of the set S is equal to

$$
\sum_{k=1}^{n-r+2} r \cdot P_{r-1}^{n+1-k} = r \cdot P_{r-1}^{r-1} + r \sum_{k=1}^{n-r+1} P_{r-1}^{n+1-k}
$$

$$
= r! + r(P_{r-1}^n + P_{r-1}^{n-1} + P_{r-1}^{n-2} + \cdots + P_{r-1}^r).
$$

The desired follows.

15. (i) There are $\binom{5}{3}$ ways to choose 3 girls out of 5 and there are $\binom{10}{6}$ ways to choose 6 boys out of 10 to form a committee that satisfies the given conditions. By (MP), the number of ways is equal to $\binom{5}{3}\binom{10}{6}$.

   (ii) There are 9! ways for the group to take up 9 different posts in the committee. By part (i) and by (MP), we see that the number of ways is equal to $\binom{5}{3}\binom{10}{6} \cdot 9!$.

16. Firstly, there are 7! ways to arrange the seven students. Then there are $\binom{8}{3}$ ways to place the 3 remaining empty chairs together with the 7 occupied chairs, such that no 2 empty chairs are adjacent. By (MP), the number of ways is equal to $7! \cdot \binom{8}{3}$.

17. Firstly, there are 5! ways to arrange the five distinct balls. Then there are $\binom{6}{3}$ ways to place the 3 remaining empty boxes along with the 5 occupied boxes, such that no 2 empty boxes are adjacent. By (MP), the number of ways is equal to $5! \cdot \binom{6}{3} = 2400$.

18. Firstly, there are $\binom{13}{7}$ ways to choose 7 students out of the remaining 13 students to stand in the front row. Then there are 10! ways to arrange the 10 students in the front row, and likewise there are 10! ways to arrange the 10 students in the back row. By (MP), the number of ways is equal to $\binom{13}{7}(10!)^2$.

19. First, there are 7! ways to arrange the 7 boys, and then there are $5 \times 2 = 10$ ways to arrange the 2 girls in such a way that the 2 girls are separated by exactly 3 boys. By (MP), the number of ways is equal to $10 \cdot 7!$.

20. (i) There are $\binom{15}{7}$ ways to choose 7 students out of 15 to form a 7-person committee, and there are $\binom{12}{7}$ ways to choose a 7-person committee, such that all of the committee members are boys. By (CP), the number of ways is equal to $\binom{15}{7} - \binom{12}{7}$.

   (ii) There are 7! ways for the group to take up 7 different posts in the committee. By part (i) and by (MP), the number of ways is equal to $7! \left\{ \binom{15}{7} - \binom{12}{7} \right\}$.

21. As there are m 0's, it follows that there are $m+1$ places to place the n 1's in such a way that no two 1's are adjacent. So the number of ways is equal to $\binom{m+1}{n}$.

22. There are $\binom{p}{2}$ ways to choose the 2 horizontal sides of the parallelogram, and there are $\binom{q}{2}$ ways to choose the 2 vertical sides of the parallelogram. By (MP), the number of ways is equal to $\binom{p}{2}\binom{q}{2}$.

23. By the conditions given in the problem, we deduce that at least 2 of

the senior students must be boys. Let us consider these cases:

Case 1: All senior students are boys. In this case, there must be 4 senior boys, 1 junior boy and 2 junior girls in the committee. There are $\binom{10}{4}$ ways to choose 4 senior boys out of 10, $\binom{15}{1}$ ways to choose 1 junior boy out of 15 and $\binom{10}{2}$ ways to choose 2 junior girls out of 10.

Case 2: 3 out of the 4 senior students are boys. In this case, there must be 3 senior boys, 1 senior girl, 2 junior boys and 1 junior girl in the committee. There are $\binom{10}{3}$ ways to choose 3 senior boys out of 10, $\binom{4}{1}$ ways to choose 1 senior girl out of 4, $\binom{15}{2}$ ways to choose 2 junior boys out of 15 and $\binom{10}{1}$ ways to choose 1 junior girl out of 10.

Case 3: 2 out of the 4 senior students are boys. In this case, there must be 2 senior boys, 2 senior girl and 3 junior boys in the committee. There are $\binom{10}{2}$ ways to choose 2 senior boys out of 10, $\binom{4}{2}$ ways to choose 2 senior girls out of 4 and $\binom{15}{3}$ ways to choose 2 junior boys out of 15.

By (AP) and (MP), there are $\binom{10}{4}\binom{15}{1}\binom{10}{2} + \binom{10}{3}\binom{4}{1}\binom{15}{2}\binom{10}{1} + \binom{10}{2}\binom{4}{2}\binom{15}{3}$ ways to choose a committee that satisfies the conditions.

24. First, let us find the number of ways to arrange 7 identical white balls and 5 identical black balls, such that the 6-th leftmost ball is white, and the 5 leftmost balls consists of 3 black balls and 2 white balls. There are $\binom{5}{2}$ ways to arrange the 3 black balls and 2 white balls in the 5 leftmost positions, and there are $\binom{6}{2}$ ways to arrange the 2 black balls and 4 white balls in the 6 rightmost positions.

By the above, we see that the probability that the 6-th ball drawn is white, while before that exactly 3 black balls are drawn, is equal to $\frac{\binom{5}{2}\binom{6}{2}}{\binom{12}{5}} = \frac{25}{132}$.

25. (i) By using a similar argument as in Example 1.5.1, we see that the number of shortest routes from O to A is equal to $\binom{5}{2}$ and the number of shortest routes from A to P is equal to $\binom{8}{3}$. By (MP), the number of shortest routes from O to P passing through the junction A is equal to $\binom{5}{2}\binom{8}{3}$.

(ii) By using a similar argument as in Example 1.5.1, we see that the number of shortest routes from O to A is equal to $\binom{5}{2}$ and the number of shortest routes from B to P is equal to $\binom{7}{3}$. By (MP), the number of shortest routes from O to P passing through the street AB is equal to $\binom{5}{2}\binom{7}{3}$.

(iii) By using a similar argument as in Example 1.5.1, we see that the number of shortest routes from O to A is equal to $\binom{5}{2}$, the number of shortest routes from A to C is equal to $\binom{4}{1}$, and the number of shortest routes from C to P is equal to $\binom{4}{2}$. By (MP), the number of shortest routes from O to P passing through the junctions A and C is equal to $\binom{5}{2}\binom{4}{1}\binom{4}{2}$.

(iv) By (CP), by part (ii), and by using a similar argument as in Example 1.5.1, we see that the number of shortest routes from O to P when street AB is closed is equal to $\binom{13}{5} - \binom{5}{2}\binom{7}{3}$.

26. (i) The number of ways to choose a group of $2k$ people from n couples, given that there are k couples in such a group, is equal to the number of ways to choose k couples out of n, which is equal to $\binom{n}{k}$.

(ii) Firstly, there are $\binom{n}{2k}$ ways to choose $2k$ couples out of n. Then for each chosen couple, there are 2 ways to choose one person out of the couple. By (MP), the number of ways is equal to $\binom{n}{2k} \cdot 2^{2k}$.

(iii) By definition, there are $\binom{2n}{2k}$ ways to choose $2k$ people out of n couples. By (CP), there are $\binom{2n}{2k} - \binom{n}{2k} \cdot 2^{2k}$ ways to ensure that at least one couple is included in such a group.

(iv) Firstly, there are $\binom{n}{2}$ ways to choose 2 couples out of n. Then there are $\binom{n-2}{2k-4}$ ways to choose $2k-4$ couples out of the remaining $n-2$ couples. Then for each of the remaining $2k-4$ couples, there are 2 ways to choose one person out of the couple. By (MP), the number of ways is equal to $\binom{n}{2}\binom{n-2}{2k-4} \cdot 2^{2k-4}$.

27. First we choose z from $\{2, 3, \ldots, n+1\}$, and write z as $k+1$ where $k \in \{1, 2, \ldots, n\}$. Then we choose x, y from $\{1, 2, \ldots, k\}$ so that $x < z$ and $y < z$, and the number of ways to do so is equal to k^2. Hence the number of points (x, y, z) in S^3 satisfying $x < z$ and $y < z$ is equal to

$$\sum_{k=1}^{n} k^2.$$

For the alternative method of counting, let us consider these cases:

Case 1: $\{x, y, z\} = \{a, b\}$, where $1 \le a < b \le n+1$. In this case, we set $x = y = a$, $z = b$, and there are $\binom{n+1}{2}$ ways to choose a and b out of S.

Case 2: $\{x, y, z\} = \{a, b, c\}$, where $1 \le a < b < c \le n+1$. In this case, we either have $x = a$, $y = b$, $z = c$, or $x = b$, $y = a$, $z = c$, and there are $\binom{n+1}{3}$ ways to choose a, b and c out of S. Hence, the number of points (x, y, z) in S^3 satisfying $x < z$ and $y < z$ is $\binom{n+1}{2} + 2\binom{n+1}{3}$, and we conclude that

$$\sum_{k=1}^{n} k^2 = |T| = \binom{n+1}{2} + 2\binom{n+1}{3}$$

as desired.

28. (i) There are $\binom{6}{2}$ ways to choose the 2 horizontal sides of the rectangle, and there are $\binom{10}{2}$ ways to choose the 2 vertical sides of the rectangle. By (MP), the number of rectangles whose vertices are in A is equal to $\binom{6}{2}\binom{10}{2} = 675$.

(ii) For $k = 1, 2, 3, 4, 5$, we see that the number of squares whose side is of length k is equal to $(10-k)(6-k)$. Hence the number of squares whose vertices are in A is equal to $9 \times 5 + 8 \times 4 + 7 \times 3 + 6 \times 2 + 5 \times 1 = 115$.

29. (i) Since P_1, P_2, P_3, P_4, P_5 are collinear and no other 3 points are collinear, we see that any two of P_1, P_2, P_3, P_4, P_5 gives rise to the same line. By (CP), the number of such lines is equal to $\binom{15}{2} - \binom{5}{2} + 1 = 96$.

(ii) Let A, B and C be 3 of the 15 points. We observe that A, B and C form a triangle if and only if A, B and C are not collinear. Also, we see that A, B and C are collinear if and only if A, B and C belong to the set $\{P_1, P_2, P_3, P_4, P_5\}$. By (CP), the number of triangles whose vertices are 3 of the 15 points is equal to $\binom{15}{3} - \binom{5}{3} = 445$.

30. We shall consider these cases:

Case 1: There are 3 different digits in the number, and 0 is one of them. Then each digit must appear exactly twice, and there are $\binom{9}{2}$ to choose the 2 remaining distinct digits. Then the leftmost digit must not be 0, and there are 2 ways to choose the leftmost digit. Subsequently, there are $\binom{5}{2}\binom{3}{2}$ ways to arrange the remaining 5 digits. There are $\binom{9}{2} \times 2 \times \binom{5}{2}\binom{3}{2}$ such numbers.

Case 2: There are 3 different digits in the number, and all digits are non-zero. Then each digit must appear exactly twice, and there are $\binom{9}{3}$ to choose the 3 distinct digits. Then there are $\binom{6}{2}\binom{4}{2}$ ways to arrange the 6 digits. There are $\binom{9}{3}\binom{6}{2}\binom{4}{2}$ such numbers.

Case 3: There are 2 different digits in the number, and 0 appears exactly twice in the number. Then there are 9 ways to choose the other digit. Also, since the leftmost digit must not be 0, it follows that there are $\binom{5}{2}$ ways to arrange the remaining 5 digits. There are $9 \times \binom{5}{2}$ such numbers.

Case 4: There are 2 different digits in the number, and 0 appears exactly three times in the number. Then there are 9 ways to choose the other digit. Also, since the leftmost digit must not be 0, it follows that there are $\binom{5}{3}$ ways to arrange the remaining 5 digits. There are $9 \times \binom{5}{3}$ such numbers.

Case 5: There are 2 different digits in the number, and 0 appears exactly four times in the number. Then there are 9 ways to choose the other digit. Also, since the leftmost digit must not be 0, it follows that there are $\binom{5}{4}$ ways to arrange the remaining 5 digits. There are $9 \times \binom{5}{4}$ such numbers.

Case 6: There are 2 different digits in the number, and both digits are non-zero and each digit appear exactly thrice. Then there are $\binom{9}{2}$ ways to choose the 2 digits, and there are $\binom{6}{3}$ ways to arrange the 6 digits. There are $\binom{9}{2}\binom{6}{3}$ such numbers.

Case 7: There are 2 different digits in the number, and both digits are non-zero and one of the digits appear exactly twice. Then there are P_2^9 ways to choose the 2 digits, and there are $\binom{6}{2}$ ways to arrange the 6 digits. There are $P_2^9 \times \binom{6}{2}$ such numbers.

Case 8: There is only 1 distinct digit in the number. In this case, there are 9 ways to choose the number.

By (AP), the total number of such numbers is equal to

$$2\binom{9}{2}\binom{5}{2}\binom{3}{2} + \binom{9}{3}\binom{6}{2}\binom{4}{2} + 9\binom{5}{2}$$
$$+ 9\binom{5}{3} + 9\binom{5}{4} + \binom{9}{2}\binom{6}{3} + P_2^9\binom{6}{2} + 9$$
$$= 11754.$$

31. We shall consider these cases:

Case 1: There are 2 different digits in the number, and 0 appears exactly thrice in the number. Then there are 9 ways to choose the other digit. Also, since the leftmost digit must not be 0, it follows that there are $\binom{6}{3}$ ways to arrange the remaining 6 digits. There are $9 \times \binom{6}{3}$ such numbers.

Case 2: There are 2 different digits in the number, and 0 appears exactly four times in the number. Then there are 9 ways to choose the other digit. Also, since the leftmost digit must not be 0, it follows that there are $\binom{6}{4}$ ways to arrange the remaining 6 digits. There are $9 \times \binom{6}{4}$ such numbers.

Case 3: There are 2 different digits in the number, and both digits are non-zero and one of the digits appear exactly thrice. Then there are

P_2^9 ways to choose the 2 digits, and there are $\binom{7}{3}$ ways to arrange the 7 digits. There are $P_2^9 \times \binom{7}{3}$ such numbers.

Case 4: There is only 1 distinct digit in the number. In this case, there are 9 ways to choose the number.

By (AP), the total number of such numbers is equal to

$$9\binom{6}{3} + 9\binom{6}{4} + P_2^9\binom{7}{3} + 9 = 2844.$$

32. For a 2-element subset $\{a, b\}$ of $\{1, 2, \ldots, 1000\}$, we note that ab is divisible by 5 if and only if at least one of a, b is divisible by 5. We observe that there are $\binom{800}{2}$ ways of choosing 2 distinct numbers a and b from $\{1, 2, \ldots, 1000\}$ such that both a and b are not divisible by 5. By (CP), the number of 2-element subsets $\{a, b\}$ of $\{1, 2, \ldots, 1000\}$ such that ab is divisible by 5 is equal to $\binom{1000}{2} - \binom{800}{2} = 179900$.

33.   (i) It is easy to see that $A = \{(0,0), (\pm 1, \pm 1), (\pm 2, 0), (0, \pm 2),$
      $(\pm 1, 0), (0, \pm 1)\}$. So $|A| = 13$.

  (ii) It is easy to verify that the following groups of points are collinear:

$$\{(-2, 0), (-1, 1), (0, 2)\}, \{(-1, -1), (0, 0), (1, 1)\},$$
$$\{(0, -2), (1, -1), (2, 0)\}, \{(0, 2), (1, 1), (2, 0)\},$$
$$\{(-1, 1), (0, 0), (1, -1)\}, \{(-2, 0), (-1, -1), (0, -2)\},$$
$$\{(-1, 1), (0, 1), (1, 1)\}, \{(-2, 0), (-1, 0), (0, 0), (1, 0), (2, 0)\},$$
$$\{(-1, -1), (0, -1), (1, -1)\}, \{(-1, 1), (-1, 0), (-1, -1)\},$$
$$\{(0, 2), (0, 1), (0, 0), (0, -1), (0, -2)\}, \{(1, -1), (1, 0), (1, 1)\}.$$

By using a similar argument as in Problem 1.29 and by (CP), we see that the number of straight lines that passes through at least 2 points in A is equal to $\binom{13}{2} - 10\binom{3}{2} - 2\binom{5}{2} + 12 = 40$.

  (iii) By using a similar argument as in Problem 1.29 and by (CP), we see that the number of triangles whose vertices are in A is equal to $\binom{13}{3} - 10\binom{3}{3} - 2\binom{5}{3} = 256$.

34. Fix a vertex A in P. Let us count the number of triangles, containing A as a vertex, whose vertices are in P and are pairwise non-adjacent. There are $\binom{n-3}{2}$ ways to choose vertices B and C in P such that B and C is not adjacent to A. Out of these $\binom{n-3}{2}$ ways, $n - 4$ of them have B and C adjacent to each other. As each such triangle is counted exactly 3 times in this process, we see that the number of triangles formed by any 3 vertices of P such that they are pairwise non-adjacent in P is equal to $\frac{n}{3}\left(\binom{n-3}{2} - (n-4)\right) = \frac{n}{3}\binom{n-4}{2}$.

35. (i) The number of ways is equal to $Q_{11}^{11} = (11 - 1)! = 10!$.

    (ii) Firstly, there are $Q_6^6 = (6 - 1)! = 5!$ ways to arrange the 6 boys around the table. Then there are P_5^6 ways to arrange the 5 girls in such a way that no 2 girls are adjacent. By (MP), the total number of ways is equal to $5!P_5^6 = 5! \times 6!$.

    (iii) Firstly, there are 5! ways to arrange the 5 girls within the block. Then there are $Q_7^7 = (7 - 1)! = 6!$ ways to arrange the 6 boys and the block of girls around the table. By (MP), the total number of ways is equal to $5!6!$.

    (iv) Firstly, there are 2 ways to arrange the boys B_1 and B_2. Then there are $Q_9^9 = (9 - 1)! = 8!$ ways to arrange the remaining 8 boys and girls, and the block consisting of B_1, B_2 and G around the table. By (MP), the total number of ways is equal to $2 \cdot 8!$.

36. Firstly, there are $\binom{n}{r} = \frac{n!}{(n-r)!r!}$ ways to choose r objects out of n distinct objects. Then there are $Q_r^r = (r - 1)!$ ways to arrange the r objects around the circle. By (MP), the number of such circular permutations is equal to

$$\frac{n!}{(n-r)!r!} \cdot (r - 1)! = \frac{n!}{(n-r)!r}.$$

37. Firstly, there are $\binom{kn}{n}$ ways to choose n people to seat at the first table. Then, there are $\binom{(k-1)n}{n}$ ways to choose n people to seat at the second table. In general, there are $\binom{(k-r)n}{n}$ ways to choose n people to seat at the $(r + 1)$-th table. Then, for each table, there are $Q_n^n = (n - 1)!$ ways to arrange the n people around the table. By (MP), the number of ways is equal to

$$\binom{kn}{n}\binom{(k-1)n}{n}\binom{(k-2)n}{n}\cdots\binom{n}{n} \cdot ((n-1)!)^k$$

$$= \frac{(kn)!}{((k-1)n)!n!} \cdot \frac{((k-1)n)!}{((k-2)n)!n!} \cdots \frac{(2n)!}{n!n!} \cdot \frac{n!}{n!0!} \cdot ((n-1)!)^k$$

$$= \frac{(kn)!}{(n!)^k} \cdot ((n-1)!)^k$$

$$= \frac{(kn)!}{n^k}.$$

38. We have

$$\frac{1}{\binom{9}{r}} - \frac{1}{\binom{10}{r}} = \frac{11}{6\binom{11}{r}}$$

$$\Rightarrow \frac{r!(9-r)!}{9!} - \frac{r!(10-r)!}{10!} = \frac{r!(11-r)!}{6 \cdot 10!}$$

$$\Rightarrow \frac{r!(9-r)!}{6\cdot 10!}(60 - 6(10-r) - (10-r)(11-r)) = 0$$
$$\Rightarrow r^2 - 27r + 110 = 0$$
$$\Rightarrow (r-5)(r-22) = 0.$$

This gives us $r = 5$ or $r = 22$. Since $0 \le r \le 9$, we must have $r = 5$.

39. (a) Let us count the number of ways to choose r people from a set of n people to form a committee consisting of a chairman and $r-1$ members, in 2 different ways. On one hand, we can first choose the r people for the committee in $\binom{n}{r}$ ways, and then choose the chairman for the committee in r ways.

On the other hand, there are exactly n ways to first choose the chairman for the committee, and subsequently there are $\binom{n-1}{r-1}$ ways to choose the $r-1$ members for the committee. Hence, we have $r\binom{n}{r} = n\binom{n-1}{r-1}$, or equivalently,

$$\binom{n}{r} = \frac{n}{r}\binom{n-1}{r-1}$$

as desired.

(b) Let us count the number of ways to choose r people from a set of n people to form a committee consisting of a chairman and $r-1$ members, in 2 different ways. On one hand, we can first choose the r people for the committee in $\binom{n}{r}$ ways, and then choose the chairman for the committee in r ways.

On the other hand, we can first choose the $r-1$ members for the committee in $\binom{n}{r-1}$ ways, and there are $n-r+1$ ways to choose the chairman for the committee. Hence, we have $r\binom{n}{r} = (n-r+1)\binom{n}{r-1}$, or equivalently,

$$\binom{n}{r} = \frac{n-r+1}{r}\binom{n}{r-1}$$

as desired.

(c) Let us count the number of ways to choose $r+1$ people from a set of n people to form a committee consisting of a chairman and r members, in 2 different ways. On one hand, we can first choose the r members for the committee in $\binom{n}{r}$ ways, and there are $n-r$ ways to choose the chairman for the committee.

On the other hand, there are exactly n ways to first choose the chairman for the committee, and subsequently there are $\binom{n-1}{r}$

ways to choose the r members for the committee. Hence, we have $(n-r)\binom{n}{r} = n\binom{n-1}{r}$, or equivalently,

$$\binom{n}{r} = \frac{n}{n-r}\binom{n-1}{r}$$

as desired.

(d) Let us count the number of ways to choose m people from a set of n people to form a committee consisting of r normal members and $m-r$ executive members, in 2 different ways. On one hand, we can first choose m people for the committee in $\binom{n}{m}$ ways, and then choose r normal members from the m committee members in $\binom{m}{r}$ ways. By (MP), we have $\binom{n}{m}\binom{m}{r}$ committees that can be formed. On the other hand, we can first choose r normal members for the committee in $\binom{n}{r}$ ways, and then choose $m-r$ executive members for the committee in $\binom{n-r}{m-r}$ ways. Hence, we have

$$\binom{n}{m}\binom{m}{r} = \binom{n}{r}\binom{n-r}{m-r}$$

as desired.

40. Let A denote the set of binary strings of length n consisting of r 0's and $n-r$ 1's, and let B denote the set of binary strings of length n consisting of $n-r$ 0's and r 1's. Then we have $|A| = \binom{n}{r}$ and $|B| = \binom{n}{n-r}$ by definition. Also, it is easy to see that the function $f: A \to B$ given by $f(\overline{a_1 a_2 \cdots a_n}) = \overline{b_1 b_2 \cdots b_n}$, where $b_i = 1 - a_i$ for all $i \in \{1, 2, \ldots n\}$, is a well-defined bijection from A to B. By (BP), we have

$$\binom{n}{r} = |A| = |B| = \binom{n}{n-r}$$

as desired.

41. It is easy to see that the function $f: \mathcal{A} \to \mathcal{B}$ given by $f(A) = X \setminus A$, is a well-defined bijection. By (BP), we have $|\mathcal{A}| = |\mathcal{B}|$ as desired.

42. We have

$$\frac{(n+1)(n+2)\cdots(n+r)}{r!} = \frac{n!(n+1)(n+2)\cdots(n+r)}{n!r!}$$
$$= \frac{(n+r)!}{n!r!}$$
$$= \binom{n+r}{r}$$

to be an integer by definition. Hence $(n+1)(n+2)\cdots(n+r)$ is divisible by $r!$ as desired.

43. By using a similar argument to that in Example 1.4.4 Solution 3, we see that the number of k-groupings of A is equal to $\frac{(kn)!}{n!(k!)^n}$, since the order in each k-element subset of A, and the order of the k-element subsets of A is immaterial.

44. First Solution: Let us fix a sitting configuration of the knights. Regarding the 25 knights as the vertices of a 25-sided regular convex polygon Q, the given problem is equivalent to finding the probability that a triangle formed by the vertices of the polygon have at least a pair of adjacent vertices in Q. Fix a vertex A in P. Let us count the number of triangles, containing A as a vertex, whose vertices are in Q and are pairwise non-adjacent.

    There are $\binom{22}{2}$ ways to choose vertices B and C in Q such that B and C are both not adjacent to A. Out of these $\binom{22}{2}$ ways, 21 of them have B and C adjacent to each other. As each such triangle is counted exactly 3 times in this process, we see that the number of triangles formed by any 3 vertices of Q such that they are pairwise non-adjacent in Q is equal to $\frac{25}{3}\left(\binom{22}{2} - 21\right) = 1750$.

    By (CP), the number of triangles formed by any 3 vertices of Q such that at least a pair of vertices of the triangle is adjacent in Q is equal to $\binom{25}{3} - 1750 = 550$. So the probability that a triangle formed by the vertices of the polygon have at least a pair of adjacent vertices in Q is $\frac{550}{\binom{25}{3}} = \frac{11}{46}$, and hence $P = \frac{11}{46}$. So the desired answer is $11 + 46 = 57$.

    Second Solution: Same as before as in the first solution, let us regard the 25 knights as the vertices of a 25-sided regular convex polygon Q. Let us label the vertices $1 - 25$, and find the probability that a triangle formed by any 3 vertices of the polygon has at least a pair of adjacent vertices in Q. To this end, let us consider these cases:

    Case 1: The triangle has two adjacent pairs of vertices in Q. There are 25 such triangles.

    Case 2: The triangle has exactly one adjacent pair of vertices in Q. There are 25 ways to choose two vertices A and B in Q such that they are adjacent to each other, and there are $25 - 4 = 21$ ways to choose a third vertex C in Q such that C is not adjacent to both A and B.

    By (AP) and (MP), the total number of triangles formed by any 3 vertices of the polygon, such that the triangle has at least a pair of adjacent vertices in Q is equal to $25 + 25 \cdot 21 = 550$. As there is a total of $\binom{25}{3} = 2300$ triangles that can be formed by any 3 vertices of the polygon without any restriction, it follows that $P = \frac{550}{2300} = \frac{11}{46}$, and hence the desired answer is $11 + 46 = 57$.

45. Since the empty set and the whole set are not admissible combinations, the total number of admissible combinations is equal to $2^{10} - 2 = 1022$. By (CP), the number of additional admissible combinations is equal to $1022 - \binom{10}{5} = 770$.

46. We can visualize the clay targets as follows: the left chain is a chain of 3 red targets, the middle chain is a chain of 2 white targets, and the right chain is a chain of 3 blue targets. Then each valid shooting order corresponds to an arrangement of the 3 red targets, 2 white targets and 3 blue targets in a row. Hence, the number of valid shooting orders is equal to $P(8; 3, 2, 3) = \frac{8!}{3!2!3!} = 560$.

47. For a sequence $a = \overline{a_1 a_2 \cdots a_k}$ of distinct digits, where $k = 1, 2, 3, 4, 5$, and $a_i \in \{1, 2, 3, 4, 5\}$ for all $i = 1, 2, \ldots, k$, we see that the number of 5-digit integers that uses the digits $1, 2, 3, 4, 5$ exactly once, and begins with the sequence a is equal to $(5 - k)!$.

  (i) Using the above fact, we see that:
   12345 is the first number,
   31245 is the $(1 + 2 \cdot 4!)$-th number,
   35124 is the $(1 + 2 \cdot 4! + 3 \cdot 3!)$-th number, and
   35412 is the $(1 + 2 \cdot 4! + 3 \cdot 3! + 2 \cdot 2!)$-th number.
   So 35421 is the $(1 + 2 \cdot 4! + 3 \cdot 3! + 2 \cdot 2! + 1)$-th $= 72$-th number.

  (ii) First, we note that $\lfloor \frac{100}{4!} \rfloor = 4$, and $100 - 4 \cdot 4! - 1 = 3 \geq 0$. Based on this, we see that the $(1 + 4 \cdot 4!)$-th $= 97$-th number in the sequence is 51234. Also, since $2! < 3 < 3!$, we see that the $(97 + 2!)$-th $= 99$-th number in the sequence is 51324. So the 100-th number in the sequence is 51342.

48. For a sequence $a = (a_1, a_2, \ldots, a_k)$ of distinct digits, where $k = 1, 2, 3$, $a_i \in \{1, 2, \ldots 9\}$, we see that the number of 4-permutations of $\{1, 2, \ldots 9\}$ that begins with the sequence a is equal to P_{4-k}^{9-k}.
   Using the above fact, we see that:
   "1234" is the first permutation,
   "4123" is the $(1 + 3 \cdot P_3^8)$-th permutation,
   "4512" is the $(1 + 3 \cdot P_3^8 + 3 \cdot P_2^7)$-th permutation,
   "4561" is the $(1 + 3 \cdot P_3^8 + 3 \cdot P_2^7 + 3 \cdot P_1^6)$-th permutation.
   So "4567" is the $(1 + 3 \cdot P_3^8 + 3 \cdot P_2^7 + 3 \cdot P_1^6 + 3)$-th $= 1156$-th permutation, that is, "4567" is in the 1156-th position.
   Using the above fact, we see that:
   "1234" is the first permutation,

"5123" is the $(1 + 4 \cdot P_3^8)$-th permutation.

So "5182" is the $(1 + 4 \cdot P_3^8 + 5 \cdot P_1^6)$-th $= 1375$-th permutation, that is, "5182" is in the 1375-th position.

49. Fix a strictly increasing sequence $(a_1, a_2, \ldots, a_k)$, where $k = 1, 2, 3$, and $a_1 \geq 1$, $a_k \leq 10$. Let $B = \{b_1, b_2, b_3, b_4\}$ be a 4-element subset of $\{1, 2, \ldots, 10\}$ with the property that $b_1 < b_2 < b_3 < b_4$, and $b_i = a_i$ for $i = 1, 2, \ldots, k$. Then we see that the number of such subsets B is equal to $\binom{10-a_k}{4-k}$.

Using the above fact, we see that:

$\{1, 2, 3, 4\}$ is the first subset, $\{2, 3, 4, 5\}$ is the $(1 + \binom{10-1}{4-1})$-th subset.

So $\{3, 4, 5, 6\}$ is the $(1 + \binom{10-1}{4-1} + \binom{10-2}{4-1})$-th $= 141$-th subset, that is, $\{3, 4, 5, 6\}$ is in the 141-th position.

Using the above fact, we see that:

$\{3, 5, 6, 7\}$ is the $(141 + \binom{10-4}{4-2})$-th subset,

$\{3, 5, 7, 8\}$ is the $(141 + \binom{10-4}{4-2} + \binom{10-6}{4-3})$-th subset.

So $\{3, 5, 7, 9\}$ is the $(141 + \binom{10-4}{4-2} + \binom{10-6}{4-3} + 1)$-th $= 161$-th subset, that is, $\{3, 5, 7, 9\}$ is in the 161-th position.

50. Let the scientists be $S_1, S_2, \ldots, S_6$, and let L be the set of locks, and let S be the set of 2-element subsets of $\{S_1, S_2, \ldots, S_6\}$. For each element $T = \{S_a, S_b\}$ in S, let L_T denote the set of locks that cannot be opened by the scientists S_a, S_b.

We first note that L_T is non-empty for each element T in S; this is because the cabinet can be opened only when 3 or more scientists are present. Furthermore; for any two distinct elements $U, V \in S$; we must have $L_U \cap L_V = \emptyset$. To see that this is the case; we note that if $\ell \in L_U \cap L_V$; then we see that lock ℓ cannot be opened by the set of scientists $U \cup V$. As the set $U \cup V$ contains at least 3 scientists, this contradicts the fact that the cabinet can be opened by any group containing at least 3 scientists.

Now, let us define a function $f : S \to L$ by $f(T) = \ell_T$, where ℓ_T is an element arbitrary chosen from the set L_T. Then it is clear from the above that f is a well-defined injection. By (IP), we have $|L| \geq |S| = \binom{6}{2} = 15$, so the minimum number of locks required is 15.

For the second part of the problem, let us consider the number of keys that S_1 needs to carry. Since the cabinet can be opened by any group containing at least 3 scientists, it follows that for every two-element subset W of $\{S_2, S_3, \ldots, S_6\}$, S_1 must possess the keys to the locks in L_W. Furthermore, we see from the above that $|L_W| \geq 1$ and

$L_U \cap L_V = \emptyset$, so the number of keys that S_1 must carry is equal to

$$\sum_{V \subseteq \{S_2, S_3, \ldots, S_6\}, |V|=2} |L_V| \geq \binom{5}{2} = 10.$$

Hence, the minimum number of keys that S_1 needs to carry is 10. Since S_1 is arbitrary, we see that each scientist needs to carry at least 10 keys.

51. (i) There are 4^{10} ways to paint the building.

(ii) Firstly, there are 4 ways to paint the first storey. Then for $k = 1, 2, \ldots, 9$, there are 3 ways to paint the $(k + 1)$-th storey since the $(k + 1)$-th storey must be of a different colour from the k-th storey. By (MP), the number of ways is equal to $4 \cdot 3^9$.

52. We observe that a multi-subset of $\{r_1 \cdot a_1, r_2 \cdot a_2, \ldots, r_n \cdot a_n\}$ is of the form

$$\{s_1 \cdot a_1, s_2 \cdot a_2, \ldots, s_n \cdot a_n\},$$

where $0 \leq s_i \leq a_i$ for all $i \in \{1, 2, \ldots, n\}$. By (MP), the number of such multi-subsets is equal to $(r_1 + 1)(r_2 + 1) \cdots (r_n + 1)$.

53. Let S denote the set of permutations of $\{1, 2, \ldots, 2n\}$ without the property $P(r)$, and T denote the set of permutations of $\{1, 2, \ldots, 2n\}$ with the property $P(r)$. Without loss of generality, let us assume that S is non-empty. Let us define a function $f \colon S \to T$ by $f(x_1 x_2 \cdots x_{2n}) = x_2 x_3 \cdots x_{s-1} x_s x_1 x_{s+1} \cdots x_{2n}$, where s is the smallest integer such that $|x_{s+1} - x_1| = r$. (Note: Depending on the value of r and x_1, there may be more than one integer s satisfying the relation $|x_{s+1} - x_1| = r$.) Then by a similar argument as in Example 1.5.4, we see that f is an injection but not a bijection, so by (IP) and (BP), we have $|S| < |T|$ as desired.

54. Let A be a set of kn elements, where k, n are positive integers. Let us define a k-ordering of A to be a sequence $(A_1, A_2, \ldots, A_n)$ of n pairwise disjoint k-element subsets of A, and count the number of such k-orderings of A. Firstly, there are $\binom{kn}{k}$ ways to choose k elements of A for A_1. Then, there are $\binom{(n-1)k}{k}$ ways to choose k elements of $A \setminus A_1$ for A_2. In general, there are $\binom{(n-r)k}{k}$ ways to choose k elements of $A \setminus (\cup_{i=1}^{r} A_i)$ for A_{r+1}, where $r \in \{1, 2, \ldots, n-1\}$. By (MP), the number of such k-orderings of A is equal to

$$\binom{kn}{k} \binom{(n-1)k}{k} \binom{(n-2)k}{k} \cdots \binom{k}{k}$$

$$= \frac{(kn)!}{((n-1)k)!k!} \cdot \frac{((n-1)k)!}{((n-2)k)!k!} \cdot \frac{((n-2)k)!}{((n-3)k)!k!} \cdots \frac{(2k)!}{k!k!} \cdot \frac{k!}{k!0!}$$

$$= \frac{(kn)!}{(k!)^n}.$$

Now,

(i) By setting $k = 3$, we see that the number of k-orderings of A is equal to $\frac{(kn)!}{(k!)^n} = \frac{(3n)!}{(3!)^n} = \frac{(3n)!}{2^n \cdot 3^n}$. Hence $\frac{(3n)!}{2^n \cdot 3^n}$ is an integer as desired.

(ii) By setting $k = 6$, we see that the number of k-orderings of A is equal to $\frac{(kn)!}{(k!)^n} = \frac{(6n)!}{(6!)^n} = \frac{(3n)!}{2^{4n} \cdot 3^{2n} \cdot 5^n}$. Hence $\frac{(3n)!}{2^{4n} \cdot 3^{2n} \cdot 5^n}$ is an integer as desired.

(iii) By setting $k = n$, we see that the number of k-orderings of A is equal to $\frac{(kn)!}{(k!)^n} = \frac{(n^2)!}{(n!)^n}$. Hence $\frac{(n^2)!}{(n!)^n}$ is an integer as desired.

Finally, by defining a n-ordering of A to be a sequence $(A_1, A_2, \ldots, A_k)$ of k pairwise disjoint n-element subsets of A, we see by a similar argument as above that the number of n-orderings of A is equal to $\frac{(kn)!}{(n!)^k}$.

(iv) Hence by setting $k = (n-1)!$, we see that the number of n-orderings of A is equal to $\frac{(kn)!}{(n!)^k} = \frac{(n!)!}{(n!)^{(n-1)!}}$. Hence $\frac{(n!)!}{(n!)^{(n-1)!}}$ is an integer as desired.

55. Let M be an r-element multi-subset of $\{1 \cdot a_1, \infty \cdot a_2, \ldots, \infty \cdot a_n\}$. If $a_1 \in M$, then there are $H_{r-1}^{n-1} = \binom{r+n-3}{r-1}$ such sets. Else, if $a_1 \notin M$, then there are $H_r^{n-1} = \binom{r+n-2}{r}$ such sets. By (AP), the number of r-elements multi-subsets of $\{1 \cdot a_1, \infty \cdot a_2, \ldots, \infty \cdot a_n\}$ is equal to $\binom{r+n-3}{r-1} + \binom{r+n-2}{r}$.

56. Firstly, there are 6! ways to arrange the 6 distinct symbols. Suppose that the order of which the symbols are transmitted is $s_1, s_2, s_3, s_4, s_5, s_6$. Let b_i denote the number of blanks transmitted between the symbols s_i and s_{i+1} for $i = 1, 2, 3, 4, 5$. Then we have $b_1 + b_2 + b_3 + b_4 + b_5 = 18$, with the condition that $b_i \geq 2$ for all $i = 1, 2, 3, 4, 5$. By writing $c_i = b_i - 2$ for $i = 1, 2, 3, 4, 5$, we see that the number of ways to insert the blanks is equal to the number of non-negative integer solutions of the equation $c_1 + c_2 + c_3 + c_4 + c_5 = 8$, which is equal to $H_8^5 = \binom{12}{4}$. By (MP), the total number of ways is equal to $\binom{12}{4} \cdot 6!$.

57. Firstly, by considering the 3 X's and 2 Y's in isolation, we see that there are exactly 3 ways $(XYXYX, XYYXX, XXYYX)$ of arrang-

ing the 3 X's and 2 Y's in such a way that every Y lies between 2 X's. Then, we see that there are 6! ways to arrange the letters A, B, C, D, E, F in a row. Finally, by identifying the X's and the Y's, we see that there are $H_5^7 = \binom{11}{5}$ ways of inserting the letters X's and Y's into the row. By (MP), the number of ways is equal to $3 \cdot \binom{11}{5} \cdot 6!$.

58. (i) The number of pairwise non-equivalent 5-digit integers is equal to the number of 5-element multi-subsets of the multi-set $\{\infty \cdot 0, \infty \cdot 1, \ldots, \infty \cdot 9\}$, which is equal to $H_5^{10} = \binom{14}{5}$.

   (ii) We shall consider these cases:

   Case 1: All of the digits $5, 7, 9$ appear in the 5-digit integer. Then the number of such pairwise non-equivalent 5-digit integers satisfying the conditions is equal to $H_2^7 = \binom{8}{2}$.

   Case 2: Exactly two of the digits $5, 7, 9$ appear in the 5-digit integer. Then the number of such pairwise non-equivalent 5-digit integers satisfying the conditions is equal to $3 \cdot H_3^7 = 3\binom{9}{3}$.

   Case 3: Exactly one of the digits $5, 7, 9$ appear in the 5-digit integer. Then the number of such pairwise non-equivalent 5-digit integers satisfying the conditions is equal to $3 \cdot H_4^7 = 3\binom{10}{4}$.

   Case 4: None of the digits $5, 7, 9$ appear in the 5-digit integer. Then the number of such pairwise non-equivalent 5-digit integers satisfying the conditions is equal to $H_5^7 = \binom{11}{5}$.

   By (AP), the number of pairwise non-equivalent 5-digit integers satisfying the conditions is equal to $\binom{11}{5} + 3\binom{10}{4} + 3\binom{9}{3} + \binom{8}{2}$.

59. (i) The number of 10-letter words is equal to 6^{10}.

   (ii) The number of such 10-letter words is equal to $\frac{10!}{3!3!}$.

   (iii) The number of such 10-letter words is equal to the number of 10-element multi-subsets of the multi-set $\{\infty \cdot a, \infty \cdot b, \ldots, \infty \cdot f\}$, which is equal to $H_{10}^6 = \binom{15}{5}$.

   (iv) Let t_α denote the number of times that the letter α appears in the 10-letter word for $\alpha = a, b, c, d, e, f$. Then we have $t_a + t_b + t_c + t_d + t_e + t_f = 10$, with the condition that $t_\alpha \geq 1$ for all $\alpha = a, b, c, d, e, f$. This implies that the number of such 10-letter words is equal to the number of non-negative integer solutions of the equation $u_a + u_b + u_c + u_d + u_e + u_f = 4$, which is equal to $H_4^6 = \binom{9}{4}$.

60. Label the boxes $B_1, B_2, \ldots, B_n$, and let the number of objects in the box B_i be equal to a_i, where $i = 1, 2, \ldots, n$. Then we have $a_1 + a_2 + \cdots + a_n = r$, with the condition that $a_i \geq k$ for all $i = 1, 2, \ldots, 5$. This implies that the number of ways of distributing is equivalent to

the number of non-negative integer solutions of the equation $b_1 + b_2 + \cdots + b_n = r - nk$, which is equal to $H_{r-nk}^n = \binom{r-n(k-1)-1}{n-1}$.

61. Firstly, by considering the letters x, y, z in isolation, we see that there are 2 ways of arranging x, y, z such that y always lie between x and z. Then, we see that there are 6! ways to arrange the letters r, s, t, u, v, w in a row. Finally, by identifying the letters x, y, z, we see that there are $H_3^7 = \binom{9}{3}$ ways of inserting the letters x, y and z into the row. By (MP), the number of ways is equal to $2 \cdot \binom{9}{3} \cdot 6! = 2 \cdot \frac{9!}{3!}$.

62. Firstly, there are 2 ways to arrange the girls A, B, C in such a way that B must lie between A and C. Next, we note that the number of ways to insert 9 identical blanks in the arrangement of the girls A, B, C, such that there are exactly 4 blanks between the girls A and B, is equal to $H_5^3 = 21$. Finally, there are 9! ways to arrange the 9 boys into the 9 blanks. By (MP), the number of ways is equal to $2 \cdot 9! \cdot 21 = 42 \cdot 9!$.

63. Firstly, let us find the number of ways of inserting 11 blanks in the arrangement of the girls, such that there must be at least 3 blanks between G_1 and G_2, and there is at most one blank between G_4 and G_5.

    We note that the number of ways of inserting 11 blanks in the arrangement of the girls, such that there must be at least 3 blanks between G_1 and G_2 is equal to $H_8^6 = 1287$, and the number of ways of inserting 11 blanks in the arrangement of the girls, such that there must be at least 3 blanks between G_1 and G_2, and there must be at least 2 blanks between G_4 and G_5 is equal to $H_6^6 = 462$. By (CP), the number of ways of inserting 11 blanks in the arrangement of the girls, such that there must be at least 3 blanks between G_1 and G_2, and there is at most one blank between G_4 and G_5 is equal to $1287 - 462 = 825$. Next, we see that the number of ways of arranging the 11 boys into the 11 blanks is equal to 11!. By (MP), the number of ways is equal to $825 \cdot 11!$.

64. Let $M(r, n)$ denote the number of ways of distributing r distinct objects into n distinct boxes, such that no box is empty, and the objects in each box are arranged in a row. Let $s_1, s_2, \ldots, s_r, t$ be $r + 1$ distinct symbols. Then we see that $M(r, n)$ is equal to the number of permutations of the multi-set $\{1 \cdot s_1, 1 \cdot s_2, \ldots, 1 \cdot s_r, (n-1) \cdot t\}$, such that each of the t's must appear in between two of the s_i's.

    Firstly, we see that there are $r!$ ways to arrange the symbols $s_1, s_2, \ldots, s_r$ in a row, and there are $\binom{r-1}{n-1}$ ways to place $n - 1$ t's within the arrangement of the s_i's, such that each of the t's appear in between two of the s_i's. By (MP), we see that $M(r,n) = r!\binom{r-1}{n-1}$. Since the boxes in the case of $L(r,n)$ is identical, we see that $L(r,n) = \frac{M(r,n)}{n!} = \frac{r!}{n!}\binom{r-1}{n-1}$.

65. (i) By setting $y_i = x_i - (i-1)$ for $i = 1,2,3,4,5,6$, we see that the number of integer solutions to the equation $x_1 + x_2 + x_3 + x_4 + x_5 + x_6 = 60$ that satisfies $x_i \geq i - 1$ for $i = 1,2,3,4,5,6$, is equal to the number of non-negative integer solutions to the equation $y_1 + y_2 + y_3 + y_4 + y_5 + y_6 = 45$, which is equal to $H_{45}^6 = \binom{50}{5}$.

   (ii) By setting $a_1 = 2$, $a_2 = 5$, $a_3 = 2$, $a_4 = 1$, $a_5 = 3$ and $a_6 = 2$, we see that the number of integer solutions to the equation $x_1 + x_2 + x_3 + x_4 + x_5 + x_6 = 60$ that satisfies $x_i \geq a_i$ for $i = 1,2,3,4,5,6$, is equal to the number of non-negative integer solutions to the equation $y_1 + y_2 + y_3 + y_4 + y_5 + y_6 = 45$, with the additional condition that $y_3 \leq 5$.

   We note that the number of non-negative integer solutions to the equation

   $$y_1 + y_2 + y_3 + y_4 + y_5 + y_6 = 45$$

   is equal to $\binom{50}{5}$ by part (i). Also, by setting $z_i = y_i$ for $i = 1,2,4,5,6$ and $z_3 = y_3 - 6$, we see that the number of non-negative integer solutions to the equation $y_1 + y_2 + y_3 + y_4 + y_5 + y_6 = 45$ such that $y_3 \geq 6$, is equal to the number of non-negative integer solutions to the equation $z_1 + z_2 + z_3 + z_4 + z_5 + z_6 = 39$, which is then equal to $H_{39}^6 = \binom{44}{5}$. By (CP), the number of integer solutions to the equation $x_1 + x_2 + x_3 + x_4 + x_5 + x_6 = 60$ that satisfies $x_i \geq a_i$ for $i = 1,2,3,4,5,6$ and $x_3 \leq 7$ is equal to $\binom{50}{5} - \binom{44}{5}$.

66. (i) The number of non-negative integer solutions to the equation $x_1 + x_2 + x_3 + x_4 = 30$ is equal to $H_{30}^4 = \binom{33}{3}$.

   (ii) By setting $a_1 = 2$ and $a_2 = a_3 = a_4 = 0$, we see that the number of integer solutions to the equation $x_1 + x_2 + x_3 + x_4 = 30$ satisfying $x_i \geq a_i$ for $i = 1,2,3,4$ and $x_1 \leq 7$ is equal to the number of non-negative integer solutions to the equation $y_1 + y_2 + y_3 + y_4 = 28$, subject to the additional condition $y_1 \leq 5$.

   We note that the number of non-negative integer solutions to the equation

   $$y_1 + y_2 + y_3 + y_4 = 28$$

is equal to $H_{28}^4 = \binom{31}{3}$. Also, by setting $z_1 = y_1 - 6$ and $z_i = y_i$ for $i = 2, 3, 4$, we see that the number of non-negative integer solutions to the equation $y_1 + y_2 + y_3 + y_4 = 28$, such that $y_1 \geq 6$, is equal to the number of non-negative integer solutions to the equation $z_1 + z_2 + z_3 + z_4 = 22$, which is then equal to $H_{22}^4 = \binom{25}{3}$. By (CP), the number of integer solutions to the equation $x_1 + x_2 + x_3 + x_4 = 30$ satisfying the conditions given in the problem is equal to $\binom{31}{3} - \binom{25}{3}$.

(iii) By setting $a_1 = -5$, $a_2 = -1$, $a_3 = 1$ and $a_4 = 2$ and $y_i = x_i - a_i$ for $i = 1, 2, 3, 4$, we see that the number of integer solutions to the equation $x_1 + x_2 + x_3 + x_4 = 30$ that satisfies $x_i \geq a_i$ for $i = 1, 2, 3, 4$, is equal to the number of non-negative integer solutions to the equation $y_1 + y_2 + y_3 + y_4 = 33$, which is equal to $H_{33}^4 = \binom{36}{3}$.

67. We note that a tuple (w, x, y, z) of non-negative integers satisfies $w + x + y + z \leq 1992$ if and only if $w + x + y + z = 1992 - v$ for some non-negative integer v. Hence, we see that the number of 4-tuples (w, x, y, z) of non-negative integers that satisfy the inequality $w + x + y + z \leq 1992$ is equal to the number of tuples (v, w, x, y, z) of non-negative integers that satisfy the equation $v + w + x + y + z = 1992$, which is equal to $H_{1992}^5 = \binom{1996}{4}$.

68. Clearly, the only values of x_1 for which the equation $5x_1 + x_2 + x_3 + x_4 = 14$ of non-negative integers admits a solution are $0, 1$ and 2. For $v = 0, 1, 2$, we see that the number of non-negative integer solutions to the equation $5x_1 + x_2 + x_3 + x_4 = 14$ with $x_1 = v$ is equal to the number of non-negative integer solutions to the equation $x_2 + x_3 + x_4 = 14 - 5v$, which is equal to $H_{14-5v}^3 = \binom{16-5v}{2}$.

By (AP), the total number of solutions to the equation $5x_1 + x_3 + x_4 = 14$ is equal to $\sum_{v=0}^{2} \binom{16-5v}{2} = \binom{16}{2} + \binom{11}{2} + \binom{6}{2}$.

69. Clearly, the only values of x_1 for which the equation $rx_1 + x_2 + \cdots + x_n = kr$ of non-negative integers admits a solution are $0, 1, \ldots, k$. For $i = 0, 1, \ldots, k$, we see that the number of non-negative integer solutions to the equation $rx_1 + x_2 + \cdots + x_n = kr$ with $x_1 = i$ is equal to the number of non-negative integer solutions to the equation $x_2 + \cdots + x_n = r(k - i)$, which is equal to $H_{r(k-i)}^{n-1} = \binom{r(k-i)+n-2}{n-2}$.

By (AP), the total number of non-negative integer solutions to the equation $rx_1 + x_2 + \cdots + x_n = kr$ is equal to $\sum_{i=0}^{k} \binom{r(k-i)+n-2}{n-2}$.

70. Clearly, the only values of (x_1, x_2) for which the equation $3x_1 +$

$5x_2 + x_3 + x_4 = 10$ of non-negative integers admits a solution are $(0,0), (0,1), (0,2), (1,0), (1,1), (2,0)$ and $(3,0)$. For a pair of integers $(a,b) = (0,0), (0,1), (0,2), (1,0), (1,1), (2,0), (3,0)$, we see that the number of non-negative integer solutions to the equation

$$5x_1 + x_2 + x_3 + x_4 = 14$$

with $x_1 = a, x_2 = b$ is equal to the number of non-negative integer solutions to the equation $x_3 + x_4 = 10 - 3a - 5b$, which is equal to

$$H^2_{10-3a-5b} = \binom{11 - 3a - 5b}{1} = 11 - 3a - 5b.$$

By (AP), the total number of solutions to the equation $3x_1 + 5x_2 + x_3 + x_4 = 10$ is equal to

$$
\begin{aligned}
&(11 - 3(0) - 5(0)) + (11 - 3(0) - 5(1)) + (11 - 3(0) - 5(2)) \\
&+ (11 - 3(1) - 5(0)) + (11 - 3(1) - 5(1)) + (11 - 3(2) - 5(0)) \\
&+ (11 - 3(3) - 5(0))
\end{aligned}
$$

$$= 36.$$

71. Since all of $x_1, x_2, x_3, y_1, y_2, y_3, y_4$ are positive, and the only positive factors of 77 are $1, 7, 11, 77$, we must have either $x_1 + x_2 + x_3 = 7$ and $y_1 + y_2 + y_3 + y_4 = 11$, or $x_1 + x_2 + x_3 = 11$ and $y_1 + y_2 + y_3 + y_4 = 7$. We note that for a given equation $a_1 + a_2 + \cdots + a_n = r$ where $r \geq n$ is a positive integer, we see that by writing $b_i = a_i - 1$ for $i = 1, 2, \ldots, n$, the number of positive integer solutions to the equation $a_1 + a_2 + \cdots + a_n = r$ is equal to the number of non-negative integer solutions to the equation $b_1 + b_2 + \cdots + b_n = r - n$, which is equal to $H^n_{r-n} = \binom{r-1}{n-1}$.

    For the former case, there are $\binom{7-1}{3-1} = \binom{6}{2}$ positive integer solutions to the equation $x_1 + x_2 + x_3 = 7$ and there are $\binom{11-1}{4-1} = \binom{10}{3}$ positive integer solutions to the equation $y_1 + y_2 + y_3 + y_4 = 11$.

    For the latter case, there are $\binom{11-1}{3-1} = \binom{10}{2}$ positive integer solutions to the equation $x_1 + x_2 + x_3 = 11$ and there are $\binom{7-1}{4-1} = \binom{6}{3}$ positive integer solutions to the equation $y_1 + y_2 + y_3 + y_4 = 7$.

    By (AP) and (MP), there is a total of $\binom{6}{2}\binom{10}{3} + \binom{10}{2}\binom{6}{3}$ positive integer solutions to the equation $(x_1 + x_2 + x_3)(y_1 + y_2 + y_3 + y_4) = 77$.

72. Since the only positive factors of p are 1 and p, we must have either $x_1 + \cdots + x_n = 1$ and $y_1 + \cdots + y_n = p$, or $x_1 + \cdots + x_n = p$ and $y_1 + \cdots + y_n = 1$.

For the former case, there are $H_1^n = \binom{n}{1} = n$ non-negative integer solutions to the equation $x_1 + \cdots + x_n = 1$ and $H_p^n = \binom{p+n-1}{p}$ non-negative integer solutions to the equation $y_1 + \cdots + y_n = p$. By (MP), the total number of solutions in this case is equal to $n\binom{p+n-1}{p}$. By symmetry, the number of solutions in the latter case is equal to $n\binom{p+n-1}{p}$.

By (AP), there are a total of $2n\binom{p+n-1}{p}$ non-negative integer solutions to the equation $(x_1 + \cdots + x_n)(y_1 + \cdots + y_n) = p$.

73. It is easy to see that the number of ways to express r as a sum of n non-negative integers in which the order counts is equal to the number of non-negative integer solutions to the equation $x_1 + \cdots + x_n = r$, which is equal to $H_r^n = \binom{r+n-1}{r}$.

74. It is easy to see that the number of ways to express r as a sum of n positive integers in which the order counts is equal to the number of positive integer solutions to the equation $x_1 + \cdots + x_n = r$, which is equal to $H_{r-n}^n = \binom{r-1}{n-1}$.

75. For the completeness of the argument here let us also assume that 0 is an ascending integer. Then it is easy to see that the number of non-negative ascending integers lesser than 10^9 is equal to the number of 9-element multi-subsets of the multi-set

$$\{\infty \cdot 0, \infty \cdot 1, \ldots, \infty \cdot 9\},$$

which is equal to $H_9^{10} = \binom{18}{9}$. Excluding 0, the number of (positive) ascending integers is equal to $\binom{18}{9} - 1$.

76. (i) The number of strictly ascending integers lesser than 10^9 is equal to the number of non-empty subsets of the set $\{1, 2, \ldots, 9\}$, which is equal to $2^9 - 1$.

(ii) The number of strictly ascending integers lesser than 10^5 is equal to the number of non-empty subsets of the set $\{1, 2, \ldots, 9\}$ with at most 5 elements. We note that the number of k-element subsets of $\{1, 2, \ldots, 9\}$ is equal to $\binom{9}{k}$. Therefore, the number of strictly ascending integers lesser than 10^5 is equal to $\sum_{k=1}^{5} \binom{9}{k} = 381$.

77. (i) Let $B \subseteq A$ in which $k \in B$ and k is the maximum number amongst the elements in B. Then it follows that $B \setminus \{k\}$ must be a subset of $\{1, 2, \ldots, k-1\}$ for $k > 1$, or $B = \{k\}$ when $k = 1$. For the former case, there are 2^{k-1} ways to choose elements for $B \setminus \{k\}$ so the number of such sets B is equal to 2^{k-1}. For the latter case, there is only $1 = 2^{1-1} = 2^{k-1}$ such set. The desired follows.

(ii) We shall count the number of non-empty subsets C of A in two different ways. On one hand, it is clear by (CP) that $|C| = 2^n - 1$, since the empty set is excluded. On the other hand, for $i \in A$, let S_i denote the set of subsets B of A in which $i \in B$ and i is the maximum number amongst the elements in B. Then it is clear that S_i and S_j are pairwise disjoint for all $i \neq j$, and $C = \bigcup\limits_{i=1}^{n} S_i$. Hence, we have

$$|C| = \left| \bigcup_{i=1}^{n} S_i \right| = \sum_{i=1}^{n} |S_i| = \sum_{i=1}^{n} 2^{i-1} = \sum_{i=0}^{n-1} 2^i.$$

Therefore, we have $\sum\limits_{i=0}^{n-1} 2^i = |C| = 2^n - 1$ as desired.

78. Let C denote the set of chords. For each chord $c \in C$, let r_c denote the number of line segments that comprise c, and let m_c denote the number of points of intersections lying on chord c. Then it is clear that $r_c = 1 + m_c$ for all $c \in C$. Furthermore, we see that since no three chords are concurrent within the interior of the circle, we see that each point of intersection within the interior of the circle is counted exactly twice in the sum $\sum\limits_{c \in C} m_c$, and hence $\sum\limits_{c \in C} m_c = 2m$. Therefore, we have

$$r = \sum_{c \in C} r_c = \sum_{c \in C} (1 + m_c) = \sum_{c \in C} 1 + \sum_{c \in C} m_c = |C| + 2m = n + 2m.$$

79. (i) The number of such chords is equal to $\binom{p}{2}$.

(ii) Since no three chords are concurrent within the interior of the circle, we see that each point of intersection within the interior of the circle is determined by a unique set of 4 distinct points on the circumference of the circle. Hence the number of points of intersection within the interior is equal to $\binom{p}{4}$.

(iii) By Problem 1.78, we see that the number of line segments is equal to $\binom{p}{2} + 2\binom{p}{4}$.

(iv) Let us count the number of triangles whose vertices are points of intersection within the interior of the circle, and whose sides lie on chords of the circle. Let T be the set of such triangles, let P denote the set of points on the circumference of the circle, and let S denote the number of 6-element subsets of P.

For a given triangle XYZ in T, the sides XY, YZ and ZX must lie on some chords AB, CD and EF respectively. Furthermore, it is easy to see that the points A, B, C, D, E, F must be pairwise distinct. This gives rise to a function $f: T \to S$ given by $f(XYZ) = \{A, B, C, D, E, F\}$ for every triangle XYZ in T. It is easy to check that f is a bijection, so by (BP), we have $|T| = |S| = \binom{p}{6}$.

80. Firstly, there are $\binom{n+1}{2}$ ways to choose 2 prizes out of $n + 1$. Then there are n ways to choose a student to get the 2 prizes. Finally, there are $(n-1)!$ ways to distribute $n-1$ prizes amongst the remaining $n-1$ students. By (MP), the number of ways is equal to $\binom{n+1}{2} \cdot n \cdot (n-1)! = \binom{n+1}{2} \cdot n!$.

81. (a) (i) The number of mappings from N_n to N_m is equal to m^n.

   (ii) Let f be a $1-1$ mapping from N_n to N_m. Then there are m choices for $f(1)$, $m-1$ choices for $f(2)$ after choosing $f(1)$, and in general, $m-r$ choices for $f(r+1)$ after choosing $f(1), \ldots, f(r)$ for $r = 1, \ldots, n-1$. By (MP), the number of $1-1$ mappings from N_n to N_m is equal to

$$m(m-1)\cdots(m-(n-1)) = P_n^m.$$

(b) The number of strictly increasing mappings from N_n to N_m is equal to the number of n-element subsets of N_m, which is equal to $\binom{m}{n}$.

(c) First, we note from definition that there are $S(n, m)$ ways to partition N_n into m non-empty subsets. Then there are $m!$ ways to assign a number from N_m to each of the m subsets. By (MP), there are $m! S(n, m)$ mappings from N_n onto N_m.

82. (i) $s(r, 1)$ is equal to the number of ways to arrange r objects around a single circle, which is equal to $Q_r^r = (r-1)!$.

   (ii) Fix an object x, and suppose that the number of objects that are in the same circle as x is equal to $k+1$, where $k \in \{0, 1, \ldots, r-2\}$. There are $\binom{r-1}{k}$ ways to choose k objects out of the remaining $r-1$ objects, $Q_{k+1}^{k+1} = k!$ ways of arranging x and the k objects around in the first circle, and $Q_{r-(k+1)}^{r-(k+1)} = (r-k-2)!$ ways of arranging the remaining $r - (k+1)$ objects around in the second circle. By (MP), there are $\binom{r-1}{k} k! (r-k-2)! = \frac{(r-1)!}{k!(r-k-1)!} \cdot k!(r-k-2)! = \frac{(r-1)!}{r-k-1}$ ways in this case, when the number of objects that are in the same circle as x is equal to $k + 1$. By (AP), we have

$$s(r, 2) = \sum_{k=0}^{r-2} \frac{(r-1)!}{r-k-1}$$

$$= (r-1)! \left(\sum_{k=0}^{r-2} \frac{1}{r-k-1} \right)$$

$$= (r-1)! \left(1 + \frac{1}{2} + \cdots + \frac{1}{r-1} \right).$$

(iii) In this case, one circle must have exactly 2 distinct objects, and the rest of the circles must have one object each. There are $\binom{r}{2}$ ways to choose 2 objects out of r for the first circle, so we have $s(r, r-1) = \binom{r}{2}$.

(iv) Let us consider these cases:

Case 1: 2 of the circles contain 2 objects each, and the rest of the circles contain one object each. There are $\binom{r}{4}$ ways to choose 4 objects for the first two circles, and there are $\frac{1}{2!} \binom{4}{2} = 3$ ways of splitting 4 objects to be placed in the first 2 circles, so that each circle has exactly 2 objects. There are $3\binom{r}{4}$ such arrangements.

Case 2: 1 of the circle contain 3 objects, and the rest of the circles contain one object each. There are $\binom{r}{3}$ ways to choose 3 objects for the first circle, and there are $Q_3^3 = 2! = 2$ ways to arrange the 3 objects in the first circle.

Hence, we have

$$s(r, r-2) = 3\binom{r}{4} + 2\binom{r}{3}$$

$$= \frac{1}{24} r(r-1)(r-2)(3(r-3)+8)$$

$$= \frac{1}{24} r(r-1)(r-2)(3r-1).$$

(v) For each $r \geq 0$, let $s_r = \sum_{n=0}^{r} s(r, n)$. We note that s_r denotes the number of ways to arrange r objects in at most r circles, such that no circle is empty. Let us proceed by induction that $s_r = r!$ for all $r \geq 0$. The base cases $r = 0$ and $r = 1$ are trivial. For $n = 2$, we have $s(2,2) = 1, s(2,1) = 1, s(2,0) = 0$ and $s_2 = \sum_{n=0}^{2} s(2,n) = 2 = 2!$. The result holds for $r = 2$. Suppose that the equation holds for some $r = k$, where $k \geq 2$. By induction hypothesis, we have $s_k = k!$. Now, let us consider s_{k+1}.

First, let us denote the objects by $a_1, a_2, \ldots, a_{k+1}$. There are s_k ways to arrange $a_1, a_2, \ldots, a_k$ in at most k circles, such that no circle is empty. Fix a particular arrangement A of the objects $a_1, a_2, \ldots, a_k$. Without loss of generality, let the objects be arranged in m identical circles, where $1 \leq m \leq k$, and without loss of generality, let us assume that the circles are labeled $C_1, C_2, \ldots, C_m$, and $a_{p_{i-1}+1}, a_{p_{i-1}+2}, \ldots, a_{p_i}$ are arranged (in that order) in circle C_i, where $0 = p_0 < p_1 < p_2 < \cdots < p_m = k$.

We see that there are $k+1$ ways to place a_{k+1} into A such that $a_1, a_2, \ldots, a_{k+1}$ is arranged in at most $k+1$ identical circles, with no empty circles. To see that this is the case, we see that either a_{k+1} must be placed in one of the circles C_i, of which there are $p_i - p_{i-1}$ ways to do so, or a_{k+1} could be placed in a new circle. Furthermore, the way in which we add a_{k+1} into the arrangement would ensure that $a_1, a_2, \ldots, a_{k+1}$ is arranged in at most $k+1$ identical circles. This implies that $s_{k+1} = (k+1)s_k = (k+1)k! = (k+1)!$, thereby completing the inductive step. We are done.

83. We shall proceed by induction on r to show that $x(x+1) \cdots (x+r-1) = \sum_{n=0}^{r} s(r,n)x^n$ for all $r \geq 1$, with the base case $r = 1$ being trivial. For $r = 2$, $x(x+1) = x + x^2$, $s(2,1) = 1$ and $s(2,2) = 1$. That is the result holds for $r = 2$. Assume that the assertion holds for some $r = k$ where $k \geq 2$. By induction hypothesis, we have

$$x(x+1) \cdots (x+k-1) = \sum_{n=0}^{k} s(k,n)x^n.$$

On the other hand, let us consider the product $x(x+1) \cdots (x+k)$. Since

$$x(x+1) \cdots (x+k) = x(x+1) \cdots (x+k-1)x + x(x+1) \cdots (x+k-1)k,$$

we have

$$\text{coefficient of } x^n \text{ in } x(x+1) \cdots (x+k)$$
$$= \text{coefficient of } x^{n-1} \text{ in } x(x+1) \cdots (x+k-1)$$
$$+k \times \text{coefficient of } x^n \text{ in } x(x+1) \cdots (x+k-1)$$
$$= s(k,n-1) + ks(k,n).$$

On the other hand, we assert that $s(k+1,n) = s(k,n-1) + ks(k,n)$. Indeed, let us consider an arrangement of objects $a_1, a_2, \ldots, a_{k+1}$ in n identical circles. Fix an arrangement A, and let us consider these cases:

Case 1: a_{k+1} is the only object in some circle. Then the objects $a_1, a_2, \ldots, a_k$ must be arranged in the remaining $n-1$ circles, and there are $s(k, n-1)$ such arrangements by definition.

Case 2: The circle containing a_{k+1} contains some object other than a_{k+1}. First, we note that there are $s(k, n)$ ways by definition to arrange the objects $a_1, a_2, \ldots, a_k$ in n identical circles, and by using a similar argument as in Problem 1.82(v), we see that for each arrangement B of the objects $a_1, a_2, \ldots, a_k$ in n identical circles, there are k ways to place a_{k+1} into B, such that $a_1, a_2, \ldots, a_{k+1}$ is arranged in n identical circles. By (MP), the number of arrangements in this case is equal to $ks(k, n)$.

This implies that $s(k + 1, n) = s(k, n - 1) + ks(k, n)$, so we have the coefficient of x^n in $x(x + 1) \cdots (x + k)$ to be equal $s(k + 1, n)$, and this completes the inductive step. We are done.

84. (i) Let the r objects be $a_1, a_2, \ldots, a_r$. Let the box that contains a_r be denoted B, and let $A = \{a_i \mid a_i \in B\}$. Since no box can be empty, we must have $A - \{a_r\} \subsetneq \{a_1, a_2, \ldots, a_{r-1}\}$. There are $2^{r-1} - 1$ valid possibilities for $A - \{a_r\}$, and hence $S(r, 2) = 2^{r-1} - 1$.

(ii) Let us first consider the distribution of r objects into three distinct boxes X, Y, Z. If there are no restrictions, then there are 3^r ways to do so. Of these 3^r ways, 3 of them involve leaving exactly two of those boxes empty. Also, let us count the number of ways of distributing r objects into three distinct boxes X, Y, Z, where exactly one of those boxes are empty. We note that there are exactly 3 ways of choosing one box to be empty, and there are $2! \cdot S(r, 2)$ ways to distribute r distinct objects into the other 2 distinct boxes, such that none of them is empty.

By (CP), the number of ways to distribute r objects into three distinct boxes X, Y, Z, such that none of them are empty, is equal to

$$3^r - 3 - 3 \cdot 2! \cdot S(r, 2) = 3(3^{r-1} - 1) - 6(2^{r-1} - 1),$$

by part (i) above. Therefore, the number of ways to distribute r objects into three distinct boxes X, Y, Z, such that none of them are empty, is equal to

$$\frac{3(3^{r-1} - 1) - 6(2^{r-1} - 1)}{3!} = \frac{1}{2}(3^{r-1} + 1) - 2^{r-1}.$$

(iii) In this case, one of the boxes must contain exactly 2 distinct objects, and the other $r-2$ boxes must contain one object. Hence, we have $S(r, r - 1) = \binom{r}{2}$.

(iv) Let us consider these cases:

Case 1: One of the boxes contain exactly 3 distinct objects, and the other $r - 3$ boxes contains exactly one object. In this case, there are $\binom{r}{3}$ ways to do so.

Case 2: Two of the boxes contain exactly 2 distinct objects each, and the other $r - 4$ boxes must contain exactly one object. In this case, there are $\binom{r}{4}$ ways to choose 4 distinct objects for the first two boxes, and there are $\frac{1}{2!}\binom{4}{2} = 3$ ways to distribute the 4 distinct objects into the first 2 boxes, such that each box has exactly 2 objects. Hence, we have $S(r, r - 2) = \binom{r}{3} + 3\binom{r}{4}$.

85. We shall proceed by induction on r. Clearly the assertion holds for $r = 0$ and $r = 1$. The examples in the problem show that the result holds for $r = 2, 3, 4$ as well. Now, let us assume that the assertion holds for some $r = k \geq 4$. By induction hypothesis, we have $x^k = \sum_{n=0}^{k} S(k, n)(x)_n$. Noting from equation (1.7.2) that

$$S(r, n) = S(r - 1, n - 1) + nS(r - 1, n),$$

we have

$$\sum_{n=0}^{k+1} S(k + 1, n)(x)_n$$

$$= \sum_{n=1}^{k+1} S(k + 1, n)(x)_n \quad \text{(since } S(k + 1, 0) = 0\text{)}$$

$$= \sum_{n=1}^{k+1} S(k, n - 1)(x)_n + \sum_{n=1}^{k+1} nS(k, n)(x)_n$$

$$= \sum_{n=0}^{k} S(k, n)(x)_{n+1} + \sum_{n=0}^{k+1} nS(k, n)(x)_n$$

$$= \sum_{n=0}^{k} S(k, n)(x)_n(x - n) + \sum_{n=0}^{k} nS(k, n)(x)_n \quad \text{(since } S(k, k + 1) = 0\text{)}$$

$$= \sum_{n=0}^{k} (x - n + n)S(k, n)(x)_n$$

$$= \sum_{n=0}^{k} xS(k, n)(x)_n$$

$$= x \sum_{n=0}^{k} S(k, n)(x)_n$$

$$= x \cdot x^k$$
$$= x^{k+1},$$

thereby completing the induction step. We are done.

86. Let us denote the m chords $\ell_1, \ell_2, \ldots, \ell_m$. We shall count the number of new regions formed as the chords are drawn in succession, and without loss of generality, let us assume that the order is $\ell_1, \ell_2, \ldots, \ell_m$. Firstly, we note that whenever a new chord is drawn, a new region will be created. Also, for each of new intersection points that is created by drawing the new chord, we note that there will be an extra region created. Thus, by defining n_i to be the number of new intersection points of the chord ℓ_i with $\ell_1, \ell_2, \ldots, \ell_{i-1}$ for $i = 2, 3, \ldots, m$, we see that the additional number of regions formed by drawing the new chord ℓ_i is equal to $1 + n_i$.

Hence, the number of regions divided by the chords in the circle is equal to

$$2 + \sum_{i=2}^{m}(1 + n_i) = 1 + m + \sum_{i=2}^{m} n_i,$$

since ℓ_1 would always cut the region into 2 new regions. As no 3 chords are concurrent, it follows that each of the intersection points in the sum $\sum_{i=2}^{m} n_i$ is counted exactly once. Hence, we must have $\sum_{i=2}^{m} n_i = n$, so the number of regions is equal to $1 + m + \sum_{i=2}^{m} n_i = 1 + m + n$ as desired.

87. First Solution: We shall proceed by induction on n. Clearly the assertion holds for $n = 4$ as we have $\binom{4}{4} + \binom{4-1}{2} = 1 + 3 = 4 = r(4)$. Now, let us assume that the assertion holds for some $k \geq 4$. By induction hypothesis, we have $r(k) = \binom{k}{4} + \binom{k-1}{2}$. Let us take a polygon with $(k+1)$ sides, and label its vertices $v_1, v_2, \ldots, v_{k+1}$ in the counterclockwise direction. By induction hypothesis, the k-gon with vertices $v_1, v_2, \ldots, v_k$ has $r(k) = \binom{k}{4} + \binom{k-1}{2}$ interior regions divided by all of its diagonals. Now, by drawing the sides $v_1 v_{k+1}$ and $v_k v_{k+1}$, we see that one extra region is created.

Firstly, we see that there will be an extra region created for each of the diagonals $v_2 v_{k+1}, v_3 v_{k+1}, \ldots, v_{k-1} v_{k+1}$ drawn. Also, it is easy to see that none of the diagonals $v_2 v_{k+1}, v_3 v_{k+1}, \ldots, v_{k-1} v_{k+1}$ taken pairwise would intersect inside the polygon. Finally, for $2 \leq \ell \leq k-1$, we note that a diagonal $v_i v_j$ with $1 \leq i < j < k+1$, $\{i, j\} \cap \{\ell, k+1\} = \emptyset$, would intersect with the diagonal $v_\ell v_{k+1}$ if and only if the vertices v_i and v_j are on the different sides of the diagonal $v_\ell v_{k+1}$, that is, $1 \leq i < \ell$ and $\ell < j < k+1$. Now, there are $\ell - 1$ choices for i and $k - \ell$ choices for j. Hence, there are $(\ell - 1)(k - \ell)$ new intersection points created when the diagonal $v_\ell v_{k+1}$ is drawn. Since a new region is created whenever there is a new intersection point created, the total number of new regions created is equal to $k - 2 + \sum_{\ell=2}^{k-1}(\ell - 1)(k - \ell)$. Therefore, we have

$$r(k+1)$$

$$= r(k) + 1 + k - 2 + \sum_{\ell=2}^{k-1}(\ell - 1)(k - \ell)$$

$$= \left[\binom{k}{4} + \binom{k-1}{2}\right] + k - 1 + \sum_{\ell=1}^{k-2}\ell(k - \ell - 1)$$

(by induction hypothesis)

$$= \frac{k(k-1)(k-2)(k-3)}{24} + \frac{(k-1)(k-2)}{2}$$

$$+ k - 1 - \sum_{\ell=1}^{k-2}\left[\ell^2 + (1-k)\ell\right]$$

$$= \frac{k(k-1)(k-2)(k-3)}{24} + \frac{k(k-1)}{2}$$

$$- \frac{(k-2)(k-1)(2k-3)}{6} - \frac{(1-k)(k-2)(k-1)}{2}$$

$$= \frac{(k-1)(k-2)[k(k-3) - 4(2k-3) - 12(1-k)]}{24} + \frac{k(k-1)}{2}$$

$$= \frac{(k+1)k(k-1)(k-2)}{24} + \frac{k(k-1)}{2}$$

$$= \binom{k+1}{4} + \binom{k}{2},$$

thereby completing the induction step. We are done.

Second Solution: Without loss of generality, we may inscribe the n-gon in a circle such that no three diagonals are concurrent. Let us label the vertices of the n-gon by $v_1, v_2, \ldots, v_n$. There are $\binom{n}{2}$ chords that can be drawn using these n vertices. For each set of 4 vertices, there is exactly one intersection point of the chords in the interior of the circle. There are $\binom{n}{4}$ intersection points in the interior of the circle. By Problem 1.86, there are $\binom{n}{2} + \binom{n}{4} + 1$ regions in the circle.

For each side $v_i v_{i+1}$ of the n-gon, we exclude the region formed by the side $v_i v_{i+1}$ and the minor arc from v_i to v_{i+1} as it is outside the n-gon. It follows that there are $\binom{n}{2} + \binom{n}{4} + 1 - n = \binom{n}{4} + \binom{n-1}{2}$ regions formed in the n-gon. The desired statement now follows.

88. Let the prime factorisation of n be $p_1^{\alpha_1} \cdot p_2^{\alpha_2} \cdots p_k^{\alpha_k}$, where $p_1, p_2, \ldots, p_k$ are pairwise distinct primes and $\alpha_1, \alpha_2, \ldots, \alpha_k$ are positive integers. Then the prime factorisation of n^2 is $p_1^{2\alpha_1} \cdot p_2^{2\alpha_2} \cdots p_k^{2\alpha_k}$. Let (x, y) be one such solution to the equation

$$\frac{xy}{x+y} = n,$$

and let $m = x - n$. Then m is a positive integer; indeed, we have

$$x = \frac{xy}{y} > \frac{xy}{x+y} = n.$$

This implies that

$$\frac{xy}{x+y} = n$$
$$\Rightarrow \frac{(m+n)y}{m+n+y} = n$$
$$\Rightarrow (m+n)y = n(m+n+y)$$
$$\Rightarrow y = n + \frac{n^2}{m}.$$

Since y, n are positive integers, we must have $\frac{n^2}{m}$ to be a positive integer, that is, $m|n^2$, or equivalently, $(x, y) \in S$, where $S = \left\{ \left(n + m, n + \frac{n^2}{m} \right) \mid m \in \mathbf{N} \,\&\, m|n^2 \right\}$. On the other hand, every element of S is also a solution to the given equation. Therefore, there exists a bijection between the solution set and S. By (BP), the total number of solutions (x, y) is equal to $|S|$, which in turn is equal to the number of positive divisors of n^2, which is then equal to $(2\alpha_1 + 1)(2\alpha_2 + 1) \cdots (2\alpha_k + 1)$.

89. (i) For $k = 0, 1, 2$, define the set $[k]$ to be $\{x \in S \mid x \equiv k \pmod{3}\}$. Then we have $|[0]| = |[1]| = |[2]| = \frac{1992}{3} = 664$. Without loss of generality, let us assume that $a \in [p]$, $b \in [q]$ and $c \in [r]$, where $0 \le p \le q \le r \le 2$, such that $3|a + b + c$. We need to have $p + q + r \equiv 0 \pmod{3}$. There are two cases to consider:

Case 1: $p = q = r = k$, $0 \le k \le 2$. There are 3 choices for k, and $\binom{664}{3}$ ways to choose a, b, c from $[k]$. There are $3\binom{664}{3}$ 3-element subsets in this case.

Case 2: $p = 0, q = 1, r = 2$. There are 664 choices each for a, b and c. There are 664^3 3-element subsets in this case.

So the total number of such 3-element subsets $\{a, b, c\}$ is equal to $3\binom{664}{3} + 664^3$.

(ii) For $k = 0, 1, 2, 3$, define the set $[k]$ to be $\{x \in S \mid x \equiv k \pmod{4}\}$. Then we have $\|[0]\| = \|[1]\| = \|[2]\| = \|[3]\| = \frac{1992}{4} = 498$. Without loss of generality, let us assume that $a \in [p]$, $b \in [q]$ and $c \in [r]$, where $0 \le p \le q \le r \le 3$, such that $4 \mid a + b + c$. We need to have $p + q + r \equiv 0 \pmod{4}$. We shall consider the following cases:

Case 1: $p = q = r = 0$. There are $\binom{498}{3}$ ways to choose a, b, c from $[0]$. There are $\binom{498}{3}$ 3-element subsets in this case.

Case 2: $p = 0$, $q = 1$, $r = 3$. There are 498 choices each for a, b and c. There are 498^3 3-element subsets in this case.

Case 3: $p = 0$, $q = r = 2$. There are 498 choices for a, and $\binom{498}{2}$ ways to choose b, c from $[2]$. There are $498 \cdot \binom{498}{2}$ 3-element subsets in this case.

Case 4: $p = q = 1$, $r = 2$. There are 498 choices for c, and $\binom{498}{2}$ ways to choose a, b from $[1]$. There are $498 \cdot \binom{498}{2}$ 3-element subsets in this case.

Case 5: $p = 2$, $q = r = 3$. There are 498 choices for a, and $\binom{498}{2}$ ways to choose b, c from $[3]$. There are $498 \cdot \binom{498}{2}$ 3-element subsets in this case.

By (AP), the total number of such 3-element subsets $\{a, b, c\}$ is equal to

$$\binom{498}{3} + 3 \cdot 498 \cdot \binom{498}{2} + 498^3.$$

90. There are 6 choices where we could draw the letter U (and subsequently the whole block $UNIVERSITY$), namely the first, second, third, fourth, fifth and sixth draws, and subsequently there are 26 choices each for the remaining 5 draws. Hence the probability where $UNIVERSITY$ comes out as a block in the sequence of 15 random draws is equal to $\frac{6 \cdot 26^5}{26^{15}} = \frac{6}{26^{10}}$.

91. Let U be the set of r-element subsets of X which are m-separated, and V be the set of r-element subsets of $\{1, 2, \ldots, n - (m-1)(r-1)\}$. Define the function $f : U \to V$ by

$$\{a_1, a_2, \ldots, a_r\} \mapsto \{a_i - (i-1)(m-1) \mid i = 1, 2, \ldots, r\}$$

for all r-element subsets $\{a_1, a_2, \ldots, a_r\}$ of U such that $a_1 < a_2 < \cdots < a_r$. To show that f is a well-defined function, it suffices to show that $f(\{a_1, a_2, \ldots, a_r\})$ has r elements for all $\{a_1, a_2, \ldots, a_r\} \in U$ such that $a_1 < a_2 < \cdots < a_r$. Indeed, we have for $i = 1, 2, \ldots, r$, we have

$$[a_i - (i-1)(m-1)] - [a_{i-1} - (i-2)(m-1)]$$
$$= a_i - a_{i-1} - (m-1)$$
$$\geq m - (m-1)$$
$$= 1.$$

This implies that the numbers $a_1, a_2 - (m-1), \ldots, a_r - (r-1)(m-1)$ are pairwise distinct and they all belong to V. So f is well-defined. It is easily verified from the definition of the function f that the function f is bijective. By (BP), we have $|U| = |V| = \binom{n-(m-1)(r-1)}{r}$.

92. We note that the number of terms in S_k is equal to $\binom{n}{k}$ and the number of terms in S_{n-k} is equal to $\binom{n}{n-k}$. Therefore, the number of terms in $S_k S_{n-k}$ is equal to $\binom{n}{k}\binom{n}{n-k} = \binom{n}{k}^2$. Next, for each $i = 1, 2, \ldots n$, let us count the number of terms in $S_k S_{n-k}$ containing a_i. We notice that the number of terms w in S_k containing a_i is equal to $\binom{n-1}{k-1}$, the number of terms x in S_{n-k} containing a_i is equal to $\binom{n-1}{n-k-1}$, the number of terms y in S_k not containing a_i is equal to $\binom{n}{k} - \binom{n-1}{k-1}$, and the number of terms z in S_{n-k} not containing a_i is equal to $\binom{n}{n-k} - \binom{n-1}{n-k-1}$. Therefore,

The number of terms in $S_k S_{n-k}$ containing a_i but not a_i^2

in their product

$$= wz + xy$$
$$= \binom{n-1}{k-1}\left[\binom{n}{n-k} - \binom{n-1}{n-k-1}\right]$$
$$+ \binom{n-1}{n-k-1}\left[\binom{n}{k} - \binom{n-1}{k-1}\right],$$

The number of terms in $S_k S_{n-k}$ containing a_i^2 in their product

$$= wy$$
$$= \binom{n-1}{k-1}\binom{n-1}{n-k-1}.$$

Therefore, the power of a_i in the product of all the terms in $S_k S_{n-k}$ is equal to

$$\binom{n-1}{k-1}\left[\binom{n}{n-k} - \binom{n-1}{n-k-1}\right]$$

$$+\binom{n-1}{n-k-1}\left[\binom{n}{k}-\binom{n-1}{k-1}\right]+2\binom{n-1}{k-1}\binom{n-1}{n-k-1}$$

$$=\binom{n-1}{k-1}\binom{n}{n-k}+\binom{n}{k}\binom{n-1}{n-k-1}$$

$$=\binom{n}{k}\left[\binom{n-1}{k-1}+\binom{n-1}{k}\right]$$

$$=\binom{n}{k}^2 \quad \text{(by (1.4.1))}.$$

Hence, the geometric mean GM of the terms in $S_k S_{n-k}$ is equal to

$$\left[a_1^{\binom{n}{k}^2}\cdot a_2^{\binom{n}{k}^2}\cdots a_n^{\binom{n}{k}^2}\right]^{\frac{1}{\binom{n}{k}^2}}=a_1 a_2\cdots a_n.$$

Also, we note that the arithmetic mean of the terms in $S_k S_{n-k}$ is equal to $\frac{1}{\binom{n}{k}^2}S_k S_{n-k}$.

Therefore, by the $AM-GM$ inequality, we have

$$\frac{1}{\binom{n}{k}^2}S_k S_{n-k}\geq a_1 a_2\cdots a_n,$$

that is,

$$S_k S_{n-k}\geq\binom{n}{k}^2 a_1 a_2\cdots a_n.$$

Remark. The $AM-GM$ inequality states that if $a_1, a_2,\ldots, a_r$ are non-negative real numbers, then the arithmetic mean of $a_1, a_2,\ldots, a_r$ is greater than or equal to the geometric mean of $a_1, a_2,\ldots, a_r$, that is,

$$\frac{a_1+a_2+\cdots+a_r}{r}\geq\sqrt[r]{a_1 a_2\cdots a_r}.$$

93. For completeness, let us define the alternating sum of an empty set to be equal to 0. Let S denote the set of subsets of $\{1,2,3,4,5,6\}$, and let T denote the set of subsets of $\{1,2,3,4,5,6,7\}$ that contains the element 7. Then it is easily seen that the function $f\colon S\to T$, $f(A)=A\cup\{7\}$ is bijective.

Now, we assert that the sum of the alternating sums of B and $f(B)$ for each non-empty $B\in S$ is equal to 7. Indeed, let $B=\{b_1,\ldots,b_r\}$ where $b_i>b_j$ for all $i<j$. Then it is easily seen that the alternating sum of B is equal to $b_1-b_2+\cdots+(-1)^{r+1}b_r$, and the alternating sum of $f(B)$ is equal to $7-b_1+b_2-\cdots+(-1)^r b_r$, and the assertion follows. Also, it is easily seen that the sum of the alternating sums of

$\emptyset$ and $f(\emptyset) = \{7\}$ is equal to 7. We note that every non-empty subset $U \subseteq \{1, 2, \ldots, 7\}$ belongs to either S or T but not both. Therefore, the sum of all such alternating sums for $n = 7$ is equal to $7 \cdot |S| = 7 \cdot 2^6 = 448$.

94. We note that the total number of possible arrangements is equal to $\frac{12!}{3!4!5!} = 27720$. Let us count the number of ways to arrange the trees such that no two birch trees are adjacent to each other. There are $\binom{3+4}{3} = 35$ ways to arrange the 3 maple and 4 oak trees, and then there are $\binom{8}{5} = 56$ ways to place the 5 birch trees in the 8 slots created by the 3 maple threes and 4 oak trees, so that no two birch trees are adjacent. Hence the probability is $\frac{35 \cdot 56}{27720} = \frac{7}{99}$, so $m+n = 7+99 = 106$.

95. Let A denote the set of the players with $|A| = n$ and let L denote the set of the 10 lowest scoring players. For every $x \in A$, let x_L denote the total points scored by x against the players in L, and $x_{A \setminus L}$ denote the total points scored by x against the players in $A \setminus L$. We note that $x_L = x_{A \setminus L}$, and the total points earned by x is equal to $x_L + x_{A \setminus L}$. Now, let us count the total scores of the players in 2 different ways. Firstly, we note that the total scores of the players is equal to the total number of games played, which is $\binom{n}{2}$. On the other hand, it is also the sum of the players' individual scores, which is equal to

$$\sum_{x \in A} (x_L + x_{A \setminus L}) = \sum_{x \in L} (x_L + x_{A \setminus L}) + \sum_{y \in A \setminus L} (y_L + y_{A \setminus L})$$

$$= 2 \sum_{x \in L} x_L + 2 \sum_{y \in A \setminus L} y_{A \setminus L}$$

$$= 2\binom{10}{2} + 2\binom{n - 10}{2},$$

where the last inequality holds by the fact that $\sum_{x \in L} x_L$ is equal to the total number of games played amongst the players in L, which is equal to $\binom{10}{2}$. The same argument applies for $\sum_{y \in A \setminus L} y_{A \setminus L}$. This gives us

$$\binom{n}{2} = 2\binom{10}{2} + 2\binom{n - 10}{2}$$

$$\Rightarrow \frac{n(n - 1)}{2} = 90 + (n - 10)(n - 11)$$

$$\Rightarrow n^2 - 41n + 400 = 0$$

$$\Rightarrow (n - 16)(n - 25) = 0,$$

which would imply that $n = 16$ or $n = 25$. As we have $\sum_{x \in L} (x_L + x_{A \setminus L}) = 2\binom{10}{2} = 90$ and $|L| = 10$, there exist some $a \in L$ such that $a_L + a_{A \setminus L} \geq 9$. As L's players are lowest scoring, we must have $y_L + y_{A \setminus L} \geq 9$ for all $y \in A \setminus L$. When $n = 16$, we have $\sum_{y \in A \setminus L} (y_L + y_{A \setminus L}) \geq |A \setminus L| \cdot 9 = 54$. However, at the same time we also have $\sum_{y \in A \setminus L} (y_L + y_{A \setminus L}) = 2\binom{16-10}{2} = 30 \geq 54$, which is a contradiction. So $n = 25$.

96. We observe that $1000000 = 2^6 \cdot 5^6$, and the divisors of 1000000 are of the form $2^x 5^y$ where $0 \leq x, y \leq 6$ and x, y are integers. Thus, we have

$$S = \sum_{0 \leq x, y \leq 6} \log\left(2^x 5^y\right) - \log 1000000 - \log 1$$

$$= \sum_{x=0}^{6} \sum_{y=0}^{6} (x \log 2 + y \log 5) - 6$$

$$= \sum_{x=0}^{6} \sum_{y=0}^{6} x \log 2 + \sum_{x=0}^{6} \sum_{y=0}^{6} y \log 5 - 6$$

$$= \sum_{x=0}^{6} 7x \log 2 + \sum_{y=0}^{6} 7y \log 5 - 6$$

$$= 147 \log 2 + 147 \log 5 - 6$$

$$= 147 \log 10 - 6$$

$$= 141,$$

so the nearest integer nearest to S is 141.

97. Let us consider each of the subsequences of two coin tosses as an operation instead; this operation takes a sequence and appends the second coin toss of the subsequence to the end of the old sequence. We shall first consider the case where the subsequence in question is HT or TH. We observe that by performing the operation HT or TH, the next coin toss of the newly obtained sequence would be different from the last coin toss of the old sequence. We note from the problem that there are three HT operations and four TH operations. Hence, it follows that our subsequence will contain a subsequence of $THTHTHTH$ (Note: these coin tosses need not be adjacent).

Now it remains to add the remaining letters to the subsequence to obtain a string that satisfy the desired conditions. There are 5 TT operations, which means that we have to add 5 T's into the sequence, with the condition that the new T's are adjacent to the existing T's. As there are already 4 T's in the subsequence, it follows the number ways of adding 5 T's into the subsequence is equal to the number of ways of distributing 5 identical balls into 4 distinct boxes, which is equal to $H_5^4 = 56$. By a similar argument, we see that there are $H_2^4 = 10$ ways of adding 2 H's into the subsequence. By (MP), there are $56 \cdot 10 = 560$ different sequences of 15 coin tosses that contains exactly 2 HH, 3 HT, 4 TH and 5 TT subsequences.

98. (i) Since no carrying over is required, the range of possible values of any digit of m is from 0 to the corresponding digit in 1492 (and the corresponding digit of n would be fixed accordingly depending on m). Thus, the number of simple ordered pairs (m, n) of non-negative integers that sum up to 1492 is $(1+1)(4+1)(9+1)(2+1) = 300$.

(ii) By a similar reasoning as in part (i), we deduce that the number of such pairs is equal to $(1 + 1)(9 + 1)(9 + 1)(2 + 1) = 600$.

99. We note that $10^{99} = 2^{99} \cdot 5^{99}$ and $10^{88} = 2^{88} \cdot 5^{88}$. A divisor of 10^{99} must be of the form $2^x 5^y$ where $0 \le x, y \le 99$ and x, y are integers. Also, an integer multiple of 10^{88} that is also a divisor of 10^{99} must be of the form $2^x 5^y$ where $88 \le x, y \le 99$ and x, y are integers. Hence the probability that a divisor of 10^{99} is also an integer multiple of 10^{88} is equal to $\frac{(99-88+1)(99-88+1)}{(99-0+1)(99-0+1)} = \frac{9}{625}$, so we have $m + n = 9 + 625 = 634$.

100. We note that a vertex of the polyhedron is also a vertex of a square, a hexagon and an octagon. Let the total number of vertices of the squares, hexagons and octagons be A, B and C respectively. Then it follows that the number of vertices in the polyhedron is equal to $\frac{1}{3}(A + B + C) = \frac{1}{3}(12 \cdot 4 + 8 \cdot 6 + 6 \cdot 8) = 48$, since each vertex in the polyhedron is the vertex of one square, one hexagon and one octagon. Also, we note that an edge of the polyhedron is also an edge of a square or a hexagon, or an octagon, and vice versa. Let the total number of edges of the squares, hexagons and octagons be D, E and F respectively. Then it follows that the number of edges in the polyhedron is equal to $\frac{1}{2}(D+E+F) = \frac{1}{2}(12 \cdot 4 + 8 \cdot 6 + 6 \cdot 8) = 72$, since each edge in the polyhedron is the edge of 2 faces in the polyhedron. Now, we see that the number of segments joining the vertices of the polyhedron is equal to $\binom{48}{2} = 1128$, the number of such segments along an edge or face of a square is equal to $12 \cdot \binom{4}{2} = 72$, the number of such segments along an edge or face of a hexagon is equal to $8 \cdot \binom{6}{2} = 120$, and the number of such segments along an edge or face of an octagon

is equal to $6 \cdot \binom{8}{2} = 168$.

Hence, the number of segments that lie in the interior of the polyhedron is equal to $1128 - 72 - 120 - 168 + 72 = 840$, since each edge of the polyhedron are subtracted twice in the computation.

101. Let $\ell \in \mathbf{Z}$, $n \geq 6$ be such that $n! = (\ell + 4)(\ell + 5) \cdots (\ell + n)$. Then $n \geq 6$ implies that $\ell \geq 1$ and

$$
\begin{aligned}
\frac{n!}{(n-3)!} &= \frac{(\ell + 4)(\ell + 5) \cdots (\ell + n)}{(n-3)!} \\
&= \frac{(\ell + n)!}{(\ell + 3)!(n-3)!} \\
&= \frac{(n+\ell)(n+\ell-1)\cdots(n+1)}{(3+\ell)(2+\ell)\cdots 4} \cdot \frac{n!}{3!(n-3)!},
\end{aligned}
$$

that is,

$$
\frac{\ell + n}{3 + \ell} \cdot \frac{\ell + n - 1}{2 + \ell} \cdots \frac{2 + n}{5} \cdot \frac{n + 1}{4!} = 1.
$$

Since $n \geq 6$, it implies that $\frac{n+i}{3+i} > 1$ for $i = 2, 3, \ldots, \ell$. If $n > 23$, then we have $\frac{n+1}{4!} > 1$, so there are no solutions for ℓ if $n > 23$. One easily checks that the only solution (ℓ, n) to the equation is $(1, 23)$.

102. Let us pick any d. Since $|A| \geq 2$, this would imply that $a + d \leq n$, that is, $a \leq n - d$. Also, since A is maximal, we must have $1 \leq a \leq d$. Hence, for a given d, there exists a unique A for every $a = 1, \ldots, \min(n-d, d)$. Therefore, for every $d = 1, 2, \ldots, n-1$, the number of A's with common difference d is $\min(n-d, d)$. Let us consider these cases:

Case 1: n is even, that is, $n = 2m$ for some positive integer m. Then the number of such sets A is equal to

$$
\begin{aligned}
&\sum_{d=1}^{n-1} \min(n - d, d) \\
&= \sum_{d=1}^{m} \min(2m - d, d) + \sum_{d=m+1}^{2m-1} \min(2m - d, d) \\
&= \sum_{d=1}^{m} d + \sum_{d=m+1}^{2m-1} (2m - d) \\
&= \frac{m(m+1)}{2} + 2m(2m - 1 - (m+1) + 1) - \sum_{d=m+1}^{2m-1} d
\end{aligned}
$$

$$= \frac{m(m+1)}{2} + 2m(m-1) - \sum_{d=1}^{2m-1} d + \sum_{d=1}^{m} d$$

$$= \frac{m(m+1)}{2} + 2m(m-1) - \frac{(2m-1)2m}{2} + \frac{m(m+1)}{2}$$

$$= m^2$$

$$= \left\lfloor \frac{n^2}{4} \right\rfloor .$$

Case 2: n is odd, that is, $n = 2m+1$ for some non-negative integer m. Then the number of such sets A is equal to

$$\sum_{d=1}^{n-1} \min(n-d, d)$$

$$= \sum_{d=1}^{m} \min(2m+1-d, d) + \sum_{d=m+1}^{2m} \min(2m+1-d, d)$$

$$= \sum_{d=1}^{m} d + \sum_{d=m+1}^{2m} (2m+1-d)$$

$$= \frac{m(m+1)}{2} + (2m+1)(2m - (m+1) + 1) - \sum_{d=m+1}^{2m} d$$

$$= \frac{m(m+1)}{2} + m(2m+1) - \sum_{d=1}^{2m} d + \sum_{d=1}^{m} d$$

$$= \frac{m(m+1)}{2} + m(2m+1) - \frac{2m(2m+1)}{2} + \frac{m(m+1)}{2}$$

$$= m^2 + m$$

$$= \left\lfloor \frac{4m^2 + 4m + 1}{4} \right\rfloor$$

$$\left(\text{since } m^2 + m < \frac{4m^2 + 4m + 1}{4} < m^2 + m + 1\right)$$

$$= \left\lfloor \frac{n^2}{4} \right\rfloor .$$

In both cases, the number of such sets A is equal to $\left\lfloor \frac{n^2}{4} \right\rfloor$.

103. It is easy to verify that $(m, n) = (3, 2), (6, 3), (10, 4)$ are solutions to the equation $1! \, 3! \cdots (2n-1)! = m!$. We shall show that there are no solutions for $n \geq 5$. Firstly, we shall prove by induction that $1! \, 3! \cdots (2n-1)! > (4n)!$ for all integers $n \geq 7$. For $n = 7$, it is

straightforward to verify that the inequality holds. Assume that the assertion holds for some integer $k \geq 7$. By induction hypothesis, we have $1!3! \cdots (2k-1)! > (4k)!$. As we have $k \geq 7$, it follows that inequality

$$(j+2)(2k-3+j) > 4k+j$$

holds for $j = 1, 2, 3, 4$. This implies that

$$
\begin{aligned}
&1!3! \cdots (2k-1)!(2(k+1)-1)! \\
&> (4k)! \cdot 1 \cdot 2 \cdot 3 \cdot 4 \cdot 5 \cdot 6 \cdots (2k-2)(2k-1)(2k)(2k+1) \\
&> (4k)! \cdot 3 \cdot 4 \cdot 5 \cdot 6 \cdot (2k-2)(2k-1)(2k)(2k+1) \\
&= (4k)! \cdot 3(2k-2) \cdot 4(2k-1) \cdot 5(2k) \cdot 6(2k+1) \\
&= (4k)!(4k+1)(4k+2)(4k+3)(4k+4) \\
&= (4k+4)!,
\end{aligned}
$$

thereby completing the induction step. We are done.

Now, suppose on the contrary that there exists an integer solution (m, n) to the equation

$$1!3! \cdots (2n-1)! = m!$$

with $n \geq 7$. Then by the assertion above, we must have $m > 4n$. By Bertrand's postulate, there exists a prime p such that $2n < p < 4n$. This would imply that p divides $m!$. However, we see that p does not divide the product $1!3! \cdots (2n-1)!$, which is a contradiction. So $n \leq 6$. If $n = 6$, then it is easy to see that $1!3! \cdots 11! > 12!$, so if m is an integer satisfying $1!3! \cdots 11! = m!$, then we have $m \geq 13$. However, 13 divides $m!$ and 13 does not divide $1!3! \cdots 11!$, a contradiction. A similar argument shows that there are no integer solutions to the equation $1!3! \cdots (2n-1)! = m!$ when $n = 5$.

Remark 1. The Bertrand's postulate states that for any integer $n > 1$, there exists some prime p such that $n < p < 2n$.

Remark 2. The solution to the above problem is given by D. Enkers, Rand Afrikaans University, Johannesburg, South Africa. We shall refer the reader to *Amer. Math. Monthly,* **94** (1987), 190 for further details.

104. By definition, we have $e = \sum_{i=0}^{\infty} \frac{1}{i!}$. Then we have

$$n!e = n! \sum_{i=0}^{\infty} \frac{1}{i!}$$

$$= n! \sum_{i=0}^{n} \frac{1}{i!} + n! \sum_{i=n+1}^{\infty} \frac{1}{i!}$$

$$= \sum_{i=0}^{n} P_{n-i}^{n} + \left[\frac{1}{n+1} + \frac{1}{(n+1)(n+2)} + \cdots \right]$$

$$< \sum_{r=0}^{n} P_{r}^{n} + \left[\frac{1}{n+1} + \frac{1}{(n+1)^2} + \cdots \right]$$

$$= \sum_{r=0}^{n} P_{r}^{n} + \frac{1}{n+1} \cdot \left[\frac{1}{1 - \frac{1}{n+1}} \right]$$

$$= \sum_{r=0}^{n} P_{r}^{n} + \frac{1}{n},$$

that is, $\sum_{r=0}^{n} P_{r}^{n} < n!e < \sum_{r=0}^{n} P_{r}^{n} + \frac{1}{n}$. Hence, we have $\lfloor n!e \rfloor = \sum_{r=0}^{n} P_{r}^{n}$.

105. For $k = 0, 1, \ldots, 4$, let us define

$$[k] = \left\{ A \subset S \mid |A| = 31 \text{ and } \sum_{a \in A} a \equiv k \pmod 5 \right\}$$

and $[5] = [0]$. Furthermore, for each $k = 0, 1, 2, 3, 4$ and $A = \{a_1, a_2, \ldots, a_{31}\} \in [k]$ such that $a_i < a_j$ for all $i < j$, let us define $f_k(A)$ as follows:

$$f_k(A) = \begin{cases} \{a_1 + 1, a_2 + 1, \ldots, a_{31} + 1\} & \text{if } a_{31} \neq 1990, \\ \{1, a_1 + 1, a_2 + 1, \ldots, a_{30} + 1\} & \text{if } a_{31} = 1990. \end{cases}$$

Firstly, we shall show that for each $k = 0, 1, 2, 3, 4$ and $A = \{a_1, a_2, \ldots, a_{31}\} \in [k]$ such that $a_i < a_j$ for all $i < j$, we have $f_k(A) \in [k+1]$. If $a_{31} \neq 1990$, then we have

$$1 \leq a_1 < a_2 < \cdots < a_{31} < 1990.$$

Thus, we have

$$2 \leq a_1 + 1 < a_2 + 1 < \cdots < a_{31} + 1 \leq 1990$$

and $\sum_{n=1}^{31} (a_n + 1) \equiv k + 1 \pmod 5$. Else, we have

$$1 \leq a_1 < a_2 < \cdots < a_{31} = 1990.$$

Thus, we have

$$1 < a_1 + 1 < a_2 + 1 < \cdots < a_{30} + 1 \leq 1990,$$

and $1 + \sum_{n=1}^{30} (a_n + 1) \equiv k + 1 \pmod 5$.

Now, it is easy to check that the function $f_k \colon [k] \to [k+1]$ defined is bijective for each $k = 0, 1, 2, 3, 4$. By (BP), we have $|[0]| = |[1]| = |[2]| = |[3]| = |[4]|$, so the number required is equal to $|[0]| = \frac{1}{5}\binom{1990}{31}$.

106. We shall count the total number of appearances of the ordered pairs in P. Let us take any $(a_1, a_2, \ldots, a_{100}) \in P$. As the a_i's are distinct, this sequence contributes a total of $\binom{100}{2}$ appearances. Also, as every ordered pair of elements of S appears in at most 1 member of P, P contributes a total of $|P| \cdot \binom{100}{2}$ appearances, and this total is less than or equal to the number of appearances that can be formed from S. Hence, we have $|P| \cdot \binom{100}{2} \le 2\binom{1990}{2}$, which would imply that $|P| \le 799\frac{34}{55} \le 800$ as desired.

107. For each $i = 1, \ldots, n$, denote b_i to be the number of r-permutations of M with a_i as the first element of the permutation, and c_i to be the number of $(r-1)$-permutations of M_i where $M_i = \{r_1 \cdot a_1, \ldots, (r_i - 1) \cdot a_i, \ldots, r_n \cdot a_n\}$. Then we have $b_i = c_i$ for all i. By (AP), the number of r-permutations of M (which is equal to $\sum_{i=1}^{n} b_i$) must be equal to the number of $(r-1)$-permutations of M (which is equal to $\sum_{i=1}^{n} c_i$).

108. Suppose on the contrary that it is possible for such a configuration to occur. Let the 7 lines be $\ell_1, \ell_2, \ldots, \ell_7$. We shall count the number of pairs of lines $\{\ell_i, \ell_j\}$, $i \ne j$ in 2 different ways. Firstly, the number of pairs is equal to the number of ways to choose 2 lines out of 7, which is equal to $\binom{7}{2} = 21$.

On the other hand, each of the 6 points where exactly 3 of the lines intersect contributes $\binom{3}{2} = 3$ pairs. Hence, these 6 points contribute a total of $6\binom{3}{2} = 18$ pairs. Also, each of the 4 points where exactly 2 of the lines intersect contributes 1 pair of line. Hence, these 4 points contribute a total of 4 pairs. Hence, the total number of pairs is at least $18 + 4 = 22 > 21$, which is a contradiction. Hence it is not possible to achieve such a configuration.

109. We first note that the largest possible value of $|x_k - k|$ is $n - 1$, and is achieved when $n_k \ne k$ and $\{n_k, k\} = \{1, n\}$. Let $(x_1, x_2, \ldots, x_n)$ be a permutation of $(1, 2, \ldots, n)$. Then it is clear that $\sum_{k=1}^{n} (x_k - k) = 0$.

Now, define the set $A = \{k \mid x_k > k\}$. Then we have $\sum_{k=1}^{n} |x_k - k| =$

$$\sum_{k=1}^{n} |x_k - k| + \sum_{k=1}^{n} (x_k - k) = \sum_{k \in A} 2(x_k - k), \text{ which is even.}$$

Now, we claim that if $n \equiv 2, 3 \pmod 4$, then no permutation $(x_1, x_2, \ldots, x_n)$ would satisfy the condition that the differences $|x_k - k|$, $1 \leq k \leq n$, are all distinct. Indeed, if $n \equiv 2 \pmod 4$, then we have $n = 4m+2$ for some non-negative integer m. Suppose on the contrary that there exists a permutation $(x_1, x_2, \ldots, x_n)$ of $(1, 2, \ldots, n)$ for which the differences $|x_k - k|$, $1 \leq k \leq n$ are all distinct. Then we have $\{|x_1 - 1|, |x_2 - 2|, \ldots, |x_n - n|\} = \{0, 1, 2, \ldots, n-1\}$, so we have

$$\sum_{k=1}^{n} |x_k - k| = \sum_{j=0}^{n-1} j = \frac{n(n-1)}{2} = (2m+1)(4m+1),$$

which is odd, a contradiction. One would also arrive at a similar contradiction if $n \equiv 3 \pmod 4$.

Next, we shall construct a permutation $(x_1, x_2, \ldots, x_n)$ of $(1, 2, \ldots, n)$ such that the differences $|x_k - k|$, $1 \leq k \leq n$ are pairwise distinct for the cases $n \equiv 0, 1 \pmod 4$. First, let us consider the case where $n \equiv 0 \pmod 4$. Then we have $n = 4m$ for some positive integer m. For each $k \in \{1, 2, \ldots, n\}$, let us define x_k as follows:

$$x_k = \begin{cases} 4m + 1 - k & \text{if } 1 \leq k \leq m, \\ 4m - k & \text{if } m + 1 \leq k \leq 2m - 1, \\ 1 & \text{if } k = 2m, \\ 4m + 1 - k & \text{if } 2m + 1 \leq k \leq 3m - 1, \\ k & \text{if } k = 3m, \text{ and} \\ 4m + 2 - k & \text{if } 3m + 1 \leq k \leq 4m. \end{cases}$$

Then we have

$$|x_k - k| = \begin{cases} 4m + 1 - 2k & \text{if } 1 \leq k \leq m, \\ 4m - 2k & \text{if } m + 1 \leq k \leq 2m - 1, \\ 2m - 1 & \text{if } k = 2m, \\ 2k - 1 - 4m & \text{if } 2m + 1 \leq k \leq 3m - 1, \\ 0 & \text{if } k = 3m, \text{ and} \\ 2k - 2 - 4m & \text{if } 3m + 1 \leq k \leq 4m. \end{cases}$$

Hence, it follows that we have

$$\{|x_k - k| \mid 1 \leq k \leq m\} = \{2m + 1, 2m + 3, \ldots, 4m - 1\},$$
$$\{|x_k - k| \mid m + 1 \leq k \leq 2m - 1\} = \{2, 4, \ldots, 2m - 2\},$$
$$\{|x_k - k| \mid 2m + 1 \leq k \leq 3m - 1\} = \{1, 3, \ldots, 2m - 3\},$$
$$\{|x_k - k| \mid 3m + 1 \leq k \leq 4m\} = \{2m, 2m + 2, \ldots, 4m - 2\}.$$

This implies that $\{|x_1 - 1|, \ldots, |x_{4m} - 4m|\} = \{0, 1, \ldots, 4m - 1\}$, so we have the differences $|x_k - k|$, $1 \le k \le n$ to be pairwise distinct as required.

Next, let us consider the case where $n \equiv 1 \pmod 4$. When $n = 5$, we see that the permutation $(5, 2, 4, 1, 3)$ of $(1, 2, 3, 4, 5)$ satisfies the property that the differences $|x_k - k|$, $1 \le k \le 5$ are pairwise distinct. Hence, let us assume that $n > 5$. Then we have $n = 4m + 1$ for some positive integer $m > 1$. For each $k \in \{1, 2, \ldots, n\}$, let us define x_k as follows:

$$x_k = \begin{cases} 4m + 2 - k & \text{if } 1 \le k \le m, \\ k & \text{if } k = m + 1, \\ 4m + 3 - k & \text{if } m + 2 \le k \le 2m + 1, \\ 1 & \text{if } k = 2m + 2, \\ 4m + 4 - k & \text{if } 2m + 3 \le k \le 3m + 2, \\ 4m + 3 - k & \text{if } 3m + 3 \le k \le 4m + 1. \end{cases}$$

Then we have

$$|x_k - k| = \begin{cases} 4m + 2 - 2k & \text{if } 1 \le k \le m, \\ 0 & \text{if } k = m + 1, \\ 4m + 3 - 2k & \text{if } m + 2 \le k \le 2m + 1, \\ 2m + 1 & \text{if } k = 2m + 2, \\ 2k - 4 - 4m & \text{if } 2m + 3 \le k \le 3m + 2, \text{ and} \\ 2k - 3 - 4m & \text{if } 3m + 3 \le k \le 4m + 1. \end{cases}$$

Hence, we have

$$\{|x_k - k| \mid 1 \le k \le m\} = \{2m + 2, 2m + 4, \ldots, 4m\},$$
$$\{|x_k - k| \mid m + 2 \le k \le 2m + 1\} = \{1, 3, \ldots, 2m - 1\},$$
$$\{|x_k - k| \mid 2m + 3 \le k \le 3m + 2\} = \{2, 4, \ldots, 2m\},$$
$$\{|x_k - k| \mid 3m + 3 \le k \le 4m + 1\} = \{2m + 3, \ldots, 4m - 1\}.$$

This implies that $\{|x_1 - 1|, |x_2 - 2|, \ldots, |x_{4m+1} - (4m + 1)|\} = \{0, 1, \ldots, 4m\}$, so we have the differences $|x_k - k|$, $1 \le k \le n$ to be pairwise distinct as required.

110. Let $f(n, m)$ denote the number of ways of choosing a team A of m boys and a team B of m girls from a group of n boys and n girls respectively, and let $g(n, m)$ denote the number of ways of choosing a team A of m boys and a team B of m girls from a group of n boys

and n girls respectively, along with a team leader from each of the 2 teams. We shall prove that $d(n,m) = f(n,m)$, thereby showing that $d(n,m)$ is an integer. Clearly $f(n,0) = f(n,n) = 1$.

We shall count $g(n,m)$, where $0 < m \leq n$, in 2 different ways. On one hand, there are $f(n,m)$ ways to choose a team of m boys and a team of m girls from a group of n boys and n girls respectively, and then that there are m choices each to choose a team leader for both teams. By (MP), we have $g(n,m) = m^2 f(n,m)$. On the other hand, let us fix a particular boy a and a particular girl b, and consider the cases where either $a \in A$ and/or $b \in B$:

Case 1: $a \in A, b \in B$. Then by definition, there are $f(n-1,m-1)$ ways of choosing a sub-team A' of the remaining $m-1$ boys and a sub-team B' of the remaining $m-1$ girls from a group of $n-1$ boys and $n-1$ girls respectively, and there are m choices each to choose a team leader for both teams. The number of ways in this case is $m^2 f(n-1,m-1)$.

Case 2: $a \notin A, b \notin B$. Then by definition, there are $f(n-1,m)$ ways of choosing a team A of m boys and a team B of m girls from a group of $n-1$ boys and $n-1$ girls respectively, and there are m choices each to choose a team leader for both teams. The number of ways in this case is $m^2 f(n-1,m)$.

Case 3: $a \in A, b \notin B$. Firstly, by definition, there are $f(n-1,m-1)$ ways of choosing a sub-team A' of the remaining $m-1$ boys and a sub-team B' of the remaining $m-1$ girls from a group of $n-1$ boys and $n-1$ girls respectively, that does not include both a and b. Next, let us choose a team leader for both A and B. Since A' along with a would have formed a team A of m people, we see that there are m ways to choose a team leader from A. For team B, we see that as there are only $m-1$ people in the sub-team B', and $b \notin B$, we would need to select a team leader from the remaining $n - (m-1) - 1 = n - m$ girls, so that we could form a team B of m girls. The number of ways in this case is $m(n-m)f(n-1,m-1)$.

Case 4: $a \notin A, b \in B$. This case is similar to Case 3.

Hence, we have

$$\begin{aligned} g(n,m) &= m^2 f(n-1,m-1) + m^2 f(n-1,m) \\ &\quad + 2m(n-m)f(n-1,m-1) \\ &= m^2 f(n-1,m) + m(2n-m)f(n-1,m-1), \end{aligned}$$

so this implies that
$$m^2 f(n, m) = m^2 f(n - 1, m) + m(2n - m)f(n - 1, m - 1).$$
By dividing m on both sides of the above equation, we have
$$mf(n, m) = mf(n - 1, m) + (2n - m)f(n - 1, m - 1)$$
for all $0 < m \le n$. Since we have
$$md(n, m) = md(n - 1, m) + (2n - m)d(n - 1, m - 1)$$
for all $0 < m \le n$ and $d(n, 0) = d(n, n) = 1$ for all positive integers n, this would imply that $d(m, n) = f(m, n)$ as desired.

111. Each pupil attempts 7 questions and hence $\binom{7}{2} = 21$ pairs of questions. There are $\binom{28}{2} = 378$ pairs of questions in total and each pair of questions is attempted by 2 pupils, so there must be $\frac{378 \cdot 2}{21} = 36$ pupils. Now, suppose n pupils solved question 1. Each pupil who solved question 1 solved $7 - 1 = 6$ pairs involving question 1, so there must be $\frac{n \cdot 6}{2} = 3n$ pairs of questions involving question 1. But there are $28 - 1 = 27$ pairs of questions involving question 1, so $n = 9$. As the same argument applies to any other question, we see that each question was solved by 9 pupils.

Suppose that all contestants solved either 1, 2 or 3 questions in part I. Let the number of questions in part I be m, and the number of pupils solving 1, 2, 3 questions in part I be a, b and c respectively. Then we have $a + b + c = 36$ and $a + 2b + 3c = 9m$. Now consider the number of pairs of questions in part I. There are $\binom{m}{2} = \frac{m(m-1)}{2}$ such pairs. We note that the pupils solving just 1 question in part I solve no pairs of questions in part I, pupils solving 2 questions in part I solve 1 pair of questions in part I and pupils that solve 3 questions in part I solve $\binom{3}{2} = 3$ pairs of questions in part I, so we have $b + 3c = m(m - 1)$ (since each pair of questions in part I is solved by exactly 2 pupils). As $b + 2c = (a + 2b + 3c) - (a + b + c) = 9m - 36$, we have
$$\begin{aligned} b &= 3(b + 2c) - 2(b + 3c) \\ &= 3(9m - 36) - 2m(m - 1) \\ &= -2m^2 + 29m - 108 \\ &= -2\left(m - \frac{29}{4}\right)^2 - \frac{23}{8} \\ &< 0, \end{aligned}$$
which is a contradiction. So there must exist some student that has solved either no questions, or at least 4 questions in part I.

Remark. The solution is courtesy of John Scholes. We refer the reader to https://mks.mff.cuni.cz/kalva/usa/usoln/usol844.html for the original solution.

112. Consider any of the 5 given points. The number of perpendiculars from the chosen point to the lines joining any 2 of the other 4 points is equal to $\binom{4}{2} = 6$. Hence, the total number of perpendiculars is equal to $5 \cdot 6 = 30$. Also, since no 2 distinct lines joining any two of the 5 given points are parallel, it follows that there is at most one point of intersection between any 2 distinct perpendiculars. This implies that the maximum number of intersection points between any of the 2 perpendiculars is equal to $\binom{30}{2} = 435$.

Firstly, fix any 2 of the 5 given points, and consider the line ℓ joining them. As no 2 lines joining the 5 points are perpendicular, we see that the $5 - 2 = 3$ perpendiculars to ℓ do not intersect. As there are $\binom{5}{2} = 10$ possibilities for ℓ and hence a total of $10 \cdot 3 = 30$ such perpendiculars, we see that this reduces the maximum number of intersection points between any of the 2 perpendiculars to $435 - 30 = 405$.

Secondly, let us consider the perpendiculars from a fixed point chosen from the 5 given points to the lines joining any two of the other 4 points. We note that there are $\binom{4}{2} = 6$ such lines. Since these perpendiculars only intersect at this fixed point, and this point of intersection has been counted $\binom{6}{2} = 15$ times in the counting of the number of intersection points, we see that each intersection point has been counted an extra $15 - 1 = 14$ times. Since there are 5 such intersection points, we see that this reduces the maximum number of intersection points between any of the 2 perpendiculars to $405 - 14 \cdot 5 = 335$.

Finally, since no 2 of the lines joining any two points of the 5 given points are coincident, we see that any 3 of the 5 given points determines a non-degenerate triangle, and hence a point of intersection, which is the intersection of the altitudes of the triangle. As this point of intersection is counted $\binom{3}{2} = 3$ times in the counting of the number of intersection points, we see that each intersection point has been counted an extra $3 - 1 = 2$ times. Since there are $\binom{5}{3} = 10$ such triangles, we see that this reduces the maximum number of intersection points between any of the 2 perpendiculars to $335 - 2 \cdot 10 = 315$.

113. Label the points $a_1, a_2, \ldots, a_n$, and let the number of points that are of unit distance from point a_i be m_i. Then the problem is equivalent to showing that $\sum_{i=1}^{n} m_i < 2 \cdot 2n^{\frac{3}{2}} = 4n^{\frac{3}{2}}$ (since each such pair is double

counted). We shall first prove the assertion for the case where $m_i \geq 2$ for all $i \in \{1, 2, \ldots, n\}$.

We note that the m_i points of unit distance from a_i must all lie on the circumference of C_i, where C_i denotes the circle of unit radius centered at a_i. We note that there are at most 2 points of intersection between any pair of distinct circles. Since there are $\binom{n}{2} = \frac{n(n-1)}{2}$ such pairs of circles, we see that the total number of such intersection points, counting multiplicity, is at most $2 \cdot \frac{n(n-1)}{2} = n(n-1)$. We also observe that point a_i lies on m_i such C_j's, so as a point of intersection, it is counted $\binom{m_i}{2} = \frac{m_i(m_i-1)}{2}$ times. Thus, all the points together are counted a total of $\sum_{i=1}^{n} \frac{m_i(m_i-1)}{2}$ times. Hence, we have $\sum_{i=1}^{n} \frac{m_i(m_i-1)}{2} \leq n(n-1)$, that is, $\sum_{i=1}^{n} m_i(m_i - 1) \leq 2n(n-1)$. Since $m_i \geq 2$, we have $m_i \leq 2(m_i - 1)$, so this implies that

$$\sum_{i=1}^{n} m_i^2 \leq 2 \sum_{i=1}^{n} m_i(m_i - 1) \leq 4n(n-1) < 4n^2.$$

By the Cauchy-Schwarz inequality, we have $\left(\sum_{i=1}^{n} m_i\right)^2 \leq n \sum_{i=1}^{n} m_i^2 < 4n^3$, and the desired result follows.

Now, let us deal with the general case. Without loss of generality, let us assume that $m_i \neq 0$ for all $i \in \{1, 2, \ldots, n\}$. If $m_i = 1$ for all $i \in \{1, 2, \ldots, n\}$ then the result is immediate. Else, let us assume without loss of generality that $m_1 = m_2 = \cdots = m_k = 1$ for some $1 \leq k < n$, and $m_i \geq 2$ for all $i > k$. Then by what we have shown earlier, we have $\sum_{i=k+1}^{n} m_i < 4(n-k)^{\frac{3}{2}}$. Thus, we have

$$\sum_{i=1}^{n} m_i = \sum_{i=1}^{k} m_i + \sum_{i=k+1}^{n} m_i = k + 4(n-k)^{\frac{3}{2}} < 4n^{\frac{3}{2}},$$

where the last inequality follows from the fact that the derivative of the function $f(x) = 4(x + a)^{\frac{3}{2}} - 4x^{\frac{3}{2}} - a$, $a \geq 1$ is always positive for all $x > 0$, and $f(0) > 0$. We are done.

Remark 1. The Cauchy-Schwarz inequality states that if $a_1, a_2, \ldots, a_n, b_1, b_2, \ldots, b_n$ are real numbers, then we have

$$\left(\sum_{i=1}^{n} a_i b_i\right)^2 \leq \left(\sum_{i=1}^{n} a_i^2\right)\left(\sum_{j=1}^{n} b_j^2\right).$$

Remark 2. The solution is courtesy of John Scholes. We refer the reader to `https://mks.mff.cuni.cz/kalva/putnam/psoln/psol786.html` for the original solution.

114. First Solution: Let $S(c,m)$ denote the set of ordered c-tuples $(n_1, \ldots, n_c)$ satisfying $n_i \in \{1, \ldots, m\}$ for all $i \in \{1, 2, \ldots, c\}$, $n_1 > n_2$ and $n_2 \le n_3 \le \cdots \le n_c$. Let $P(c,m)$ denote the set of permutations of the multiset $\{c \cdot x, (m-2) \cdot y, 1 \cdot z\}$ satisfying the following conditions: 1. z must lie between 2 (not necessarily adjacent) x's, and not be preceded immediately by any y's. We shall construct a function from $S(c,m)$ to $P(c,m)$, and show that the function is bijective. Firstly, for each $i \in \{1, \ldots, m\}$, let us define $f_i \colon S(c,m) \to \mathbf{N}$ by $f_i((n_1, \ldots, n_c)) = a_i$, where $a_i = |\{j \mid n_j = i\}|$. Also, let us define $g \colon S(c,m) \to \mathbf{N}$ as follows: $g((n_1, n_2, \ldots, n_c)) = N$, where $N = \max\{i \mid i < n_1, a_i \ne 0\}$. Finally, let us define $f \colon S(c,m) \to P(c,m)$ by $f((n_1, n_2, \ldots, n_c)) = k_1 k_2 \cdots k_{c+m-1}$, where

$$
k_p = \begin{cases}
y, & \text{if } p = j + \sum_{i=1}^{j} a_i, j \in \{1, 2, \ldots, m-1\} \setminus \{N\}, \\
z, & \text{if } p = N + \sum_{i=1}^{N} a_i \\
x, & \text{otherwise.}
\end{cases}
$$

Clearly, f is well-defined. It remains to show that f is bijective. Let us first show that f is injective. Let $(n_1, \ldots, n_c), (n'_1, \ldots, n'_c) \in S(c,m)$ be distinct c-tuples, and let $f((n_1, \ldots, n_c)) = k_1 \cdots k_{c+m-1}$, $f((n'_1, \ldots, n'_c)) = k'_1 \cdots k'_{c+m-1}$. We need to show that $k_1 \cdots k_{c+m-1} \ne k'_1 \cdots k'_{c+m-1}$. For each $i \in \{1, 2, \ldots, m\}$, let us define $a_i = f_i((n_1, \ldots, n_c))$ and $a'_i = f_i((n'_1, \ldots, n'_c))$, and let

$$
N = g((n_1, \ldots, n_c)), N' = g((n'_1, \ldots, n'_c)).
$$

Then either we have $a_i \ne a'_i$ for some $i \in \{1, 2, \ldots, m-1\}$, or $n_1 \ne n'_1$. If it is the former, then let $q = \min\{i \mid a_i \ne a'_i\}$, and without loss of generality, let us assume that $a_q > a'_q$. Then by the definition given above, we have $k_p = x$ and $k'_p = y$ or $k'_p = z$, where $p = q + \sum_{i=1}^{q} a'_i$, which implies that $k_p \ne k'_p$, and hence $k_1 \cdots k_{c+m-1} \ne k'_1 \cdots k'_{c+m-1}$. Else, we must have $a_i = a'_i$ for all $i \in \{1, 2, \ldots, m-1\}$ and $n_1 \ne n'_1$. Without loss of generality, let us assume that $n_1 > n'_1$. Then we must have $a_{n'_1} = a'_{n'_1} \ne 0$, and hence $n'_1 \in \{i \mid i < n_1, a_i \ne 0\}$. This implies that $N \ge n'_1$. By definition, we have $n'_1 > N'$ so this implies that $N' < N$. This implies that $k_p = z$ and $k'_p = x$ or

$k'_p = y$, where $p = N + \sum_{i=1}^{N} a'_i$, which implies that $k_p \neq k'_p$, and hence $k_1 \cdots k_{c+m-1} \neq k'_1 \cdots k'_{c+m-1}$.

Next, let us show that f is surjective. Given any $k_1 \cdots k_{c+m-1} \in P(c, m)$, we let $\{i \mid k_i \neq x\} = \{b_1, b_2, \ldots, b_{m-1}\}$, where $b_i < b_j$ for all $i < j$, and let b_N be the unique integer for which $k_{b_N} = z$, where $N \in \{1, 2, \ldots, m-1\}$. Let $a_1 = b_1 - 1$, $a_i = b_{i+1} - b_i - 1$ for all $i = 2, \ldots, m-1$, $a_m = c + m - 1 - b_{m-1}$, and n be the smallest integer for which $n > N$ and $a_i \neq 0$ (Note: since z must lie between 2 x's, there always exists some $i > N$ for which $a_i \neq 0$). Finally, define $c_i = a_i$ for all $i \neq n$, and $c_n = a_n - 1$. Then the sequence $(n_1, n_2, \ldots, n_c) \in S(c, m)$, defined by $n_1 = n$, and $n_i = \min\left\{ k \mid \sum_{j=1}^{k} c_j \geq i - 1 \right\}$, for $i = 2, \ldots, c$ clearly satisfies $f_i((n_1, n_2, \ldots, n_c)) = a_i$ for all $i \in \{1, 2, \ldots, m\}$ and $g((n_1, n_2, \ldots, n_c)) = N$, and $j + \sum_{i=1}^{j} a_i = b_j$ for $j \in \{1, 2, \ldots, m-1\}$, so this shows that $f((n_1, n_2, \ldots, n_c)) = k_1 k_2 \cdots k_{c+m-1}$.

Hence, f is bijective, so we have $n(c, m) = |S(c, m)| = |P(c, m)|$ by (BP). Now, we need to find $|P(c, m)|$. We note that there are $\binom{c+m-2}{c}$ ways to permute c x's and $m - 2$ y's. Also, there are $(c + m - 2) - (m - 1) = c - 1$ ways to place the z so that z lies between 2 (not necessarily adjacent) x's, and not be preceded immediately by any y's. This implies that $|P(c, m)| = (c-1)\binom{c+m-2}{c}$ by (MP), so we have

$$n(c, m) = (c - 1)\binom{c + m - 2}{c}.$$

Second Solution: Let $S(c, m)$ denote the set of ordered c-tuples $(n_1, n_2, \ldots, n_c)$ that satisfy $n_i \in \{1, 2, \ldots, m\}$ for all $i = 1, 2, \ldots, c$, $n_2 < n_1$, and $n_2 \leq n_3 \leq \cdots \leq n_c$. Let $(n_1, n_2, \ldots, n_c) \in S(c, m)$ with $n_2 = k$. Clearly, we have $k \in \{1, 2, \ldots, m-1\}$ since $n_2 < n_1 \leq m$. We notice that $\{n_3, \ldots, n_c\}$ is a $(c-2)$-element multi-subset of the multiset $M_k = \{\infty \cdot k, \infty \cdot (k+1), \ldots, \infty \cdot m\}$. Conversely, any $(c-2)$-element multi-subset of M_k gives rise to exactly one $(c-1)$-tuple $(n_2, n_3, \ldots, n_c)$ with $k = n_2 \leq n_3 \leq \cdots \leq n_c$. There are $H_{c-2}^{m-k+1} = \binom{m-k+c-2}{c-2}$ such multi-subsets and hence $(c-1)$-tuples where $n_2 = k$. As $n_1 > n_2 = k$, there are $m - k$ choices for n_1. The number of ordered c-tuples $(n_1, n_2, \ldots, n_c) \in S(c, m)$ with $n_2 = k$ is equal to $(m - k)\binom{m-k+c-2}{c-2}$. Thus, by (AP), we have

$$n(c, m) = |S(c, m)|$$

$$= \sum_{k=1}^{m-1} (m-k) \binom{m-k+c-2}{c-2}$$

$$= \sum_{u=1}^{m-1} u \binom{u+r}{r} \quad \text{(by setting } m-k=u \text{ and } c-2=r\text{)}$$

$$= \sum_{u=1}^{m-1} u \frac{(u+r)!}{u!r!}$$

$$= \sum_{u=1}^{m-1} (r+1) \frac{(u+r)!}{(u-1)!(r+1)!}$$

$$= (r+1) \sum_{u=1}^{m-1} \binom{u+r}{r+1}$$

$$= (r+1) \binom{m+r}{r+2} \quad \text{(by (2.5.1))}$$

$$= (c-1) \binom{c+m-2}{c} \quad \text{(since } c-2=r\text{).}$$

115. (i) We note that $A(x_i, \cdot)$ and $A(x_j, \cdot)$ are pairwise disjoint for all $i \neq j$ by definition. As we have

$$\bigcup_{i=1}^{m} A(x_i, \cdot) = \bigcup_{i=1}^{m} ((\{x_i\} \times Y) \cap A)$$

$$= \left(\bigcup_{i=1}^{m} (\{x_i\} \times Y) \right) \cap A$$

$$= (X \times Y) \cap A$$

$$= A,$$

it follows that we have

$$\sum_{i=1}^{m} |A(x_i, \cdot)| = \left| \bigcup_{i=1}^{m} A(x_i, \cdot) \right| = |A|.$$

Similarly, we have $\sum_{j=1}^{n} |A(\cdot, y_j)| = |A|$.

(ii) Denote the set of n given points by A, and let a and b be two distinct points in A for which the distance between a and b is minimal. Denote ℓ to be the line joining points a and b, and let us fix the x- and y-axes of the plane in such a way that the x-axis is parallel to ℓ. Let X be the set of all possible x-coordinates of the points in A, and let Y be the set of all possible y-coordinates of the points in A. Then we have $A \subseteq X \times Y$.

Since any 3 distinct points in A form a non-degenerate triangle, it follows that no 3 distinct points in A are collinear, and hence we have $|A(x_i, \cdot)| \leq 2$ for all $x_i \in X$. Furthermore, let ℓ_a and ℓ_b be lines passing through a and b respectively that are perpendicular to the x-axis, and without loss of generality, let us assume that ℓ_a is on the left of ℓ_b. Since all triangles formed by any 3 distinct points in A are right-angled, it follows that each triangle must contain one right angle and two acute angles. In particular, this implies that no point in A can lie to the right of ℓ_b or to the left of ℓ_a. Indeed, if there exists some point d that lies on the right of ℓ_b, then it is visibly seen that the triangle formed by the vertices a, b and d has an obtuse angle at the vertex b, which is a contradiction. The same argument applies when d lies on the left of ℓ_a instead. Now, let c be a point in A distinct from a and b, and let ℓ_c be the line passing through c perpendicular to the x-axis. By the above assertion, ℓ_c cannot lie on the right of ℓ_b or on the left of ℓ_a. We want to show that $\ell_c = \ell_a$ or $\ell_c = \ell_b$. Suppose on the contrary that ℓ_c does not coincide with ℓ_a and ℓ_b. Then we must have ℓ_c to lie strictly in between ℓ_a and ℓ_b. Since the triangle formed by the vertices a, b and c is right-angled, it follows that the right angle must be at vertex c. By Pythagoras' Theorem, this implies that the distance between a and c is strictly smaller than that of a and b, which contradicts the minimality of the distance between a and b. Hence, we must have ℓ_c to coincide with either ℓ_a or ℓ_b, that is, c lies on ℓ_a or ℓ_b. This implies that $|X| = 2$, so we have $|A| = \sum_{x \in X} |A(x, \cdot)| \leq 2|X| = 4$. Indeed, equality is achieved only when the vertices in A form a rectangle. So the maximum value of n is 4.

Solutions to Exercise 2

1. We note that the number of ways to write n as a sum of r positive integers with the order being taken into account is equal to the number of positive integer solutions to the equation $x_1 + x_2 + \cdots + x_r = n$, which is equal to $H_{n-r}^r = \binom{n-r+r-1}{n-r} = \binom{n-1}{r-1}$. Since the possible values of r are $1, 2, \ldots, n$, we see that the total number of ways that n can be expressed is equal to

$$\sum_{r=1}^{n} \binom{n-1}{r-1} = \sum_{r=0}^{n-1} \binom{n-1}{r} = 2^{n-1}.$$

2. Let A be a $2n$-digit binary sequence such that the number of 0's in the first n digits is equal to the number of 1's in the last n digits. Let the number of 0's in the first n digits of A be m. Then the number of 1's in the last n digits of A is equal to m, and hence the number of 0's in the last n digits of A is equal to $n - m$. Hence, the number of 0's in the binary representation of A is equal to n.

Conversely, if B is a $2n$-digit binary sequence whose number of 0's in the binary representation is equal to n, then let the number of 0's in the first n digits of B be p. Then the number of 0's in the last n digits of B is equal to $n - p$, and hence the number of 1's in the last n digits of B is equal to p. Hence, the number of 0's in the first n digits of B is equal to the number of 1's in the last n digits of B.

Therefore, we see that the number of $2n$-digit binary sequences whose the number of 0's in the first n digits is equal to the number of 1's in the last n digits, is equal to the number of $2n$-digit binary sequences whose number of 0's in the binary representation is equal to n, which is then equal to $\binom{2n}{n}$.

3. (i) Let N be an r-element multi-subset of M, and let $p = |\{i \mid a_i \in N\}|$. Since $p \le r$, $r \le m$ and $r \le n$ it follows that the set of

possible values of p is $\{0, 1, \ldots, r\}$. For each $j \in \{0, 1, \ldots, r\}$, there are $\binom{n}{j}$ such r-element multi-subset N of M for which $|\{i \mid a_i \in N\}| = j$. By (AP), the number of r-element multi-subsets of M is equal to $\sum_{j=0}^{r} \binom{n}{j}$.

(ii) Let N be an r-element multi-subset of M, and let $p = |\{i \mid a_i \in N\}|$. Since $p \leq r$, $r \leq m$ and $r \geq n$ it follows that the set of possible values of p is $\{0, 1, \ldots, n\}$. For each $j \in \{0, 1, \ldots, n\}$, there are $\binom{n}{j}$ such r-element multi-subset N of M for which $|\{i \mid a_i \in N\}| = j$. By (AP), the number of r-element multi-subsets of M is equal to $\sum_{j=0}^{n} \binom{n}{j} = 2^n$.

(iii) Let N be an r-element multi-subset of M, and let $p = |\{i \mid a_i \in N\}|$. Since $p \leq r$, $r \leq n$ and $r \geq m$ it follows that the set of possible values of p is $\{r - m, r - m + 1, \ldots, r\}$. For each $j \in \{r - m, r - m + 1, \ldots, r\}$, there are $\binom{n}{j}$ such r-element multi-subset N of M for which $|\{i \mid a_i \in N\}| = j$. By (AP), the number of r-element multi-subsets of M is equal to $\sum_{j=r-m}^{r} \binom{n}{j}$.

4. We note that for each $n \in \{3, 4, \ldots, 10\}$, and any n-element subset of the ten vertices, there is exactly one convex polygon that can be drawn using all of the elements in the subset as vertices. As the number of distinct convex n-sided polygons that can be drawn using some, or all of the ten points as vertices is equal to $\binom{10}{n}$, we see that the number of distinct convex polygons of three or more sides that can be drawn using some, or all of the ten points as vertices is equal to

$$\sum_{n=3}^{10} \binom{10}{n} = \sum_{n=0}^{10} \binom{10}{n} - \sum_{n=0}^{2} \binom{10}{n} = 2^{10} - 1 - 10 - 45 = 968.$$

5. We note that a term in the expansion of $(1 + x + x^2)^8$ is of the form $1^a x^b (x^2)^c = x^{b+2c}$, where a, b and c are non-negative integers satisfying $a + b + c = 8$. Also, we see that the coefficient of the term x^{b+2c} in the expansion of $(1 + x + x^2)^8$ is equal to $\binom{8}{8-b-c, b, c}$. Finally, we see that the pairs (b, c) of non-negative integers satisfying $b + c \leq 8$ and $b + 2c = 5$ are $(1, 2)$, $(3, 1)$ and $(5, 0)$. Therefore, the coefficient of x^5 in the expansion of $(1 + x + x^2)^8$ is

$$\binom{8}{8 - 1 - 2, 1, 2} + \binom{8}{8 - 3 - 1, 3, 1} + \binom{8}{8 - 5 - 0, 5, 0}$$

$$= \binom{8}{5, 1, 2} + \binom{8}{4, 3, 1} + \binom{8}{3, 5, 0}$$

$$= \frac{8!}{5!1!2!} + \frac{8!}{4!3!1!} + \frac{8!}{3!5!0!}$$
$$= 504.$$

6. We note that a term in the expansion of $(1 + x + x^2)^9$ is of the form $1^a x^b (x^2)^c = x^{b+2c}$, where a, b and c are non-negative integers satisfying $a + b + c = 9$. Also, we see that the coefficient of the term x^{b+2c} in the expansion of $(1 + x + x^2)^9$ is equal to $\binom{9}{9-b-c, b, c}$. Finally, we see that the pairs (b, c) of non-negative integers satisfying $b + c \le 9$ and $b + 2c = 6$ are $(0, 3)$, $(2, 2)$, $(4, 1)$ and $(6, 0)$. Therefore, the coefficient of x^6 in the expansion of $(1 + x + x^2)^9$ is equal to

$$\binom{9}{9-0-3, 0, 3} + \binom{9}{9-2-2, 2, 2}$$
$$+ \binom{9}{9-4-1, 4, 1} + \binom{9}{9-6-0, 6, 0}$$
$$= \binom{9}{6, 0, 3} + \binom{9}{5, 2, 2} + \binom{9}{4, 4, 1} + \binom{9}{3, 6, 0}$$
$$= \frac{9!}{6!0!3!} + \frac{9!}{5!2!2!} + \frac{9!}{4!4!1!} + \frac{9!}{3!6!0!}$$
$$= 1554.$$

7. We note that a term in the expansion of $(1 + x^3 + x^5 + x^7)^{100}$ is of the form $x^{3a+5b+7c}$, where a, b and c are non-negative integers satisfying $a + b + c \le 100$. Also, we see that the coefficient of the term $x^{3a+5b+7c}$ in the expansion of $(1+x^3+x^5+x^7)^{100}$ is equal to $\binom{100}{100-a-b-c, a, b, c}$. Finally, we see that the triples (a, b, c) of non-negative integers satisfying $a+b+c \le 100$ and $3a + 5b + 7c = 18$ are $(6, 0, 0)$, $(2, 1, 1)$ and $(1, 3, 0)$. Therefore, the coefficient of x^{18} in the expansion of $(1 + x^3 + x^5 + x^7)^{100}$ is

$$\binom{100}{6} + \binom{100}{96, 2, 1, 1} + \binom{100}{96, 1, 3}$$
$$= \binom{100}{6} + \frac{100 \cdot 99 \cdot 98 \cdot 97}{2!} + \frac{100 \cdot 99 \cdot 98 \cdot 97}{3!}$$
$$= \binom{100}{6} + 97 \cdot 98 \cdot \frac{100 \cdot 99}{2!} + 97 \cdot \frac{100 \cdot 99 \cdot 98}{3!}$$
$$= \binom{100}{6} + 97 \cdot 98 \cdot \binom{100}{2} + 97 \cdot \binom{100}{3}.$$

8. We note that a term in the expansion of $(1 + x^5 + x^7 + x^9)^{1000}$ is of the form $x^{5a+7b+9c}$, where a, b and c are non-negative integers satisfying

$a + b + c \leq 1000$. Also, we see that the coefficient of the term $x^{5a+7b+9c}$ in the expansion of $(1 + x^5 + x^7 + x^9)^{1000}$ is equal to $\binom{1000}{1000-a-b-c,a,b,c}$. Finally, we see that the triples (a, b, c) of non-negative integers satisfying $a + b + c \leq 100$ and $5a + 7b + 9c = 29$ are $(3, 2, 0)$ and $(4, 0, 1)$. Therefore, the coefficient of x^{29} in the expansion of $(1+x^5+x^7+x^9)^{1000}$ is equal to

$$
\begin{aligned}
&\binom{1000}{995, 3, 2} + \binom{1000}{995, 4, 1} \\
=\ & \frac{1000 \cdot 999 \cdot 998 \cdot 997 \cdot 996}{3!2!} + \frac{1000 \cdot 999 \cdot 998 \cdot 997 \cdot 996}{4!1!} \\
=\ & \frac{1000 \cdot 999 \cdot 998}{3!} \cdot \frac{997 \cdot 996}{2!} + \frac{1000 \cdot 999 \cdot 998 \cdot 997}{4!} \cdot \frac{996}{1!} \\
=\ & \binom{1000}{3}\binom{997}{2} + \binom{1000}{4}\binom{996}{1}.
\end{aligned}
$$

9. We note that a term in the expansion of $\left(\sum_{i=0}^{10} x^i\right)^3$ is of the form $x^{n_1 + 2n_2 + \cdots + 10n_{10}}$, where $n_1, n_2, \ldots, n_{10}$ are non-negative integers satisfying $\sum_{i=1}^{10} n_i \leq 3$. Also, we see that the coefficient of the term $x^{n_1 + 2n_2 + \cdots + 10n_{10}}$ in the expansion of $\left(\sum_{i=0}^{10} x^i\right)^3$ is equal to $\binom{3}{3 - \sum_{i=1}^{10} n_i, n_1, n_2, \ldots, n_{10}}$.

(i) We see that the 10-tuples $(n_1, n_2, \ldots, n_{10})$ of non-negative integers satisfying $\sum_{i=1}^{10} n_i \leq 3$ and $\sum_{i=1}^{10} i n_i = 5$ are

$$
\begin{aligned}
&(0, 0, 0, 0, 1, 0, 0, 0, 0, 0), (1, 0, 0, 1, 0, 0, 0, 0, 0, 0), \\
&(0, 1, 1, 0, 0, 0, 0, 0, 0, 0), (2, 0, 1, 0, 0, 0, 0, 0, 0, 0), \\
&(1, 2, 0, 0, 0, 0, 0, 0, 0, 0).
\end{aligned}
$$

Therefore, the coefficient of x^5 in the expansion of $\left(\sum_{i=0}^{10} x^i\right)^3$ is equal to

$$
3\binom{3}{2} + 2\binom{3}{1, 1, 1} = 21.
$$

(ii) We see that the 10-tuples $(n_1, n_2, \ldots, n_{10})$ of non-negative integers satisfying $\sum\limits_{i=1}^{10} n_i \leq 3$ and $\sum\limits_{i=1}^{10} in_i = 8$ are

$$(0,0,0,0,0,0,0,1,0,0), \ (1,0,0,0,0,0,1,0,0,0),$$
$$(0,1,0,0,0,1,0,0,0,0), \ (2,0,0,0,0,1,0,0,0,0),$$
$$(0,0,1,0,1,0,0,0,0,0), \ (1,1,0,0,1,0,0,0,0,0),$$
$$(0,0,0,2,0,0,0,0,0,0), \ (1,0,1,1,0,0,0,0,0,0),$$
$$(0,2,0,1,0,0,0,0,0,0), \ (0,1,2,0,0,0,0,0,0,0).$$

Therefore, the coefficient of x^8 in the expansion of $\left(\sum\limits_{i=0}^{10} x^i \right)^3$ is equal to

$$5\binom{3}{2} + 5\binom{3}{1,1,1} = 45.$$

10. Fix an element $x \in X$. Define a function $f : \mathcal{O} \to \mathcal{E}$ as follows: $f(A) = A - \{x\}$ if $x \in A$, and $f(A) = A \cup \{x\}$ if $x \notin A$. Then it is easily verified that f is a bijection. Hence $|\mathcal{O}| = |\mathcal{E}|$.

11. Let M_r be a $(m+r)$-permutation of the multi-set $\{m \cdot 1, n \cdot 2\}$ that contains the m 1's, where $0 \leq r \leq n$. We note that there are $\binom{m+r}{m}$ such permutations M_r. By (AP), the number of permutations of the multi-set $\{m \cdot 1, n \cdot 2\}$ that contains the m 1's is equal to $\sum\limits_{r=0}^{n} \binom{m+r}{m} = \binom{m+n+1}{m+1} = \binom{m+n+1}{n}$ by (2.5.1).

12. Let A_m be an r-element subset of the set $\{1, 2, \ldots, n\}$ with largest member m, where $r \leq m \leq n$. Then it follows that $A_m - \{m\} \subseteq \{1, 2, \ldots, m-1\}$, and hence there are $\binom{m-1}{r-1}$ such sets A_m. Since there are $\binom{n}{r}$ r-element subsets of $\{1, 2, \ldots, n\}$, we have

$$
\begin{aligned}
H(n,r) &= \frac{1}{\binom{n}{r}} \sum_{m=r}^{n} m \binom{m-1}{r-1} \\
&= \frac{1}{\binom{n}{r}} \sum_{m=r}^{n} r \binom{m}{r} \quad \text{(by (2.1.2))} \\
&= \frac{r}{\binom{n}{r}} \sum_{m=r}^{n} \binom{m}{r} \\
&= \frac{r}{\binom{n}{r}} \binom{n+1}{r+1} \\
&= \frac{r \cdot r! \cdot (n-r)!}{n!} \cdot \frac{(n+1)!}{(r+1)!(n-r)!}
\end{aligned}
$$

$$= \frac{r(n+1)}{r+1}.$$

13. Let us denote the $n+1$ lines obtained from the n-th subdivision that are parallel to the side BC, and whose two endpoints lie on the sides AB and AC be denoted $\ell_1, \ell_2, \ldots, \ell_{n+1}$, from the top to the bottom. Without loss of generality, we may assume that vertex Y is on the left of vertex Z. Firstly, we note that for $i \in \{1, 2, \ldots, n+1\}$, there are $i+1$ points on the line ℓ_i, and hence there are $\binom{i+1}{2}$ ways to choose two points on ℓ_i to form the side YZ such that the vertices of the side YZ lies on ℓ_i. Furthermore, we see that after fixing the vertices Y and Z of the triangle XYZ, we see that by the condition that X and A must be on the same side of YZ that we must have the side XY to be parallel to AB and the side XZ to be parallel to AC. This implies that the vertex X is uniquely determined by the vertices Y and Z. Therefore, by (2.5.1), we have

$$\Delta(n) = \sum_{i=1}^{n+1} \binom{i+1}{2} = \binom{n+3}{3}.$$

14. For $i \in \{0, 1, \ldots, n\}$, we note that the coefficient of x^{n-i} in the binomial expansion of $(1+x)^{2n-i}$ is equal to $\binom{2n-i}{n-i} = \binom{2n-i}{n}$. Hence, the coefficient of x^n in the expansion of $\sum_{i=0}^{n} x^i (1+x)^{2n-i}$ is equal to

$$\sum_{i=0}^{n} \binom{2n-i}{n} = \sum_{i=n}^{2n} \binom{i}{n} = \binom{2n+1}{n+1} = \binom{2n+1}{n},$$

where the second equality follows from (2.5.1). Similarly, we see that the coefficient of x^{n+r-i} in the binomial expansion of $(1+x)^{2n-i}$ is equal to $\binom{2n-i}{n+r-i} = \binom{2n-i}{n-r}$. Therefore, the coefficient of x^n in the expansion of $\sum_{i=0}^{n} x^i (1+x)^{2n-i}$ is equal to

$$\sum_{i=0}^{n} \binom{2n-i}{n-r} = \sum_{i=n}^{2n} \binom{i}{n-r}$$

$$= \sum_{i=n-r}^{2n} \binom{i}{n-r} - \sum_{i=n-r}^{n-1} \binom{i}{n-r}$$

$$= \binom{2n+1}{n-r+1} - \binom{n}{n-r+1} \quad \text{(by (2.5.1))}$$

$$= \binom{2n+1}{n+r} - \binom{n}{r-1}.$$

15. For $i \in \{1, 2, \ldots, n\}$, we note that the coefficient of x^{n+i} in the expansion of $\left(\sum_{r=1}^{n} r x^r \right)^2$ is equal to

$$\sum_{r=i}^{n} r(n+i-r)$$

$$= \sum_{r=1}^{n} r(n+i-r) - \sum_{r=1}^{i-1} r(n+i-r)$$

$$= (n+i) \sum_{r=1}^{n} r - \sum_{r=1}^{n} r^2 - (n+i) \sum_{r=1}^{i-1} r + \sum_{r=1}^{i-1} r^2$$

$$= \frac{n(n+1)(n+i)}{2} - \frac{n(n+1)(2n+1)}{6}$$

$$- \frac{i(i-1)(n+i)}{2} + \frac{i(i-1)(2i-1)}{6}$$

$$= \frac{n(n+1)(n+3i-1)}{6} - \frac{i(i-1)(3n+i+1)}{6}.$$

This implies that $a_{n+i} = \frac{n(n+1)(n+3i-1)}{6} - \frac{i(i-1)(3n+i+1)}{6}$, and therefore we have

$$\sum_{i=n+1}^{2n} a_i$$

$$= \sum_{j=1}^{n} a_{n+j} \quad \text{(by setting } j = i - n\text{)}$$

$$= \sum_{j=1}^{n} \left(\frac{n(n+1)(n+3j-1)}{6} - \frac{j(j-1)(3n+j+1)}{6} \right)$$

$$= \sum_{j=1}^{n} \frac{n(n+1)(n-1)}{6} + \frac{n(n+1)}{2} \sum_{j=1}^{n} j$$

$$- n \sum_{j=1}^{n} \frac{j(j-1)}{2} - \sum_{j=1}^{n} \frac{j(j-1)(j+1)}{6}$$

$$= \frac{n^2(n+1)(n-1)}{6} + \frac{n^2(n+1)^2}{4} - n \left(\frac{n(n+1)(2n+1)}{12} - \frac{n(n+1)}{4} \right)$$

$$- \left(\frac{n^2(n+1)^2}{24} - \frac{n(n+1)}{12} \right)$$

$$= \frac{n(n+1)(4n^2 - 4n + 6n^2 + 6n - 4n^2 - 2n + 6n - n^2 - n + 2)}{24}$$

$$= \frac{n(n+1)(5n^2+5n+2)}{24}.$$

16. We have

$$P_r^r + P_r^{r+1} + \cdots + P_r^{2r} = \sum_{i=r}^{2r} P_r^i$$

$$= \sum_{i=r}^{2r} r!\binom{i}{r}$$

$$= r! \sum_{i=r}^{2r} \binom{i}{r}$$

$$= r!\binom{2r+1}{r+1} \quad \text{(by (2.5.1))}$$

$$= r!\binom{2r+1}{r}$$

$$= P_r^{2r+1}.$$

17. We have

$$\sum_{i=n}^{m+n}\binom{i}{r} = \sum_{i=r}^{m+n}\binom{i}{r} - \sum_{i=r}^{n-1}\binom{i}{r} = \binom{m+n+1}{r+1} - \binom{n}{r+1}$$

by (2.5.1). Hence, we have

$$P_r^n + P_r^{n+1} + \cdots + P_r^{n+m}$$

$$= \sum_{i=n}^{n+m} P_r^i$$

$$= \sum_{i=n}^{n+m} r!\binom{i}{r}$$

$$= r!\sum_{i=n}^{n+m}\binom{i}{r}$$

$$= r!\left(\sum_{i=r}^{n+m}\binom{i}{r} - \sum_{i=r}^{n-1}\binom{i}{r}\right)$$

$$= r!\left(\binom{m+n+1}{r+1} - \binom{n}{r+1}\right) \quad \text{(by (2.5.1))}$$

$$= \frac{1}{r+1}\left((r+1)!\binom{m+n+1}{r+1} - (r+1)!\binom{n}{r+1}\right)$$

$$= \frac{1}{r+1}(P_{r+1}^{m+n+1} - P_{r+1}^n).$$

18. Let r and n be non-negative integers such that $n > 0$ and $r \leq n$. We note that
$$\frac{\binom{n}{r-1}}{\binom{n}{r}} = \frac{n!}{(r-1)!(n-r+1)!} \cdot \frac{r!(n-r)!}{n!} = \frac{r}{n-r+1}.$$
Then it is clear that $\frac{\binom{n}{r-1}}{\binom{n}{r}} < 1$ if and only if $r < \frac{n+1}{2}$.

   (i) If n is even, then we see from the above that $\binom{n}{r-1} < \binom{n}{r}$ if and only if $r \leq \frac{n}{2}$. This implies that $\binom{n}{i} < \binom{n}{j}$ if $0 \leq i < j \leq \frac{n}{2}$. Also, if $\frac{n}{2} \leq i < j \leq n$, then this implies that $0 \leq n-j < n-i \leq \frac{n}{2}$, and hence we have
   $$\binom{n}{i} = \binom{n}{n-i} > \binom{n}{n-j} = \binom{n}{j}.$$

   (ii) If n is odd, then we see from the above that $\binom{n}{r-1} < \binom{n}{r}$ if and only if $r \leq \frac{n-1}{2}$. This implies that $\binom{n}{i} < \binom{n}{j}$ if $0 \leq i < j \leq \frac{n-1}{2}$. Also, if $\frac{n+1}{2} \leq i < j \leq n$, then this implies that $0 \leq n-j < n-i \leq \frac{n-1}{2}$, and hence we have
   $$\binom{n}{i} = \binom{n}{n-i} > \binom{n}{n-j} = \binom{n}{j}.$$

19. (i) First Solution: We have
$$\begin{aligned}
\binom{2n}{n+1} + 2\binom{2n}{n} + \binom{2n}{n-1} &= \frac{2(2n)!}{(n+1)!(n-1)!} + \frac{2 \cdot (2n)!}{(n!)^2} \\
&= \frac{2(2n)!}{((n+1)!)^2}(n(n+1) + (n+1)^2) \\
&= \frac{(2n)!}{((n+1)!)^2}(n+n+1)(2(n+1)) \\
&= \frac{(2n)!}{((n+1)!)^2}(2n+1)(2n+2) \\
&= \frac{(2n+2)!}{((n+1)!)^2} \\
&= \binom{2(n+1)}{n+1}.
\end{aligned}$$
Second Solution: We have
$$\begin{aligned}
\binom{2(n+1)}{n+1} &= \binom{2n+1}{n} + \binom{2n+1}{n+1} \quad \text{(by (2.1.4))} \\
&= \binom{2n}{n-1} + \binom{2n}{n} + \binom{2n}{n} + \binom{2n}{n+1} \quad \text{(by (2.1.4))} \\
&= \binom{2n}{n+1} + 2\binom{2n}{n} + \binom{2n}{n-1}.
\end{aligned}$$

Third Solution: The identity holds trivially for the case where $n = 0$, so let us assume that $n > 0$. We shall count the number of ways to choose a team of $n + 1$ people from a group of $2(n + 1)$ people in two different ways. On one hand, the number of ways is equal to $\binom{2(n+1)}{n+1}$ by definition. On the other hand, let us fix persons A and B from the group of $2(n + 1)$ people, and let us consider these cases:

Case 1: Both A and B are chosen for the team. There are $\binom{2n}{n-1}$ ways to choose the remaining $n - 1$ people for the team out of the remaining $2n$ people in the group.

Case 2: Exactly one of A and B is chosen for the team. Then there are two ways to choose a person out of A and B, and then there are $\binom{2n}{n}$ ways to choose the remaining n people for the team out of the remaining $2n$ people in the group.

Case 3: Both A and B are not chosen for the team. Then there are $\binom{2n}{n+1}$ ways to choose the remaining $n + 1$ people for the team out of the remaining $2n$ people in the group.

By (AP), the total number of ways is equal to $\binom{2n}{n+1} + 2\binom{2n}{n} + \binom{2n}{n-1}$, so we must have

$$\binom{2(n + 1)}{n + 1} = \binom{2n}{n + 1} + 2\binom{2n}{n} + \binom{2n}{n - 1}$$

as desired.

(ii) First Solution: We have

$$\binom{n}{m - 1} + \binom{n - 1}{m} + \binom{n - 1}{m - 1}$$

$$= \frac{n!}{(m - 1)!(n - m + 1)!} + \frac{(n - 1)!}{m!(n - m - 1)!} + \frac{(n - 1)!}{(m - 1)!(n - m)!}$$

$$= \frac{(n - 1)!}{m!(n - m + 1)!}(nm + (n - m + 1)(n - m) + m(n - m + 1))$$

$$= \frac{(n - 1)!}{m!(n - m + 1)!}(n + 1)n$$

$$= \frac{(n + 1)!}{m!(n - m + 1)!}$$

$$= \binom{n + 1}{m}.$$

Second Solution: We have

$$\binom{n+1}{m} = \binom{n}{m-1} + \binom{n}{m} \quad \text{(by (2.1.4))}$$

$$= \binom{n}{m-1} + \binom{n-1}{m-1} + \binom{n-1}{m} \quad \text{(by (2.1.4))}.$$

Third Solution: Let us assume that $n > 0$. We shall count the number of ways to choose a team of m people from a group of $n+1$ people in two different ways. On one hand, the number of ways is equal to $\binom{n+1}{m}$ by definition. On the other hand, let us fix persons A and B from the group of $n+1$ people, and let us consider these cases:

Case 1: A is chosen for the team. Then there are $\binom{n}{m-1}$ ways to choose the remaining $m-1$ people for the team out of the remaining n people in the group.

Case 2: A is not chosen for the team, but B is chosen for the team. Then there are $\binom{n-1}{m-1}$ ways to choose the remaining $m-1$ people for the team out of the remaining $n-1$ people in the group.

Case 3: Both A and B are not chosen for the team. Then there are $\binom{n-1}{m}$ ways to choose the remaining m people for the team out of the remaining $n+1$ people in the group.

By (AP), the total number of ways is equal to $\binom{n}{m-1} + \binom{n-1}{m-1} + \binom{n-1}{m}$, so we must have

$$\binom{n+1}{m} = \binom{n}{m-1} + \binom{n-1}{m-1} + \binom{n-1}{m}$$

as desired.

20. Let us count the number of ways to choose m people from a set of n people to form a committee consisting of r normal members and $m-r$ executive members, in 2 different ways. On one hand, we can first choose m people for the committee in $\binom{n}{m}$ ways, and choose r normal members from the m committee members in $\binom{m}{r}$ ways.

On the other hand, we can first choose r normal members for the committee in $\binom{n}{r}$ ways, and choose $m-r$ executive members for the committee in $\binom{n-r}{m-r}$ ways. Hence, we have

$$\binom{n}{m}\binom{m}{r} = \binom{n}{r}\binom{n-r}{m-r}$$

as desired.

21. We have

$$\sum_{r=0}^{n} \frac{(2n)!}{(r!)^2((n-r)!)^2} = \sum_{r=0}^{n} \left(\frac{(2n)!}{(n!)^2} \cdot \frac{(n!)^2}{(r!)^2((n-r)!)^2} \right)$$

$$= \binom{2n}{n} \sum_{r=0}^{n} \binom{n}{r}^2$$

$$= \binom{2n}{n}\binom{2n}{n} \quad \text{(by (2.3.6))}$$

$$= \binom{2n}{n}^2.$$

22. Let us find the coefficient of x^{2m} in the expansion of $(1 - x^2)^n = (1 + x)^n(1 - x)^n$ in two different ways. On one hand, we see that the coefficient of x^{2m} in the expansion of $(1 - x^2)^n$ is equal to $(-1)^m\binom{n}{m}$. On the other hand, we see that for each $i \in \{0, 1, \ldots, 2m\}$, we have the coefficient of x^i in the expansion of $(1 - x)^n$ to be equal to $(-1)^i\binom{n}{i}$, and we have the coefficient of x^{2m-i} in the expansion of $(1 + x)^n$ to be equal to $\binom{n}{2m-i}$. By (AP), the coefficient of x^{2m} in the expansion of $(1 + x)^n(1 - x)^n$ is equal to $\sum_{i=0}^{2m}(-1)^i\binom{n}{i}\binom{n}{2m-i}$, so we must have

$$\sum_{i=0}^{2m}(-1)^i\binom{n}{i}\binom{n}{2m-i} = (-1)^m\binom{n}{m}. \tag{2.1}$$

Next, let us compute the coefficient of x^{2m+1} in the binomial expansion of $(1 - x^2)^n = (1 + x)^n(1 - x)^n$ in two different ways. On one hand, we see that the coefficient of x^{2m+1} in the expansion of $(1 - x^2)^n$ is equal to 0. On the other hand, we see that for each $i \in \{0, 1, \ldots, 2m + 1\}$, we have the coefficient of x^i in the expansion of $(1 - x)^n$ to be equal to $(-1)^i\binom{n}{i}$, and we have the coefficient of x^{2m+1-i} in the expansion of $(1 + x)^n$ to be equal to $\binom{n}{2m+1-i}$. By (AP), the coefficient of x^{2m} in the expansion of $(1 + x)^n(1 - x)^n$ is equal to $\sum_{i=0}^{2m+1}(-1)^i\binom{n}{i}\binom{n}{2m+1-i}$, so we must have

$$\sum_{i=0}^{2m+1}(-1)^i\binom{n}{i}\binom{n}{2m+1-i} = 0.$$

When n is even, we have $n = 2k$ for some non-negative integer k. As

$k \leq n$, we have

$$\sum_{i=0}^{n}(-1)^i \binom{n}{i}^2 = \sum_{i=0}^{n}(-1)^i \binom{n}{i}\binom{n}{n-i}$$

$$= \sum_{i=0}^{2k}(-1)^i \binom{n}{i}\binom{n}{2k-i}$$

$$= (-1)^k \binom{n}{k} \quad \text{(by equation (2.1))}$$

$$= (-1)^{\frac{n}{2}} \binom{n}{\frac{n}{2}}.$$

When n is odd, we have $n = 2k + 1$ for some non-negative integer k. As $k \leq n$, we have

$$\sum_{i=0}^{n}(-1)^i \binom{n}{i}^2 = \sum_{i=0}^{n}(-1)^i \binom{n}{i}\binom{n}{n-i}$$

$$= \sum_{i=0}^{2k+1}(-1)^i \binom{n}{i}\binom{n}{2k+1-i}$$

$$= 0.$$

23. We have

$$S = m! + \frac{(m+1)!}{1!} + \frac{(m+2)!}{2!} + \cdots + \frac{(m+n)!}{n!}$$

$$= \sum_{i=0}^{n} \frac{(m+i)!}{i!}$$

$$= m! \sum_{i=0}^{n} \frac{(m+i)!}{i!m!}$$

$$= m! \sum_{i=0}^{n} \binom{m+i}{m}$$

$$= m! \binom{m+n+1}{m+1} \quad \text{(by (2.5.1))}$$

$$= m! \binom{m+n+1}{n}.$$

24. By setting $x = 1$ and $y = 3$ in the Binomial Theorem, we get

$$\sum_{r=0}^{n} 3^r \binom{n}{r} = (1+3)^n = 4^n.$$

25. We have

$$\sum_{r=0}^{n}(r+1)\binom{n}{r} = \sum_{r=0}^{n}r\binom{n}{r} + \sum_{r=0}^{n}\binom{n}{r}$$

$$= n\cdot 2^{n-1} + 2^n \quad \text{(by (2.3.1) and (2.3.4))}$$

$$= (n+2)2^{n-1}.$$

26. We have

$$\sum_{r=0}^{n}\frac{1}{r+1}\binom{n}{r} = \sum_{r=0}^{n}\frac{1}{n+1}\binom{n+1}{r+1} \quad \text{(by (2.1.2))}$$

$$= \frac{1}{n+1}\left(\sum_{r=0}^{n+1}\binom{n+1}{r} - \sum_{r=0}^{0}\binom{n+1}{r}\right)$$

$$= \frac{1}{n+1}(2^{n+1}-1) \quad \text{(by (2.3.1)).}$$

27. We have

$$\sum_{r=0}^{n}\frac{(-1)^r}{r+1}\binom{n}{r} = -\sum_{r=0}^{n}\frac{(-1)^{r+1}}{n+1}\binom{n+1}{r+1} \quad \text{(by (2.1.2))}$$

$$= -\frac{1}{n+1}\left(\sum_{r=0}^{n+1}(-1)^r\binom{n+1}{r} - \sum_{r=0}^{0}(-1)^r\binom{n+1}{r}\right)$$

$$= -\frac{1}{n+1}(0-1) \quad \text{(by (2.3.2))}$$

$$= \frac{1}{n+1}.$$

28. We have

$$\sum_{r=m}^{n}\binom{n}{r}\binom{r}{m} = \sum_{r=m}^{n}\binom{n}{m}\binom{n-m}{r-m} \quad \text{(by (2.1.5))}$$

$$= \binom{n}{m}\sum_{r=m}^{n}\binom{n-m}{r-m}$$

$$= \binom{n}{m}\sum_{r=0}^{n-m}\binom{n-m}{r}$$

$$= \binom{n}{m}2^{n-m} \quad \text{(by (2.3.1)).}$$

29. The result follows immediately from (2.3.2) when $m = n$, so let us

assume that $m < n$. Then we have

$$\sum_{r=0}^{m}(-1)^r\binom{n}{r}$$

$$= \binom{n}{0} + \sum_{r=1}^{m}(-1)^r\left(\binom{n-1}{r} + \binom{n-1}{r-1}\right) \quad \text{(by (2.1.4))}$$

$$= \binom{n}{0} - \left(\binom{n-1}{1} + \binom{n-1}{0}\right) + \left(\binom{n-1}{2} + \binom{n-1}{1}\right)$$

$$+ \cdots + (-1)^m\left(\binom{n-1}{m} + \binom{n-1}{m-1}\right)$$

$$= (-1)^m\binom{n-1}{m}.$$

30. We have

$$\sum_{r=0}^{m}(-1)^{m-r}\binom{n}{r} = (-1)^m\sum_{r=0}^{m}(-1)^{-r}\binom{n}{r}$$

$$= (-1)^m\sum_{r=0}^{m}(-1)^r\binom{n}{r}$$

$$= (-1)^m \cdot (-1)^m\binom{n-1}{m} \quad \text{(by Problem 2.29)}$$

$$= \binom{n-1}{m}.$$

31. We have

$$\sum_{r=0}^{n}(-1)^r r\binom{n}{r} = \sum_{r=1}^{n}(-1)^r n\binom{n-1}{r-1} \quad \text{(by (2.1.2))}$$

$$= -n\sum_{r=0}^{n-1}(-1)^r\binom{n-1}{r}$$

$$= 0 \quad \text{(by (2.3.2))}.$$

32. Since $\binom{2n-1}{r} = \binom{2n-1}{2n-1-r}$ for all $r \in \{0, 1, \ldots, n-1\}$, we have

$$\sum_{r=0}^{n-1}\binom{2n-1}{r} = \sum_{r=0}^{n-1}\binom{2n-1}{2n-1-r} = \sum_{r=n}^{2n-1}\binom{2n-1}{r}.$$

By (2.3.1), we have

$$2^{2n-1} = \sum_{r=0}^{2n-1}\binom{2n-1}{r}$$

$$= \sum_{r=0}^{n-1} \binom{2n-1}{r} + \sum_{r=n}^{2n-1} \binom{2n-1}{r}$$

$$= 2 \sum_{r=0}^{n-1} \binom{2n-1}{r},$$

so by the above equation, we have

$$\sum_{r=0}^{n-1} \binom{2n-1}{r} = \frac{1}{2} \cdot 2^{2n-1} = 2^{2n-2}.$$

33. Since $\binom{2n}{r} = \binom{2n}{2n-r}$ for all $r \in \{0, 1, \ldots, n-1\}$, we have

$$\sum_{r=0}^{n-1} \binom{2n}{r} = \sum_{r=0}^{n-1} \binom{2n}{2n-r} = \sum_{r=n+1}^{2n} \binom{2n}{r}.$$

By (2.3.1), we have

$$2^{2n} = \sum_{r=0}^{2n} \binom{2n}{r}$$

$$= \sum_{r=0}^{n-1} \binom{2n}{r} + \sum_{r=n+1}^{2n} \binom{2n}{r} + \binom{2n}{n}$$

$$= 2 \sum_{r=0}^{n-1} \binom{2n}{r} + \binom{2n}{n},$$

so by the previous equation, we have

$$\sum_{r=0}^{n} \binom{2n}{r} = \sum_{r=0}^{n-1} \binom{2n}{r} + \binom{2n}{n}$$

$$= \frac{1}{2} \left[2^{2n} - \binom{2n}{n} \right] + \binom{2n}{n}$$

$$= \frac{1}{2} \left[2^{2n} + \binom{2n}{n} \right].$$

34. We have

$$\sum_{r=0}^{n} r \binom{2n}{r} = \sum_{r=1}^{n} 2n \binom{2n-1}{r-1} \quad \text{(by (2.1.2))}$$

$$= 2n \sum_{r=1}^{n} \binom{2n-1}{r-1}$$

$$= 2n \sum_{r=0}^{n-1} \binom{2n-1}{r}$$

$$= 2n \cdot 2^{2n-2} \quad \text{(by Problem 2.32)}$$

$$= n \cdot 2^{2n-1}.$$

35. By (2.5.1), we have

$$\sum_{r=k}^{m}\binom{n+r}{n} = \sum_{r=0}^{m}\binom{n+r}{n} - \sum_{r=0}^{k-1}\binom{n+r}{n} = \binom{n+m+1}{n+1} - \binom{n+k}{n+1}.$$

36. By (2.3.5), we have

$$\sum_{r=1}^{n}\binom{n}{r}\binom{n-1}{r-1} = \sum_{r=0}^{n-1}\binom{n}{r+1}\binom{n-1}{r}$$

$$= \sum_{r=0}^{n-1}\binom{n}{n-r-1}\binom{n-1}{r}$$

$$= \binom{2n-1}{n-1}.$$

37. If $m = n$, then the sum is equal to $(-1)^m \binom{n}{m}\binom{m}{m} = (-1)^m$. Else, if $m < n$, then we have

$$\sum_{r=m}^{n}(-1)^r\binom{n}{r}\binom{r}{m} = \sum_{r=m}^{n}(-1)^r\binom{n}{m}\binom{n-m}{r-m} \quad \text{(by (2.1.5))}$$

$$= \binom{n}{m}\sum_{r=m}^{n}(-1)^r\binom{n-m}{r-m}$$

$$= \binom{n}{m}\sum_{r=0}^{n-m}(-1)^{r+m}\binom{n-m}{r}$$

$$= (-1)^m\binom{n}{m}\sum_{r=0}^{n-m}(-1)^r\binom{n-m}{r}$$

$$= 0 \quad \text{(by (2.3.2))}.$$

38. We have

$$\sum_{r=1}^{n-1}(n-r)^2\binom{n-1}{n-r}$$

$$= \sum_{k=1}^{n-1}k^2\binom{n-1}{k}$$

$$= \sum_{k=1}^{n-1}k(n-1)\binom{n-2}{k-1} \quad \text{(by (2.1.2))}$$

$$= (n-1)\left[\sum_{k=1}^{n-1}(k-1)\binom{n-2}{k-1} + \sum_{k=1}^{n-1}\binom{n-2}{k-1}\right]$$

$$= (n-1) \left[\sum_{k=2}^{n-1} (k-1) \binom{n-2}{k-1} + \sum_{k=0}^{n-2} \binom{n-2}{k} \right]$$

$$= (n-1) \left[\sum_{k=2}^{n-1} (n-2) \binom{n-3}{k-2} + \sum_{k=0}^{n-2} \binom{n-2}{k} \right] \quad \text{(by (2.1.2))}$$

$$= (n-1) \left[(n-2) \sum_{k=2}^{n-1} \binom{n-3}{k-2} + 2^{n-2} \right] \quad \text{(by (2.3.2))}$$

$$= (n-1) \left[(n-2) \sum_{k=0}^{n-3} \binom{n-3}{k} + 2^{n-2} \right]$$

$$= (n-1) \left[(n-2)2^{n-3} + 2^{n-2} \right] \quad \text{(by (2.3.2))}$$

$$= n(n-1)2^{n-3}.$$

39. Since $\binom{2n}{r} = \binom{2n}{2n-r}$ for all $r \in \{0, 1, \ldots, n-1\}$, we have

$$\sum_{r=0}^{n-1} \binom{2n}{r}^2 = \sum_{r=0}^{n-1} \binom{2n}{2n-r}^2 = \sum_{r=n+1}^{2n} \binom{2n}{r}^2.$$

By (2.3.6), we have

$$\binom{4n}{2n} = \sum_{r=0}^{2n} \binom{2n}{r}^2$$

$$= \sum_{r=0}^{n-1} \binom{2n}{r}^2 + \sum_{r=n+1}^{2n} \binom{2n}{r}^2 + \binom{2n}{n}^2$$

$$= 2 \sum_{r=0}^{n-1} \binom{2n}{r}^2 + \binom{2n}{n}^2$$

$$= 2 \sum_{r=0}^{n} \binom{2n}{r}^2 - \binom{2n}{n}^2.$$

Hence, we have

$$\sum_{r=0}^{n} \binom{2n}{r}^2 = \frac{1}{2} \left[\binom{4n}{2n} + \binom{2n}{n}^2 \right].$$

40. We have

$$\sum_{r=0}^{n} \binom{n}{r}^2 \binom{r}{n-k} = \sum_{r=0}^{n} \binom{n}{r} \binom{n}{n-k} \binom{k}{r-n+k} \quad \text{(by (2.1.5))}$$

$$= \binom{n}{n-k} \sum_{r=0}^{n} \binom{n}{r}\binom{k}{n-r}$$

$$= \binom{n}{k}\binom{n+k}{n} \quad \text{(by (2.3.5))}.$$

41. We have

$$\sum_{r=0}^{n} \binom{n}{r}\binom{m+r}{n}$$

$$= \sum_{r=0}^{n} \binom{n}{r} \sum_{s=0}^{n} \binom{m}{s}\binom{r}{n-s} \quad \text{(by (2.3.5))}$$

$$= \sum_{s=0}^{n} \binom{m}{s} \sum_{r=0}^{n} \binom{n}{r}\binom{r}{n-s}$$

$$= \sum_{s=0}^{n} \binom{m}{s}\binom{n}{s} \sum_{r=0}^{n} \binom{s}{r-n+s} \quad \text{(by (2.1.5))}$$

$$= \sum_{s=0}^{n} \binom{m}{s}\binom{n}{s} \sum_{r=n-s}^{n} \binom{s}{r-n+s}$$

(since $r - n + s < 0$ for all $r < n - s$)

$$= \sum_{s=0}^{n} \binom{m}{s}\binom{n}{s} \sum_{t=0}^{s} \binom{s}{t} \quad \text{(by setting } t = r - n + s)$$

$$= \sum_{s=0}^{n} \binom{m}{s}\binom{n}{s} 2^s \quad \text{(by (2.3.1))}.$$

42. We have

$$\sum_{r=0}^{m} \binom{m}{r}\binom{n}{r}\binom{p+r}{m+n}$$

$$= \sum_{r=0}^{m} \binom{m}{r}\binom{n}{r} \sum_{s=0}^{m+n} \binom{p}{m+n-s}\binom{r}{s} \quad \text{(by (2.3.5))}$$

$$= \sum_{r=0}^{m} \sum_{s=0}^{m+n} \binom{n}{r}\binom{p}{m+n-s}\binom{m}{s}\binom{m-s}{r-s} \quad \text{(by (2.1.5))}$$

$$= \sum_{s=0}^{m+n} \binom{p}{m+n-s}\binom{m}{s} \sum_{r=0}^{m} \binom{n}{r}\binom{m-s}{m-r}$$

$$= \sum_{s=0}^{m+n} \binom{p}{m+n-s}\binom{m}{s}\binom{n+m-s}{m} \quad \text{(by (2.3.5))}$$

$$= \sum_{s=0}^{m+n} \binom{m}{s}\binom{p}{m}\binom{p-m}{n-s} \quad \text{(by (2.1.5))}$$

$$= \binom{p}{m} \sum_{s=0}^{n} \binom{m}{s}\binom{p-m}{n-s} \quad \text{(since } n-s<0 \text{ for all } s>n\text{)}$$

$$= \binom{p}{m}\binom{p}{n} \quad \text{(by (2.3.5))}.$$

43. We have

$$\sum_{r=0}^{m} \binom{m}{r}\binom{n}{r}\binom{p+m+n-r}{m+n}$$

$$= \sum_{r=0}^{m} \binom{m}{r}\binom{n}{r} \sum_{s=0}^{m+n} \binom{p+m}{m+n-s}\binom{n-r}{s} \quad \text{(by (2.3.5))}$$

$$= \sum_{r=0}^{m} \sum_{s=0}^{m+n} \binom{m}{r}\binom{p+m}{m+n-s}\binom{n}{n-r}\binom{n-r}{s}$$

$$= \sum_{r=0}^{m} \sum_{s=0}^{m+n} \binom{m}{r}\binom{p+m}{m+n-s}\binom{n}{s}\binom{n-s}{n-r-s} \quad \text{(by (2.1.5))}$$

$$= \sum_{s=0}^{m+n} \binom{p+m}{m+n-s}\binom{n}{s} \sum_{r=0}^{m} \binom{m}{m-r}\binom{n-s}{r}$$

$$= \sum_{s=0}^{m+n} \binom{p+m}{m+n-s}\binom{n}{s}\binom{m+n-s}{m} \quad \text{(by (2.3.5))}$$

$$= \sum_{s=0}^{m+n} \binom{n}{s}\binom{p+m}{m}\binom{p}{n-s} \quad \text{(by (2.1.5))}$$

$$= \binom{p+m}{m} \sum_{s=0}^{n} \binom{n}{s}\binom{p}{n-s} \quad \text{(since } n-s<0 \text{ for all } s>n\text{)}$$

$$= \binom{p+m}{m}\binom{p+n}{n} \quad \text{(by (2.3.5))}.$$

44. (i) Let ℓ denote the straight line in the $x-y$ plane whose endpoints are $(0,n)$ and $(n,0)$. Let us count the number of shortest routes from the origin O to any lattice point on the line ℓ.

On one hand, we note that any lattice point on the line ℓ must be of the form $(r, n-r)$ where r is an integer and $0 \le r \le n$. Furthermore, the number of shortest routes from the origin O to

$(r, n - r)$ for any integer r such that $0 \le r \le n$ must be equal to $\binom{n}{r}$. This implies that the number of such routes is equal to $\sum_{r=0}^{n} \binom{n}{r}$.

On the other hand, by using a '0' to denote a path from (x, y) to $(x + 1, y)$ and '1' to denote a path from (x, y) to $(x, y + 1)$ for any integers x and y, we see that any shortest route from the origin O to any lattice point on the line ℓ can be uniquely represented as a binary sequence of length n. Thus the number of such routes is equal to 2^n, and therefore we have

$$\sum_{r=0}^{n} \binom{n}{r} = 2^n$$

as desired.

(ii) Let us count the number of shortest routes from O to $(n, r + 1)$ in two different ways. On one hand, the number of such routes is equal to $\binom{n+r+1}{r+1}$ by definition. Every shortest route from O to $(n, r + 1)$ must pass through exactly one of the segments joining (k, r) to $(k, r + 1)$ for $k = 0, 1, \ldots, n$. We note that there is only one route to go from $(k, r + 1)$ to $(n, r + 1)$. For any integer k such that $0 \le k \le n$, the number of such routes passing through both (k, r) and $(k, r + 1)$ is equal to $\binom{k+r}{r}$ by definition. By (AP), the number of such routes is equal to $\sum_{k=0}^{n} \binom{k+r}{r}$, and therefore we have

$$\sum_{k=0}^{n} \binom{k+r}{r} = \binom{n+r+1}{r+1}$$

as desired.

45. Let us count the number of shortest routes from O to $(p - q + r + 1, q)$ in two different ways. On one hand, the number of such routes is equal to $\binom{p+r+1}{q}$ by definition. On the other hand, for any integer i such that $0 \le i \le q$, the number of such routes passing through both (r, i) and $(r + 1, i)$ is equal to $\binom{r+i}{i}\binom{p-i}{q-i}$ by definition. We note that every shortest route from O to $(p - q + r + 1, q)$ must pass through exactly one of the segments joining (r, i) and $(r + 1, i)$ for some $i = 0, 1, \ldots, q$. By (MP) and (AP), the number of such routes is equal to $\sum_{i=0}^{q} \binom{r+i}{i}\binom{p-i}{q-i}$, and therefore we have

$$\sum_{i=0}^{q} \binom{r+i}{i}\binom{p-i}{q-i} = \binom{p+r+1}{q}$$

as desired.

46. We shall count the number of ways to choose a team consisting of at least one member, along with a team leader, from a group of n people in two different ways. On one hand, for any integer r such that $1 \le r \le n$, there are $\binom{n}{r}$ ways to choose a team consisting of r members, and subsequently there are r ways to choose a team leader out of the r members of the team. By (MP) and (AP), this implies that there are a total of $\sum_{r=1}^{n} r\binom{n}{r}$ ways to choose a team consisting of at least one member, along with a team leader, from a group of n people. On the other hand, there are n ways to first choose a leader for the team, and subsequently there are $\sum_{s=0}^{n-1} \binom{n-1}{s} = 2^{n-1}$ ways to choose the remaining s members of the team out of the remaining $n-1$ people. By (MP) again, we must have

$$\sum_{r=1}^{n} r\binom{n}{r} = n \cdot 2^{n-1}$$

as desired.

47. (i) The equation clearly holds for $n = 1$, so let's us assume that $n > 1$. We have

$$\sum_{r=1}^{n} r^2 \binom{n}{r} = \sum_{r=1}^{n} rn \binom{n-1}{r-1} \quad \text{(by (2.1.2))}$$

$$= n \sum_{r=1}^{n} (r-1+1)\binom{n-1}{r-1}$$

$$= n \left(\sum_{r=2}^{n} (r-1)\binom{n-1}{r-1} + \sum_{r=1}^{n} \binom{n-1}{r-1} \right)$$

$$= n \left(\sum_{r=1}^{n-1} r\binom{n-1}{r} + \sum_{r=0}^{n-1} \binom{n-1}{r} \right)$$

$$= n \left((n-1)2^{n-2} + 2^{n-1} \right) \quad \text{(by (2.3.1) and (2.3.4))}$$

$$= n(n+1)2^{n-2}.$$

(ii) The equation clearly holds for $n = 1$, so let's us assume that $n > 1$. We have

$$\sum_{r=1}^{n} r^3 \binom{n}{r}$$

$$= \sum_{r=1}^{n} r^2 n \binom{n-1}{r-1} \quad \text{(by (2.1.2))}$$

$$= n \sum_{r=1}^{n} ((r-1)^2 + 2(r-1) + 1) \binom{n-1}{r-1}$$

$$= n \left(\sum_{r=2}^{n} (r^2 - 1) \binom{n-1}{r-1} + \sum_{r=0}^{n-1} \binom{n-1}{r} \right)$$

$$= n \left(\sum_{r=1}^{n-1} (r^2 + 2r) \binom{n-1}{r} + 2^{n-1} \right) \quad \text{(by (2.3.1))}$$

$$= n \left(n(n-1)2^{n-3} + 2(n-1)2^{n-2} + 2^{n-1} \right)$$

(by part (i) and (2.3.4))

$$= n^2 (n+3) 2^{n-3}.$$

(iii) The equation clearly holds for $n = 1$, so let us assume that $n > 1$. We have

$$\sum_{r=1}^{n} r^4 \binom{n}{r}$$

$$= \sum_{r=1}^{n} r^3 n \binom{n-1}{r-1} \quad \text{(by (2.1.2))}$$

$$= n \left(\sum_{r=1}^{n} ((r-1)^3 + 3(r-1)^2 + 3(r-1) + 1) \binom{n-1}{r-1} \right)$$

$$= n \left(\sum_{r=2}^{n} (r^3 - 1) \binom{n-1}{r-1} + \sum_{r=0}^{n-1} \binom{n-1}{r} \right)$$

$$= n \left(\sum_{r=1}^{n-1} (r^3 + 3r^2 + 3r) \binom{n-1}{r} + 2^{n-1} \right)$$

(by (2.3.1))

$$= n(n-1) \left((n-1)(n+2)2^{n-4} + 3n2^{n-3} + 3 \cdot 2^{n-2} \right) + n2^{n-1}$$

(by parts (i) and (ii) and (2.3.1))

$$= n(n+1)(n^2 + 5n - 2)2^{n-4}.$$

48. (i) We have

$$r^k = [(r-1) + 1]^k = \sum_{i=0}^{k} \binom{k}{i} (r-1)^{k-i} 1^i = \sum_{i=0}^{k} \binom{k}{i} (r-1)^{k-i}.$$

(ii) We observe that we have $R(n,k) = \sum_{r=0}^{n} r^k \binom{n}{r}$ for all non-negative integers n, k, noting that we have $0^0 := 1$ by definition. Then for all positive integers n, k, we have

$$R(n,k) = \sum_{m=0}^{n} m^k \binom{n}{m}$$

$$= \sum_{m=1}^{n} m^k \binom{n}{m}$$

$$= \sum_{r=0}^{n-1} \binom{n}{r+1} (1+r)^k \quad \text{(by setting } r = m-1)$$

$$= \sum_{r=0}^{n-1} \frac{n}{r+1} \binom{n-1}{r} (1+r)^k \quad \text{(by (2.1.2))}$$

$$= n \cdot \sum_{r=0}^{n-1} \binom{n-1}{r} (1+r)^{k-1}$$

$$= n \cdot \sum_{r=0}^{n-1} \binom{n-1}{r} \sum_{j=0}^{k-1} \binom{k-1}{j} r^j$$

$$= n \cdot \sum_{j=0}^{k-1} \binom{k-1}{j} \sum_{r=0}^{n-1} \binom{n-1}{r} r^j$$

$$= n \cdot \sum_{j=0}^{k-1} \binom{k-1}{j} R(n-1, j)$$

as desired.

49. We have

$$\sum_{r=1}^{n} \frac{1}{r} \binom{n}{r} = \sum_{r=1}^{n} \sum_{k=r-1}^{n-1} \frac{1}{r} \binom{k}{r-1} \quad \text{(by (2.5.1))}$$

$$= \sum_{r=1}^{n} \sum_{k=r-1}^{n-1} \frac{1}{k+1} \binom{k+1}{r} \quad \text{(by (2.1.2))}$$

$$= \sum_{r=1}^{n} \sum_{k=r}^{n} \frac{1}{k} \binom{k}{r}$$

$$= \sum_{k=1}^{n} \sum_{r=1}^{k} \frac{1}{k} \binom{k}{r} \quad \text{(since } 1 \le r \le k \le n)$$

$$= \sum_{k=1}^{n} \frac{1}{k} \sum_{r=1}^{k} \binom{k}{r}$$

$$= \sum_{k=1}^{n} \frac{1}{k} \left(\sum_{r=0}^{k} \binom{k}{r} - \binom{k}{0} \right)$$

$$= \sum_{k=1}^{n} \frac{1}{k} \left(2^k - 1 \right) \quad \text{(by (2.3.1))}.$$

50. **First Solution:** We have

$$\sum_{r=1}^{n} r \binom{n}{r}^2 = \sum_{r=1}^{n} r \binom{n}{r} \binom{n}{r}$$

$$= \sum_{r=1}^{n} n \binom{n-1}{r-1} \binom{n}{r} \quad \text{(by (2.1.2))}$$

$$= n \cdot \sum_{r=0}^{n-1} \binom{n-1}{r} \binom{n}{r+1}$$

$$= n \cdot \sum_{r=0}^{n-1} \binom{n-1}{r} \binom{n}{n-1-r}$$

$$= n \binom{2n-1}{n-1} \quad \text{(by (2.3.5))}.$$

Second Solution: Let us count the number of ways to choose a team of n members, inclusive of a team leader, out of a group of n men and n women, such that the team leader is a male. On one hand, there are n ways to choose a male team leader, and $\binom{2n-1}{n-1}$ ways to choose the remaining $n-1$ team members out of the remaining $2n-1$ people. By (MP), the total number of such ways is equal to $n\binom{2n-1}{n-1}$. On the other hand, suppose that the team comprises r men and $n-r$ women, where $1 \le r \le n$. There are $\binom{n}{r}$ ways to choose r men for the team, $\binom{n}{n-r} = \binom{n}{r}$ ways to choose $n-r$ women for the team, and r ways to choose a male team leader. By (MP) and (AP), the total number of ways is equal to $\sum_{r=1}^{n} \binom{n}{r} \cdot \binom{n}{r} \cdot r = \sum_{r=1}^{n} r \binom{n}{r}^2$. Thus we must have

$$\sum_{r=1}^{n} r \binom{n}{r}^2 = n \binom{2n-1}{n-1}$$

as desired.

51. By (2.1.2), we have $\binom{p}{r} = \frac{p}{r}\binom{p-1}{r-1}$, or equivalently, $r\binom{p}{r} = p\binom{p-1}{r-1}$. As $1 \leq r \leq p-1$, we must have $(p,r) = 1$, and hence r must divide $\binom{p-1}{r-1}$. Therefore, $\frac{1}{r}\binom{p-1}{r-1}$ is necessarily an integer, which implies that $\binom{p}{r} = p \cdot \frac{1}{r}\binom{p-1}{r-1}$ is a multiple of p, or equivalently, $\binom{p}{r} \equiv 0 \pmod{p}$ as desired. Consequently, we have

$$(1+x)^p - (1+x^p) = \sum_{r=0}^{p}\binom{p}{r}x^r - \binom{p}{0} - \binom{p}{p}x^p$$

$$= \sum_{r=1}^{p-1}\binom{p}{r}x^r$$

$$\equiv 0 \pmod{p},$$

or equivalently, $(1+x)^p \equiv (1+x^p) \pmod{p}$ as desired.

Remark. Let $f(x) = a_n x^n + a_{n-1}x^{n-1} + \cdots + a_0$ and $g(x) = b_n x^n + b_{n-1}x^{n-1} + \cdots + b_0$ be two polynomials with integer coefficients. We say that $f(x)$ and $g(x)$ are congruent to each other modulo p (and write $f(x) \equiv g(x) \pmod{p}$) if p divides $a_i - b_i$ for all $i = 0, 1, \ldots, n$. Equivalently, $f(x) \equiv g(x) \pmod{p}$ if there exists a polynomial $h(x)$ with integer coefficients, such that $f(x) - g(x) = ph(x)$. We note that this congruence relation is an equivalence relation, and this definition extends easily for polynomials in more than one variable.

52. By (2.3.6), we have $\binom{2p}{p} = \sum_{r=0}^{p}\binom{p}{r}^2$, and by the previous problem, we have

$$\binom{2p}{p} - 2 = \sum_{r=0}^{p}\binom{p}{r}^2 - \binom{p}{0}^2 - \binom{p}{p}^2 = \sum_{r=1}^{p-1}\binom{p}{r}^2 \equiv 0 \pmod{p},$$

or equivalently, $\binom{2p}{p} \equiv 2 \pmod{p}$ as desired.

53. (i) Let $f(\mathbf{N}_n) = X$. Firstly, there are $S(n,p)$ p-partitions of $\mathbf{N}_n$ by definition. Next, there are $\binom{m}{p}$ choices for X such that $|X| = p$, and subsequently there are $p!$ ways to assign a unique integer $n_A \in X$ to each subset A of $\mathcal{P}$ for each p-partition $\mathcal{P}$ of $\mathbf{N}_n$, such that $f(x) = n_A$ for all $x \in A$, and $f(y) \neq n_A$ for all $y \notin A$. By (MP), the number of mappings $f \colon \mathbf{N}_n \to \mathbf{N}_m$ such that $|f(\mathbf{N}_n)| = p$ is equal to $\binom{m}{p} \cdot p! \cdot S(n,p)$.

(ii) Let $f(\mathbf{N}_n) = X$ and $|X| = k$. As $\mathbf{N}_p \subseteq X \subseteq \mathbf{N}_m$ we must have $p \leq k \leq m$. Firstly, there are $S(n,k)$ k-partitions of $\mathbf{N}_n$ by definition. Next, there are $\binom{m-p}{k-p}$ choices for X such that $\mathbf{N}_p \subseteq X$, and subsequently there are $k!$ ways to assign a unique integer

$n_A \in X$ to each subset A of $\mathcal{P}$ for each k-partition $\mathcal{P}$ of $\mathbf{N}_n$, such that $f(x) = n_A$ for all $x \in A$, and $f(y) \neq n_A$ for all $y \notin A$. By (MP) and (AP), the number of mappings $f : \mathbf{N}_n \to \mathbf{N}_m$ such that $\mathbf{N}_p \subseteq f(\mathbf{N}_n)$ is equal to $\sum_{k=p}^{m} \binom{m-p}{k-p} \cdot k! \cdot S(n, k)$.

54. (a) By (2.1.3), we have

$$H_r^n = \binom{r+n-1}{r} = \frac{(r+n-1) - r + 1}{r} \binom{r+n-1}{r-1} = \frac{n}{r} H_{r-1}^{n+1}.$$

(b) By (2.1.2), we have

$$H_r^n = \binom{r+n-1}{r} = \frac{r+n-1}{r} \binom{r+n-2}{r-1} = \frac{n+r-1}{r} H_{r-1}^n.$$

(c) By (2.1.4), we have

$$H_r^n = \binom{r+n-1}{r} = \binom{r+n-2}{r} + \binom{r+n-2}{r-1} = H_r^{n-1} + H_{r-1}^n.$$

(d) By (2.5.2), we have

$$\sum_{k=0}^{r} H_k^n = \sum_{k=0}^{r} \binom{k+n-1}{k} = \binom{r+n}{r} = H_r^{n+1}.$$

(e) By parts (a) and (d), we have

$$\sum_{k=1}^{r} k H_k^n = \sum_{k=1}^{r} n H_{k-1}^{n+1} = n \sum_{k=0}^{r-1} H_k^{n+1} = n H_{r-1}^{n+2}.$$

(f) Let us count the number of shortest routes from O to $(m+n-1, r)$ in two different ways. On one hand, the number of such routes is equal to $\binom{r+m+n-1}{r} = H_r^{m+n}$ by definition. On the other hand, for any integer k such that $0 \leq k \leq r$, the number of such routes passing through both $(m-1, k)$ and (m, k) is equal to $\binom{m+k-1}{k}\binom{n-1+r-k}{r-k} = H_k^m H_{r-k}^n$ by definition. By (MP) and (AP), the number of such routes is equal to $\sum_{k=0}^{r} H_k^m H_{r-k}^n$, and therefore we have

$$\sum_{k=0}^{r} H_k^m H_{r-k}^n = H_r^{m+n}$$

as desired.

55. (i) It is easy to see that $d_{\min}(n) \geq 0$ for all $n \geq 2$ by definition. By (2.1.3), we have for all $k \in \{1, 2, \ldots, n\}$ that

$$d_k(n) = \left| \binom{n}{k} - \binom{n}{k-1} \right|$$
$$= \left| \frac{n-k+1}{k} \binom{n}{k-1} - \binom{n}{k-1} \right|$$
$$= \left| \frac{n-2k+1}{k} \right| \binom{n}{k-1}.$$

This implies that

$$d_{\min}(n) = 0$$
$$\Leftrightarrow d_k(n) = 0 \text{ for some } k \in \{1, 2, \ldots, n\}$$
$$\Leftrightarrow \left| \frac{n-2k+1}{k} \right| \binom{n}{k-1} = 0 \text{ for some } k \in \{1, 2, \ldots, n\}$$
$$\Leftrightarrow \frac{n-2k+1}{k} = 0 \text{ for some } k \in \{1, 2, \ldots, n\}$$
$$\Leftrightarrow n - 2k + 1 = 0 \text{ for some } k \in \{1, 2, \ldots, n\}$$
$$\Leftrightarrow n = 2k - 1 \text{ for some } k \in \{1, 2, \ldots, n\}$$
$$\Leftrightarrow n \text{ is odd.}$$

(ii) By the proof of part (i), we see that $d_k(n) = 0$ if and only if $n - 2k + 1 = 0$, or equivalently, $k = \frac{n+1}{2}$.

(iii) It is easy to verify that the equation holds for $n = 2$ and $n = 3$. Henceforth, let us assume that $n \geq 5$. Firstly, we note that $d_1(n) = \left| \binom{n}{1} - \binom{n}{0} \right| = n - 1$. Next, it is easy to check for all $k \in \{1, 2, \ldots, n\}$ that

$$d_k(n) = \left| \binom{n}{k} - \binom{n}{k-1} \right|$$
$$= \left| \binom{n}{n-k} - \binom{n}{n-k+1} \right|$$
$$= d_{n-k+1}(n).$$

As the set $\left\{ d_k(n) \mid 1 \leq k \leq n, k \neq \frac{n+1}{2} \right\}$ consists of all integers of the form $d_k(n)$ where $1 \leq k \leq n, k \neq \frac{n+1}{2}$, it suffices to show by symmetry that we have $d_k(n) \geq n - 1$ for all integers k such that $2 \leq k \leq \frac{n}{2}$, from which the desired conclusion would hold. When $k = 2$, we have by part (i) that

$$d_k(n) = \left| \frac{n-2k+1}{k} \right| \binom{n}{k-1} = \frac{n-3}{2} \cdot \binom{n}{1} \geq \frac{5-3}{2} \cdot n > n - 1.$$

When $k > 2$ and $k \leq \frac{n}{2}$, we have

$$
\begin{aligned}
d_k(n) &= \left| \frac{n - 2k + 1}{k} \right| \binom{n}{k-1} \quad \text{(by part (i))} \\
&= \frac{n - 2k + 1}{k} \binom{n}{k-1} \quad \left(\text{since } k < \frac{n+1}{2} \right) \\
&\geq \frac{n - 2 \cdot \frac{n}{2} + 1}{\frac{n}{2}} \cdot \binom{n}{2} \quad \text{(by Problem 2.18)} \\
&= \frac{2}{n} \cdot \frac{n(n-1)}{2} \\
&= n - 1,
\end{aligned}
$$

and equality holds if and only if $k = \frac{n}{2}$ and $k = 3$ (since $\binom{n}{\ell} > \binom{n}{2}$ for all integers ℓ such that $3 \leq \ell \leq \frac{n}{2}$ by Problem 2.18), or equivalently, $n = 6$. This implies that $d_k(n) \geq n - 1$ for all integers k such that $2 \leq k \leq \frac{n}{2}$. We are done.

(iv) It is easy to verify that the result holds for $n = 2$, $n = 3$, $n = 4$ and $n = 5$. Henceforth, let us assume that $n \geq 7$. We have shown in part (iii) that $d_1(n) = n - 1$ for all $n \geq 2$ and $d_k(n) = d_{n-k+1}(n)$ for all $k \in \{1, 2, \ldots, n\}$. Hence, it follows that if $k = 1$ or $k = n$, then we have $d_k(n) = n - 1$.

Conversely, suppose that $k \in \{1, 2, \ldots, n\}$ and $k \neq 1, n$. By symmetry and the fact that $d_{\frac{n+1}{2}}(n) = 0$ when n is odd, we may assume without loss of generality that $2 \leq k \leq \frac{n}{2}$. By what we have shown earlier in part (iii), it follows that $d_k(n) > n - 1$ (since $n > 6$). The desired result follows.

(v) If $k = 1, 3, 4, 6$ then it is easy to verify that $d_k(6) = 5$, and if $k = 2$ or 5, then we have $d_k(6) = 9$. The desired statement follows.

56. (i) We have

$$
\begin{aligned}
\binom{\binom{n}{2}}{2} &= \frac{\binom{n}{2}\left(\binom{n}{2} - 1\right)}{2} \\
&= \frac{\frac{n(n-1)}{2}\left(\frac{n(n-1)}{2} - 1\right)}{2} \\
&= \frac{n(n-1)(n^2 - n - 2)}{8} \\
&= 3 \cdot \frac{n(n-1)(n+1)(n-2)}{24} \\
&= 3\binom{n+1}{4}.
\end{aligned}
$$

(ii) We have

$$\binom{\binom{n}{2}}{3} - \binom{\binom{n}{3}}{2}$$

$$= \frac{\binom{n}{2}\left(\binom{n}{2} - 1\right)\left(\binom{n}{2} - 2\right)}{6} - \frac{\binom{n}{3}\left(\binom{n}{3} - 1\right)}{2}$$

$$= \frac{\frac{n(n-1)}{2}\left(\frac{n(n-1)}{2} - 1\right)\left(\frac{n(n-1)}{2} - 2\right)}{6}$$

$$- \frac{\frac{n(n-1)(n-2)}{6}\left(\frac{n(n-1)(n-2)}{6} - 1\right)}{2}$$

$$= \frac{n(n-1)(n^2 - n - 2)(n^2 - n - 4)}{48}$$

$$- \frac{n(n-1)(n-2)(n^3 - 3n^2 + 2n - 6)}{72}$$

$$= \frac{n(n-1)(n-2)(3(n+1)(n^2 - n - 4) - 2(n^3 - 3n^2 + 2n - 6))}{144}$$

$$= \frac{n(n-1)(n-2)(n(n^2 + 6n - 19))}{144}$$

$$= \frac{n^2(n-1)(n-2)((n+3)^2 - 28)}{144}$$

$$\geq \frac{3^2(3-1)(3-2)((3+3)^2 - 28)}{144}$$

$$> 0.$$

(iii) Firstly, it is easy to see that we have $\epsilon_j = \frac{(-1)^{j+1}+1}{2}$ for all positive integers j. This implies that for all positive integers j, we have

$$\binom{\binom{r}{j} + \epsilon_j}{2} = \frac{1}{2}\left(\binom{r}{j} + \epsilon_j\right)\left(\binom{r}{j} + \epsilon_j - 1\right)$$

$$= \frac{1}{2}\left(\binom{r}{j}^2 + (2\epsilon_j - 1)\binom{r}{j} + \epsilon_j^2 - \epsilon_j\right)$$

$$= \frac{1}{2}\left(\binom{r}{j}^2 + (-1)^{j+1}\binom{r}{j}\right).$$

Consequently, this gives us

$$\sum_{j=1}^{r} \binom{\binom{r}{j} + \epsilon_j}{2}\binom{n+r-j}{2r}$$

$$= \sum_{j=1}^{r} \frac{1}{2}\left(\binom{r}{j}^2 + (-1)^{j+1}\binom{r}{j}\right)\binom{n+r-j}{2r}$$

$$= \frac{1}{2} \sum_{j=0}^{r} \left(\binom{r}{j}^2 + (-1)^{j+1} \binom{r}{j} \right) \binom{n+r-j}{2r}$$

$$= \frac{1}{2} \sum_{j=0}^{r} \left(\binom{r}{r-j}^2 - (-1)^{j} \binom{r}{r-j} \right) \binom{n+r-j}{2r}.$$

We have

$$\sum_{j=0}^{r} (-1)^j \binom{r}{r-j} \binom{n+r-j}{2r}$$

$$= \sum_{j=0}^{r} (-1)^j \binom{r}{r-j} \sum_{k=0}^{2r} \binom{n}{k} \binom{r-j}{2r-k} \quad \text{(by (2.3.5))}$$

$$= \sum_{k=0}^{2r} \binom{n}{k} \sum_{j=0}^{r} (-1)^j \binom{r}{k-r} \binom{k-r}{k-r-j} \quad \text{(by (2.1.5))}$$

$$= \sum_{k=r}^{2r} \binom{n}{k} \binom{r}{k-r} \sum_{j=0}^{r} (-1)^j \binom{k-r}{j}$$

(since $k - r < 0$ for all $k < r$)

$$= \sum_{i=0}^{r} \binom{n}{i+r} \binom{r}{i} \sum_{j=0}^{r} (-1)^j \binom{i}{j} \quad \text{(by setting } k - r = i)$$

$$= \sum_{i=0}^{r} \binom{n}{i+r} \binom{r}{i} \sum_{j=0}^{i} (-1)^j \binom{i}{j}$$

(since $\binom{i}{j} = 0$ for all $j > i$, and $i \leq r$)

$$= \binom{n}{r},$$

where the last equality follows from the fact that we have $\sum_{j=0}^{i} (-1)^j \binom{i}{j} = 0$ for all $i > 0$ by (2.3.2). Next, we have

$$\sum_{j=0}^{r} \binom{r}{r-j}^2 \binom{n+r-j}{2r}$$

$$= \sum_{j=0}^{r} \binom{r}{r-j}^2 \sum_{k=0}^{2r} \binom{n}{k} \binom{r-j}{2r-k} \quad \text{(by (2.3.5))}$$

$$= \sum_{k=0}^{2r} \sum_{j=0}^{r} \binom{r}{r-j} \binom{n}{k} \binom{r}{r-j} \binom{r-j}{2r-k}$$

$$= \sum_{k=0}^{2r} \sum_{j=0}^{r} \binom{r}{r-j} \binom{n}{k} \binom{r}{2r-k} \binom{k-r}{k-r-j} \quad \text{(by (2.1.5))}$$

$$= \sum_{k=r}^{2r} \binom{n}{k} \binom{r}{2r-k} \sum_{j=0}^{r} \binom{r}{r-j} \binom{k-r}{j}$$

(since $2r - k > r$ for all $k < r$)

$$= \sum_{k=r}^{2r} \binom{n}{k} \binom{r}{k-r} \binom{k}{r} \quad \text{(by (2.3.5))}$$

$$= \sum_{i=0}^{r} \binom{n}{i+r} \binom{r}{i} \binom{i+r}{r} \quad \text{(by setting } k - r = i)$$

$$= \sum_{i=0}^{r} \binom{n}{r} \binom{n-r}{i} \binom{r}{i} \quad \text{(by (2.1.5))}$$

$$= \binom{n}{r} \sum_{i=0}^{r} \binom{n-r}{i} \binom{r}{r-i}$$

$$= \binom{n}{r}^2 \quad \text{(by (2.3.5))}.$$

Therefore, we have

$$\sum_{j=1}^{r} \left(\frac{\binom{r}{j} + \epsilon_j}{2} \right) \binom{n+r-j}{2r}$$

$$= \frac{1}{2} \sum_{j=0}^{r} \left(\binom{r}{r-j}^2 - (-1)^j \binom{r}{r-j} \right) \binom{n+r-j}{2r}$$

$$= \frac{1}{2} \left(\binom{n}{r}^2 - \binom{n}{r} \right)$$

$$= \frac{1}{2} \cdot \binom{n}{r} \left(\binom{n}{r} - 1 \right)$$

$$= \binom{\binom{n}{r}}{2}$$

as desired

(iv) Let us count the number of ways that we can choose 2 distinct r-element subsets X and Y of $\mathbf{N}_n$. On one hand, the number of ways is equal to $\binom{\binom{n}{r}}{2}$ by definition. On the other hand, let $|X \cap Y| = r - j$. This implies that $|X \cup Y| = |X| + |Y| - |X \cap Y| = r + j$. As $X \neq Y$, it follows that we have $1 \leq r < n$.

Now, there are $\binom{n}{r+j}$ ways to choose $r+j$ elements out of the set $\mathbf{N}_n$ for the elements of $X \cup Y$, $\binom{r+j}{r-j} = \binom{r+j}{2j}$ ways to choose $r+j$ elements out of the set $X \cup Y$ for the elements of $X \cap Y$, and $\frac{1}{2}\binom{2j}{j} = \binom{2j-1}{j-1} = \binom{2j-1}{j}$ to partition $(X \cup Y) \setminus (X \cap Y)$ into two sets A and B, such that $A \cap B = \emptyset$, $|A| = |B| = j$, and $X = A \cup (X \cap Y)$, $Y = B \cup (X \cap Y)$. By (MP) and (AP), the total number of ways is equal to $\sum_{j=1}^{r} \binom{2j-1}{j}\binom{r+j}{2j}\binom{n}{r+j}$, and hence we have

$$\binom{\binom{n}{r}}{2} = \sum_{j=1}^{r}\binom{2j-1}{j}\binom{r+j}{2j}\binom{n}{r+j}$$

as desired.

57. For all positive integers n, we have

$$\begin{aligned}
a_n &= 6^n + 8^n \\
&= (-1+7)^n + (1+7)^n \\
&= \sum_{i=0}^{n}\binom{n}{i}(-1)^{n-i}7^i + \sum_{i=0}^{n}\binom{n}{i}7^i \\
&= \sum_{i=0}^{n}\binom{n}{i}(1+(-1)^{n-i})7^i \\
&= \binom{n}{0}(1+(-1)^n) + 7\binom{n}{1}(1+(-1)^{n-1}) \\
&\quad + \sum_{i=2}^{n}\binom{n}{i}(1+(-1)^{n-i})7^i \\
&= 1 + (-1)^n + 7n(1+(-1)^{n-1}) + \sum_{i=2}^{n}\binom{n}{i}(1+(-1)^{n-i})7^i.
\end{aligned}$$

This implies that

$$\begin{aligned}
a_{83} &= 1 + (-1)^{83} + 7 \cdot 83 \cdot (1+(-1)^{83-1}) + \sum_{i=2}^{83}\binom{83}{i}(1+(-1)^{83-i})7^i \\
&= 1162 + \sum_{i=2}^{83}\binom{83}{i}(1+(-1)^{83-i})7^i.
\end{aligned}$$

As 49 divides each of $\binom{83}{i}(1+(-1)^{83-i})7^i$ for each $i = 2, 3, \ldots, 83$, it follows that the remainder of a_{83} divided by 49 is equal to the remainder of 1162 divided by 49, which is equal to 35.

58. Let $T = \{a_0, a_1, a_2, \ldots\}$ denote the set of all non-negative integers which are either powers of 3, sums of distinct powers of 3 or zero, where $a_i < a_j$ for all $i < j$. Then it follows that all integers in T are of the form $\sum\limits_{n \in S} 3^n$ where S is a finite subset of $\mathbf{N}^*$. When S is the empty subset, the sum is taken to be 0. We note that the first few elements are $a_0 = 0, a_1 = 1, a_2 = 3, a_3 = 4$. Furthermore, for each positive integer m, let us define $T_m := \{a_n \mid a_n < 3^m\}$. Let us first show that $T_m = \{a_0, a_1, \ldots, a_{2^m - 1}\}$ for all positive integers m. Indeed, we have $\sum\limits_{i=0}^{m-1} 3^i = \frac{3^m - 1}{3 - 1} < 3^m$. This implies that for all subsets U of $\{0, 1, \ldots, m - 1\}$, we have $\sum\limits_{i \in U} 3^i \le \sum\limits_{i=0}^{m-1} 3^i < 3^m$ (noting that an empty sum is equal to 0). As there are 2^m subsets of U, we see that $|T_m| \ge 2^m$ and $a_{2^m - 1} = \sum\limits_{i=0}^{m-1} 3^i = \frac{1}{2}(3^m - 1)$. The next term, $a_{2^m} = 3^m$, and all elements in $T_{m+1} \setminus T_m$ are necessarily of the form $3^m + a$, where $a \in T_m$, or equivalently, all elements in $T_{m+1} \setminus T_m$ are necessarily of the form $a_{2^m} + a_k$ where $k \in \{0, 1, \ldots, 2^m - 1\}$. Now we have

$$
\begin{aligned}
a_{100} &= a_{2^6} + a_{36} \quad \text{(since } 2^6 = 64 < 100 < 127 = 2^7 - 1) \\
&= a_{2^6} + a_{2^5} + a_4 \quad \text{(since } 2^5 = 32 < 36 < 63 = 2^6 - 1) \\
&= a_{2^6} + a_{2^5} + a_{2^2} + a_0 \quad \text{(since } 2^2 = 4 = 4 + 0 < 7 = 2^3 - 1) \\
&= 3^6 + 3^5 + 3^2 + 0 \\
&= 981.
\end{aligned}
$$

So the 100-th term of the sequence is 981.

59. As $x = y - 1$, we have

$$
\begin{aligned}
&1 - x + x^2 - x^3 + \cdots + x^{16} - x^{17} \\
&= 1 - (y - 1) + (y - 1)^2 - (y - 1)^3 + \cdots + (y - 1)^{16} - (y - 1)^{17} \\
&= 1 + (1 - y) + (1 - y)^2 + \cdots + (1 - y)^{17} \\
&= a_0 + a_1 y + a_2 y^2 + \cdots + a_{17} y^{17}.
\end{aligned}
$$

For all $n = 2, 3, \ldots, 17$, we note from the Binomial Theorem that the coefficient of y^2 in the expansion of $(1 - y)^n$ is equal to $\binom{n}{2} 1^{n-2} (-1)^2 = \binom{n}{2}$. Hence, by (2.5.1), we have

$$
a_2 = \binom{2}{2} + \binom{3}{2} + \cdots + \binom{17}{2} = \binom{18}{3} = 816.
$$

60. Let us consider these cases:

Case 1: Letter 9 has been typed by the secretary before lunch. In this case, for each $i = 1, 2, 3, 4, 5, 6, 7$, letter i may or may not have been typed before lunch. Hence, the number of after-lunch typing orders in this case is equal to the number of subsets of the set $\{1, 2, 3, 4, 5, 6, 7\}$ which is $2^7 = 128$. We note that if there are r letters to be typed after lunch, say letters, $i_1, i_2, \ldots, i_r$ with $i_1 < i_2 < \cdots < i_r$ then these letters must be in the order $i_1, i_2, \ldots, i_r$ with i_1 at the bottom and i_r at the top.

Case 2: Letter 9 has yet to be typed by the secretary before lunch. Without loss of generality, we may assume that letter 9 would only be delivered to the secretary after lunch, Let S be the set of letters amongst letters $1, 2, 3, 4, 5, 6, 7$ that have yet to be typed by the secretary before lunch, and $|S| = r$. Then it follows that $0 \leq r \leq 7$. It is easy to see that there are $r + 1$ possibilities in which letter 9 could be delivered to the secretary. We note that letter 9 could be delivered after exactly j letters in the set S have been typed, where $j \in \{0, 1, \ldots, r\}$. Furthermore, for each $r \in \{0, 1, 2, 3, 4, 5, 6, 7\}$, there are $\binom{7}{r}$ ways to choose S such that $|S| = r$. By (MP) and (AP), the number of after-lunch typing orders in this case is equal to

$$\sum_{r=0}^{7}(r+1)\binom{7}{r} = \sum_{r=1}^{7} r\binom{7}{r} + \sum_{r=0}^{7}\binom{7}{r}$$

$$= 7 \cdot 2^{7-1} + 2^7 \quad \text{(by (2.3.1) and (2.3.4))}$$

$$= 576.$$

Combining Cases 1 and 2, we see that the number of after-lunch typing orders is equal to $128 + 576 = 704$.

61. By (2.1.3), we have

$$A_k = \binom{1000}{k}(0.2)^k$$

$$= \frac{1000 - k + 1}{k} \cdot \binom{1000}{k-1} \cdot 0.2 \cdot (0.2)^{k-1}$$

$$= \frac{1001 - k}{5k} A_{k-1}.$$

This implies that $\frac{A_{k-1}}{A_k} = \frac{5k}{1001-k}$ for all $k \in \{1, 2, \ldots, 1,000\}$. Now, we have

$$A_{k-1} < A_k \Leftrightarrow \frac{5k}{1001 - k} < 1$$

$$\Leftrightarrow 5k < 1001 - k$$
$$\Leftrightarrow 6k < 1001$$
$$\Leftrightarrow k < 166\frac{5}{6},$$

and equality holds if and only if $k = 166\frac{5}{6}$. This implies that

$$A_0 < A_1 < \cdots < A_{165} < A_{166} > A_{167} > \cdots > A_{1000}.$$

So the desired value of k is 166.

62. We note that all terms in the expansion of $(x_1 + x_2 + \cdots + x_m)^n$ are of the form $x_1^{n_1} x_2^{n_2} \cdots x_m^{n_m}$ where $n_1, n_2, \ldots, n_m$ are non-negative integers and $n_1 + n_2 + \cdots + n_m = n$. Thus, the number of distinct terms in the expansion of $(x_1 + x_2 + \cdots + x_m)^n$ is equal to H_n^m by definition.

63. First Solution: We have

$$\binom{n}{n_1, n_2, \ldots, n_m} = \frac{n!}{n_1! n_2! \cdots n_m!}$$

$$= \frac{n(n-1)!}{n_1! n_2! \cdots n_m!}$$

$$= \frac{(n-1)!}{n_1! n_2! \cdots n_m!}(n_1 + n_2 + \cdots + n_m)$$

$$= \sum_{i=1}^{m} \frac{(n-1)!}{n_1! \cdots n_{i-1}!(n_i - 1)! n_{i+1}! \cdots n_m!}$$

$$= \sum_{i=1}^{m} \binom{n-1}{n_1, \ldots, n_{i-1}, n_i - 1, n_{i+1}, \ldots, n_m}.$$

Second Solution: Let us count the number of ways to distribute n distinct objects into m boxes $B_1, B_2, \ldots, B_m$, such that there are n_i objects in box B_i for $i = 1, 2, \ldots, m$, and $n = n_1 + n_2 + \cdots + n_m$ in two different ways. On one hand, the number of ways is equal to $\binom{n}{n_1, n_2, \ldots, n_m}$ by definition. On the other hand, let us fix an object O, and consider the box that it is placed in. If O is placed in box B_i, then there are $\binom{n-1}{n_1, \ldots, n_{i-1}, n_i - 1, n_{i+1}, \ldots, n_m}$ ways to distribute the remaining $n - 1$ objects into the boxes $B_1, B_2, \ldots, B_m$ such that there are n_i objects in box B_i for $i = 1, 2, \ldots, m$. By (AP), the number of ways is equal to $\sum_{i=1}^{m} \binom{n-1}{n_1, \ldots, n_{i-1}, n_i - 1, n_{i+1}, \ldots, n_m}$, and hence we must have

$$\binom{n}{n_1, n_2, \ldots, n_m} = \sum_{i=1}^{m} \binom{n-1}{n_1, \ldots, n_{i-1}, n_i - 1, n_{i+1}, \ldots, n_m}$$

as desired.

64. Let us count the number of surjective functions $f: \mathbf{N}_n \to \mathbf{N}_m$ in two different ways.

   For the first method of counting, we first note that there are $S(n, m)$ m-partitions of $\mathbf{N}_n$ by definition. Next, there are $m!$ ways to assign a unique integer $n_A \in \mathbf{N}_m$ to each subset A of $\mathcal{P}$ for each m-partition $\mathcal{P}$ of $\mathbf{N}_n$, such that $f(x) = n_A$ for all $x \in A$, and $f(y) \neq n_A$ for all $y \notin A$. By (MP), the number of surjective functions $f: \mathbf{N}_n \to \mathbf{N}_m$ is equal to $m! \cdot S(n, m)$.

   For the second method of counting, let T denote the set of all m-tuples $(n_1, \ldots, n_m)$ of positive integers such that $\sum_{i=1}^{m} n_i = n$. We note that for each surjective function $f: \mathbf{N}_n \to \mathbf{N}_m$, we must have $|f^{-1}(\{i\})| \neq 0$ for each $i \in \mathbf{N}_m$, where $f^{-1}(\{i\}) := \{x \in \mathbf{N}_n \mid f(x) = i\}$. Based on this, we note that for each $(n_1, \ldots, n_m) \in T$, there are $\binom{n}{n_1, \ldots, n_m}$ surjective functions $f: \mathbf{N}_n \to \mathbf{N}_m$ such that $|f^{-1}(\{i\})| = n_i$ for each $i \in \mathbf{N}_m$. By (AP), the number of surjective functions $f: \mathbf{N}_n \to \mathbf{N}_m$ is equal to

$$\sum_{(n_1, \ldots, n_m) \in T} \binom{n}{n_1, \ldots, n_m}.$$

   Hence, we must have

$$\sum_{(n_1, \ldots, n_m) \in T} \binom{n}{n_1, \ldots, n_m} = m! \cdot S(n, m)$$

   as desired.

65. Let T denote the set of all m-tuples $(n_1, \ldots, n_m)$ of non-negative integers such that $\sum_{i=1}^{m} n_i = n$. By the Multinomial Theorem, we have

$$(x_1 + x_2 + \cdots + x_m)^n = \sum_{(n_1, \ldots, n_m) \in T} \binom{n}{n_1, \ldots, n_m} x_1^{n_1} x_2^{n_2} \cdots x_m^{n_m}.$$

   The desired conclusion follows by setting $x_i = 1$ for all odd integers i and $x_j = -1$ for all even integers j.

66. Let us count the number of ways to assign p boys and q girls into m rooms $R_1, R_2, \ldots, R_m$, such that there are k_i people in room R_i for $i = 1, 2, \ldots, m$, and $p + q = k_1 + k_2 + \cdots + k_m$ in two different ways. On one hand, the number of ways is equal to $\binom{p+q}{k_1, k_2, \ldots, k_m}$ by definition. On the other hand, let T denote the set of all m-tuples $(j_1, \ldots, j_m)$ of non-negative integers such that $\sum_{i=1}^{m} j_i = p$. For each tuple $(j_1, \ldots, j_m) \in T$, we note that there are $\binom{p}{j_1, j_2, \ldots, j_m}$ ways to assign the p

boys into the rooms $R_1, R_2, \ldots, R_m$, such that there are j_i boys in room R_i for $i = 1, 2, \ldots, m$, and consequently, there are $\binom{q}{k_1 - j_1, k_2 - j_2, \ldots, k_m - j_m}$ ways to assign the q girls into the rooms $R_1, R_2, \ldots, R_m$, such that there are k_i people in room R_i for $i = 1, 2, \ldots, m$. By (MP) and (AP), the total number of ways is equal to

$$\sum_{(j_1,\ldots,j_m) \in T} \binom{p}{j_1, j_2, \ldots, j_m} \binom{q}{k_1 - j_1, k_2 - j_2, \ldots, k_m - j_m},$$

and hence we must have

$$\binom{p+q}{k_1, k_2, \ldots, k_m}$$
$$= \sum_{(j_1,\ldots,j_m) \in T} \binom{p}{j_1, j_2, \ldots, j_m} \binom{q}{k_1 - j_1, k_2 - j_2, \ldots, k_m - j_m}$$

as desired.

67. By (2.8.7), we may assume without loss of generality that $p > n_1 \geq 1$. By (2.8.4) and Problem 2.51, we have

$$\binom{p}{n_1, n_2, \ldots, n_m} = \binom{p}{n_1}\binom{p - n_1}{n_2} \cdots \binom{p - (n_1 + n_2 + \cdots + n_{m-1})}{n_m}$$
$$\equiv 0 \pmod{p}.$$

Now, let T denote the set of all m-tuples $(n_1, \ldots, n_m)$ of non-negative integers such that $\sum_{i=1}^{m} n_i = p$, and S to be the subset of T consisting of all m-tuples $(n_1, \ldots, n_m)$ satisfying $n_i \neq p$ for $i = 1, 2, \ldots, m$. As a consequence, we have

$$\left(\sum_{i=1}^{m} x_i \right)^p - \sum_{i=1}^{m} x_i^p$$
$$= \sum_{(n_1,\ldots,n_m) \in T} \binom{p}{n_1, n_2, \ldots, n_m} x_1^{n_1} x_2^{n_2} \cdots x_m^{n_m} - \sum_{i=1}^{m} x_i^p$$
$$= \sum_{(n_1,\ldots,n_m) \in S} \binom{p}{n_1, n_2, \ldots, n_m} x_1^{n_1} x_2^{n_2} \cdots x_m^{n_m}$$
$$\equiv 0 \pmod{p},$$

or equivalently, $\left(\sum_{i=1}^{m} x_i \right)^p \equiv \sum_{i=1}^{m} x_i^p \pmod{p}$ as desired.

68. It is easy to see that if $f(x_1, \ldots, x_n)$ and $g(x_1, \ldots, x_n)$ are polynomials in $x_1, \ldots, x_n$ with integer coefficients such that $f(x_1, \ldots, x_n) \equiv g(x_1, \ldots, x_n) \pmod{p}$, then the number of terms of $f(x_1, \ldots, x_n)$ whose coefficients are not divisible by p is equal to the number of terms of $g(x_1, \ldots, x_n)$ whose coefficients are not divisible by p. For each $i = 0, 1, \ldots, k$, let T_i denote the set of all m-tuples $(n_{i,1}, \ldots, n_{i,m})$ of non-negative integers such that $\sum_{j=1}^{m} n_{i,j} = n_i$. By Problem 2.67, we have

$$(x_1 + \cdots + x_m)^n$$
$$= (x_1 + \cdots + x_m)^{n_0} \cdots (x_1 + \cdots + x_m)^{n_k p^k}$$
$$\equiv (x_1 + \cdots + x_m)^{n_0} \cdots \left(x_1^{p^k} + \cdots + x_m^{p^k} \right)^{n_k} \pmod{p}$$
$$\equiv \prod_{i=0}^{k} \sum_{(n_{i,1}, \ldots, n_{i,m}) \in T_i} \binom{n_i}{n_{i,1}, \ldots, n_{i,m}} x_1^{n_{i,1} p^i} \cdots x_m^{n_{i,m} p^i} \pmod{p}$$
$$\equiv \sum_{\substack{(n_{i,1}, \ldots, n_{i,m}) \in T_i \\ i=0,1,\ldots,k}} \left[\prod_{i=0}^{k} \binom{n_i}{n_{i,1}, \ldots, n_{i,m}} \right]$$
$$\times x_1^{n_{0,1} + \cdots + n_{k,1} p^k} \cdots x_m^{n_{0,m} + \cdots + n_{k,m} p^k} \pmod{p}.$$

By noting that

$$m_0 + m_1 p + \cdots + m_s p^s \leq (p-1)(1 + p + \cdots + p^s) = (p-1) \cdot \frac{p^{s+1} - 1}{p - 1} < p^{s+1}$$

for all $m_0, m_1, \ldots, m_s \in \{0, 1, \ldots, p-1\}$, it follows that the two terms

$$x_1^{n_{0,1} + \cdots + n_{k,1} p^k} \cdots x_m^{n_{0,m} + \cdots + n_{k,m} p^k}$$

and

$$x_1^{n'_{0,1} + \cdots + n'_{k,1} p^k} \cdots x_m^{n'_{0,m} + \cdots + n'_{k,m} p^k}$$

are the same if and only if $n_{i,j} = n'_{i,j}$ for all $i = 0, 1, \ldots, k$ and $j = 1, 2, \ldots, m$. As there are $H_{n_i}^m = \binom{n_i + m - 1}{m - 1}$ elements in the set T_i for all $i = 0, 1, \ldots, k$, it follows that the number of terms in the polynomial

$$\sum_{\substack{(n_{i,1}, \ldots, n_{i,m}) \in T_i \\ i=0,1,\ldots,k}} \left[\prod_{i=0}^{k} \binom{n_i}{n_{i,1}, \ldots, n_{i,m}} \right]$$
$$\times x_1^{n_{0,1} + \cdots + n_{k,1} p^k} \cdots x_m^{n_{0,m} + \cdots + n_{k,m} p^k}$$

is equal to $\prod\limits_{i=0}^{k} \binom{n_i+m-1}{m-1}$. It remains to show that every term of the polynomial

$$\sum_{\substack{(n_{i,1},\ldots,n_{i,m})\in T_i \\ i=0,1,\ldots,k}} \left[\prod_{i=0}^{k} \binom{n_i}{n_{i,1},\ldots,n_{i,m}} \right]$$

$$\times x_1^{n_{0,1}+\cdots+n_{k,1}p^k} \cdots x_m^{n_{0,m}+\cdots+n_{k,m}p^k}$$

is not divisible by p, from which we could conclude henceforth by our earlier observation that the number of terms in the expansion of $(x_1 + \cdots + x_m)^n$ whose coefficients is not divisible by p is equal to $\prod\limits_{i=0}^{k} \binom{n_i+m-1}{m-1}$. Indeed, since $n_i \leq p - 1$ for all $i = 0, 1, \ldots, k$, we have $p \nmid n_i!$ for all $i = 0, 1, \ldots, k$. This implies that $p \nmid \binom{n_i}{n_{i,1},\ldots,n_{i,m}}$ for all $i = 0, 1, \ldots, k$ and $(n_{i,1},\ldots,n_{i,m}) \in T_i$, and hence we have $p \nmid \prod\limits_{i=0}^{k} \binom{n_i}{n_{i,1},\ldots,n_{i,m}}$ for all $i = 0, 1, \ldots, k$ and $(n_{i,1},\ldots,n_{i,m}) \in T_i$. This completes the proof, and we are done.

69. By Problem 2.43, we have

$$\sum_{r=0}^{n} \binom{n}{r}\binom{q}{r}\binom{m+n+q-r}{n+q} = \binom{m+n}{n}\binom{m+q}{q}.$$

The desired equation then follows by setting $q = n$.

70. We have

$$\sum_{r=1}^{n} \frac{(-1)^{r-1}}{r}\binom{n}{r} = -\sum_{r=1}^{n}\sum_{k=r-1}^{n-1} \frac{(-1)^r}{r}\binom{k}{r-1} \quad \text{(by (2.5.1))}$$

$$= -\sum_{r=1}^{n}\sum_{k=r-1}^{n-1} \frac{(-1)^r}{k+1}\binom{k+1}{r} \quad \text{(by (2.1.2))}$$

$$= -\sum_{r=1}^{n}\sum_{k=r}^{n} \frac{(-1)^r}{k}\binom{k}{r}$$

$$= -\sum_{k=1}^{n}\sum_{r=1}^{k} \frac{(-1)^r}{k}\binom{k}{r} \quad \text{(since } 1 \leq r \leq k \leq n)$$

$$= -\sum_{k=1}^{n} \frac{1}{k}\sum_{r=1}^{k}(-1)^r\binom{k}{r}$$

$$= -\sum_{k=1}^{n} \frac{1}{k}\left(\sum_{r=0}^{k}(-1)^r\binom{k}{r} - \binom{k}{0}\right)$$

$$= \sum_{k=1}^{n} \frac{1}{k} \quad \text{(by (2.3.2))}.$$

71. For any positive integer n, let us define $u_n = \frac{1}{(n+r-1)(n+r-2)\cdots n}$ and $v_n = \frac{1}{(n+r-2)\cdots n}$. Then it is easy to see that $v_{n+1} - v_n = (1-r)u_n$ and $\lim_{n\to\infty} v_{n-r+2} = \lim_{n\to\infty} \frac{1}{n\cdots(n-r+2)} = 0$ for all positive integers n. Now, for all integers $m \geq r$, we have

$$\sum_{n=r}^{m} \frac{1}{\binom{n}{r}} = \sum_{n=r}^{m} \frac{r!}{n(n-1)\cdots(n-r+1)}$$

$$= \sum_{n=r}^{m} r! u_{n-r+1}$$

$$= r! \sum_{n=r}^{m} \left(\frac{v_{n-r+2} - v_{n-r+1}}{1-r} \right)$$

$$= \frac{r!}{1-r}(v_{m-r+2} - v_1)$$

$$= \frac{r!}{1-r}\left(v_{m-r+2} - \frac{1}{(r-1)!} \right).$$

Hence, we have

$$\sum_{n=r}^{\infty} \frac{1}{\binom{n}{r}} = \lim_{m\to\infty} \sum_{n=r}^{m} \binom{n}{r}^{-1}$$

$$= \lim_{m\to\infty} \frac{r!}{1-r}\left(v_{m-r+2} - \frac{1}{(r-1)!} \right)$$

$$= \frac{r!}{1-r} \lim_{m\to\infty} v_{m-r+2} - \frac{r!}{1-r} \lim_{m\to\infty} \frac{1}{(r-1)!}$$

$$= \frac{r!}{r-1} \cdot \frac{1}{(r-1)!}$$

$$= \frac{r}{r-1}.$$

72. Fix an $r \in \{1, 2, \ldots, n\}$. We note that $M(A) = r$ if and only if $r \in A$ and $A \subseteq \{1, 2, \ldots, r\}$. Hence, the number of subsets S of $\{1, 2, \ldots n\}$ such that $M(S) = r$ is equal to 2^{r-1}. Also, we note that $m(A) = n-r+1$ if and only if $n-r+1 \in A$ and $A \subseteq \{n-r+1, n-r+2, \ldots, n\}$. Hence, the number of subsets T of $\{1, 2, \ldots n\}$ such that $m(T) = n-r+1$ is

equal to 2^{r-1}. Hence, the desired arithmetic mean is equal to

$$\frac{1}{2^n - 1} \sum_{\substack{A \in \mathbf{N}_n, \\ A \neq \emptyset}} \alpha(A) = \frac{1}{2^n - 1} \left(\sum_{\substack{A \in \mathbf{N}_n \\ A \neq \emptyset}} M(A) + \sum_{\substack{A \in \mathbf{N}_n \\ A \neq \emptyset}} m(A) \right)$$

$$= \frac{1}{2^n - 1} \left(\sum_{r=1}^{n} r 2^{r-1} + \sum_{r=1}^{n} (n - r + 1) 2^{r-1} \right)$$

$$= \frac{n+1}{2^n - 1} \left(\sum_{r=1}^{n} 2^{r-1} \right)$$

$$= \frac{n+1}{2^n - 1} \cdot (2^n - 1)$$

$$= n + 1.$$

73. Firstly, it is easy to see that $a_n \geq \binom{n}{0}^{-1} + \binom{n}{n}^{-1} = 2$ for all positive integers n. Next, we shall that for all positive integers $n \geq 4$ and all positive integers $k \in \{2, 3, \ldots, n-2\}$, we have $\binom{n}{k} \geq \binom{n}{2}$. If $2 \leq k \leq \frac{n}{2}$, then we must have $\binom{n}{k} \geq \binom{n}{2}$ by Problem 2.18. Else, we must have $\frac{n}{2} < k \leq n - 2$, and by Problem 2.18 again we have $\binom{n}{k} \geq \binom{n}{n-2} = \binom{n}{2}$. Hence, for all positive integers $n \geq 4$, we have

$$a_n = \sum_{k=0}^{n} \binom{n}{k}^{-1}$$

$$= \binom{n}{0}^{-1} + \binom{n}{1}^{-1} + \binom{n}{n-1}^{-1} + \binom{n}{n}^{-1} + \sum_{k=2}^{n-2} \binom{n}{k}^{-1}$$

$$\leq 2 + \frac{2}{n} + \sum_{k=2}^{n-2} \binom{n}{2}^{-1}$$

$$\leq 2 + \frac{2}{n} + (n - 3) \cdot \frac{2}{n(n-1)}$$

$$\leq 2 + \frac{2}{n} + (n - 1) \cdot \frac{2}{n(n-1)}$$

$$= 2 + \frac{4}{n}.$$

As $2 \leq a_n \leq 2 + \frac{4}{n}$ and $\lim_{n \to \infty} \left(2 + \frac{4}{n} \right) = 2$, it follows from the Squeeze Theorem that $\lim_{n \to \infty} a_n = 2$.

Remark. The Squeeze Theorem (for sequences) asserts that if $(x_n)_{n=1}^{\infty}$, $(y_n)_{n=1}^{\infty}$ and $(z_n)_{n=1}^{\infty}$ are sequences of real numbers such that there exists some positive integer N for which $x_n \leq y_n \leq z_n$ for all $n \geq N$, and the sequences $(x_n)_{n=1}^{\infty}$ and $(z_n)_{n=1}^{\infty}$ are convergent with $\lim_{n \to \infty} x_n = \lim_{n \to \infty} z_n = L$, then the sequence $(y_n)_{n=1}^{\infty}$ is convergent with $\lim_{n \to \infty} y_n = L$.

74. We shall prove by induction that $(x + y)_n = \sum_{i=0}^{n} \binom{n}{i} (x)_i (y)_{n-i}$ for all positive integers n, with the base case $n = 1$ being trivial. Suppose that the equation holds true for some positive integer m. By induction hypothesis, we have $(x + y)_m = \sum_{i=0}^{m} \binom{m}{i} (x)_i (y)_{m-i}$. This implies that

$$(x + y)_{m+1}$$
$$= (x + y)_m (x + y - m)$$
$$= \sum_{i=0}^{m} \binom{m}{i} (x)_i (y)_{m-i} (x + y - m)$$
$$= \sum_{i=0}^{m} \binom{m}{i} (x)_i (y)_{m-i} (x - i + y - m + i)$$
$$= \sum_{i=0}^{m} \binom{m}{i} (x)_i (x - i)(y)_{m-i} + \sum_{i=0}^{m} \binom{m}{i} (x)_i (y)_{m-i} (y - m + i)$$
$$= \sum_{i=0}^{m} \binom{m}{i} (x)_{i+1} (y)_{m-i} + \sum_{i=0}^{m} \binom{m}{i} (x)_i (y)_{m-i+1}$$
$$= (x)_{m+1} (y)_0 + (x)_0 (y)_{m+1}$$
$$\quad + \sum_{i=0}^{m-1} \binom{m}{i} (x)_{i+1} (y)_{m-i} + \sum_{i=1}^{m} \binom{m}{i} (x)_i (y)_{m-i+1}$$
$$= (x)_{m+1} (y)_0 + (x)_0 (y)_{m+1}$$
$$\quad + \sum_{i=1}^{m} \binom{m}{i-1} (x)_i (y)_{m-i+1} + \sum_{i=1}^{m} \binom{m}{i} (x)_i (y)_{m-i+1}$$
$$= (x)_{m+1} (y)_0 + (x)_0 (y)_{m+1} + \sum_{i=1}^{m} \left(\binom{m}{i-1} + \binom{m}{i} \right) (x)_i (y)_{m-i+1}$$
$$= (x)_{m+1} (y)_0 + (x)_0 (y)_{m+1} + \sum_{i=1}^{m} \binom{m+1}{i} (x)_i (y)_{m-i+1} \quad \text{(by (2.1.4))}$$
$$= \sum_{i=0}^{m+1} \binom{m+1}{i} (x)_i (y)_{m-i+1},$$

thereby completing the inductive step. So we are done.

75. Let $a_1 \cdots a_n$ be a permutation of $\{1, \ldots, n\}$, and for each $k = 1, \ldots, n$, let b_k be the unique integer such that $a_{b_k} = k$. We shall show that $a_1 \cdots a_n$ satisfies the condition that for each integer i with $2 \le i \le n$, there exists some $j < i$, such that $|a_i - a_j| = 1$, if and only if for all integers $p, q \in \{1, \ldots, n\}$ such that $a_1 \le p < q$ or $q < p \le a_1$, we have $b_p < b_q$.

Firstly, let us suppose that for all integers $p, q \in \{1, 2, \ldots, n\}$ such that $a_1 \le p < q$ or $q < p \le a_1$, we have $b_p < b_q$. We name this property P. Let us take any $i \in \{2, 3, \ldots, n\}$, and show that there exists some $j < i$, such that $|a_i - a_j| = 1$. We consider the following cases.

Case 1: We suppose that $a_i > a_1$. If $a_i = a_1 + 1$, then a_1 is to the left of a_i and $|a_i - a_1| = 1$. If $a_i > a_1 + 1$, then $a_1 < a_i - 1 < a_i$. By property P, $b_{a_i - 1} < b_{a_i}$. That is, $a_i - 1$ is to the left of a_i and $|(a_i - 1) - a_i| = 1$.

Case 2: We suppose that $a_i < a_1$. If $a_i = a_1 - 1$, then a_1 is to the left of a_i and $|a_i - a_1| = 1$. If $a_i < a_1 - 1$, then $a_i < a_i + 1 < a_1$. By property P, $b_{a_i + 1} < b_{a_i}$. That is, $a_i + 1$ is to the left of a_i and $|(a_i + 1) - a_i| = 1$. Hence the claim.

Conversely, let us suppose that for each $i = 2, 3, \ldots, n$, there exists some $j < i$, such that $|a_i - a_j| = 1$. Let us show that for all integers $p, q \in \{1, 2, \ldots, n\}$ such that $a_1 \le p < q$ or $q < p \le a_1$, we have $b_p < b_q$. To this end, let us first prove by induction that $b_{s+1} < b_s$ for all $s < a_1$. If $a_1 = 1$, then there is nothing to prove, so let us assume that $a_1 > 1$. By assumption, there exists some $c < b_1$, such that $|a_c - a_{b_1}| = |a_c - 1| = 1$. This implies that $a_c = 2$, and hence $b_2 = c < b_1$, so the base case holds. Assume that the proposition holds for some positive integer $k < a_1$. By induction hypothesis, we have $b_{k+1} < b_k$. Without loss of generality, we may assume that $k + 1 < a_1$. By assumption, there exists some $d < b_{k+1}$, such that $|a_d - a_{b_{k+1}}| = |a_d - (k+1)| = 1$. This implies that $a_d = k$ or $a_d = k + 2$, or equivalently, $d = b_k$ or $d = b_{k+2}$. As we have $b_{k+1} < b_k$, we must have $a_d = k + 2$, and hence $b_{k+2} = d < b_{k+1}$, thereby completing the inductive step. By a similar proof, we have $b_r < b_{r+1}$ for all $r > a_1$.

Now, for each $m = 1, 2, \ldots, n$, let us count the number of permutations $a_1 a_2 \cdots a_n$ of $\{1, 2, \ldots, n\}$ with $a_1 = m$ that satisfies the condition that for each $i = 2, 3, \ldots, n$, there exists some $j < i$, such that $|a_i - a_j| = 1$. Now, there are $\binom{n-1}{m-1}$ ways to choose $\{b_1, \ldots, b_{m-1}\} \subseteq \{2, 3, \ldots, n\}$, and exactly one way to ensure that $b_{m+1} < b_{m+2} < \cdots < b_n$ and $b_{m-1} < b_{m-2} < \cdots < b_1$. Hence by (AP), the total number of permutations that satisfy the condition is equal to

$$\sum_{m=1}^{n}\binom{n-1}{m-1} = \sum_{m=0}^{n-1}\binom{n-1}{m} = 2^{n-1}.$$

76. We have

$$\sum_{r=0}^{\lfloor\frac{n-1}{2}\rfloor}\left[\frac{n-2r}{n}\binom{n}{r}\right]^{2}$$

$$= \binom{n}{0}^{2} + \sum_{r=1}^{\lfloor\frac{n-1}{2}\rfloor}\left[\binom{n}{r} - \frac{2r}{n}\binom{n}{r}\right]^{2}$$

$$= \binom{n-1}{0}^{2} + \sum_{r=1}^{\lfloor\frac{n-1}{2}\rfloor}\left[\binom{n}{r} - 2\binom{n-1}{r-1}\right]^{2} \quad \text{(by (2.1.4))}$$

$$= \binom{n-1}{0}^{2} + \sum_{r=1}^{\lfloor\frac{n-1}{2}\rfloor}\left[\binom{n-1}{r} + \binom{n-1}{r-1} - 2\binom{n-1}{r-1}\right]^{2}$$

(by (2.1.2))

$$= \binom{n-1}{0}^{2} + \sum_{r=1}^{\lfloor\frac{n-1}{2}\rfloor}\binom{n-1}{r}^{2}$$

$$+ \sum_{r=1}^{\lfloor\frac{n-1}{2}\rfloor}\binom{n-1}{n-r}^{2} - 2\sum_{r=1}^{\lfloor\frac{n-1}{2}\rfloor}\binom{n-1}{r}\binom{n-1}{n-r}.$$

Let us consider the following cases:

Case 1: n is even, that is, $n = 2k$ for some positive integer k. We have

$$\sum_{r=0}^{\lfloor\frac{n-1}{2}\rfloor}\left[\frac{n-2r}{n}\binom{n}{r}\right]^{2}$$

$$= \binom{n-1}{0}^{2} + \sum_{r=1}^{k-1}\binom{n-1}{r}^{2}$$

$$+ \sum_{r=1}^{k-1}\binom{n-1}{n-r}^{2} - 2\sum_{r=1}^{k-1}\binom{n-1}{r}\binom{n-1}{n-r}$$

$$= \sum_{r=0}^{k-1}\binom{n-1}{r}^{2} + \sum_{r=k+1}^{n-1}\binom{n-1}{r}^{2} - 2\sum_{r=1}^{k-1}\binom{n-1}{r}\binom{n-1}{n-r}$$

(since $n = 2k$)

$$= \sum_{r=0}^{n-1} \binom{n-1}{r}^2 - \binom{n-1}{k}^2 - 2\sum_{r=1}^{k-1} \binom{n-1}{r}\binom{n-1}{n-r}$$

$$= \binom{2n-2}{n-1} - \binom{n-1}{k}^2 - 2\sum_{r=1}^{k-1} \binom{n-1}{r}\binom{n-1}{n-r} \quad \text{(by (2.3.6))}$$

$$= \binom{2n-2}{n-1} - \binom{n-1}{k}^2 - \sum_{r=1}^{k-1} \binom{n-1}{r}\binom{n-1}{n-r}$$

$$\quad - \sum_{t=k+1}^{n-1} \binom{n-1}{n-t}\binom{n-1}{t}$$

(by setting $t = n - r$)

$$= \binom{2n-2}{n-1} - \sum_{r=1}^{n-1} \binom{n-1}{r}\binom{n-1}{n-r}$$

$$= \binom{2n-2}{n-1} - \sum_{r=0}^{n} \binom{n-1}{r}\binom{n-1}{n-r} \quad \left(\text{since } \binom{n-1}{n} = 0\right)$$

$$= \binom{2n-2}{n-1} - \binom{2n-2}{n} \quad \text{(by (2.3.5))}$$

$$= \binom{2n-2}{n-1} - \frac{2n-2-n+1}{n}\binom{2n-2}{n-1} \quad \text{(by (2.1.3))}$$

$$= \frac{1}{n}\binom{2n-2}{n-1}.$$

Case 2: n is odd, that is, $n = 2k+1$ for some non-negative integer k. We have

$$\sum_{r=0}^{\lfloor \frac{n-1}{2} \rfloor} \left[\frac{n-2r}{n}\binom{n}{r}\right]^2$$

$$= \binom{n-1}{0}^2 + \sum_{r=1}^{k} \binom{n-1}{r}^2 + \sum_{r=1}^{k} \binom{n-1}{n-r}^2 - 2\sum_{r=1}^{k} \binom{n-1}{r}\binom{n-1}{n-r}$$

$$= \sum_{r=0}^{k} \binom{n-1}{r}^2 + \sum_{r=k+1}^{n-1} \binom{n-1}{r}^2 - 2\sum_{r=1}^{k} \binom{n-1}{r}\binom{n-1}{n-r}$$

$$= \sum_{r=0}^{n-1} \binom{n-1}{r}^2 - 2\sum_{r=1}^{k} \binom{n-1}{r}\binom{n-1}{n-r}$$

$$= \binom{2n-2}{n-1} - 2\sum_{r=1}^{k} \binom{n-1}{r}\binom{n-1}{n-r} \quad \text{(by (2.3.6))}$$

$$= \binom{2n-2}{n-1} - \sum_{r=1}^{n-1} \binom{n-1}{r}\binom{n-1}{n-r}$$

$$= \frac{1}{n}\binom{2n-2}{n-1} \quad \text{(by Case 1)}.$$

77. We have

$$\sum_{r=1}^{n} r\sqrt{\binom{n}{r}}$$

$$= \sum_{r=1}^{n} r\sqrt{\frac{n}{r}\binom{n-1}{r-1}} \quad \text{(by (2.1.2))}$$

$$= \sqrt{n}\sum_{r=1}^{n}\sqrt{r}\sqrt{\binom{n-1}{r-1}}$$

$$\leq \sqrt{n}\sqrt{\sum_{r=1}^{n}r}\sqrt{\sum_{r=1}^{n}\binom{n-1}{r-1}} \quad \text{(by Cauchy-Schwarz inequality)}$$

$$= \sqrt{n}\sqrt{\frac{n(n+1)}{2}}\sqrt{\sum_{r=0}^{n-1}\binom{n-1}{r}}$$

$$< \sqrt{n}\sqrt{n^2}\sqrt{2^{n-1}} \quad \text{(since } n > 1)$$

$$= \sqrt{2^{n-1}n^3}.$$

78. By (2.3.5), we have for all $j = 0, 1, \ldots, r$ that

$$\binom{n+j}{r} = \sum_{i=0}^{r}\binom{j}{i}\binom{n}{r-i} = \sum_{i=0}^{j}\binom{j}{i}\binom{n}{r-i},$$

noting that $\binom{j}{i} = 0$ for all $i > j$. We shall prove by strong induction that k divides $\binom{n}{r-j}$ for all $j = 0, 1, \ldots, r$, with the base case being trivial. Suppose that there exists a non-negative integer $s < r$ such that k divides $\binom{n}{r-j}$ for all integers j satisfying $0 \leq j \leq s$. We note that

$$\binom{n+s+1}{r} = \sum_{i=0}^{s+1}\binom{s+1}{i}\binom{n}{r-i}$$

$$= \binom{n}{r-(s+1)} + \sum_{i=0}^{s}\binom{s+1}{i}\binom{n}{r-i},$$

so that we have

$$\binom{n}{r-(s+1)} = \binom{n+s+1}{r} - \sum_{i=0}^{s}\binom{s+1}{i}\binom{n}{r-i}.$$

By the given assumption, we have k to divide $\binom{n+s+1}{r}$, and by induction hypothesis, we have k to divide $\binom{n}{r-j}$ for all integers j satisfying $0 \leq j \leq s$. So k divides $\binom{n}{r-(s+1)}$, and this completes the inductive step, and we are done. Since k divides $\binom{n}{r-r}$, and $\binom{n}{r-r} = 1$, we must have $k = 1$.

79. Arguing by contradiction, suppose on the contrary that there exist four consecutive binomial coefficients $\binom{n}{r}, \binom{n}{r+1}, \binom{n}{r+2}, \binom{n}{r+3}$ which are in an arithmetic progression. As $\binom{n}{r}, \binom{n}{r+1}, \binom{n}{r+2}$ are in an arithmetic progression, we must have $\binom{n}{r} + \binom{n}{r+2} = 2\binom{n}{r+1}$. By (2.1.3), we have $\binom{n}{r+2} = \frac{n-r-1}{r+2}\binom{n}{r+1}$ and $\binom{n}{r+1} = \frac{n-r}{r+1}\binom{n}{r}$. This implies that

$$\binom{n}{r} + \binom{n}{r+2} = 2\binom{n}{r+1}$$
$$\Rightarrow \frac{r+1}{n-r}\binom{n}{r+1} + \frac{n-r-1}{r+2}\binom{n}{r+1} = 2\binom{n}{r+1}$$
$$\Rightarrow (r+1)(r+2) + (n-r-1)(n-r) = 2(n-r)(r+2)$$
$$\Rightarrow r^2 - (n-2)r + \frac{n^2 - 5n + 2}{4} = 0.$$

Likewise, since $\binom{n}{r+1}, \binom{n}{r+2}, \binom{n}{r+3}$ are in an arithmetic progression, we must have $\binom{n}{r+1} + \binom{n}{r+3} = 2\binom{n}{r+2}$. This also implies that

$$(r+1)^2 - (n-2)(r+1) + \frac{n^2 - 5n + 2}{4} = 0.$$

Hence, we must have $(r+1)^2 - (n-2)(r+1) = r^2 - (n-2)r$, from which we get $n = 2r+3$.

On the other hand, it follows from (2.7.1) and (2.7.2) that $\binom{n}{r}, \binom{n}{r+1}, \binom{n}{r+2}, \binom{n}{r+3}$ cannot be in a constant arithmetic progression, that is, with the common difference equal to 0, so we must have

$$\binom{n}{r+2} - \binom{n}{r+1} = \frac{n-r-1}{r+2}\binom{n}{r+1} - \binom{n}{r+1}$$
$$= \frac{n-2r-3}{r+2}\binom{n}{r+1}$$
$$\neq 0.$$

This implies that $n - 2r - 3 \neq 0$, which contradicts the fact that $n = 2r + 3$. Consequently, there does not exist four consecutive binomial coefficients

$$\binom{n}{r}, \binom{n}{r+1}, \binom{n}{r+2}, \binom{n}{r+3}$$

which are in an arithmetic progression.

80. We have

$$\sum_{i=0}^{n-1} \binom{2n}{2i+1} = \sum_{i=0}^{n-1} \left[\binom{2n-1}{2i+1} + \binom{2n-1}{2i} \right] \quad \text{(by (2.1.4))}$$

$$= \sum_{i=0}^{n-1} \binom{2n-1}{2i+1} + \sum_{i=0}^{n-1} + \binom{2n-1}{2i}$$

$$= \sum_{i=0}^{2n-1} \binom{2n-1}{i}$$

$$= 2^{2n-1} \quad \text{(by (2.3.1))}.$$

Let d be the greatest common divisor of $\binom{2n}{1}, \binom{2n}{3}, \ldots, \binom{2n}{2n-1}$. We write $n = 2^k \cdot p$, where k and p are integers and p is odd. As $\sum_{i=0}^{n-1} \binom{2n}{2i+1} = 2^{2n-1}$, it follows that $d | 2^{2n-1}$. For all $i = 0, 1, \ldots, n-1$, we have by (2.1.2) that

$$(2i+1)\binom{2n}{2i+1} = 2n\binom{2n-1}{2i} = 2^{k+1}p\binom{2n-1}{2i}.$$

As we have $(2^{k+1}, 2i+1) = 1$, it follows that $2i+1 | p\binom{2n-1}{2i}$, which implies that $\frac{p\binom{2n-1}{2i}}{2i+1}$ is an integer. As we have $\frac{1}{2^{k+1}}\binom{2n}{2i+1} = \frac{p\binom{2n-1}{2i}}{2i+1}$, an integer, we have $2^{k+1} | \binom{2n}{2i+1}$, and hence $2^{k+1} | d$ by the definition of d. Next, since $d | \binom{2n}{1} = 2n = 2^{k+1}p$ and $(d, p) = 1$ (since $d | 2^{2n-1}$), we must have $d | 2^{2k+1}$, and hence we have $d = 2^{k+1}$.

81. Firstly, let us suppose that $\binom{n}{r}$ is odd for all $r \in \{0, 1, \ldots, n\}$. Let k be the smallest positive integer for which $n < 2^k$. This would imply that $2^{k-1} \leq n$. By (2.1.3), we have $2^{k-1}\binom{n}{2^{k-1}} = (n - 2^{k-1} + 1)\binom{n}{2^{k-1}-1}$. As $\binom{n}{2^{k-1}-1}$ is odd, we must have 2^{k-1} and $\binom{n}{2^{k-1}-1}$ to be coprime, and hence we must have $2^{k-1} | (n - 2^{k-1} + 1)$. This implies that $n - 2^{k-1} + 1 \geq 2^{k-1}$, or equivalently, $n \geq 2^k - 1$. As $n < 2^k$, we must have $n = 2^k - 1$. Conversely, suppose that $n = 2^k - 1$ for some positive integer k. We shall prove by induction that $\binom{n}{r}$ is odd for all $r \in \{0, 1, \ldots, n\}$, with the base cases $r = 0$ and $r = 1$ being trivial. Suppose that $\binom{n}{r}$ is odd

for some $k \geq 0$. We note that $\binom{n}{k+1} = \frac{n-k}{k+1}\binom{n}{k}$ by (2.1.3). If k is even, that is, $k = 2m$ for some non-negative integer m, then we have $\binom{n}{k+1} = \frac{n-2m}{2m+1}\binom{n}{k} = \frac{2^k-1-2m}{2m+1}\binom{n}{k}$. As $2^k - 1 - 2m$ and $2m + 1$ are both odd integers, $\binom{n}{k+1}$ is necessarily odd as well. Else, if k is odd, that is, $k = 2m - 1$ for some positive integer m, then let us write $m = 2^p \cdot r$ where r is odd. If $k = n$, then there is nothing to show, so let us assume that $k < n$. We first note that

$$2m - 1 < n \Rightarrow 2m < n + 1 = 2^k$$
$$\Rightarrow 2^{p+1}r < 2^k$$
$$\Rightarrow 1 \leq r < 2^{k-p-1}.$$

This implies that

$$\binom{n}{k+1} = \frac{n - 2m + 1}{2m}\binom{n}{k} = \frac{2^k - 2^{p+1}r}{2^{p+1}r}\binom{n}{k} = \frac{2^{k-p-1} - r}{r}\binom{n}{k}.$$

As $2^{k-p-1} > 1$, we must have 2^{k-p-1} to be even, so $2^{k-p-1} - r$ is odd, and hence $\binom{n}{k+1}$ is necessarily odd as well. This completes the inductive step and we are done.

82. We have

$$\frac{1}{2^{n-1}} \sum_{k<\frac{n}{2}} (n - 2k)\binom{n}{k}$$

$$= \frac{n}{2^{n-1}} \sum_{k<\frac{n}{2}} \binom{n}{k} - \frac{n}{2^{n-1}} \sum_{k<\frac{n}{2}} 2\binom{n-1}{k-1} \quad \text{(by (2.1.2))}$$

$$= \frac{n}{2^{n-1}} \sum_{k<\frac{n}{2}} \left(\binom{n-1}{k-1} + \binom{n-1}{k}\right) - \frac{n}{2^{n-1}} \sum_{k<\frac{n}{2}} 2\binom{n-1}{k-1}$$

$$\text{(by (2.1.4))}$$

$$= \frac{n}{2^{n-1}} \sum_{k<\frac{n}{2}} \binom{n-1}{k} - \frac{n}{2^{n-1}} \sum_{k<\frac{n}{2}} \binom{n-1}{k-1}$$

$$= \frac{n}{2^{n-1}} \sum_{k<\frac{n}{2}} \binom{n-1}{k} - \frac{n}{2^{n-1}} \sum_{k<\frac{n-2}{2}} \binom{n-1}{k}$$

$$= \frac{n}{2^{n-1}} \sum_{\frac{n-2}{2} \leq k < \frac{n}{2}} \binom{n-1}{k}.$$

It is easy to see that the only integer that satisfies the inequality $\frac{n-2}{2} \leq k < \frac{n}{2}$ is $\lfloor \frac{n-1}{2} \rfloor$. Therefore, we have

$$\frac{1}{2^{n-1}} \sum_{k<\frac{n}{2}} (n - 2k)\binom{n}{k} = \frac{n}{2^{n-1}} \sum_{\frac{n-2}{2} \leq k < \frac{n}{2}} \binom{n-1}{k} = \frac{n}{2^{n-1}} \binom{n-1}{\lfloor \frac{n-1}{2} \rfloor}.$$

83. We shall prove by induction on a that $\binom{pa}{pb} \equiv \binom{a}{b}$ (mod p) for all non-negative integers a and non-negative integers b such that $b \leq a$, with the base cases $a = 0, 1$ being clear. Suppose that the proposition holds for some non-negative integer $a = n$. By induction hypothesis, we have $\binom{pn}{pb} \equiv \binom{n}{b}$ (mod p) for all non-negative integers b such that $b \leq n$. Let y be a non-negative integer satisfying $y \leq n + 1$. We have

$$\binom{p(n+1)}{py}$$

$$= \sum_{k=0}^{p} \binom{p}{k} \binom{pa}{py-k} \quad \text{(by (2.3.6))}$$

$$= \binom{pn}{py} + \binom{pn}{p(y-1)} + \sum_{k=1}^{p-1} \binom{p}{k} \binom{pa}{py-k}$$

$$\equiv \binom{pn}{py} + \binom{pn}{p(y-1)} \quad \text{(mod p)} \quad \text{(by Problem 2.51)}$$

$$\equiv \binom{n}{y} + \binom{n}{y-1} \quad \text{(mod p)} \quad \text{(by induction hypothesis)}$$

$$\equiv \binom{n+1}{y} \quad \text{(mod p)} \quad \text{(by (2.1.4))}.$$

This completes the inductive step, and we are done.

84. Let $S = \{a_1, a_2, \ldots, a_n\}$, and for each non-empty subset A of S, let us denote the geometric mean of the elements in A by $g(A)$. Let us fix an $i \in \{1, 2, \ldots, n\}$. For each $k = 1, 2, \ldots, n$, we note that there are $\binom{n-1}{k-1}$ subsets B of S that contains a_i and $|B| = k$. Furthermore, we note that for each subset C of S that contains a_i and $|C| = k$, the power of a_i in the geometric mean $g(C)$ of the elements in C is equal to $\frac{1}{k}$ by definition. Hence, the power of a_i in the product $\prod_{\substack{A \subseteq S, \\ A \neq \emptyset}} g(A)$ is equal to

$$\sum_{k=1}^{n} \frac{1}{k} \binom{n-1}{k-1} = \sum_{k=1}^{n} \frac{1}{n} \binom{n}{k} \quad \text{(by (2.1.2))}$$

$$= \frac{1}{n} \left(\sum_{k=0}^{n} \binom{n}{k-1} - \binom{n}{0} \right)$$

$$= \frac{2^n - 1}{n} \quad \text{(by (2.3.1))}.$$

This implies that $\prod_{\substack{A \subseteq S, \\ A \neq \emptyset}} g(A) = (a_1 a_2 \cdots a_n)^{\frac{2^n - 1}{n}}$. Consequently, the geometric mean of all of the $g(A)$'s, where A is a non-empty subset of

S, is equal to

$$\left((a_1 a_2 \cdots a_n)^{\frac{2^n-1}{n}} \right)^{\frac{1}{2^n-1}} = (a_1 a_2 \cdots a_n)^{\frac{1}{n}},$$

which is the geometric mean of all the elements in S. We are done.

85. (i) We have

$$\sum_{k=0}^{m-1} \binom{m}{k} S_k(n) = \sum_{k=0}^{m-1} \binom{m}{k} \sum_{j=1}^{n} j^k$$

$$= \sum_{k=0}^{m-1} \sum_{j=1}^{n} \binom{m}{k} j^k$$

$$= \sum_{j=1}^{n} \sum_{k=0}^{m-1} \binom{m}{k} j^k$$

$$= \sum_{j=1}^{n} \left[\sum_{k=0}^{m} \binom{m}{k} j^k - j^m \right]$$

$$= \sum_{j=1}^{n} [(1+j)^m - j^m]$$

$$= (n+1)^m - 1.$$

(ii) We have

$$S_m(n) - \sum_{k=0}^{m} (-1)^{m-k} \binom{m}{k} S_k(n)$$

$$= \sum_{j=1}^{n} j^m - \sum_{k=0}^{m} (-1)^m (-1)^{-k} \binom{m}{k} \sum_{j=1}^{n} j^k$$

$$= \sum_{j=1}^{n} j^m - \sum_{k=0}^{m} (-1)^m \sum_{j=1}^{n} (-1)^k \binom{m}{k} j^k$$

$$= \sum_{j=1}^{n} j^m - \sum_{j=1}^{n} (-1)^m \sum_{k=0}^{m} \binom{m}{k} (-j)^k$$

$$= \sum_{j=1}^{n} j^m - \sum_{j=1}^{n} (-1)^m (1-j)^m$$

$$= \sum_{j=1}^{n} j^m - \sum_{j=1}^{n} (j-1)^m$$

$$= n^m.$$

86. We first recall that if $f(x)$ is a real polynomial of degree n, and there exist pairwise distinct real numbers $x_1, \ldots, x_{n+1}$ such that $f(x_i) = 0$ for all $i = 1, \ldots, n+1$, then we must have $f(x) = 0$ for all real numbers x. As an immediate corollary, it follows that if $f(x)$ and $g(x)$ are polynomials of degree n and there exist pairwise distinct real numbers $x_1, \ldots, x_{n+1}$ such that $f(x_i) = g(x_i)$ for all $i = 1, \ldots, n+1$, then we must have $f(x) = g(x)$ for all real numbers x.

Let us define

$$Q(x) = 2\left(1 + \sum_{q=1}^{n} \frac{(x-1)(x-2)\cdots(x-q)}{q!}\right) = 2\left(1 + \sum_{q=1}^{n}\binom{x-1}{q}\right).$$

Then it is readily seen that $Q(x)$ is a polynomial of degree n. Next, let us compute $Q(k)$ for $k = 1, 2, \ldots, n+1$. We have

$$Q(k) = 2\left(1 + \sum_{q=1}^{n}\binom{k-1}{q}\right)$$

$$= 2\left(1 + \sum_{q=1}^{k-1}\binom{k-1}{q}\right)$$

$$= 2(1 + 2^{k-1} - 1)$$

$$= 2^k$$

$$= P(k).$$

Thus, by our earlier corollary, we must have $Q(x) = P(x)$ for all real numbers x. In particular, we have

$$P(n+2) = Q(n+2)$$

$$= 2\left(1 + \sum_{q=1}^{n}\binom{n+1}{q}\right)$$

$$= 2\left(1 + \sum_{q=1}^{n+1}\binom{n+1}{q} - \binom{n+1}{n+1}\right)$$

$$= 2(2^{n+1} - 1) \quad \text{(by (2.3.1))}$$

$$= 2^{n+2} - 2.$$

87. Let $\mathcal{B} = \{B \subseteq X \mid |B| = 5\}$ and $\mathcal{C} = \{A \in \mathcal{A} \mid f(A) \notin A\}$. Furthermore, let $\mathcal{F}$ denote the collection all subsets F of X that is of the form $A \cup \{f(A)\}$, where $A \in \mathcal{C}$. We first note that $\mathcal{F} \subseteq \mathcal{B}$, and $|\mathcal{F}| = |\mathcal{C}| \leq |\mathcal{A}| = \binom{10}{4} = 210$. As $|\mathcal{B}| = \binom{10}{5} = 252 > 210 = |\mathcal{F}|$, it

follows that there exists some element $S \in \mathcal{B} \setminus \mathcal{F}$. Let us show that $f(S \setminus \{r\}) \neq r$ for all $r \in S$.

Arguing by contradiction, suppose there exists some $r \in S$, such that $f(S \setminus \{r\}) = r$. By setting $T = S \setminus \{r\}$, it follows that $F(T) = r \notin S \setminus \{r\} = T$, so this implies that $T \in \mathcal{C}$. As we have $S = (S \setminus \{r\}) \cup \{r\} = T \cup \{f(T)\}$, this implies that $S \in \mathcal{F}$, which is a contradiction. We are done.

88.   (i) By the AM-GM inequality, we have

$$\frac{1}{n} \sum_{r=1}^{n} \binom{n+1}{r} \geq \left(\prod_{r=1}^{n} \binom{n+1}{r} \right)^{\frac{1}{n}}$$

$$\Rightarrow \sum_{r=0}^{n+1} \binom{n+1}{r} - \binom{n+1}{0} - \binom{n+1}{n+1} \geq n \left(\prod_{r=1}^{n} \binom{n+1}{r} \right)^{\frac{1}{n}}$$

$$\Rightarrow 2^{n+1} - 2 \geq n \left(\prod_{r=1}^{n} \binom{n+1}{r} \right)^{\frac{1}{n}}$$

$$\Rightarrow (2^{n+1} - 2)^n \geq n^n \prod_{r=1}^{n} \binom{n+1}{r}.$$

(ii) Let us first prove by induction that $\prod_{r=1}^{n} r^{n+1-r} = \prod_{r=1}^{n} r!$ for all positive integers n, with the base case $n = 1$ being trivial. Suppose that the equation holds for some positive integer $n = m$. By induction hypothesis, we have

$$\prod_{r=1}^{m+1} r^{m+2-r} = \left(\prod_{r=1}^{m+1} r \right) \left(\prod_{r=1}^{m+1} r^{m+1-r} \right)$$

$$= (m+1)! \cdot \prod_{r=1}^{m} r!$$

$$= \prod_{r=1}^{m+1} r!,$$

thereby completing the inductive step. Hence, we have

$$(n!)^{n+1} = \prod_{r=1}^{n} r^{n+1} = \left(\prod_{r=1}^{n} r^r \right) \left(\prod_{r=1}^{n} r^{n+1-r} \right) = \left(\prod_{r=1}^{n} r^r \right) \left(\prod_{r=1}^{n} r! \right).$$

(iii) We have

$$\left(\frac{n(n+1)!}{2^{n+1}-2}\right)^{\frac{n}{2}} = \left(\frac{n^n((n+1)!)^n}{(2^{n+1}-2)^n}\right)^{\frac{1}{2}}$$

$$\leq \left(\frac{n^n((n+1)!)^n}{n^n \prod_{r=1}^{n} \binom{n+1}{r}}\right)^{\frac{1}{2}} \quad \text{(by part (i))}$$

$$= \left(\frac{((n+1)!)^n}{\prod_{r=1}^{n} \frac{(n+1)!}{r!(n+1-r)!}}\right)^{\frac{1}{2}}$$

$$= \left(\frac{((n+1)!)^n}{((n+1)!)^n \prod_{r=1}^{n} \frac{1}{r!} \prod_{r=1}^{n} \frac{1}{(n+1-r)!}}\right)^{\frac{1}{2}}$$

$$= \left(\prod_{r=1}^{n} r! \prod_{k=1}^{n} k!\right)^{\frac{1}{2}} \quad \text{(by setting } k = n+1-r)$$

$$= \prod_{r=1}^{n} r!$$

$$= \frac{(n!)^{n+1}}{\prod_{r=1}^{n} r^r} \quad \text{(by part (ii)).}$$

(iv) Equality holds in part (iii) if and only if $(2^{n+1}-2)^n = n^n \prod_{r=1}^{n} \binom{n+1}{r}$.
In turn, by the AM-GM inequality, we have the equality

$$(2^{n+1} - 2)^n = n^n \prod_{r=1}^{n} \binom{n+1}{r}$$

to hold if any only if $\binom{n+1}{1} = \binom{n+1}{2} = \cdots = \binom{n+1}{n}$. By the unimodal property of the binomial coefficients, the equality $\binom{n+1}{1} = \binom{n+1}{2} = \cdots = \binom{n+1}{n}$ holds if and only if $n = 1$ or $n = 2$. We are done.

89. Let a be a positive integer whose base-n representation consists of distinct digits with the property that, except for the leftmost digit, every digit differs by ± 1 from some digit further to the left. Let the base n-representation of a be $a_1 a_2 \cdots a_m$, and let $a_{\min} = \min\{a_i \mid i = 1, 2, \ldots, m\}$ and $a_{\max} = \max\{a_i \mid i = 1, 2, \ldots, m\}$. We note that we must have $a_i \in \{0, 1, \ldots, n-1\}$ for all $i = 1, 2, \ldots, m$, and $a_1 \neq 0$. Also, by assumption, it follows that for each integer i that satisfies $2 \leq i \leq m$ there exists some integer j with $1 \leq j < i$ such that $|a_i - a_j| = 1$.

Let us show that for each integer r such that $a_1 < r \leq a_{\max}$, there exists some integer b_r such that $a_{b_r} = r$. If $a_1 = a_{\max}$, then there is nothing to show, so let us assume that $a_1 < a_{\max}$. Now, by assumption, either $a_{\max}+1$ or $a_{\max}-1$ must appear on the left of the digit $a_{\max}$. As $a_{\max} = \max\{a_i \mid i = 1, \ldots, m\}$, the former case cannot occur. Hence, $a_{\max} - 1$ must appear on the left of the digit $a_{\max}$. Now, by the same assumption, either $a_{\max}$ or $a_{\max}-2$ must appear on the left of the digit $a_{\max}-1$. As $a_{\max}$ appears on the right of $a_{\max}-1$, the latter case must occur. The desired statement now follows by applying this argument repeatedly. In particular, this argument also shows that $r-1$ must appear on the left of r for each integer r that satisfies $a_1 < r \leq a_{\max}$. Likewise, for each integer s such that $a_{\min} \leq s < a_1$, there exists some integer b_s such that $a_{b_s} = s$. This implies that $\{a_1, \ldots, a_m\} = \{a_{\min}, a_{\min} + 1, \ldots, a_{\max}\}$. In particular, we must have $\{a_1, \ldots, a_m\} = \{b, b+1, \ldots, b+m-1\}$ for some integer $b \in \{0, 1, \ldots, n-m+1\}$.

Now, for each $b \in \{0, 1, \ldots, n-m\}$, the number of permutations $a_1 a_2 \cdots a_m$ of the set $\{b, b+1, \ldots, b+m-1\}$ that satisfy the property that for each integer i that satisfies $2 \leq i \leq m$, there exists some integer j with $1 \leq j < i$ such that $|a_i - a_j| = 1$ is equal to 2^{m-1} by Problem 2.75. We note that there are $n-m+1$ choices for $b \in \{0, 1, \ldots, n-m\}$. Furthermore, by our earlier argument, it follows that the only permutation $a_1 a_2 \cdots a_m$ of the set $\{0, 1, \ldots, m-1\}$ that satisfy $a_1 = 0$ and the aforementioned property is $01 \cdots (m-1)$. Thus, the number of positive integers whose base-n representation satisfy the property stated in the problem is equal to

$$\sum_{m=1}^{n} ((n-m+1)2^{m-1} - 1)$$

$$= \sum_{m=0}^{n-1} (n-m)2^m - n$$

$$= -n + n \sum_{m=0}^{n-1} 2^m - \sum_{m=1}^{n-1} m2^m$$

$$= -n + n(2^n - 1) - \sum_{m=1}^{n-1} m2^{m+1} + \sum_{m=1}^{n-1} m2^m$$

$$= n(2^n - 2) + \sum_{m=1}^{n-1} 2^{m+1} - \sum_{m=1}^{n-1} (m+1)2^{m+1} + \sum_{m=1}^{n-1} m2^m$$

$$= n(2^n - 2) + 2^2 \sum_{m=0}^{n-2} 2^m - n2^n - \sum_{m=2}^{n-1} m2^m + 2 + \sum_{m=2}^{n-1} m2^m$$

$$= n(2^n - 2) + 2^2(2^{n-1} - 1) - n2^n + 2$$

$$= 2^{n+1} - 2n - 2.$$

90. We note that $S_n = \sum_{k=0}^{n} \binom{3n}{3k} \leq \sum_{k=0}^{3n} \binom{3n}{k} = 2^{3n}$, which implies that
$(S_n)^{\frac{1}{3n}} \leq 2$ for all $n \in \mathbf{N}$. On the other hand, let us show that $S_n \geq \frac{1}{8n} 2^{3n}$ for all $n \geq 5$. To this end, let us consider these cases:

Case 1: n is even. By the unimodal property of the binomial coefficients, it follows that $\binom{3n}{\frac{3n}{2}} \geq \binom{3n}{j}$ for all $j \in \{0, 1, \ldots, 3n\}$. In particular, we have $\binom{3n}{\frac{3n}{2}} \geq \binom{3n}{3k}$ for all $k \in \{0, 1, \ldots, n\}$. Hence, we have

$$8nS_n = 8n \sum_{k=0}^{n} \binom{3n}{3k} \geq 8n \binom{3n}{\frac{3n}{2}} \geq (3n+1) \binom{3n}{\frac{3n}{2}} \geq \sum_{k=0}^{3n} \binom{3n}{k} = 2^{3n},$$

or equivalently, $S_n \geq \frac{1}{8n} 2^{3n}$ for all positive even integers $n \geq 2$.

Case 2: n is odd, that is, $n = 2m+1$ for some positive integer $m \geq 2$. By the unimodal property of the binomial coefficients, it follows that $\binom{3n}{\frac{3n-1}{2}} \geq \binom{3n}{j}$ for all $j \in \{0, 1, \ldots, 3n\}$. In particular, we have $\binom{3n}{\frac{3n-1}{2}} \geq \binom{3n}{3k}$ for all $k \in \{0, 1, \ldots, n\}$. Furthermore, we note from identity (2.1.3) that $\binom{2j+1}{j} = \frac{2j+1-j+1}{j} \binom{2j+1}{j-1} = \frac{j+2}{j} \binom{2j+1}{j-1}$ for all positive integers j. As we have $\frac{j+2}{j} \leq 2$ for all positive integers $j \geq 2$, it follows that we have $2\binom{2j+1}{j-1} \geq \frac{j+2}{j} \binom{2j+1}{j-1} = \binom{2j+1}{j}$ for all positive integers $j \geq 2$. Hence, for all integers $n \geq 5$ (or equivalently, $m \geq 2$), we have

$$8nS_n = 8n \sum_{k=0}^{n} \binom{3n}{3k}$$

$$\geq 8n \binom{3n}{\frac{3(n-1)}{2}}$$

$$= 8n \binom{2(3m+1)+1}{3m}$$

$$\geq 4n \binom{2(3m+1)+1}{3m+1}$$

$$= 4n \binom{3n}{\frac{3n-1}{2}}$$

$$\geq (3n+1) \binom{3n}{\frac{3n-1}{2}}$$

$$\geq \sum_{k=0}^{3n} \binom{3n}{k}$$

$$= 2^{3n},$$

or equivalently, $S_n \geq \frac{1}{8n} 2^{3n}$ for all positive odd integers $n \geq 5$. This implies that $S_n \geq \frac{1}{8n} 2^{3n}$ for all $n \geq 5$, or equivalently, $(S_n)^{\frac{1}{3n}} \geq 2 \left(\frac{1}{8n}\right)^{\frac{1}{3n}}$ for all $n \geq 5$. As we have

$$\lim_{n\to\infty} 2 \left(\frac{1}{8n}\right)^{\frac{1}{3n}} = 2 \lim_{n\to\infty} \left(\frac{1}{8}\right)^{\frac{1}{3n}} \lim_{n\to\infty} n^{-\frac{1}{3n}}$$

$$= 2 \left(\lim_{n\to\infty} \left(\frac{1}{2}\right)^{\frac{1}{n}}\right) \left(\lim_{n\to\infty} n^{\frac{1}{n}}\right)^{-\frac{1}{3}}$$

$$= 2,$$

it follows from the Squeeze Theorem that $\lim_{n\to\infty} (S_n)^{\frac{1}{3n}} = 2$.

91. (i) Firstly, let us count the number of n-digit numbers that contains exactly k 0's. We note that there are $\binom{n-1}{k}$ ways to place k 0's in the rightmost $n-1$ digits, and there are 9 choices for each of the remaining $n-k$ non-zero digits. This implies that there are $\binom{n-1}{k} 9^{n-k}$ n-digit numbers that contains exactly k 0's. Furthermore, each number that has k 0's contributes 2^k to the sum S. Hence, we have

$$S = \sum_{n=1}^{10} \left(\sum_{k=0}^{n-1} \binom{n-1}{k} 9^{n-k} 2^k \right)$$

$$= 9 \sum_{n=1}^{10} \left(\sum_{k=0}^{n-1} \binom{n-1}{k} 9^{n-1-k} 2^k \right)$$

$$= 9 \sum_{n=1}^{10} 11^{n-1}$$

$$= 9 \cdot \frac{11^{10} - 1}{11 - 1}$$

$$= \frac{9}{10} (11^{10} - 1)$$

$$= 23343682140.$$

(ii) Firstly, let us count the number of positive integers whose base $b+1$ representation has d digits, of which exactly k of them are 0's. We note that there are $\binom{d-1}{k}$ ways to place k 0's in the rightmost $d-1$

digits, and there are b choices for each of the remaining $d - k$ non-zero digits. This implies that there are $\binom{d-1}{k}b^{d-k}$ positive integers whose base $b + 1$ representation has d digits, of which exactly k of them are 0's. Furthermore, each number that has k 0's in their base $b + 1$ representation contributes a^k to the sum S_n, and n has m digits in its base $b + 1$ representation. Hence, we have

$$
\begin{aligned}
S_n &= \sum_{d=1}^{m} \left(\sum_{k=0}^{d-1} \binom{d-1}{k} b^{d-k} a^k \right) \\
&= b \sum_{d=1}^{m} \left(\sum_{k=0}^{d-1} \binom{d-1}{k} b^{d-1-k} a^k \right) \\
&= b \sum_{d=1}^{m} (a + b)^{d-1} \\
&= \begin{cases} bm & \text{if } a + b = 1, \\ b\frac{(a+b)^m - 1}{a+b-1} & \text{if } a + b \neq 1. \end{cases}
\end{aligned}
$$

92. Let S denote the set of all binary sequences of length $n + 2$ that contain exactly m occurrences of "01" and $m + 1$ occurrences of "10", and let T denote the set of all binary sequences of length n that contain exactly m occurrences of "01". We shall construction a bijection from S to T. To this end, let us define a function $f \colon S \to T$ by $f(a_1 a_2 \cdots a_{n+2}) = a_2 a_3 \cdots a_{n+1}$. Let us show that f is a well-defined bijection. Indeed, for all $a_1 a_2 \cdots a_{n+2} \in S$, let $\{k_1, k_2, \ldots, k_{m+1}\}$ denote the set of integers in $\{1, 2, \ldots, n + 1\}$ that satisfy $a_{k_i} a_{k_i+1} = 10$. Furthermore, let us assume without loss of generality that $k_i < k_j$ for all $i < j$. As there are exactly m occurrences of "01" in $a_1 a_2 \cdots a_{n+2}$, there must exist integers $j_1, j_2, \ldots, j_m$ such that $k_i < j_i < k_{i+1}$ for all $i = 1, 2, \ldots, m$, and $a_{j_i} a_{j_i+1} = 01$ for all $i = 1, 2, \ldots, m$. This necessarily implies that $a_1 a_2 \neq 01$ and $a_{n+1} a_{n+2} \neq 01$, so there are exactly m occurrences of "01" in the binary sequence $a_2 a_3 \cdots a_{n+1}$. This shows that f is well-defined.

Next, let us show that f is injective. Suppose $a_1 a_2 \cdots a_{n+2}$ and $b_1 b_2 \cdots b_{n+2}$ are two sequences in S that satisfy $f(a_1 a_2 \cdots a_{n+2}) = f(b_1 b_2 \cdots b_{n+2})$. Clearly, by the definition of f, we must have $a_i = b_i$ for all $i = 2, 3, \ldots, n - 1$. Also, by an earlier argument in the previous paragraph, we must have $a_1 = 1$, $a_{n+2} = 0$, $b_1 = 1$, $b_{n+2} = 0$, which implies that $a_1 = b_1$ and $a_{n+2} = b_{n+2}$. So this shows that $a_1 a_2 \cdots a_{n+2} = b_1 b_2 \cdots b_{n+2}$, and hence f is injective.

Finally, let us show that f is surjective. Indeed, let us take any $c_1c_2\cdots c_n \in T$, and let us define $c_0 = 1$, and $c_{n+1} = 0$. We shall show that $c_0c_1\cdots c_{n+1} \in S$. Let $\{p_1, p_2, \ldots, p_m\}$ denote the set of integers in $\{1, 2, \ldots, n-1\}$ that satisfy $c_{p_i}c_{p_i+1} = 01$. Furthermore, let us assume without loss of generality that $p_i < p_j$ for all $i < j$. Necessarily, there must exist integers $q_1, q_2, \ldots, q_{m-1}$ such that $p_i < q_i < p_{i+1}$ for all $i = 1, 2, \ldots, m-1$, and $c_{q_i}c_{q_i+1} = 10$ for all $i = 1, 2, \ldots, m-1$. This implies that there are exactly m occurrences of "01" and $m-1$ occurrences of "10" in the sequence $c_{p_1}c_{p_1+1}\cdots c_{p_m+1}$. Now, let us show that there is exactly one occurrence of "10" and no occurrences of "01" in the sequence $c_0c_1\cdots c_{p_1}$. We first note that $c_0 = 1$ and $c_{p_1} = 0$. Suppose on the contrary that there exists some occurrence of "01" in the sequence $c_0c_1\cdots c_{p_1}$. As $c_0 = 1$, it follows that "01" must in fact occur in $c_1\cdots c_{p_1}$, which contradicts the fact that $c_1c_2\cdots c_n$ contain exactly m occurrences of "01". Necessarily, this implies that there must exist some integer $s \in \{1, 2, \ldots, p_1\}$, such that $c_\ell = 1$ for all $\ell < s$, and $c_\ell = 0$ for all $\ell \geq s$. Consequently, there is exactly one occurrence of "10" in the sequence $c_0c_1\cdots c_{p_1}$, and this completes the claim. Likewise, there is exactly one occurrence of "10" and no occurrences of "01" in the sequence $c_{p_m+1}c_{p_m+2}\cdots c_{n+1}$. So there are exactly m occurrences of "01" and $m+1$ occurrences of "10" in the sequence $c_0c_1\cdots c_{n+1}$, and hence we have $c_0c_1\cdots c_{n+1} \in S$ as claimed. Consequently, it is easy to see that $f(c_0c_1\cdots c_{n+1}) = c_1\cdots c_n$, and hence f is surjective, so f is bijective.

Now, let us take any $a_1a_2\cdots a_{n+2} \in S$, and let $\{k_1, k_2, \ldots, k_{2m+1}\}$ denote the set of integers in $\{1, 2, \ldots, n+1\}$ that satisfy $a_{k_i} \neq a_{k_i+1}$ for all $i = 1, 2, \ldots, 2m+1$, and without loss of generality let's assume that $k_i < k_j$ for all $i < j$. We note that there are $\binom{n+1}{2m+1}$ ways to choose the set $\{k_1, \ldots, k_{2m+1}\}$. Furthermore, as there are exactly m occurrences of "01" and $m+1$ occurrences of "10" in the sequence $a_1a_2\cdots a_{n+2}$, we must have $a_{k_i} = 1$ for all odd i and $a_{k_j} = 0$ for all even j. Finally, it is easy to see that there is exactly one choice for a_t for all $t \in \{1, 2, \ldots, n+2\} \setminus \{k_1, k_2, \ldots, k_{2m+1}\}$. So there are $\binom{n+1}{2m+1}$ such sequences in S, and hence there are $\binom{n+1}{2m+1}$ binary sequences of length n that contain exactly m occurrences of "01".

93. It is easy to verify that the result holds for $n = 1$ and $n = 2$, and the assumptions cannot be satisfied for $n = 3$. Henceforth, we assume $n \geq 4$. We label the given assumptions as follows:

Assumption A: Every 2 strangers have exactly 2 common friends.

Assumption B: Every 2 acquaintances have no common friends.

Let the n people be $a_1, a_2, \ldots, a_n$. For each $i = 1, 2, \ldots, n$, let A_i denote the set of acquaintances of a_i. As every 2 acquaintances have no acquaintances in common, it follows that for each $i = 1, 2, \ldots, m$, there must exist some $j_i \in \{1, 2, \ldots, m\} \setminus \{i\}$ such that $a_{j_i} \notin A_i$. This implies that a_i and a_{j_i} must have exactly two acquaintances in common, so we must have $|A_i| \geq 2$ for all $i = 1, 2, \ldots, n$.

Let us first show that if a_m and a_p are acquainted, then $|A_m| = |A_p|$. Firstly, we note that $a_p \in A_m$ and $a_m \in A_p$, and let us define $N_m = A_m \setminus \{a_p\}$ and $N_p = A_p \setminus \{a_m\}$. Next, let us take any $a_i \in N_m$. Then a_i and a_p are not acquainted to each other by assumption B, so a_i and a_p must have exactly two acquaintances in common, of which one of them must be a_m, and the other, say a_q, must be in N_p. We note that a_q is unique in that if there is another person, say $a_r \in N_p \setminus \{a_q\}$ that is acquainted to a_i, we have a violation of Assumption A as a_i and a_p who are strangers now have at least three common friends, a_m, a_q, a_r. We also note that a_q is unique to $a_i \in N_m$ in that there is no other element $a_j \in N_m \setminus \{a_i\}$ such that a_j is acquainted to a_q (*). Also, if some $a_j \in N_m \setminus \{a_i\}$ is acquainted with a_q, then we have a violation of Assumption A as a_m and a_q are strangers with at least three common friends a_p, a_i, a_j. This implies that every person in N_m must have exactly one acquaintance in N_p and $|N_m| \leq |N_p|$ by statement (*). By switching the roles of N_m and N_p, we have $|N_m| = |N_p|$, and consequently $|A_m| = |A_p|$ as desired.

It remains to show that any two strangers a_s and a_t must have the same number of acquaintances. By assumption, there must exist some $u \in \{1, 2, \ldots, n\}$ for which a_u is an acquaintance of both a_s and a_t. This implies that $|A_s| = |A_u|$ and $|A_t| = |A_u|$, and hence we must have $|A_s| = |A_t|$ as desired. Consequently, everyone must have the same number of acquaintances at the gathering.

94. Suppose that n is odd, that is, $n = 2m + 1$ for some non-negative integer m. Let us first compute the coefficient of $x^{n+1} = x^{2m+2}$ in the expansion of $(1 - x^2)^n$ in two different ways. On one hand, the coefficient of $x^{n+1} = (x^2)^{m+1}$ in the expansion of $(1 - x^2)^n$ is equal to $(-1)^{m+1}\binom{n}{m+1} = (-1)^{m+1}\binom{2m+1}{m+1}$ by definition. On the other hand, we first observe that we have $(1 - x^2)^n = (1 - x)^n(1 + x)^n$. For each $k = 1, \ldots, n$, we note that the coefficient of x^k in the expansion of $(1 - x)^n$ is equal to $(-1)^k\binom{n}{k}$, and the coefficient of x^{n+1-k} in the

expansion of $(1 + x)^n$ is equal to $\binom{n}{n+1-k} = \binom{n}{k-1}$. Thus, we have $\sum_{k=1}^{n} (-1)^k \binom{n}{k}\binom{n}{k-1} = (-1)^{m+1}\binom{2m+1}{m+1}$, or equivalently,

$$\sum_{k=1}^{2m+1} (-1)^k \binom{2m+1}{k}\binom{2m+1}{k-1} = (-1)^{m+1}\binom{2m+1}{m+1}.$$

This implies that

$$(-1)^{m+1}\binom{2m+1}{m+1}$$

$$= \sum_{k=1}^{2m+1} (-1)^k \binom{2m+1}{k}\binom{2m+1}{k-1}$$

$$= \sum_{k=1}^{m} (-1)^k \binom{2m+1}{k}\binom{2m+1}{k-1} + (-1)^{m+1}\binom{2m+1}{m+1}\binom{2m+1}{m}$$

$$\quad + \sum_{k=m+2}^{2m+1} (-1)^k \binom{2m+1}{k}\binom{2m+1}{k-1}$$

$$= \sum_{k=1}^{m} (-1)^k \binom{2m+1}{k}\binom{2m+1}{k-1} + (-1)^{m+1}\binom{2m+1}{m+1}^2$$

$$\quad + \sum_{j=1}^{m} (-1)^{2m+2-j} \binom{2m+1}{2m+2-j}\binom{2m+1}{2m+1-j}$$

(by setting $j = 2m + 2 - k$)

$$= \sum_{k=1}^{m} (-1)^k \binom{2m+1}{k}\binom{2m+1}{k-1} + (-1)^{m+1}\binom{2m+1}{m+1}^2$$

$$\quad + \sum_{j=1}^{m} (-1)^j \binom{2m+1}{j-1}\binom{2m+1}{j}$$

$$= 2\sum_{k=1}^{m} (-1)^k \binom{2m+1}{k}\binom{2m+1}{k-1} + (-1)^{m+1}\binom{2m+1}{m+1}^2,$$

or equivalently,

$$2\sum_{k=1}^{m} (-1)^k \binom{2m+1}{k}\binom{2m+1}{k-1}$$

$$= (-1)^{m+1}\binom{2m+1}{m+1} + (-1)^m\binom{2m+1}{m+1}^2.$$

Hence, we have

$$A_2(n)$$

$$= \sum_{k=0}^{m} (-1)^k \left[\binom{2m+1}{k} - \binom{2m+1}{k-1} \right]^2$$

$$= \sum_{k=0}^{m} (-1)^k \left(\binom{2m+1}{k}^2 + \binom{2m+1}{k-1}^2 - 2\binom{2m+1}{k}\binom{2m+1}{k-1} \right)$$

$$= \sum_{k=0}^{m} (-1)^k \binom{2m+1}{k}^2 + \sum_{k=0}^{m} (-1)^k \binom{2m+1}{k-1}^2$$

$$\quad - 2 \sum_{k=0}^{m} (-1)^k \binom{2m+1}{k}\binom{2m+1}{k-1}$$

$$= \sum_{k=0}^{m} (-1)^k \binom{2m+1}{k}^2 + \sum_{k=1}^{m} (-1)^k \binom{2m+1}{k-1}^2$$

$$\quad - (-1)^{m+1} \binom{2m+1}{m+1} - (-1)^m \binom{2m+1}{m+1}^2$$

$$= \sum_{k=0}^{m} (-1)^k \binom{2m+1}{k}^2 - \sum_{k=1}^{m} (-1)^{k-1} \binom{2m+1}{k-1}^2$$

$$\quad + (-1)^m \binom{2m+1}{m+1} - (-1)^m \binom{2m+1}{m+1}^2$$

$$= \sum_{k=0}^{m} (-1)^k \binom{2m+1}{k}^2 - \sum_{k=0}^{m-1} (-1)^k \binom{2m+1}{k}^2$$

$$\quad + (-1)^m \binom{2m+1}{m+1} - (-1)^m \binom{2m+1}{m+1}^2$$

$$= (-1)^m \binom{2m+1}{m}^2 + (-1)^m \binom{2m+1}{m+1} - (-1)^m \binom{2m+1}{m+1}^2$$

$$= (-1)^m \binom{2m+1}{m+1}.$$

On the other hand, we have

$$nA_1(n)$$

$$= (2m+1) \sum_{k=0}^{m} (-1)^k \left[\binom{2m+1}{k} - \binom{2m+1}{k-1} \right]$$

$$= (2m+1) \left[\sum_{k=0}^{m} (-1)^k \binom{2m+1}{k} + \sum_{k=1}^{m} (-1)^{k-1} \binom{2m+1}{k-1} \right]$$

$$= (2m+1) \left[\sum_{k=0}^{m} (-1)^k \binom{2m+1}{k} + \sum_{k=0}^{m-1} (-1)^k \binom{2m+1}{k} \right]$$

$$= (2m+1) \left[(-1)^m \binom{2m}{m} + (-1)^{m-1} \binom{2m}{m-1} \right] \quad \text{(by Problem 2.29)}$$

$$= (-1)^m (2m+1) \left[\binom{2m}{m} - \binom{2m}{m-1} \right]$$

$$= (-1)^m (2m+1) \left[\frac{2m-m+1}{m} \binom{2m}{m-1} - \binom{2m}{m-1} \right] \quad \text{(by (2.1.3))}$$

$$= (-1)^m \frac{2m+1}{m} \binom{2m}{m-1}$$

$$= (-1)^m \binom{2m+1}{m} \quad \text{(by (2.1.2))}$$

$$= (-1)^m \binom{2m+1}{m+1},$$

and hence we have $A_2(n) = nA_1(n)$ for all odd n as desired.

95. Let us consider these cases.

Case 1: n is even, that is, $n = 2m$ for some positive integer m. Then we have

$$B_2(n)$$

$$= \sum_{k=0}^{m} \left[\binom{2m}{k} - \binom{2m}{k-1} \right]^2$$

$$= \sum_{k=0}^{m} \binom{2m}{k}^2 + \sum_{k=1}^{m} \binom{2m}{k-1}^2 - 2 \sum_{k=1}^{m} \binom{2m}{k} \binom{2m}{k-1}$$

$$= \sum_{k=0}^{m} \binom{2m}{k}^2 + \sum_{j=m+1}^{2m} \binom{2m}{2m-j}^2 - \sum_{k=1}^{m} \binom{2m}{k} \binom{2m}{2m+1-k}$$

$$\quad - \sum_{k=1}^{m} \binom{2m}{k} \binom{2m}{2m+1-k} \quad \text{(by setting } j = 2m+1-k\text{)}$$

$$= \sum_{k=0}^{2m} \binom{2m}{k}^2 - \sum_{k=1}^{m} \binom{2m}{k} \binom{2m}{2m+1-k}$$

$$\quad - \sum_{i=m+1}^{2m} \binom{2m}{2m+1-i} \binom{2m}{i} \quad \text{(by setting } i = 2m+1-k\text{)}$$

$$= \binom{4m}{2m} - \sum_{k=1}^{2m} \binom{2m}{k} \binom{2m}{2m+1-k} \quad \text{(by (2.3.6))}$$

$$= \binom{4m}{2m} - \sum_{k=0}^{2m+1} \binom{2m}{k} \binom{2m}{2m+1-k} \quad \left(\text{since } \binom{2m}{2m+1} = 0\right)$$

$$= \binom{4m}{2m} - \binom{4m}{2m+1} \quad (\text{by } (2.3.5))$$

$$= \binom{2n}{n} - \binom{2n}{n+1}$$

$$= \binom{2n}{n} - \frac{2n-(n+1)+1}{n+1}\binom{2n}{n} \quad (\text{by } (2.1.3))$$

$$= \frac{1}{n+1}\binom{2n}{n}.$$

Case 2: n is odd, that is, $n = 2m+1$ for some non-negative integer m. We have

$$B_2(n)$$

$$= \sum_{k=0}^{m} \left[\binom{2m+1}{k} - \binom{2m+1}{k-1} \right]^2$$

$$= \sum_{k=0}^{m} \binom{2m+1}{k}^2 + \sum_{k=1}^{m} \binom{2m+1}{k-1}^2 - 2\sum_{k=1}^{m} \binom{2m+1}{k}\binom{2m+1}{k-1}$$

$$= \sum_{k=0}^{m} \binom{2m+1}{k}^2 + \sum_{j=m+2}^{2m+1} \binom{2m+1}{2m+1-j}^2$$

$$\quad - \sum_{k=1}^{m} \binom{2m+1}{k}\binom{2m+1}{2m+2-k}$$

$$\quad - \sum_{k=1}^{m} \binom{2m+1}{k}\binom{2m+1}{2m+2-k} \quad (\text{by setting } j = 2m+2-k)$$

$$= \sum_{k=0}^{2m+1} \binom{2m+1}{k}^2 - \binom{2m+1}{m+1}^2 - \sum_{k=1}^{m} \binom{2m+1}{k}\binom{2m+1}{2m+2-k}$$

$$\quad - \sum_{i=m+2}^{2m+1} \binom{2m+1}{2m+2-i}\binom{2m+1}{i} \quad (\text{by setting } i = 2m+2-k)$$

$$= \binom{4m+2}{2m+1} - \binom{2m+1}{m+1}^2$$

$$\quad - \left(\sum_{k=1}^{2m+1} \binom{2m+1}{k}\binom{2m+1}{2m+2-k} - \binom{2m+1}{m+1}^2 \right) \quad (\text{by } (2.3.6))$$

$$= \binom{4m+2}{2m+1} - \sum_{k=0}^{2m+2} \binom{2m+1}{k}\binom{2m+1}{2m+2-k}$$

$$\text{(since } \binom{2m+1}{2m+2} = 0)$$

$$= \binom{4m+2}{2m+1} - \binom{4m+2}{2m+2} \quad \text{(by (2.3.5))}$$

$$= \binom{2n}{n} - \binom{2n}{n+1}$$

$$= \frac{1}{n+1}\binom{2n}{n}.$$

In both cases, we have $B_2(n) = \frac{1}{n+1}\binom{2n}{n}$.

96. (i) We have

$$\binom{2n-1}{r}^{-1} = \binom{2n-1}{2n-1-r}^{-1}$$

$$= \frac{2n}{2n-r}\binom{2n}{2n-r}^{-1} \quad \text{(by (2.1.2))}$$

$$= \frac{2n}{2n+1}\frac{2n+1}{2n-r}\binom{2n}{r}^{-1}$$

$$= \frac{2n}{2n+1}\left(1 + \frac{r+1}{2n-r}\right)\binom{2n}{r}^{-1}$$

$$= \frac{2n}{2n+1}\left[\binom{2n}{r}^{-1} + \frac{r+1}{2n-r}\binom{2n}{r}^{-1}\right]$$

$$= \frac{2n}{2n+1}\left[\binom{2n}{r}^{-1} + \binom{2n}{r+1}^{-1}\right] \quad \text{(by (2.1.3))}.$$

(ii) We have

$$\sum_{r=1}^{2n-1}(-1)^{r-1}\binom{2n-1}{r}^{-1}\sum_{j=1}^{r}\frac{1}{j}$$

$$= \sum_{r=1}^{2n-1}\sum_{j=1}^{r}(-1)^{r-1}\frac{2n}{2n+1}\left[\binom{2n}{r}^{-1} + \binom{2n}{r+1}^{-1}\right]\frac{1}{j}$$

$$= \frac{2n}{2n+1}\sum_{j=1}^{2n-1}\sum_{r=j}^{2n-1}(-1)^{r-1}\left[\binom{2n}{r}^{-1} + \binom{2n}{r+1}^{-1}\right]\frac{1}{j}$$

$$= \frac{2n}{2n+1} \sum_{j=1}^{2n-1} \frac{1}{j} \left[\sum_{r=j}^{2n-1} (-1)^{r-1} \binom{2n}{r}^{-1} + \sum_{r=j+1}^{2n} (-1)^r \binom{2n}{r}^{-1} \right]$$

$$= \frac{2n}{2n+1} \sum_{j=1}^{2n-1} \frac{1}{j} \left[\sum_{r=j}^{2n-1} (-1)^{r-1} \binom{2n}{r}^{-1} - \sum_{r=j+1}^{2n} (-1)^{r-1} \binom{2n}{r}^{-1} \right]$$

$$= \frac{2n}{2n+1} \sum_{j=1}^{2n-1} \frac{1}{j} \left[(-1)^{j-1} \binom{2n}{j}^{-1} - (-1)^{2n-1} \binom{2n}{2n}^{-1} \right]$$

$$= \frac{2n}{2n+1} \sum_{j=1}^{2n-1} \frac{1}{j} \left[(-1)^{j-1} \binom{2n}{j}^{-1} + 1 \right]$$

$$= \frac{2n}{2n+1} \left[\sum_{j=1}^{2n-1} \frac{(-1)^{j-1}}{j} \binom{2n}{j}^{-1} + \sum_{j=1}^{2n-1} \frac{1}{j} \right]$$

$$= \frac{2n}{2n+1} \left[\sum_{j=1}^{2n-1} \frac{(-1)^{j-1}}{2n} \binom{2n-1}{j-1}^{-1} \right] + \frac{2n}{2n+1} \sum_{j=1}^{2n-1} \frac{1}{j}$$

(by (2.1.2))

$$= \frac{2n}{2n+1} \left[\frac{1}{2n} \sum_{j=0}^{2n-2} (-1)^j \binom{2n-1}{j}^{-1} \right] + \frac{2n}{2n+1} \sum_{j=1}^{2n-1} \frac{1}{j}$$

$$= \frac{2n}{2n+1} \left[\frac{1}{2n} + \frac{1}{2n} \sum_{j=1}^{2n-2} (-1)^j \binom{2n-1}{j}^{-1} \right] + \frac{2n}{2n+1} \sum_{j=1}^{2n-1} \frac{1}{j}$$

$$= \frac{1}{2n+1} \sum_{j=1}^{2n-2} (-1)^j \binom{2n-1}{j}^{-1} + \frac{2n}{2n+1} \sum_{j=1}^{2n} \frac{1}{j}.$$

It remains to show that

$$\sum_{j=1}^{2n-2} (-1)^j \binom{2n-1}{j}^{-1} = 0.$$

We have

$$\sum_{j=1}^{2n-2} (-1)^j \binom{2n-1}{j}^{-1}$$

$$= \sum_{j=1}^{n-1} (-1)^j \binom{2n-1}{j}^{-1} + \sum_{j=n}^{2n-2} (-1)^j \binom{2n-1}{j}^{-1}$$

$$= \sum_{j=1}^{n-1}(-1)^j\binom{2n-1}{j}^{-1} + \sum_{j=n}^{2n-2}(-1)^j\binom{2n-1}{2n-1-j}^{-1}$$

$$= \sum_{j=1}^{n-1}(-1)^j\binom{2n-1}{j}^{-1} + \sum_{k=1}^{n-1}(-1)^{2n-1-k}\binom{2n-1}{k}^{-1}$$

$$= \sum_{j=1}^{n-1}(-1)^j\binom{2n-1}{j}^{-1} - \sum_{k=1}^{n-1}(-1)^k\binom{2n-1}{k}^{-1}$$

$$= 0.$$

Hence, we have

$$\sum_{r=1}^{2n-1}(-1)^{r-1}\binom{2n-1}{r}^{-1}\sum_{j=1}^{r}\frac{1}{j} = \frac{2n}{2n+1}\sum_{j=1}^{2n}\frac{1}{j}.$$

97. We claim that the double sum is equal to $2^\ell\binom{m-n}{\ell}$. To see that this is indeed the case, we suppose that there are m students in the department, of which n of them are male, and all m students wish to take a combinatorics class that is limited to ℓ students, and each student is graded pass or fail. Then it is easy to see that if the class is restricted to female students only, then there are a total of $2^\ell\binom{m-n}{\ell}$ possible grade reports; indeed, there are $\binom{m-n}{\ell}$ to choose ℓ females out of $m-n$ total female students in the department, and for each student in the class, there are 2 ways to grade the student.

On the other hand, we note that the term $\binom{m-i}{m-\ell}\binom{n}{j}\binom{m-n}{i-j}$ refers to the number of grade reports in which i students pass the class, of which j of them are male; indeed, there are $\binom{n}{j}$ ways to choose j male students out of n total male students to pass the class, there are $\binom{m-n}{i-j}$ ways to choose $i-j$ female students out of $m-n$ total female students to pass the class, and there are $\binom{m-i}{\ell-i} = \binom{m-i}{m-\ell}$ ways to choose the remaining $\ell-i$ students out of $m-i$ remaining students to fail the class.

Let us denote the set of score reports in which there are an even number of male students who pass the class by $\mathcal{E}$, and the set of score reports in which there are an odd number of male students who pass the class by $\mathcal{O}$. Furthermore, let us denote the set of score reports in which there is at least one male student in the class, and an even number of male students who pass the class by $\mathcal{E}^+$. Then it is easy to see that $\mathcal{E}\setminus\mathcal{E}^+$ refers to set of score reports in which all of the students in the class are female, and the double sum is equal to $|\mathcal{E}| - |\mathcal{O}|$. As $|\mathcal{E}| = |\mathcal{E}^+| + |\mathcal{E}\setminus\mathcal{E}^+| = |\mathcal{E}^+| + 2^\ell\binom{m-n}{\ell}$, it remains to show that $|\mathcal{E}^+| = |\mathcal{O}|$.

To complete the proof, let us define a function $f\colon \mathcal{O} \to \mathcal{E}^+$, where for each grade report $R \in \mathcal{O}$, $f(R)$ is defined to the grade report where the grades of all students, save for the eldest male student taking the class, is left unchanged. Then it is easy to see that f is a well-defined bijection, and hence we have $|\mathcal{E}^+| = |\mathcal{O}|$ as desired.

Remark. The solution to the above problem is given by Irl C. Bivens, Davidson College, Davidson, NC. We shall refer the reader to *Amer. Math. Monthly*, **97** (1990), 428–429 for further details. The reader is also encouraged to refer to an alternative solution using the Binomial Theorem by Richard Stong, Harvard University, Cambridge, MA, in the same issue.

98. Let us consider the number of distinct committees consisting of $m + n$ men and $s + n$ women that can be formed from n married couples, q single men and p single women, under the restriction that no couples are allowed on any committee.

    Firstly, we note that for each committee, there must be r married women that are not on the committee, where $r = 0, 1, \ldots, n$. In the case where there are r married women that are not on the committee, we note that the $m + n$ men can be chosen from the q single men and the r married men whose wives are not on the committee, of which there are $\binom{q+r}{m+n}$ ways to do so. Next, we note that the $n - r$ married women on the committee can be chosen in $\binom{n}{n-r} = \binom{n}{r}$ ways, and the remaining $(n + s) - (n - r) = r + s$ single women on the committee can be chosen in $\binom{p}{r+s}$ ways. By (MP) and (AP), the total number of such committees is equal to

$$\sum_{r=0}^{n} \binom{n}{r}\binom{p}{r+s}\binom{q+r}{m+n}.$$

By switching the roles of the men and women, we see in a similar manner that the total number of such committees is equal to

$$\sum_{r=0}^{n} \binom{n}{r}\binom{q}{m+r}\binom{p+r}{n+s}.$$

Hence, we must have

$$\sum_{r=0}^{n} \binom{n}{r}\binom{p}{r+s}\binom{q+r}{m+n} = \sum_{r=0}^{n} \binom{n}{r}\binom{q}{m+r}\binom{p+r}{n+s}$$

as desired.

Remark. The above solution is given by R. C. Lyness. We refer the reader to *Crux Mathematicorum*, **9** (1983), 194–198 for further details (along with other solutions to the given problem).

Solutions to Exercise 3

1. Let ABC be an equilateral triangle of unit side length, and let S be any set of 5 points that lie in triangle ABC. Furthermore, let D, E, F be the midpoints of AB, BC and CA respectively. Then it follows that the triangles ADF, BDE, CEF and DEF are all equilateral triangles, each of $\frac{1}{2}$ unit side length. As each point in S is contained in one of the equilateral triangles ADF, BDE, CEF and DEF, it follows from (PP) that there is a equilateral triangle of $\frac{1}{2}$ unit side length, say triangle DEF, that contains at least two points of S. Let these two points be P_1 and P_2. Now, it is easy to verify that the distance between P_1 and P_2 does not exceed the length of any side of triangle DEF, which is equal to $\frac{1}{2}$ units.

2. It is easy to see that the statement holds for $n = 2$, so let's assume that $n > 2$. Let Γ be a unit circle with centre O, and let $P_1, \ldots, P_n$ be distinct points lying on the circumference of Γ satisfying $\angle P_i O P_{i+1} = \angle P_1 O P_n$ for all $i = 1, \ldots, n-1$. This implies that $\angle P_1 O P_n = \frac{2\pi}{n}$. As each point in C lies on one of the arcs $\overset{\frown}{P_1 P_2}, \overset{\frown}{P_2 P_3}, \ldots, \overset{\frown}{P_{n-1} P_n}, \overset{\frown}{P_n P_1}$, it follows from (PP) that there is an arc, say $\overset{\frown}{P_n P_1}$, that contains at least two points of C, say A and B. Necessarily, we have $\theta = \angle AOB \leq \angle P_1 O P_n = \frac{2\pi}{n}$, so that we have $\frac{\theta}{2} \leq \frac{\pi}{n} \leq \frac{\pi}{2}$. It remains to show that $AB \leq 2 \sin \frac{\pi}{n}$. Let M be the mid-point of AB; then OM is perpendicular to AB, and

$$AB = 2AM = 2OA \sin \angle AOM = 2 \sin \frac{\theta}{2} \leq 2 \sin \frac{\pi}{n},$$

where the last inequality follows from the fact that the sine function is strictly increasing on $\left[0, \frac{\pi}{2}\right]$. We are done.

3. Let us first show that the largest possible area of a triangle contained in a square $ABCD$ of side length d is equal to $\frac{1}{2}d^2$. Let XYZ be a

triangle with the largest possible area within square $ABCD$. We first note that X, Y and Z must lie on the boundary of the square. Indeed, suppose on the contrary that the point X (say) is not on the boundary of the square. Let X' denote the foot of the perpendicular from X to the side YZ. Extend $X'X$ so that it meets the boundary of the square at X''. Now $X''YZ$ is a triangle within the square with a larger area, which contradicts the choice of XYZ. So the vertices of triangle XYZ must all be on the boundary of the square.

Next, we assert that at least two of the three vertices of XYZ must be at the corners of the square. Let us consider the following cases:

Case 1: Two of the three vertices of triangle XYZ lie on the same side of the square, say on side AB. Then it is clear that they must be at the corners A and B in order to attain the largest area. The remaining point must be on the opposite side of the square giving triangle XYZ an area of $\frac{1}{2}d^2$.

Case 2: Each vertex of the triangle lie on a different side of the square. Without loss of generality, suppose X, Y and Z lie on the sides AB, BC and CD respectively. If Z is not at the corner D, we will arrive at a contradiction since triangle XYD has a bigger area than triangle XYZ. If X is not at corner A, then it is now easy to see that Y has to be at corner C and the area of XYZ is again equal to $\frac{1}{2}d^2$.

In both cases, we see that at least two of the three vertices of XYZ must be at the corners of the square. Now, if two or three of the vertices of triangle XYZ are at the vertices of the square $ABCD$, it is easy to see that the largest area that XYZ can have is $\frac{1}{2}d^2$, and the desired statement follows.

Let $ABCD$ be a square of unit side length. Let E, F, G, H be the midpoints of AB, BC, CD and DA respectively, and let I be the point of intersection of EG and FH. Then it follows that the squares $AHIE$, $BEIF$, $CFIG$ and $DGIH$ all have side length of $\frac{1}{2}$ unit each. As each point in S is contained in one of the squares $AHIE$, $BEIF$, $CFIG$ and $DGIH$, it follows from (PP) that there is a square of side length $\frac{1}{2}$ unit, say square $AHIE$, that contains at least three points of S. Let these three points be P_1, P_2 and P_3. By our earlier assertion, the area of the triangle $P_1 P_2 P_3$ is less than or equal to the area of triangle AHI, which is equal to $\frac{1}{8}$ square units.

4. Let S be any set of 5 numbers. For each $i = 0, 1, 2$, let us define
$$A_i = \{s \in S \mid s \equiv i \pmod{3}\}.$$
If A_i is non-empty for all $i = 0, 1, 2$, then let us pick some $a_i \in A_i$ for

all $i = 0, 1, 2$. Then we have

$$a_0 + a_1 + a_2 \equiv 0 + 1 + 2 \pmod{3} \equiv 0 \pmod{3},$$

which implies that $3 \mid a_0 + a_1 + a_2$. Else, there must exist some $j \in \{0, 1, 2\}$, such that A_j is empty. As each element of S is contained in A_k for some $k \in \{0, 1, 2\} \setminus \{j\}$, it follows from (PP) that there must exist some $m \in \{0, 1, 2\} \setminus \{j\}$, such that A_m contains at least three elements of S. Let these three numbers be b_1, b_2 and b_3. Now, we have

$$b_0 + b_1 + b_2 \equiv m + m + m \pmod{3} \equiv 0 \pmod{3},$$

which implies that $3 \mid b_1 + b_2 + b_3$. The desired statement follows.

5. For each $i = 0, 1, \ldots, n - 1$, let us define $A_i = \{a \in A \mid a \equiv i \pmod{n}\}$. As each element of A is contained in A_i for some $i \in \{0, 1, \ldots, n - 1\}$, it follows from (PP) that there must exist some $k \in \{0, 1, \ldots, n - 1\}$, such that A_k contains at least two (distinct) elements of A. Let these two numbers be a and b. Now, we have

$$a - b \equiv k - k \pmod{n} \equiv 0 \pmod{n},$$

which implies that $n \mid a - b$.

6. By (PP), there must exist $k + 1$ elements of A, say $a_1, a_2, \ldots, a_{k+1}$, that are of the same parity. Observe that we must have $\{a_1, a_2, \ldots, a_{k+1}\} \cap \{a_{i_1}, a_{i_2}, \ldots, a_{i_{k+1}}\} \neq \emptyset$; else this would imply that $|\{a_1, a_2, \ldots, a_{k+1}\} \cup \{a_{i_1}, a_{i_2}, \ldots, a_{i_{k+1}}\}| = 2k + 2$, a contradiction. Hence, there must exist some $m, n \in \{1, 2, \ldots, k + 1\}$, such that $a_{i_n} = a_m$. This implies that $a_{i_n} - a_n = a_m - a_n$ is even (since a_m and a_n have the same parity). The desired statement follows.

7. For each $k = 1, 2, \ldots, n$, we let $A_k = \{2^i(2k - 1) \mid i \in \mathbf{N}, 2^i(2k - 1) \leq 2n\}$. As each element of A is contained in A_k for some $k \in \{1, 2, \ldots, n\}$, it follows from (PP) that there must exist some $m \in \{1, 2, \ldots, n\}$, such that A_m contains at least two (distinct) elements of A. Let these two numbers be a and b. Then we must have $a = 2^q(2m - 1)$ and $b = 2^r(2m - 1)$ for some positive integers q and r. As a and b are distinct, we must have $q \neq r$. Without loss of generality, let us assume that $q < r$. Then we have $b = 2^r(2m - 1) = 2^{r-q}2^q(2m - 1) = 2^{r-q}a$, which implies that $a \mid b$ as desired.

8. For each $k = 1, 2, \ldots, n$, let us define $A_k = \{2k - 1, 2k\}$. As each element of A is contained in A_k for some $k \in \{1, 2, \ldots, n\}$, it follows from (PP) that there must exist some $m \in \{1, 2, \ldots, n\}$, such that A_m contains at least two (distinct) elements of A. Let these two numbers be a and b, and we may assume without loss of generality that $a < b$.

Then we must have $a = 2m - 1$ and $b = 2m$, and it is easy to check that a and b are coprime. We are done.

9. Let $A = \{a_1, a_2, \ldots, a_n\}$ be a group of n people, and for each $i = 1, 2, \ldots, n$, let us denote the number of people that a_i knows by $f(i)$. Then it is clear that $0 \le f(i) \le n-1$ for all $i = 1, 2, \ldots, n$. We consider the following cases.

Case 1: There exist distinct $i, j \in \{1, 2, \ldots, n\}$ such that $f(i) = f(j) = 0$. The result holds trivially in this case.

Case 2: There is exactly one $i \in \{1, 2, \ldots, n\}$ such that $f(i) = 0$. In this case, we have $1 \le f(j) \le n - 2$ for all $j \in \{1, 2, \ldots, n\} \setminus \{i\}$. By (PP), we have distinct $j, k \in \{1, 2, \ldots, n\} \setminus \{i\}$ such that $f(j) = f(k)$.

Case 3: We have $f(i) > 0$ for all $i = 1, 2, \ldots, n$. In this case, we have $1 \le f(i) \le n - 1$ for all $i = 1, 2, \ldots, n$, and the (PP) shows that there are distinct $i, j \in \{1, 2, \ldots, n\}$ such that $f(i) = f(j)$.

This completes the proof.

10. For $i = 1, \ldots, n$, we set $I_i = [\frac{i-1}{n}, \frac{i}{n})$. Then it is clear that each element of C is contained in I_i for some $i \in \{1, \ldots, n\}$. We have $n + 1$ points in C and n intervals I_i. By (PP), there exists an interval I_k where $k \in \{1, \ldots, n\}$ that contains at least two points, say r_p and r_q. Since the length of I_k is less than $\frac{1}{n}$, we have $|r_p - r_q| < \frac{1}{n}$.

11. Firstly, we note that $\tan \frac{\pi}{6} = \frac{\sqrt{3}}{3}$. Let us show that $\alpha := \tan \frac{\pi}{12} = 2 - \sqrt{3}$. To this end, we first note from the tangent double angle formula that $\tan 2x = \frac{2 \tan x}{1 - \tan^2 x}$ for all real numbers x. This implies that

$$\frac{\sqrt{3}}{3} = \tan \frac{\pi}{6} = \frac{2\alpha}{1 - \alpha^2},$$

or equivalently, $\alpha^2 + 2\sqrt{3}\alpha - 1 = 0$. Solving for the roots of this quadratic equation yields $\alpha = 2 - \sqrt{3}$ or $\alpha = -2 - \sqrt{3}$. As $\alpha > 0$, we must have $\alpha = 2 - \sqrt{3}$ as desired.

Now, for each $a \in A$, let θ_a denote the unique real number in $\left(-\frac{\pi}{2}, \frac{\pi}{2}\right)$ that satisfy $\tan \theta_a = a$ (note that such a θ_a exists since the tangent function is strictly increasing on $\left(-\frac{\pi}{2}, \frac{\pi}{2}\right)$ and $\tan \left(\left(-\frac{\pi}{2}, \frac{\pi}{2}\right)\right) = (-\infty, \infty)$). Then it is clear that for each $a \in A$, θ_a is contained in $\left[\frac{n\pi}{12}, \frac{(n+1)\pi}{12}\right)$ for some $n \in \{-6, -5, \ldots, 4, 5\}$. By (PP), there must exist some $m \in \{-6, -5, \ldots, 4, 5\}$ and two (distinct) elements of A, say x and y, such that $\left[\frac{m\pi}{12}, \frac{(m+1)\pi}{12}\right)$ contains θ_x and θ_y. Without loss of generality, let us assume that $\theta_x - \theta_y > 0$. Then we must have

$\theta_x - \theta_y < \frac{(m+1)\pi}{12} - \frac{m\pi}{12} = \frac{\pi}{12}$. Hence, we have

$$0 < \frac{x-y}{1+xy} = \frac{\tan\theta_x - \tan\theta_y}{1+\tan\theta_x \tan\theta_y} = \tan(\theta_x - \theta_y) < \tan\frac{\pi}{12} = 2 - \sqrt{3}$$

as desired. (Here, we have used the tangent angle sum formula in the above inequality.)

12. Let us proceed by induction on n that at least one triangle is formed when there are $2n$ points in space, $n > 1$, and they are joined by $n^2 + 1$ segments. To this end, let us first show that the assertion holds for the base case $n = 2$. Let p_1, p_2, p_3, p_4 be points in space. For each point $i = 1, 2, 3, 4$, let $\#(p_i)$ denote the number of points that is joined to p_i by a line segment. Each line segment joins two points. We must have

$$\#(p_1) + \#(p_2) + \#(p_3) + \#(p_4) = 5 \times 2 = 10.$$

By (PP), we must have $\#(p_i) \geq 3$ for some $i \in \{1, 2, 3, 4\}$. Without loss of generality, we may assume that $\#(p_1) \geq 3$. As we have only four points in total, p_1 can be joined to at most the other three points, thus we must have $\#(p_1) = 3$. This implies that p_1 is connected to p_2, p_3 and p_4 by line segments. Since there are five line segments, it is easy to see that there must exist two points amongst p_2, p_3 and p_4, say p_2 and p_3, that are connected by a line segment, so $p_1 p_2 p_3$ is our desired triangle. This completes the proof of our base case.

Now, suppose that the statement holds for some positive integer $n = k > 1$. By induction hypothesis, at least one triangle is formed when there are $2k$ points in space, and they are joined by $k^2 + 1$ segments. Now, let S be any set of $2(k + 1)$ points in space, and let us take any $q_1, q_2 \in S$ that are connected by a line segment. Let $s(q_1)$ and $s(q_2)$ denote the set of points in $S \setminus \{q_1, q_2\}$ that are connected to q_1 and q_2 by line segments respectively. If $|s(q_1)| + |s(q_2)| > 2k$, then we must have $s(q_1) \cap s(q_2) \neq \emptyset$. Let us take any $q_3 \in s(q_1) \cap s(q_2)$; then $q_1 q_2 q_3$ is our desired triangle. Else, if $|s(q_1)| + |s(q_2)| \leq 2k$, then the number of line segments that connect any two points in $S \setminus \{q_1, q_2\}$ must be greater than or equal to $(k+1)^2 + 1 - (|s(q_1)| + |s(q_2)| + 1) = (k+1)^2 + 1 - (2k+1) = k^2 + 1$, so by induction hypothesis, there must exist three points in $S \setminus \{q_1, q_2\}$, say r_1, r_2 and r_3, such that r_1, r_2 and r_3 are pairwise connected to each other by line segments. Then $r_1 r_2 r_3$ is our desired triangle in this case. This completes the induction step and we are done.

Finally, we shall show that it is possible to have $2n$ points in space that are joined by n^2 segments without any triangles being formed. To this end, let $S = \{a_1, \ldots, a_n, b_1, \ldots, b_n\}$ be a set of $2n$ points in space, and we draw a line segment between a_i and b_j for any $i, j \in \{1, 2, \ldots, n\}$. Then S is joined by n^2 line segments, but there are no triangles formed. Indeed, any such triangle must contain at least two points in either $\{a_1, \ldots, a_n\}$ or $\{b_1, \ldots, b_n\}$ by (PP), and by construction, there are no pairs of points in both $\{a_1, \ldots, a_n\}$ and $\{b_1, \ldots, b_n\}$ that are joined by a line segment, which renders the possibility of such a triangle impossible.

13. Let S be any set of 9 lattice points, and let A denote the set of 3-tuples (a_1, a_2, a_3) that satisfy $a_i \in \{0, 1\}$ for all $i = 1, 2, 3$. Then it is easy to see that $|A| = 2^3 = 8$. Furthermore, for each $(a_1, a_2, a_3) \in A$, let us define

$$S_{(a_1, a_2, a_3)} = \{(s_1, s_2, s_3) \in S \mid s_i \equiv a_i \pmod{2} \text{ for all } i = 1, 2, 3\}.$$

As each element of S is contained in $S_{(a_1, a_2, a_3)}$ for some $(a_1, a_2, a_3) \in A$, it follows from (PP) that there must exist some $(b_1, b_2, b_3) \in A$, such that $S_{(b_1, b_2, b_3)}$ contains at least two (distinct) lattice points of S. Let these two lattice points be (t_1, t_2, t_3) and (u_1, u_2, u_3). Then for all $i = 1, 2, 3$, we have $t_i + u_i \equiv b_i + b_i \pmod{2} \equiv 0 \pmod{2}$, which implies that $\left(\frac{t_1 + u_1}{2}, \frac{t_2 + u_2}{2}, \frac{t_3 + u_3}{2}\right)$ is a lattice point. Furthermore, the lattice point $\left(\frac{t_1 + u_1}{2}, \frac{t_2 + u_2}{2}, \frac{t_3 + u_3}{2}\right)$ is the midpoint of the line segment that joins (t_1, t_2, t_3) and (u_1, u_2, u_3), and this completes the proof.

14. (i) Let A denote the set of 2-tuples (a_1, a_2) that satisfy $a_i \in \{0, 1\}$ for all $i = 1, 2$. Then it is easy to see that $|A| = 2^2 = 4$. Furthermore, for each $(a_1, a_2) \in A$, let us define

$$S_{(a_1, a_2)} = \{(s_1, s_2) \in L_2 \mid s_i \equiv a_i \pmod{2} \text{ for all } i = 1, 2\}.$$

As each element of L_2 is contained in $S_{(a_1, a_2)}$ for some $(a_1, a_2) \in L_2$, it follows from (PP) that there must exist some $(b_1, b_2) \in A$, such that $S_{(b_1, b_2)}$ contains at least two (distinct) lattice points of L_2. Let these two lattice points be (t_1, t_2) and (u_1, u_2). Then for all $i = 1, 2$, we have

$$t_i + u_i \equiv b_i + b_i \pmod{2} \equiv 0 \pmod{2},$$

which implies that $\left(\frac{t_1 + u_1}{2}, \frac{t_2 + u_2}{2}\right)$ is a lattice point. Furthermore, the lattice point $\left(\frac{t_1 + u_1}{2}, \frac{t_2 + u_2}{2}\right)$ is the midpoint of the line segment that joins (t_1, t_2) and (u_1, u_2), and this completes the proof.

(ii) Let A denote the set of n-tuples $(a_1, \ldots, a_n)$ that satisfy $a_i \in \{0, 1\}$ for all $i = 1, \ldots, n$. Then it is easy to see that $|A| = 2^n$. Furthermore, for each $(a_1, \ldots, a_n) \in A$, let us define

$$S_{(a_1, \ldots, a_n)} = \{(s_1, \ldots, s_n) \in L_n \mid s_i \equiv a_i \pmod{2} \text{ for all } i\}.$$

As each element of L_n is contained in $S_{(a_1,\ldots,a_n)}$ for some $(a_1,\ldots,a_n) \in L_n$, it follows from (PP) that there must exist some $(b_1,\ldots,b_n) \in A$, such that $S_{(b_1,\ldots,b_n)}$ contains at least two (distinct) lattice points of L_n. Let these two lattice points be $(t_1,\ldots,t_n)$ and $(u_1,\ldots,u_n)$. Then for all $i = 1,\ldots,n$, we have

$$t_i + u_i \equiv b_i + b_i \pmod 2 \equiv 0 \pmod 2,$$

which implies that $\left(\frac{t_1+u_1}{2},\ldots,\frac{t_n+u_n}{2}\right)$ is a lattice point. Furthermore, the lattice point $\left(\frac{t_1+u_1}{2},\ldots,\frac{t_n+u_n}{2}\right)$ is the midpoint of the line segment that joins $(t_1,\ldots,t_n)$ and $(u_1,\ldots,u_n)$, and this completes the proof.

15. Firstly, we note that $3i + 1 = 103 - 3i$ if and only if $i = 17$. For each $i = 1,2,\ldots,16$, let us define $A_i = \{3i+1, 103-3i\}$. Furthermore, let us define $A_0 = \{1\}$ and $A_{17} = \{52\}$. Then it is immediately clear that $\bigcup_{i=0}^{17} A_i = \{1,4,7,\ldots,100\}$. As each element of A is contained in A_k for some $k \in \{0,1,\ldots,17\}$, it follows from (PP) that there must exist some $m \in \{0,1,\ldots,17\}$, such that A_m contains at least two (distinct) elements of A. Then we must have $m \in \{1,2,\ldots,16\}$. Let these two numbers be a and b, and we may assume without loss of generality that $a = 3m+1$ and $b = 103-3m$. Then we have $a+b = 3m+1+103-3m = 104$, and this completes the proof.

16. Let us denote the points in A by $P_1, P_2, P_3, P_4, P_5, P_6$, and for each pair of distinct integers $i, j \in \{1,2,3,4,5,6\}$, let us denote the length of the line segment P_iP_j by $\ell(P_iP_j)$. Without loss of generality, we may assume that $\max_{\substack{i,j\in\{1,\ldots,6\} \\ i\neq j}} \ell(P_iP_j) = \ell(P_1P_2)$. Let Γ_1 and Γ_2 denote circles of radius $\ell(P_1P_2)$ whose centres are at P_1 and P_2 respectively. By the definition of $\ell(P_1P_2)$, it follows that P_3, P_4, P_5 and P_6 must lie in the region of overlap R between Γ_1 and Γ_2.

Now, let us consider the angles $\angle P_iP_1P_j$ in the region R for each pair of distinct integers $i, j \in \{2,3,4,5,6\}$. By construction, we must have $\angle P_iP_1P_j < 180°$ for each pair of distinct integers $i, j \in \{2,3,4,5,6\}$. If there exist distinct $r, s \in \{2,3,4,5,6\}$ such that $\angle P_rP_1P_s > 120°$, then P_1, P_r and P_s are our required points, since one of $\angle P_1P_rP_s$ and $\angle P_1P_sP_r$ must be smaller than $30°$.

Else, there must exist points A_1 and A_5 on the circumference of Γ_1, such that $\angle A_1P_1A_5 = 120°$, and each of the points P_2, P_3, P_4, P_5, P_6 lie in the circular sector S bounded by the lines P_1A_1, P_1A_5 and the minor arc $\overset{\frown}{A_1 A_5}$. Furthermore, let A_2, A_3, A_4 be points on the minor arc $\overset{\frown}{A_1 A_5}$ such that $\angle A_i P_1 A_{i+1} = 30°$ for all $i = 1, 2, 3, 4$, and let us denote the circular sector bounded by the lines $P_1 A_i$, $P_1 A_{i+1}$ and the minor arc $\overset{\frown}{A_i A_{i+1}}$ by R_i for $i = 1, 2, 3, 4$. As each of the points P_2, P_3, P_4, P_5, P_6 must lie in R_j for some $j \in \{1, 2, 3, 4\}$, it follows from (PP) that there must exist some $k \in \{1, 2, 3, 4\}$, such that R_k contains at least two distinct points from the set $\{P_2, P_3, P_4, P_5, P_6\}$. Let these two points be P_i and P_j. Then P_1, P_i and P_j are our required points, for we must have $\angle P_i P_1 P_j \leq \angle A_k P_1 A_{k+1} = 30°$, and this completes the proof.

17. Let $A = \{2^1 - 1, 2^2 - 1, \ldots, 2^{n-1} - 1\}$. Firstly, we note that $(2^k, n) = 1$ for all positive integers k. This implies that $2^k \not\equiv 0 \pmod{n}$ and consequently we must have $2^k - 1 \not\equiv n - 1 \pmod{n}$ for all positive integers k. Now, suppose on the contrary that n does not divide $2^k - 1$ for all $k = 1, 2, \ldots, n - 1$. For each $i = 0, 1, \ldots, n - 1$, let us define $A_i = \{a \in A \mid a \equiv i \pmod{n}\}$. As each element of A is contained in A_k for some $k \in \{1, \ldots, n - 2\}$, it follows from (PP) that there must exist some $j \in \{1, \ldots, n - 2\}$, such that A_j contains at least two (distinct) elements of A. Let these two numbers be $2^p - 1$ and $2^q - 1$, and without loss of generality, let us assume that $p > q$. Now, we have

$$2^q(2^{p-q} - 1) = (2^p - 1) - (2^q - 1) \equiv j - j \pmod{n} \equiv 0 \pmod{n},$$

which implies that $n | 2^q(2^{p-q} - 1)$. As $(2^q, n) = 1$, we must have n to divide $2^{p-q} - 1$, which is a contradiction. So there must exists some element in set A that is divisible by n.

18. Firstly, it is easy to check that $\frac{n-3}{2} > \lfloor \frac{n}{2} \rfloor - 2$ for all positive integers n. Let $S = \{a_1, a_2, \ldots, a_n\}$ denote the set of people at the party, and for each $k = 1, 2, \ldots, n$, let us denote the number of people that a_k knows by $p(k)$. Furthermore, for all integers $i, j \in \{1, 2, \ldots, n\}$ that satisfy $i < j$, let us define $S_{i,j}$ to be the set of people at the party that either knows both or neither of a_i and a_j. We note that there are $\binom{n}{2}$ such subsets $S_{i,j}$. Let us first compute a lower bound for the sum $\sum_{1 \leq i < j \leq n} |S_{i,j}|$. For each $i, j, k \in \{1, 2, \ldots, n\}$ that satisfy $i < j$, we note that $a_k \in S_{i,j}$ if and only if a_k either knows both or neither of a_i and a_j. This implies that the number of sets $S_{i,j}$ for which a_k appears is equal to

$$\binom{p(k)}{2} + \binom{n - 1 - p(k)}{2} = p(k)^2 - (n - 1)p(k) + \frac{(n - 1)(n - 2)}{2}$$

$$= \left(p(k) - \frac{n-1}{2}\right)^2 + \frac{(n-1)(n-3)}{4}$$

$$\geq \frac{(n-1)(n-3)}{4}$$

$$> \frac{n-1}{2}\left(\left\lfloor\frac{n}{2}\right\rfloor - 2\right).$$

Hence, we have

$$\sum_{1 \leq i < j \leq n} |S_{i,j}| = \sum_{k=1}^{n} \left(\binom{p(k)}{2} + \binom{n-1-p(k)}{2}\right) > \binom{n}{2}\left(\left\lfloor\frac{n}{2}\right\rfloor - 2\right).$$

By (PP), there must exist integers $q, r \in \{1, 2, \ldots, n\}$ with $q < r$ such that $|S_{q,r}| \geq \left\lfloor\frac{n}{2}\right\rfloor - 1$. This implies that the number of people that either knows both or neither of a_q and a_r is equal to $\left\lfloor\frac{n}{2}\right\rfloor - 1$, and we are done.

19. It suffices to show that for any 6-element subset T of $\{1, 2, \ldots, 15\}$, there exist 2 disjoint subsets non-empty B and C of S such that $s(B) = s(C)$. Firstly, we note that there are $\binom{6}{1} + \binom{6}{2} + \binom{6}{3} + \binom{6}{4} = 56$ subsets D of T such that $1 \leq |D| \leq 4$. Furthermore, we note that if D is a subset of T that satisfies $1 \leq |D| \leq 4$, then we must have $1 \leq s(D) \leq 12 + 13 + 14 + 15 = 54$. By (PP), there must exist two distinct subsets P and Q of T, such that $1 \leq |P|, |Q| \leq 4$ and $s(P) = s(Q)$. If P and Q are disjoint, then we are done. Else, let us set $R = P \cap Q$, and $P' = P \setminus R$, $Q' = Q \setminus R$. Then we have

$$s(P') = s(P) - s(R) = s(Q) - s(R) = s(Q').$$

Furthermore, P' and Q' are necessarily disjoint and non-empty, and this completes the proof.

20. Let us denote the elements in the rearranged array by $b_{ij}, 1 \leq i \leq m, 1 \leq j \leq n$. Suppose on the contrary that there exists a row in the rearranged array, such that the difference between the maximum and minimum elements in the row exceeds d. Then there must exist some row i and columns j and k such that $b_{ij} - b_{ik} > d$. As we have $b_{xj} \geq b_{ij}$ and $b_{ik} \geq b_{yk}$ for all $x = 1, \ldots, i$ and $y = i, i+1, \ldots, m$, this implies that $b_{xj} - b_{yk} \geq b_{ij} - b_{ik} > d$ for all $x = 1, \ldots, i$ and $y = i, i+1, \ldots, m$. As we have $|\{b_{1j}, b_{2j}, \ldots, b_{ij}, b_{ik}, b_{(i+1)k}, \ldots, b_{mk}\}| = m + 1 > m$, it follows from (PP) that there must exist some $r \in \{1, 2, \ldots, i\}$ and $s \in \{i, i+1, \ldots, m\}$, such that b_{rj} and b_{sk} are in the same row of the original array, say row t. As we have $b_{rj} - b_{sk} > d$, it follows that the difference between the maximum and minimum elements in row t also exceeds d, which is a contradiction. So the desired statement holds.

21. Let $P_1 P_2 \cdots P_{10}$ be a regular decagon, and let us consider the diagonals $P_1 P_4$, $P_2 P_6$ and $P_3 P_9$. Let A_1 be the intersection of $P_1 P_4$ and $P_2 P_6$, A_2 be the intersection of $P_1 P_4$ and $P_3 P_9$ and A_3 be the intersection of $P_2 P_6$ and $P_3 P_9$. Then it is easy to check that no other diagonals in the decagon, apart from $P_1 P_4$, $P_2 P_6$ and $P_3 P_9$, passes through any of the points A_1, A_2 and A_3.

    Now, let the number of times that the signs of all the numbers along a given side or a given diagonal ℓ have changed after n operations be $s(n, \ell)$. By (PP), it follows that for any positive integer n, there must exist two diagonals in $\{P_1 P_4, P_2 P_6, P_3 P_9\}$, say $P_1 P_4$ and $P_2 P_6$, such that $s(n, P_1 P_4)$ and $s(n, P_2 P_6)$ have the same parity. This implies that the sign of the number at A_1 (which is the intersection point of $P_1 P_4$ and $P_2 P_6$) have been changed an even number of times, so the sign of the number at A_1 is positive after n operations. Consequently, this shows that after any number of operations, at least 1 of the points in $\{A_1, A_2, A_3\}$ have the sign of its number to be positive. Hence, it is not possible to change the signs of all numbers in the decagon to negative after a certain number of operations.

22. Arguing by contradiction, suppose otherwise that no two teams finished with the same amount of points. We note that there are $\binom{28}{2} = 378$ total matches being played, so there are most $\lfloor 0.25 \times 378 \rfloor = 94$ matches that are not draws. Let the set of teams be $T = \{t_1, t_2, \ldots, t_{28}\}$, and for each $i = 1, 2, \ldots, 28$, let $s(i)$ denote the number of points that team t_i has, and let T_i denote the set of matches that team t_i has won. Then we note that T_i and T_j are disjoint for all distinct i and j. As at most one team has 27 points, it follows from (PP) that there are at least 14 teams in T, say $t_1, t_2, \ldots, t_{14}$, such that we have either $s(i) > 27$ or $s(i) < 27$ for all $i = 1, 2, \ldots, 14$. Without loss of generality, let us assume the former. As $s(1) - 27, s(2) - 27, \ldots, s(14) - 27$ are pairwise distinct positive integers, it follows that we have $\sum_{i=1}^{14} [s(i) - 27] \geq \sum_{i=1}^{14} i = 105$. Let us show that for each $i = 1, 2, \ldots, 14$, team t_i has won at least $s(i) - 27$ matches. Suppose on the contrary that team t_i has won at most $s(i) - 28$ matches. Let w and d denote the number of matches that team t_i has won and drawn respectively. It follows that

$$s(i) = 2w + d \leq 2w + 27 - w = 27 + w \leq 27 + s(i) - 28 = s(i) - 1,$$

a contradiction, so team t_i must have won at least $s(i) - 27$ matches.

Hence, we have

$$\sum_{i=1}^{14} |T_i| \geq \sum_{i=1}^{14} [s(i) - 27] \geq 105,$$

which implies that at least 105 matches are not draws, a contradiction. So there were two teams who finished with the same amount of points.

23. By representing a correct answer with a 1 and a wrong answer with 0, we can represent an answer pattern by a binary sequence of length 15, with the added restriction that none of the 1's are consecutive. Then it follows that there are at most 8 1's in any such sequence. If there are n 1's then they must appear as $n - 1$ (10) blocks and a single 1 in the answer pattern. Now, if there are n 1's in the binary sequence, then there are $16 - n$ slots available for the n 1's. Consequently, this implies that there are $\binom{16-2n+n}{n} = \binom{16-n}{n}$ binary sequences with n 1's, such that none of the 1's are consecutive. Hence, the total number of possible answer patterns is equal to $\sum_{n=0}^{8} \binom{16-n}{n} = 1597$. By (PP), there must exist two candidates whose answer patterns are identical.

24. Firstly, we note that we must have $a_1 \leq 2$; if $a_1 > 2$, then this would imply that $a_i > 2$ for all $i = 1, 2, \ldots, n$, which implies that $a_1 + a_2 + \cdots + a_n > 2n$, a contradiction. Let $S = \{a_1, a_1 + a_2, \ldots, a_1 + a_2 + \cdots + a_{n-1}, -a_n + a_1\}$. As $-a_n < 0$, it follows that we have $|S| = n$. Let us consider the following cases:

Case 1: There exist some $x \in S$ such that $x \equiv 0 \pmod{n}$. If $x = a_1 + a_2 + \cdots + a_m$ for some $m \in \{1, 2, \ldots, n - 1\}$, then we can take $K = \{1, 2, \ldots, m\}$, for we must have

$$0 < a_1 + a_2 + \cdots + a_m < a_1 + a_2 + \cdots + a_n = 2n.$$

That is, we have $a_1 + \cdots + a_m = n$.

Else, we must have $x = a_1 - a_n$. Then we must have $a_n - a_1 = \ell n$ for some non-negative integer ℓ. If $\ell \geq 2$, then this would imply that

$$a_n = \ell n + a_1 \geq 2n + a_1 > 2n,$$

which would then imply that $a_1 + a_2 + \cdots + a_n > 2n$, a contradiction. If $\ell = 1$, then we have $a_n = n + a_1$. As $a_n \neq n + 1$, we must have $a_1 = 2$ and $a_n = n + 2$. This implies that

$$a_1 + a_2 + \cdots + a_n = a_1 + a_2 + \cdots + a_{n-1} + n + 2 \geq n + 2 + 2(n - 1) = 3n,$$

a contradiction. So we must have $\ell = 0$, and consequently, this implies that $a_i = 2$ for all $i = 1, 2, \ldots, n$. As n is even, we can take K to be the set $\{1, 2, \ldots, \frac{n}{2}\}$.

Case 2: $x \not\equiv 0 \pmod{n}$ for all $x \in S$. For each $i = 0, 1, \ldots, n-1$, let us define $S_i = \{s \in S \mid a \equiv i \pmod{n}\}$. As each element of S is contained in S_k for some $k \in \{1, \ldots, n-1\}$, it follows from (PP) that there must exist some $j \in \{1, \ldots, n-1\}$, such that S_j contains at least two (distinct) elements of S. Let these two numbers be x and y. Without loss of generality, we may assume that $x > y$. Necessarily, x must be of the form $a_1 + a_2 + \cdots + a_m$ for some $m \in \{1, 2, \ldots, n-1\}$ (since $-a_n + a_1 < 0$). If $y = a_1 + a_2 + \cdots + a_p$ for some $p \in \{1, 2, \ldots, n-1\}$, then we must have $p < m$. Consequently, we have

$$a_{p+1} + a_{p+2} + \cdots + a_m = x - y \equiv j - j \pmod{n} \equiv 0 \pmod{n}.$$

Then we can take $K = \{p+1, p+2, \ldots, m\}$, for we must have

$$0 < a_{p+1} + a_{p+2} + \cdots + a_m < 2n.$$

or equivalently, $a_{p+1} + \cdots + a_m = n$.

Else, we must have $y = -a_n + a_1$. Consequently, we have

$$a_2 + a_3 + \cdots + a_m + a_n = x - y \equiv j - j \pmod{n} \equiv 0 \pmod{n}.$$

Then we can take $K = \{2, 3, \ldots, m, n\}$, for we must have

$$0 < a_2 + a_3 + \cdots + a_m + a_n < 2n,$$

and $a_2 + \cdots + a_n = n$.

In both cases, we have found some subset K of $\{1, 2, \ldots, n\}$ for which $\sum_{i \in K} a_i = n$, so we have proven the statement when n is even. Now, if n is odd and we make the additional assumption that $a_n \neq 2$, then by proceeding in a similar argument as before, we notice that the subcase $-a_n + a_1 \equiv 0 \pmod{n}$ in Case 1 cannot occur since $a_n \neq 2$. The argument in the latter case is independent of the parity of n, which implies that the argument in Case 2 would still hold for odd n. Hence, the same statement would still hold when n is odd, provided that we make the additional assumption that $a_n \neq 2$.

25. Let $X = \{a_1, a_2, \ldots, a_n\}$ and $C = \{1, 2, \ldots, p\}$ Let us first show that $p \leq n$. Indeed, if $p > n$, then let us colour each subset A of X by colour $|A| + 1$, so that each subset of X is coloured from the set $\{1, 2, \ldots, n+1\}$. it is easy to see that for any distinct subsets B and C of X, at least two of the sets B, C, $B \cup C$ and $B \cap C$ must have distinct colours. If not, then we must have $|C| = |B \cup C| = |B \cap C|$. The first equality would imply that $C = B \cup C$, while the second equality would imply that $C = B \cap C$. Consequently, this would imply that $B = C$, which is a contradiction.

Next, let us show that $p = n$. Let us take any arbitrary colouring of the subsets of X with n colours $\{1, 2, \ldots, n\}$. If there exist any non-empty subset A of X that is coloured with the same colour as that of the empty set $\emptyset$, then we notice that $\emptyset$, A, $\emptyset \cap A = \emptyset$ and $\emptyset \cup A = A$ all have the same colour. Else, let us assume that the empty set $\emptyset$ is coloured with colour n, and all non-empty subsets of X is coloured with some colour from $\{1, 2, \ldots, n-1\}$. Now, let us define $X_k = \{a_1, a_2, \ldots, a_k\}$ for all $k = 1, 2, \ldots, n$. By (PP), there exist distinct integers $i, j \in \{1, 2, \ldots, n\}$, such that X_i and X_j are coloured with the same colour from $\{1, 2, \ldots, n-1\}$. Without loss of generality, let us assume that $i < j$. Then we notice that X_i, X_j, $X_i \cap X_j = X_i$ and $X_i \cup X_j = X_j$ all have the same colour. This shows that we can find two distinct subsets A and B of X, such that A, B, $A \cap B$ and $A \cup B$ have the same colour for any arbitrary colouring of the subsets of X. So we must have $p = n$.

26. Let A denote the set of q-tuples $(y_1, \ldots, y_q)$ that satisfy $|y_i| \leq p$ for all $i = 1, \ldots, q$. Furthermore, let us define $f_k(y) = a_{k1}y_1 + \cdots + a_{kq}y_q$ for all $k = 1, \ldots, p$ and q-tuples $y = (y_1, \ldots, y_q)$ of integers, and let us define S to be the set of p-tuples of the form $(f_1(y), \ldots, f_p(y))$ for some $y \in A$. We first note that $|A| = (2p+1)^q$. Furthermore, for each $k = 1, \ldots, p$ and $y = (y_1, \ldots, y_q) \in A$, we have

$$\begin{aligned}
|f_k(y)| &= |a_{k1}y_1 + a_{k2}y_2 + \cdots + a_{kq}y_q| \\
&\leq |a_{k1}y_1| + |a_{k2}y_2| + \cdots + |a_{kq}y_q| \\
&\leq |y_1| + |y_2| + \cdots + |y_q| \\
&\leq pq.
\end{aligned}$$

This implies that $|S| \leq (2pq+1)^p$. Now, we observe that

$$\begin{aligned}
|A| &= (2p+1)^q \\
&= (2p+1)^{2p} \\
&= (4p^2 + 4p + 1)^p \\
&\geq (4p^2 + 1)^p \\
&\geq (2pq + 1)^p \\
&\geq |S|.
\end{aligned}$$

By (PP), there must exist distinct $y = (y_1, \ldots, y_q), y' = (y'_1, \ldots, y'_q) \in A$ such that $(f_1(y'), \ldots, f_p(y')) = (f_1(y'), \ldots, f_p(y'))$. By setting $x_i = y_i - y'_i$ for all $i \in \{1, \ldots, q\}$ and setting $x = (x_1, \ldots, x_q)$, it is easy to see that x_i is an integer, and

$$f_i(x) = f_i(y - y') = f_i(y) - f_i(y') = 0$$

for all $i \in \{1, \ldots, q\}$, which implies that $x = (x_1, \ldots, x_q)$ is a solution to the system of equations. Furthermore, since $y \neq y'$, it follows that there must exist some $j \in \{1, \ldots, q\}$ for which $x_j \neq 0$. Finally, we have

$$|x_j| = |y_j - y'_j| \leq |y_j| + |y'_j| \leq 2p = q$$

for all $j \in \{1, \ldots, q\}$, and this completes the proof.

27. Let the countries be $C_1, C_2, C_3, C_4, C_5, C_6$. As $1978 > 1974 = 6 \times 329$, it follows from (PP) that there are at least 330 members that come from the same country, say C_1. Let their numbers be $a_1, a_2, \ldots, a_{330}$, and without loss of generality, let us also assume that $a_1 < a_2 < \cdots < a_{330}$. Let us consider the differences $a_2 - a_1, a_3 - a_1, \ldots, a_{330} - a_1$. If there are any members from country C_1 whose number is of the form $a_i - a_1$ for some $i \in \{2, 3, \ldots, 330\}$, then we are done. Else, all members whose number is of the form $a_i - a_1$ for some $i \in \{2, 3, \ldots, 330\}$ must come from countries C_2, C_3, C_4, C_5, C_6.

As $329 > 325 = 5 \times 65$, it follows from (PP) that there are at least 66 members whose number is of the form $a_i - a_1$ for some $i \in \{2, 3, \ldots, 330\}$ and come from the same country, say C_2. Let their numbers be $b_1, b_2, \ldots, b_{66}$, and without loss of generality, let us also assume that $b_1 < b_2 < \cdots < b_{66}$. Let us consider the differences $b_2 - b_1, b_3 - b_1, \ldots, b_{66} - b_1$. If there are any members from country C_2 whose number is of the form $b_i - b_1$ for some $i \in \{2, 3, \ldots, 66\}$, then we are done. If there are any members from country C_1 whose number is of the form $b_i - b_1$ for some $i \in \{2, 3, \ldots, 66\}$, then we are also done. To see this, we note that $b_i - b_1 = (a_m - a_1) - (a_n - a_1) = a_m - a_n$ for some distinct m, n with $1 \leq m, n \leq 330$. Now $a_m - a_n, a_n, a_m \in C_1$ with $(a_m - a_n) + a_n = a_m$. We note that if $a_m - a_n = a_n$, then a_m is twice a_n and the result holds. Else, all members whose number is of the form $b_i - b_1$ for some $i \in \{2, 3, \ldots, 66\}$ must come from countries C_3, C_4, C_5, C_6.

As $65 > 64 = 4 \times 16$, it follows from (PP) that there are at least 17 members whose number is of the form $b_i - b_1$ for some $i \in \{2, 3, \ldots, 66\}$ and come from the same country, say C_3. Let their numbers be $c_1, c_2, \ldots, c_{17}$, and without loss of generality, let us also assume that $c_1 < c_2 < \cdots < c_{17}$. Let us consider the differences $c_2 - c_1, c_3 - c_1, \ldots, c_{17} - c_1$. By a similar argument as above, if there are any members from countries C_1, C_2, C_3 whose number is of the form $c_i - c_1$ for some $i \in \{2, 3, \ldots, 17\}$, then we are done. Else, all members whose number is of the form $c_i - c_1$ for some $i \in \{2, 3, \ldots, 17\}$ must come from countries C_4, C_5, C_6.

As $16 > 15 = 3 \times 5$, it follows from (PP) that there are at least 6 members whose number is of the form $c_i - c_1$ for some $i \in \{2, 3, \ldots, 17\}$ and come from the same country, say C_4. Let their numbers be $d_1, \ldots, d_6$, and without loss of generality, let us also assume that $d_1 < \cdots < d_6$. Let us consider the differences $d_2 - d_1, d_3 - d_1, \ldots, d_6 - d_1$. By a similar argument as above, if there are any members from countries C_1, C_2, C_3, C_4 whose number is of the form $d_i - d_1$ for some $i \in \{2, 3, \ldots, 6\}$, then we are done. Else, all members whose number is of the form $d_i - d_1$ for some $i \in \{2, 3, \ldots, 6\}$ must come from countries C_5 and C_6.

As $5 > 4 = 2 \times 2$, it follows from (PP) that there are at least 3 members whose number is of the form $d_i - d_1$ for some $i \in \{2, 3, \ldots, 6\}$ and come from the same country, say C_5. Let their numbers be e_1, e_2 and e_3, and without loss of generality, let us also assume that $e_1 < e_2 < e_3$. Let us consider the differences $e_2 - e_1$ and $e_3 - e_1$. By a similar argument as above, if there are any members from countries C_1, C_2, C_3, C_4, C_5 whose number is of the form $e_i - e_1$ for some $i \in \{2, 3\}$, then we are done. Else, all members whose number is of the form $e_i - e_1$ for some $i \in \{2, 3\}$ must come from country C_6. If $e_3 - e_2 \in C_i$ for any $i \in \{1, 2, 3, 4, 5\}$ then the result holds using a similar argument used earlier. So we may assume $e_3 - e_2 \in C_6$. If $e_3 - e_2 \neq e_2 - e_1$, then $(e_2 - e_1), (e_3 - e_2), (e_3 - e_1)$ are distinct members of C_6 with the last number being the sum the previous two. If $e_3 - e_2 = e_2 - e_1$, then $e_3 - e_1 = 2(e_2 - e_1)$, and the result holds. This completes the proof and we are done.

28. Let $(a_i)_{i=1}^{R(p,q)}$ be any sequence of $R(p, q)$ distinct integers, and let $\{v_1, v_2, \ldots, v_{R(p,q)}\}$ be the set of vertices of the $R(p, q)$-clique $K_{R(p,q)}$. For each pair of integers $i, j \in \{1, 2, \ldots, R(p, q)\}$ satisfying $i < j$, let us colour the edge $v_i v_j$ blue if $a_i < a_j$ and red otherwise. By the definition of $R(p, q)$, there exists either a blue p-clique or a red q-clique. If we have a blue p-clique whose vertex set is $\{v_{i_1}, v_{i_2}, \ldots, v_{i_p}\}$ where $i_1 < i_2 < \cdots < i_p$, then we must have $a_{i_1} < a_{i_2} < \cdots < a_{i_p}$, which implies that $(a_{i_j})_{j=1}^{p}$ is an increasing sequence of p terms. Likewise, if we have a red q-clique instead, then we would get a decreasing sequence of q terms. We are done.

29. Let $(a_i)_{i=1}^{pq+1}$ be any sequence of $pq + 1$ distinct integers, and let us define $\ell\left((a_{i_j})_{j=1}^{m}\right) = m$ for all subsequences $(a_{i_j})_{j=1}^{m}$ of $(a_i)_{i=1}^{pq+1}$. Fur-

thermore, for each $k = 1, 2, \ldots, pq + 1$, let us define A_k to be the set of increasing subsequences $(a_{i_j})_{j=1}^{m}$ of $(a_i)_{i=1}^{pq+1}$, such that $a_{i_1} = a_k$, and let us define $n(k) = \max\limits_{(a_{i_j})_{j=1}^{m} \in A_k} \ell\left((a_{i_j})_{j=1}^{m}\right)$. That is, $n(k)$ is the length of the longest increasing subsequence starting with a_k. By definition, we must have $n(k) \geq 1$ for all $k = 1, 2, \ldots, pq + 1$. If there exists some $k \in \{1, 2, \ldots, pq + 1\}$ for which $n(k) \geq p + 1$, then by the definition of $n(k)$, this would imply that there exists an increasing subsequence of $p + 1$ terms. Henceforth, let us assume that $1 \leq n(k) \leq p$ for all $k = 1, 2, \ldots, pq + 1$.

By (PP), there exist some $r \in \{1, \ldots, p\}$, such that $|\{k \mid n(k) = r\}| \geq q + 1$. Let $k_1, k_2, \ldots, k_{q+1}$ be integers in $\{1, 2, \ldots, pq + 1\}$ satisfying $n(k_i) = r$ for all $i = 1, 2, \ldots, q + 1$ and $k_1 < k_2 < \cdots < k_{q+1}$. Let us show that for all integers $i, j \in \{1, 2, \ldots, q+1\}$ satisfying $i < j$, we have $a_{k_i} > a_{k_j}$. If there exist some pair of integers $s, t \in \{1, 2, \ldots, q + 1\}$ satisfying $s < t$ and $a_{k_s} < a_{k_t}$, then let $(a_{i_j})_{j=1}^{n(k_t)}$ be an increasing subsequence of $(a_i)_{i=1}^{pq+1}$, such that $a_{i_1} = a_{k_t}$ (which exists by the definition of $n(k_t)$). Then it is easy to check that $(a_{k_s}, a_{i_1}, a_{i_2}, \ldots, a_{i_{n(k_t)}})$ is an increasing subsequence consisting of $n(k_t) + 1$ terms, so this implies that $n(k_s) \geq n(k_t) + 1 = n(k_s) + 1$, a contradiction. Hence, we have $a_{k_i} > a_{k_j}$ for all integers $i, j \in \{1, 2, \ldots, q + 1\}$ satisfying $i < j$. This implies that there exists a decreasing subsequence of $q + 1$ terms, and we are done.

30. (a) Let us take any 2-colouring, red and blue, of the edges of the $R(q, p)$-clique, and obtain a new 2-colouring of the edges of the $R(q, p)$-clique by swapping the colours of each of the edges in the old 2-colouring of the edges of the $R(q, p)$-clique. By the definition of $R(q, p)$, there exists a blue q-clique or a red p-clique in the new 2-colouring of the edges of the $R(q, p)$-clique. This implies that there exists a red q-clique or a blue p-clique in the original 2-colouring of the edges of the $R(q, p)$-clique. By the definition of $R(p, q)$, we must have $R(p, q) \leq R(q, p)$. By symmetry, we must have $R(q, p) \leq R(p, q)$, so we have $R(p, q) = R(q, p)$ as desired.

(b) Firstly, let us take any 2-colouring of the edges of the q-clique. If any of the edges of the q-clique is coloured blue, then we have a blue 2-clique. Else, all of the edges of the q-clique must be coloured red and we thereby obtain a red q-clique. So we have $R(2, q) \leq q$. Next consider a 2-colouring of the edges of a $(q - 1)$-clique where all of the edges of the $(q - 1)$-clique are coloured red. Then it is easy to see that neither a blue 2-clique nor a red q-clique exists in this 2-colouring of the edges of the $(q-1)$-clique. So we must have $R(2,q) > q-1$, and hence we have $R(2,q) = q$ as desired.

31. (i) Let us take any 2-colouring of the $R(p,q)$-clique. Then there exists a blue p-clique or a red q-clique in the 2-colouring of the edges of the $R(p,q)$-clique. This implies that there exists a blue p'-clique or a red q'-clique in the 2-colouring of the edges of the $R(p,q)$-clique. By the definition of $R(p',q')$, we must have $R(p',q') \leq R(p,q)$.

(ii) Consider any 2-colouring of the edges of the $(R(p,q)-1)$-clique. Let us add a vertex v to the vertex set of the $(R(p,q)-1)$-clique, add edges connecting v to the vertices of the $(R(p,q)-1)$-clique to obtain an $R(p,q)$-clique, and colour these new edges blue to obtain a 2-colouring of the edges of the newly formed $R(p,q)$-clique. Now, if the 2-colouring of the edges of the $R(p,q)$-clique contains a red q-clique, then this would imply that the 2-colouring of the edges of a $(R(p,q)-1)$-clique must contain a red q-clique. Else, the 2-colouring of the edges of the $R(p,q)$-clique must contain a blue p-clique. If the vertex set of the blue p-clique does not contain v, then this would imply that the 2-colouring of the edges of a $(R(p,q)-1)$-clique must contain a blue p-clique (and hence a blue $(p-1)$-clique). Else, this would imply that the 2-colouring of the edges of the $R(p,q)$-clique must contain a blue $(p-1)$-clique whose vertex set does not contain v, and consequently, the 2-colouring of the edges of a $(R(p,q)-1)$-clique must contain a blue $(p-1)$-clique. So we must have $R(p-1,q) \leq R(p,q)-1$ as desired.

(iii) Clearly, if $p' = p$ and $q' = q$, then we must have $R(p',q') = R(p,q)$. Conversely, suppose we have $p' \neq p$ or $q' \neq q$. Then we must have $p' < p$ or $q' < q$. If we have $p' < p$, then we have by part (ii) that

$$R(p',m) \leq R(p'+1,m) - 1 \leq \cdots \leq R(p,m) - (p-p') < R(p,m)$$

for all positive integers m. Likewise, if $q' < q$, then this would imply that

$$R(n,q') = R(q',n) < R(q,n) = R(n,q)$$

for all positive integers n. Together, both of these inequalities would imply that $R(p',q') < R(p,q)$ if we have either $p' < p$ or $q' < q$. This completes the proof and we are done.

32. We shall prove by induction on all integers $n \geq 2$ that for all positive integers p,q that satisfy $p+q = n$, we have $R(p,q) \leq \binom{p+q-2}{p-1}$, with the

base case $n = 2$ being trivial. Suppose that we have $R(p,q) \leq \binom{p+q-2}{p-1}$ for all positive integers p, q that satisfy $p + q = k$ for some positive integer $k \geq 2$. Now, let us take any pair of positive integers a, b that satisfy $a + b = k + 1$. If we have $a = 1$ or $a = 2$, then we are done; indeed, we have

$$R(1, b) = 1 = \binom{b-1}{0} = \binom{a+b-2}{a-1}, \text{ and}$$

$$R(2, b) = b = \binom{b}{1} = \binom{a+b-2}{a-1}.$$

Likewise, we are also done if we have $b = 1$ or $b = 2$. So let us assume that $a, b \geq 3$. Then by Theorem 3.5.1 and induction hypothesis, we have

$$
\begin{aligned}
R(a, b) &\leq R(a-1, b) + R(a, b-1) \\
&\leq \binom{a+b-3}{a-2} + \binom{a+b-3}{a-1} \\
&= \binom{a+b-2}{a-1},
\end{aligned}
$$

where the last equality holds from identity (2.1.2). This completes the induction step, and we are done.

33. We shall prove by induction that $R(3, q) \leq \frac{1}{2}(q^2 + 3)$ for all positive integers q, with the base case $q = 1$ being trivial. Suppose that the inequality holds for some positive integer k. By induction hypothesis, we have $R(3, k) \leq \frac{1}{2}(k^2 + 3)$. Let us consider these cases:

Case 1: k is even. Then $\frac{1}{2}(k^2+2)$ is an integer. As $R(3, k)$ is an integer, we must have $R(3, k) \leq \frac{1}{2}(k^2 + 2)$. By Theorem 3.5.1, we have

$$R(3, k+1) \leq R(2, k+1) + R(3, k) \leq (k+1) + \frac{1}{2}(k^2+2) \leq \frac{1}{2}((k+1)^2 + 3).$$

Case 2: k is odd. Then $R(2, k + 1) = k + 1$ is even, and $\frac{1}{2}(k^2 + 3)$ is an integer. If $R(3, k) < \frac{1}{2}(k^2 + 3)$, then we must have $R(3, k) \leq \frac{1}{2}(k^2 + 3) - 1$. By Theorem 3.5.1, we have

$$R(3, k+1) \leq R(2, k+1) + R(3, k) \leq (k+1) + \frac{1}{2}(k^2+3) - 1 \leq \frac{1}{2}((k+1)^2 + 2).$$

Else, if $R(3, k) = \frac{1}{2}(k^2 + 3)$, then we must have $R(3, k)$ to be an even integer (since k is odd, we have $k^2 + 3 \equiv 0 \pmod 4$, which implies that $k^2 + 3 = 4m$ for some integer m). As both $R(2, k + 1)$ and $R(3, k)$ are both even, it follows from Theorem 3.5.2 that

$$R(3, k + 1) \leq R(2, k + 1) + R(3, k) - 1$$

$$\leq (k+1) + \frac{1}{2}(k^2 + 3) - 1$$

$$\leq \frac{1}{2}((k+1)^2 + 2).$$

In both cases, we have $R(3, k+1) \leq \frac{1}{2}((k+1)^2 + 3)$. This completes the induction step, and we are done.

34. By Problem 3.33, we have $R(3, 5) \leq \frac{1}{2}(5^2 + 3) = 14$. Next, let us show that $R(3, 5) > 13$. To this end, let the vertex set of the 13-clique be $\{v_0, v_1, \ldots, v_{12}\}$, and let us colour the edge $v_i v_j$ of the 13-clique blue if $|i - j| = 1, 5, 8, 12$, and red otherwise. Then it is easy to see that the 2-colouring of the edges of the 13-clique do not contain any blue 3-clique. It remains to show that this colouring does not contain any red 5-clique as well.

To this end, let us take any 5-element subset X of the vertex set $\{v_0, v_1, \ldots, v_{12}\}$, and consider the clique induced by X. Let us show that the clique induced by X contains at least one blue edge. By symmetry, we may assume without loss of generality that $v_0 \in X$. If any of v_1, v_5, v_8, v_{12} is in X, we are done, since $v_0 v_1, v_0 v_5, v_0 v_8$ and $v_0 v_{12}$ are all blue edges. So let us assume that $X \setminus \{v_0\} \subseteq \{v_2, v_3, v_4, v_6, v_7, v_9, v_{10}, v_{11}\}$. Let us define $A_1 = \{v_2, v_3, v_4\}$, $A_2 = \{v_6, v_7\}$ and $A_3 = \{v_9, v_{10}, v_{11}\}$. By (PP), at least two of the vertices in $X \setminus \{v_0\}$ must be in the same set A_i for some $i = 1, 2, 3$. Let these two vertices be v_k and v_m. If $|k - m| = 1$ then we are done, since $v_k v_m$ is a blue edge. So we must have either $\{v_k, v_m\} = \{v_2, v_4\}$ or $\{v_k, v_m\} = \{v_9, v_{11}\}$. By symmetry, it suffices to consider the case where $\{v_k, v_m\} = \{v_2, v_4\}$. If any of v_7, v_9, v_{10} are in X as well, then we are done, since $v_2 v_7, v_4 v_9$ and $v_2 v_{10}$ are all blue edges. Otherwise, we must have v_6 and v_{11} to be in X. But $v_6 v_{11}$ is a blue edge as well, so we are done.

Remark. The idea behind the solution is found in the paper *Combinatorial Relations and Chromatic Graphs*, *Canadian Journal of Mathematics*, **7** (1955), 1–7 by R. E. Greenwood and A. M. Gleason.

35. (a) By Problem 3.33, we have $R(3, 4) \leq \frac{1}{2}(4^2 + 3) = 9\frac{1}{2}$. So $R(3, 4) \leq 9$ as it is an integer. By Theorem 3.5.1, we have

$$R(4, 4) \leq R(3, 4) + R(4, 3) = 2R(3, 4) \leq 2 \times 9 = 18 \quad \text{(from above)}$$

as desired.

(b) By Problem 3.33, we have $R(3, 6) \leq \frac{1}{2}(6^2 + 3) = \frac{39}{2}$. As $R(3, 6)$ is an integer, we must have $R(3, 6) \leq \lfloor \frac{39}{2} \rfloor = 19$ as desired.

36. (a) As a 1-clique is vacuously defined to be of all colours, it follows that

$$R(p_1, p_2, \ldots, p_k) = 1$$

if $p_i = 1$ for some $i \in \{1, 2, \ldots, k\}$.

(b) Let the permissible colours be $\{c_1, c_2, \ldots, c_n\}$. Firstly, let us colour each of the edges of the $(p-1)$-clique with c_1. Then it is easy to check that the n-colouring does not contain any c_1 p-clique, and it does not contain any c_i 2-clique for all $i \in \{2, 3, \ldots, n\}$. So we must have $R(p, 2, \ldots, 2) \geq p$. Next, let us colour each of the edges of the p-clique with any of the possible colours in $\{c_1, c_2, \ldots, c_n\}$. If all of the edges of the p-clique are coloured with c_1, then we must have a c_1 p-clique. Else, one of the edges must be coloured with c_i for some $i \in \{2, 3, \ldots, n\}$. Then we have a c_i 2-clique. So we have $R(p, 2, \ldots, 2) \leq p$ and hence $R(p, 2, \ldots, 2) = p$ as desired.

37. Consider any colouring of the edges of the $R(p_1, p_2, \ldots, p_k)$-clique with colours $c_1, c_2, \ldots, c_{k+1}$. If there are any edges coloured c_{k+1} then we immediately have a c_{k+1} 2-clique. Else, each of the edges of the $R(p_1, p_2, \ldots, p_k)$-clique must be coloured with $c_1, c_2, \ldots, c_k$. By definition, there must exist some c_j p_j-clique for some $j \in \{1, 2, \ldots, k\}$. So we must have $R(p_1, p_2, \ldots, p_k, 2) \leq R(p_1, p_2, \ldots, p_k)$. Conversely, consider any colouring of the edges of the $R(p_1, p_2, \ldots, p_k, 2)$-clique with colours $c_1, c_2, \ldots, c_k$. Then this colouring is also a colouring of the $R(p_1, p_2, \ldots, p_k, 2)$-clique with colours $c_1, c_2, \ldots, c_{k+1}$. As there are no c_{k+1} 2-cliques (since the colouring involves only colours $c_1, c_2, \ldots, c_k$), it follows from definition that there must exist some c_j p_j-clique for some $j \in \{1, 2, \ldots, k\}$. So we must have $R(p_1, p_2, \ldots, p_k) \leq R(p_1, p_2, \ldots, p_k, 2)$ and we have $R(p_1, p_2, \ldots, p_k) = R(p_1, p_2, \ldots, p_k, 2)$ as desired.

38. For each $i = 1, \ldots, k$, we let $R_i = R(p_1, \ldots, p_{i-1}, p_i - 1, p_{i+1}, \ldots, p_k)$. Let $n = \sum_{i=1}^{k} R_i - (k-2) = \sum_{i=1}^{k} (R_i - 1) + 2$. Let a colouring of the edges of the n-clique with colours $c_1, c_2, \ldots, c_k$ be given. In order to show that $R(p_1, p_2, \ldots, p_k) \leq n$, it suffices to find a p_i-clique of colour c_i for some $i \in \{1, 2, \ldots, k\}$. To this end, let us take any vertex v of the vertex set of the n-clique. We note that v is incident to $n - 1 = \sum_{i=1}^{k} (R_i - 1) + 1$ edges in the n-clique. By (GPP), there exists some $j \in \{1, 2, \ldots, k\}$ such that at least R_j of the $n - 1$ edges are coloured c_j. Let X be the set of vertices u for which uv is of colour c_j. Now, by the definition of

R_j, there exists either a $(p_j - 1)$-clique of colour c_j, or a p_m-clique of colour c_m for some $m \in \{1, 2, \ldots, k\} \setminus \{j\}$. If the latter case occurs then we are done. Else, suppose that we have a $(p_j - 1)$-clique of colour c_j. Then it is easy to see that all of the edges of the clique induced by the vertex set $X \cup \{v\}$ is of colour c_j, and we thereby obtain a p_j-clique of colour c_j. So we are done.

39. We shall prove by induction on all integers $p \geq 2$ that for all non-negative integers $p_1, \ldots, p_k$ that satisfy $p = p_1 + \cdots + p_k$, we have $R(p_1 + 1, p_2 + 1, \ldots, p_k + 1) \leq \frac{p!}{p_1! p_2! \cdots p_k!}$. We note that $k \geq 2$. The base cases $p = 0$ and 1 are clear. Suppose that we have

$$R(p_1 + 1, p_2 + 1, \ldots, p_k + 1) \leq \frac{p!}{p_1! p_2! \cdots p_k!}$$

for all non-negative integers $p_1, p_2, \ldots, p_k$ that satisfy $p = \sum_{i=1}^{k} p_i$ for some non-negative integer p. Now, let us take any k-tuple of non-negative integers $(q_1, \ldots, q_k)$ that satisfy $p + 1 = \sum_{i=1}^{k} q_i$. If there exists some $j \in \{1, 2, \ldots, k\}$ for which $q_j = 0$, then we are done; indeed, we have $1 = R(q_1 + 1, q_2 + 1, \ldots, 1, \ldots, q_k + 1) \leq \frac{(p+1)!}{q_1! q_2! \cdots q_k!}$. So let us assume that $q_i > 0$ for all $i \in \{1, 2, \ldots, k\}$. Then we have

$$R(q_1 + 1, q_2 + 1, \ldots, q_k + 1)$$
$$\leq \sum_{i=1}^{k} R(q_1 + 1, \ldots, q_{i-1} + 1, q_i, q_{i+1} + 1, \ldots, q_k) - (k - 2)$$
(by Problem 3.38)
$$\leq \sum_{i=1}^{k} R(q_1 + 1, \ldots, q_{i-1} + 1, q_i, q_{i+1} + 1, \ldots, q_k)$$
$$\leq \sum_{i=1}^{k} \frac{p!}{q_1! \cdots q_{i-1}! (q_i - 1)! q_{i+1}! \cdots q_k!} \quad \text{(by induction hypothesis)}$$
$$= \frac{(p+1)!}{q_1! q_2! \cdots q_k!} \quad \text{(by Problem 2.63)}.$$

This completes the induction step, and we are done.

40.(a) (i) Let $n = k(R_{k-1} - 1) + 2$. Let us take any colouring of the edges of the n-clique with colours $c_1, c_2, \ldots, c_k$. In order to show that $R_k \leq n$, it suffices to find a 3-clique of colour c_i for some $i \in \{1, 2, \ldots, k\}$. To this end, let us take any vertex v of the vertex set of the n-clique. We note that v is incident to

$n - 1 = k(R_{k-1} - 1) + 1$ edges in the n-clique. By (GPP), there exists some $j \in \{1, 2, \ldots, k\}$ such that at least R_{k-1} of the $n - 1$ edges are coloured c_j. Let X be the set of vertices u for which uv is of colour c_j. Now, if there exist distinct vertices $w, x \in X$ such that wx is of colour c_j, then the 3-clique induced by the vertices v, w and x is of colour c_j. Else, each of the edges of the R_{k-1}-clique induced by X must be coloured by the colours $c_1, \ldots, c_{j-1}, c_{j+1}, \ldots, c_k$. By the definition of R_{k-1}, there exists 3 vertices $v_1, v_2, v_3 \in X$, such that 3-clique induced by the vertices v_1, v_2 and v_3 is of colour c_i for some $i \in \{1, 2, \ldots, k\} \setminus \{j\}$. So we are done.

(ii) We shall prove by induction on k that we have $R_k \leq \lfloor k!e \rfloor + 1$ for all positive integers $k \geq 2$, with the base case $k = 2$ being trivial. Suppose that the inequality $R_n \leq \lfloor n!e \rfloor + 1$ hold for some positive integer $n \geq 2$. This implies that

$$
\begin{aligned}
R_{n+1} &\leq (n+1)(R_n - 1) + 2 \quad \text{(by part (i))} \\
&\leq (n+1)\lfloor n!e \rfloor + 2 \\
&= (n+1)\sum_{r=0}^{n} \frac{n!}{r!} + 2 \quad \text{(by Problem 1.104)} \\
&= \sum_{r=0}^{n} \frac{(n+1)!}{r!} + \frac{(n+1)!}{(n+1)!} + 1 \\
&= \sum_{r=0}^{n+1} \frac{(n+1)!}{r!} + 1 \\
&= \lfloor (n+1)!e \rfloor + 1 \quad \text{(by Problem 1.104)}.
\end{aligned}
$$

This completes the induction step and we are done.

(iii) By part (i), we have $R_4 \leq 4(R_3 - 1) + 2 = 4R_3 - 2$ and $R_3 \leq 3(R_2 - 1) + 2 = 3R_2 - 1$. This implies that

$$
R_4 \leq 4R_3 - 2 \leq 4(3R_2 - 1) - 2 = 12R_2 - 6 = 12 \cdot 6 - 6 = 66
$$

as desired.

(b) Let us show that $R_k > 2^k$. To this end, let the vertex set of the 2^k-clique be $\{v_1, v_2, \ldots, v_{2^k}\}$, and let us consider a colouring of the 2^k-clique by the colours $c_0, c_1, \ldots, c_{k-1}, c_k$ by colouring the edge $v_i v_j$ of the 2^k-clique c_m if $2^m | i - j$ and $2^{m+1} \nmid i - j$. We note that none of the edges are coloured c_k since 2^k does not divide any of the integers in $\{\pm 1, \pm 2, \ldots, \pm(2^k - 1)\}$, so the colouring so defined is a colouring of the 2^k-clique by the colours $c_0, c_1, \ldots, c_{k-1}$. Let us show that there are no monochromatic 3-cliques. Suppose on the contrary that there exist integers $p, q, r \in \{1, 2, \ldots, 2^k\}$, such that the clique induced by the vertices v_p, v_q, v_r is monochromatic. This implies that there exist some integer $m \in \{0, 1, \ldots, k-1\}$, such that 2^m divides each of $p - q, q - r$ and $p - r$, and 2^{m+1} does not divide any of the integers $p - q, q - r$ and $p - r$. This implies that there exist odd integers s, t, u, such that $p - q = 2^m \cdot s$, $q - r = 2^m \cdot t$ and $p - r = 2^m \cdot u$. The first two equations imply that

$$p - r = (p - q) + (q - r) = 2^m(s + t) = 2^{m+1}\frac{s+t}{2}.$$

As s and t are both odd, it follows that $\frac{s+t}{2}$ is an integer, and hence 2^{m+1} divides $p - r$, which is a contradiction. So there are no monochromatic 3-cliques, and hence we must have $R_k \geq 2^k + 1$ as desired.

41. Let the vertex set of the n-clique be $\{v_1, v_2, \ldots, v_n\}$, and let us consider a colouring of the n-clique by the colours $c_1, c_2, \ldots, c_k$ by colouring the edge $v_i v_j$ of the n-clique c_m if $|i - j| \in S_m$. By the definition of n, there exist some $i \in \{1, 2, \ldots, k\}$ and integers $p, q, r \in \{1, 2, \ldots, n\}$, such that the clique induced by the vertices v_p, v_q and v_r is of colour c_i. Without loss of generality, let us assume that $p > q > r$. Then we have $p - q = |p - q| \in S_i$, $q - r = |q - r| \in S_i$ and $p - r = |p - r| \in S_i$. The desired follows by taking $a = p - q$, $b = q - r$ and $c = p - r$.

42. (i) By Problem 3.37, we have $R(3, 3, 2) = R(3, 3) = 6$.

(ii) By Problem 3.40(i), we have

$$R(3, 3, 3) \leq 3(R(3, 3) - 1) + 2 = 3(6 - 1) + 2 = 17.$$

43. (a) Let us take any colouring of the edges of the 6-clique by two colours: blue or red, and count the maximum possible number of non-monochromatic 3-cliques. To this end, let us denote the vertex set and edge set of the 6-clique by V and E respectively, and count the maximum possible number of triples (v_1, v_2, v_3) of distinct vertices of the 6-clique, such that the edges $v_1 v_2$ and $v_2 v_3$ are of different colours. For each edge $e \in E$, let us define $c(e) = 0$ if e is blue, and 1 if e is red. Furthermore, let us define

$$S = \{(v_1, v_2, v_3) \mid v_1, v_2, v_3 \in V, \text{ and } c(v_1 v_2) = 0, c(v_2 v_3) = 1\},$$

and $S_v = \{(v_1, v_2, v_3) \in S \mid v_2 = v\}$ for each $v \in V$. We note that each non-monochromatic 3-clique correspond to exactly two of these triples in S. For any vertex v, let the number of blue edges incident to it be k. Then the number of red edges incident to it is equal to $5 - k$. Consequently, we must have $S_v = k(5 - k)$. Now, we note that $k(5 - k) = \frac{25}{4} - \left(k - \frac{5}{2}\right)^2 \leq \frac{25}{4}$. This implies that $|S_v| \leq \lfloor \frac{25}{4} \rfloor = 6$, and hence we have $|S| = \sum_{v \in V} |S_v| \leq 6|V| = 36$. Consequently, there are at most $\frac{36}{2} = 18$ non-monochromatic 3-cliques. Hence, there are at least $\binom{6}{3} - 18 = 2$ monochromatic 3-cliques.

(b) Let us denote the vertex set of the 6-clique by $\{v_1, v_2, v_3, v_4, v_5, v_6\}$, and for each pair of distinct integers $i, j \in \{1, 2, 3, 4, 5, 6\}$, let us colour the edge $v_i v_j$ blue if $i - j$ is even, and red if $i - j$ is odd. Let us show that there are no red 3-cliques. Suppose on the contrary that there exist distinct $k, m, n \in \{1, 2, 3, 4, 5, 6\}$, such that the 3-clique induced by $\{v_k, v_m, v_n\}$ is red. By (PP), at least two of $k - m$, $m - n$ and $k - n$ must have the same parity, and without loss of generality, we may assume that $k - m$ and $m - n$ have the same parity. Then $k - n = (k - m) + (m - n)$ must be even, so the edge $v_k v_n$ is blue, a contradiction. So there are no red 3-cliques. Now, it is easy to see that the only blue 3-cliques are the cliques induced by the sets $\{v_1, v_3, v_5\}$ and $\{v_2, v_4, v_6\}$.

44. Let us take any colouring of the edges of the 7-clique by two colours: blue or red, and count the maximum possible number of non-monochromatic 3-cliques. To this end, let us denote the vertex set and edge set of the 7-clique by V and E respectively, and count the maximum possible number of triples (v_1, v_2, v_3) of distinct vertices of the 7-clique, such that the edges $v_1 v_2$ and $v_2 v_3$ are of different colours. For each edge $e \in E$, let us define $c(e) = 0$ if e is blue, and 1 if e is red. Furthermore, let us define $S = \{(v_1, v_2, v_3) \mid v_1, v_2, v_3 \in V$, and $c(v_1 v_2) = 0, c(v_2 v_3) = 1\}$, and $S_v = \{(v_1, v_2, v_3) \in S \mid v_2 = v\}$ for each $v \in V$. We note that each non-monochromatic 3-clique correspond to exactly two of these triples in S. For any vertex v, let the number of blue edges incident to it be k. Then the number of red edges incident to it is equal to $6 - k$. Consequently, we have $S_v = k(6 - k)$. Now, we have $k(6 - k) = 9 - (k - 3)^2 \leq 9$. This implies that $|S_v| \leq 9$, and hence we have $|S| \leq 9|V| = 63$. Consequently, there are at most $\lfloor \frac{63}{2} \rfloor = 31$ non-monochromatic 3-cliques. Hence, there are at least $\binom{7}{3} - 31 = 4$ monochromatic 3-cliques.

45. Let us take any colouring of the edges of the n-clique by two

colours: blue or red, and count the maximum possible number of non-monochromatic 3-cliques. To this end, let us denote the vertex set and edge set of the n-clique by V and E respectively, and count the maximum possible number of triples (v_1, v_2, v_3) of distinct vertices of the n-clique, such that the edges $v_1 v_2$ and $v_2 v_3$ are of different colours. For each edge $e \in E$, let us define $c(e) = 0$ if e is blue, and 1 if e is red. Furthermore, let us define $S = \{(v_1, v_2, v_3) \mid v_1, v_2, v_3 \in V$, and $c(v_1 v_2) = 0, c(v_2 v_3) = 1\}$, and $S_v = \{(v_1, v_2, v_3) \in S \mid v_2 = v\}$ for each $v \in V$. We note that each non-monochromatic 3-clique correspond to exactly two of these triples in S. For any vertex v, let the number of blue edges incident to it be j. Then the number of red edges incident to it is equal to $n - 1 - j$. Consequently, we must have $S_v = j(n - 1 - j)$. Now, we note that $j(n-1-j) = \frac{(n-1)^2}{4} - \left(j - \frac{(n-1)}{2}\right)^2 \leq \frac{(n-1)^2}{4}$. This implies that $|S_v| \leq \left\lfloor \frac{(n-1)^2}{4} \right\rfloor$, and hence we have $|S| = \sum_{v \in V} |S_v| \leq n \left\lfloor \frac{(n-1)^2}{4} \right\rfloor$. Consequently, there are at most $\left\lfloor \frac{n}{2} \left\lfloor \frac{(n-1)^2}{4} \right\rfloor \right\rfloor$ non-monochromatic 3-cliques. Hence, we have $T(n) \geq \binom{n}{3} - \left\lfloor \frac{n}{2} \left\lfloor \frac{(n-1)^2}{4} \right\rfloor \right\rfloor$. Now, let us consider these cases:

Case 1: $n = 2k$ for some positive integer k. Then we have

$$\left\lfloor \frac{(n-1)^2}{4} \right\rfloor = \left\lfloor \frac{(2k-1)^2}{4} \right\rfloor = \left\lfloor \frac{4k^2 - 4k + 1}{4} \right\rfloor = k^2 - k.$$

Hence, we have $\left\lfloor \frac{n}{2} \left\lfloor \frac{(n-1)^2}{4} \right\rfloor \right\rfloor = \lfloor k \cdot (k^2 - k) \rfloor = k^2(k-1)$. This implies that

$$T(n) \geq \binom{2k}{3} - k^2(k - 1) = \frac{1}{3}k(k - 1)(k - 2).$$

Case 2: $n = 4k + 1$ for some positive integer k. Then we have

$$\left\lfloor \frac{(n-1)^2}{4} \right\rfloor = \left\lfloor \frac{(4k)^2}{4} \right\rfloor = 4k^2.$$

Hence, we have $\left\lfloor \frac{n}{2} \left\lfloor \frac{(n-1)^2}{4} \right\rfloor \right\rfloor = \lfloor \frac{4k+1}{2} \cdot 4k^2 \rfloor = 2k^2(4k + 1)$. This implies that

$$T(n) \geq \binom{4k + 1}{3} - 2k^2(4k + 1) = \frac{2}{3}k(k - 1)(4k + 1).$$

Case 3: $n = 4k + 3$ for some positive integer k. Then we have

$$\left\lfloor \frac{(n-1)^2}{4} \right\rfloor = \left\lfloor \frac{(4k+2)^2}{4} \right\rfloor = 4k^2 + 4k + 1.$$

Hence, we have

$$\left\lfloor \frac{n}{2} \left\lfloor \frac{(n-1)^2}{4} \right\rfloor \right\rfloor = \left\lfloor \frac{4k+3}{2} \cdot (4k^2 + 4k + 1) \right\rfloor = 8k^3 + 14k^2 + 8k + 1.$$

This implies that

$$T(n) \geq \binom{4k+3}{3} - (8k^3 + 14k^2 + 8k + 1) = \frac{2}{3}k(k+1)(4k-1).$$

46. Firstly, from the example following Theorem 3.5.2, we note that $R(3,4) = 9$. This implies that the colouring stated in the problem defines a 2-colouring of the edges of an $R(3,4)$-clique. By assumption, there are no blue 3-cliques in the 2-colouring of the edges of the $R(3,4)$-clique (since each of the triangle determined by 3 of the 9 points contains at least one red side), so there must exist a red 4-clique in the 2-colouring of the edges of the $R(3,4)$-clique. This implies that there are four points such that the $6 (= \binom{4}{2})$ segments connecting them are red, and we are done.

Solutions to Exercise 4

1. Let C, E and M be the set of students who passed Chinese, English, and Mathematics respectively. By (4.2.1), we have

$$|C \cap E \cap M|$$
$$= |C \cup E \cup M| + |C \cap E| + |C \cap M| + |E \cap M| - |C| - |E| - |M|$$
$$\leq 102 + 65 + 54 + 48 - 92 - 75 - 63$$
$$= 39.$$

So the largest possible number of students that could have passed all three subjects is 39.

2. (a) (i) By (4.1.1), we have

$$|B| = |(A \cup \bar{A}) \cap B|$$
$$= |(A \cap B) \cup (\bar{A} \cap B)|$$
$$= |A \cap B| + |\bar{A} \cap B| - |(A \cap B) \cap (\bar{A} \cap B)|$$
$$= |A \cap B| + |\bar{A} \cap B|.$$

Hence, we have $|\bar{A} \cap B| = |B| - |A \cap B|$ as desired.

(ii) By part (i), we have

$$|\bar{A} \cap \bar{B} \cap C| = |\bar{B} \cap C| - |A \cap \bar{B} \cap C|$$
$$= |C| - |B \cap C| - |\bar{B} \cap A \cap C|$$
$$= |C| - |B \cap C| - (|A \cap C| - |B \cap A \cap C|)$$
$$= |C| - |A \cap C| - |B \cap C| + |A \cap B \cap C|.$$

(b) Let us denote the set $\{1, 2, \ldots, 10^3\}$ by S, and for each positive integer k, let us define $B_k = \{x \in S \mid k \text{ divides } x\}$. Then we have

$$|B_3| = \left\lfloor \frac{1000}{3} \right\rfloor = 333,$$

$$|B_5 \cap B_3| = |B_{15}| = \left\lfloor \frac{1000}{15} \right\rfloor = 66,$$

$$|B_7 \cap B_3| = |B_{21}| = \left\lfloor \frac{1000}{21} \right\rfloor = 47,$$

$$|B_7 \cap B_5 \cap B_3| = |B_{105}| = \left\lfloor \frac{1000}{105} \right\rfloor = 9.$$

By part (i), the number of integers in S which are not divisible by 5 nor 7, but are divisible by 3, is equal to

$$|\bar{B}_5 \cap \bar{B}_7 \cap B_3| = |B_3| - |B_5 \cap B_3| - |B_7 \cap B_3| + |B_5 \cap B_7 \cap B_3|$$

$$= 333 - 66 - 47 + 9$$

$$= 229.$$

3. Let $S = \{1, 2, \ldots, 120\}$, and for each $k = 2, 3, 5, 7$, let P_k denote the property of being divisible by k for the elements of S. Then we have

$$\omega(P_2) = \left\lfloor \frac{120}{2} \right\rfloor = 60,$$

$$\omega(P_3) = \left\lfloor \frac{120}{3} \right\rfloor = 40,$$

$$\omega(P_5) = \left\lfloor \frac{120}{5} \right\rfloor = 24,$$

$$\omega(P_7) = \left\lfloor \frac{120}{7} \right\rfloor = 17.$$

This implies that $\omega(1) = 60 + 40 + 24 + 17 = 141$. Next, we have

$$\omega(P_2 P_3) = \left\lfloor \frac{120}{2 \cdot 3} \right\rfloor = 20,$$

$$\omega(P_2 P_5) = \left\lfloor \frac{120}{2 \cdot 5} \right\rfloor = 12,$$

$$\omega(P_2 P_7) = \left\lfloor \frac{120}{2 \cdot 7} \right\rfloor = 8,$$

$$\omega(P_3 P_5) = \left\lfloor \frac{120}{3 \cdot 5} \right\rfloor = 8,$$

$$\omega(P_3 P_7) = \left\lfloor \frac{120}{3 \cdot 7} \right\rfloor = 5,$$

$$\omega(P_5 P_7) = \left\lfloor \frac{120}{5 \cdot 7} \right\rfloor = 3.$$

This implies that $\omega(2) = 20 + 12 + 8 + 8 + 5 + 3 = 56$. Next, we have

$$\omega(P_2 P_3 P_5) = \left\lfloor \frac{120}{2 \cdot 3 \cdot 5} \right\rfloor = 4,$$

$$\omega(P_2 P_3 P_7) = \left\lfloor \frac{120}{2 \cdot 3 \cdot 7} \right\rfloor = 2,$$

$$\omega(P_2 P_5 P_7) = \left\lfloor \frac{120}{2 \cdot 5 \cdot 7} \right\rfloor = 1,$$

$$\omega(P_3 P_5 P_7) = \left\lfloor \frac{120}{3 \cdot 5 \cdot 7} \right\rfloor = 1.$$

This implies that $\omega(3) = 4 + 2 + 1 + 1 = 8$. Finally, we have

$$\omega(4) = \omega(P_2 P_3 P_5 P_7) = \left\lfloor \frac{120}{2 \cdot 3 \cdot 5 \cdot 7} \right\rfloor = 0.$$

Hence, we have

$$\begin{aligned}
E(0) &= \omega(0) - \omega(1) + \omega(2) - \omega(3) + \omega(4) \\
&= 120 - 141 + 56 - 8 + 0 \\
&= 27,
\end{aligned}$$

$$\begin{aligned}
E(1) &= \omega(1) - \binom{2}{1}\omega(2) + \binom{3}{1}\omega(3) - \binom{4}{1}\omega(4) \\
&= 141 - 2 \times 56 + 3 \times 8 - 4 \times 0 \\
&= 53,
\end{aligned}$$

$$\begin{aligned}
E(2) &= \omega(2) - \binom{3}{2}\omega(3) + \binom{4}{2}\omega(4) \\
&= 56 - 3 \times 8 + 6 \times 0 \\
&= 32,
\end{aligned}$$

$$\begin{aligned}
E(3) &= \omega(3) - \binom{4}{3}\omega(4) \\
&= 8 - 4 \times 0 \\
&= 8,
\end{aligned}$$

$$E(4) = \omega(4) = 0.$$

Now, we claim that if $n \in S$ is composite, then it must be divisible by either one of $2, 3, 5, 7$. Indeed, if n is composite, then we must have $n = km$ for some positive integers $k, m > 2$. Without loss of generality, let us assume that $k \leq m$. Then we have $k^2 \leq km < 121 = 11^2$, which implies that $k < 11$. It is easy to check that k must be divisible by either one of $2, 3, 5, 7$, which completes the proof. Consequently, it is easy to see that $E(0)$ is equal to the number of non-composite integers in S that is not divisible by any of $2, 3, 5, 7$. As 1 is not a prime and $2, 3, 5, 7$ are primes, it follows that the number of primes that does not exceed 120 is equal to $27 - 1 + 4 = 30$.

4. For each positive integer k, let us denote the set of divisors of k by A_k. We first observe that we have $A_m \cap A_n = A_{(m,n)}$ for all positive integers m and n, where (m, n) denotes the greatest common divisor of m and n. Also, we have $10^{40} = 2^{40}5^{40}$ and $20^{30} = 4^{30}5^{30} = 2^{60}5^{30}$. This implies that $|A_{10^{40}}| = 41 \times 41 = 1681$, $|A_{20^{30}}| = 61 \times 31 = 1891$ and $|A_{10^{40}} \cap A_{20^{30}}| = |A_{2^{40}5^{30}}| = 41 \times 31 = 1271$. By (4.1.1), the number of positive integers n such that n is a divisor of at least one of the numbers $10^{40}, 20^{30}$ is equal to

$$\begin{aligned} |A_{10^{40}} \cup A_{20^{30}}| &= |A_{10^{40}}| + |A_{20^{30}}| - |A_{10^{40}} \cap A_{20^{30}}| \\ &= 1681 + 1891 - 1271 \\ &= 2301. \end{aligned}$$

5. For each positive integer k, let us denote the set of divisors of k by A_k. We have $10^{60} = 2^{60}5^{60}$, $20^{50} = 4^{50}5^{50} = 2^{100}5^{50}$ and $30^{40} = 2^{40}3^{40}5^{40}$. This implies that

$$\begin{aligned} |A_{10^{60}}| &= 61 \times 61 = 3721, \\ |A_{20^{50}}| &= 101 \times 51 = 5151, \\ |A_{30^{40}}| &= 41 \times 41 \times 41 = 68921, \\ |A_{10^{60}} \cap A_{20^{50}}| &= |A_{2^{60}5^{50}}| = 61 \times 51 = 3111, \\ |A_{10^{60}} \cap A_{30^{40}}| &= |A_{2^{40}5^{40}}| = 41 \times 41 = 1681, \\ |A_{20^{50}} \cap A_{30^{40}}| &= |A_{2^{40}5^{40}}| = 41 \times 41 = 1681, \\ |A_{10^{60}} \cap A_{20^{50}} \cap A_{30^{40}}| &= |A_{2^{40}5^{40}}| = 41 \times 41 = 1681. \end{aligned}$$

By (4.2.1), the number of positive integers n such that n is a divisor of at least one of the numbers $10^{60}, 20^{50}, 30^{40}$ is equal to

$$\begin{aligned} &|A_{10^{60}} \cup A_{20^{50}} \cup A_{30^{40}}| \\ &= |A_{10^{60}}| + |A_{20^{50}}| + |A_{30^{40}}| \\ &\quad - (|A_{10^{60}} \cap A_{20^{50}}| + |A_{10^{60}} \cap A_{30^{40}}| + |A_{20^{50}} \cap A_{30^{40}}|) \\ &\quad + |A_{10^{60}} \cap A_{20^{50}} \cap A_{30^{40}}| \\ &= 3721 + 5151 + 68921 - (3111 + 1681 + 1681) + 1681 \\ &= 73001. \end{aligned}$$

6. (i) Firstly, we note that $21^3 < 10^4 < 22^3$ and $4^6 < 10^4 < 5^6$. Let $S = \{1, \ldots, 10^4\}$, and for $k = 2, 3, 6$, let P_k denote the property of being of the form m^k for some positive integer m for the elements of S. This implies that $\omega(P_2) = 100$, $\omega(P_3) = 21$ and $\omega(P_2 P_3) = \omega(P_6) = 4$. Hence, the number of integers in S that are not of the form n^2 or n^3, where n is an integer, is equal to

$$E(0) = \omega(0) - \omega(1) + \omega(2) = 10000 - (100 + 21) + 4 = 9883.$$

(ii) Firstly, we note that $31^2 < 10^3 < 32^2$, $21^3 < 10^4 < 22^3$, $3^6 < 10^3 < 4^6$ and $4^6 < 10^4 < 5^6$. Let $S = \{10^3, \ldots, 10^4\}$, and for $k = 2, 3, 6$, let P_k denote the property of being of the form m^k for some positive integer m for the elements of S. This implies that $\omega(P_2) = 100 - 32 + 1 = 69$, $\omega(P_3) = 21 - 10 + 1 = 12$ and $\omega(P_2 P_3) = \omega(P_6) = 4 - 4 + 1 = 1$. Hence, the number of integers in S that are not of the form n^2 or n^3, where n is an integer, is equal to

$$E(0) = \omega(0) - \omega(1) + \omega(2) = (10000 - 1000 + 1) - (69 + 12) + 1 = 8921.$$

7. (i) Let us prove by induction on q that we have

$$|A_1 \cup A_2 \cup \cdots \cup A_q|$$
$$= \sum_{i=1}^{q} |A_i| - \sum_{i<j} |A_i \cap A_j| + \sum_{i<j<k} |A_i \cap A_j \cap A_k|$$
$$- \cdots + (-1)^{q+1} |A_1 \cap A_2 \cap \cdots \cap A_q|,$$

for any finite sets $A_1, A_2, \ldots, A_q$ and positive integer $q \geq 2$, with the base case $q = 2$ being trivial. Suppose that the identity holds for some positive integer $q = m \geq 2$, and finite sets $A_1, A_2, \ldots, A_m$. Let $B_1, B_2, \ldots, B_{m+1}$ be finite sets. By induction hypothesis, we have

$$|B_1 \cup B_2 \cup \cdots \cup B_{m+1}|$$
$$= |(B_1 \cup B_2 \cup \cdots \cup B_m) \cup B_{m+1}|$$
$$= |B_1 \cup B_2 \cup \cdots \cup B_m| + |B_{m+1}|$$
$$- |(B_1 \cup B_2 \cup \cdots \cup B_m) \cap B_{m+1}|$$
$$= |B_1 \cup B_2 \cup \cdots \cup B_m| + |B_{m+1}|$$
$$- |(B_1 \cap B_{m+1}) \cup \cdots \cup (B_m \cap B_{m+1})|.$$

By induction hypothesis once more, we have

$$
\begin{aligned}
&|B_1 \cup B_2 \cup \cdots \cup B_m| \\
&= \sum_{i=1}^{m} |B_i| - \sum_{i<j\leq m} |B_i \cap B_j| + \sum_{i<j<k\leq m} |B_i \cap B_j \cap B_k| \\
&\quad - \cdots + (-1)^{m+1} |B_1 \cap B_2 \cap \cdots \cap B_m|.
\end{aligned}
$$

At the same time, we also have

$$
\begin{aligned}
&|(B_1 \cap B_{m+1}) \cup \cdots \cup (B_m \cap B_{m+1})| \\
&= \sum_{i=1}^{m} |B_i \cap B_{m+1}| - \sum_{i<j\leq m} |(B_i \cap B_{m+1}) \cap (B_j \cap B_{m+1})| \\
&\quad + \sum_{i<j<k\leq m} |(B_i \cap B_{m+1}) \cap (B_j \cap B_{m+1}) \cap (B_k \cap B_{m+1})| \\
&\quad + \cdots + (-1)^{m} |(B_1 \cap B_{m+1}) \cap \cdots \cap (B_m \cap B_{m+1})| \\
&= \sum_{i=1}^{m} |B_i \cap B_{m+1}| - \sum_{i<j\leq m} |B_i \cap B_j \cap B_{m+1}| \\
&\quad + \sum_{i<j<k\leq m} |B_i \cap B_j \cap B_k \cap B_{m+1}| \\
&\quad - \cdots + (-1)^{m} |B_1 \cap B_2 \cap \cdots \cap B_m \cap B_{m+1}|.
\end{aligned}
$$

Combining the above two equations, we have

$$
\begin{aligned}
&|B_1 \cup B_2 \cup \cdots \cup B_{m+1}| \\
&= \sum_{i=1}^{m+1} |B_i| - \sum_{i<j} |B_i \cap B_j| + \sum_{i<j<k} |B_i \cap B_j \cap B_k| \\
&\quad - \cdots + (-1)^{m+2} |B_1 \cap B_2 \cap \cdots \cap B_{m+1}|,
\end{aligned}
$$

which completes the induction step. We are done.

(ii) By Corollary 2 to Theorem (4.3.1), we have

$$
\begin{aligned}
&|A_1 \cup A_2 \cup \cdots \cup A_q| \\
&= |S \setminus (\bar{A}_1 \cap \bar{A}_2 \cap \cdots \cap \bar{A}_q)| \\
&= |S| - |\bar{A}_1 \cap \bar{A}_2 \cap \cdots \cap \bar{A}_q| \\
&= \sum_{i=1}^{q} |A_i| - \sum_{i<j} |A_i \cap A_j| + \sum_{i<j<k} |A_i \cap A_j \cap A_k| \\
&\quad - \cdots + (-1)^{q+1} |A_1 \cap A_2 \cap \cdots \cap A_q|.
\end{aligned}
$$

8. Let $S = \{1000, 1001, \ldots, 3000\}$. For $k = 4, 100, 400$, let us define $S_k = \{x \in S \mid k \text{ divides } x\}$. We note that $S_{400} \subseteq S_{100} \subseteq S_4$, so that we have $\overline{S_{100}} \cap S_{400} = \emptyset$ and $S_4 \cap S_{100} = S_{100}$. Now, we have

$$|S_4| = \left\lfloor \frac{3000}{4} \right\rfloor - \left\lfloor \frac{999}{4} \right\rfloor = 501,$$

$$|S_{100}| = \left\lfloor \frac{3000}{100} \right\rfloor - \left\lfloor \frac{999}{100} \right\rfloor = 21,$$

$$|S_{400}| = \left\lfloor \frac{3000}{400} \right\rfloor - \left\lfloor \frac{999}{400} \right\rfloor = 5.$$

By Problem 4.2(a)(i) and (4.1.1), the number of leap years between 1000 and 3000 inclusive is equal to

$$\begin{aligned}
|(S_4 \cap \overline{S_{100}}) \cup S_{400}| &= |S_4 \cap \overline{S_{100}}| + |S_{400}| - |S_4 \cap \overline{S_{100}} \cap S_{400}| \\
&= |S_4| - |S_4 \cap S_{100}| + |S_{400}| \\
&= |S_4| - |S_{100}| + |S_{400}| \\
&= 501 - 21 + 5 \\
&= 485.
\end{aligned}$$

9. Let the set of boys be $S = \{b_1, b_2, \ldots, b_n\}$, and T denote the set of all possible groupings. For each $i = 1, 2, \ldots, n$, let P_i denote the property of b_i being grouped together with both of his parents for the elements of T. Firstly, we note that $\omega(0) = (n!)^2$, since there are $n!$ ways each to assign a male parent and a female parent to each boy. Let us compute $\omega(k)$ for $k = 1, 2, \ldots, n$. For all k-tuples $(i_1, i_2, \ldots, i_k)$ that satisfy $1 \leq i_1 < i_2 < \cdots < i_k \leq n$, we have $\omega(P_{i_1} P_{i_2} \cdots P_{i_k}) = [(n-k)!]^2$, since there are $(n-k)!$ ways each to assign a male parent and a female parent to each remaining boy that has not been grouped. This implies that $\omega(k) = \binom{n}{k}[(n-k)!]^2$ for all $k = 0, 1, \ldots, n$. Hence, the number of groupings that satisfy the property that no boy is with both of his parents in his group is equal to

$$E(0) = \sum_{k=0}^{n} (-1)^k \omega(k) = \sum_{k=0}^{n} (-1)^k \binom{n}{k} [(n-k)!]^2.$$

10. Let the set of friends be $S = \{f_1, f_2, \ldots, f_6\}$, and T denote the set of all dinner gatherings. For each $i = 1, 2, \ldots, 6$, let P_i denote the property of the man meeting f_i in the dinner gathering for the elements of T. By the assumptions given in the problem, we have $\omega(1) = \binom{6}{1} \times 12 = 72$,

$w(2) = \binom{6}{2} \times 6 = 90$, $w(3) = \binom{6}{3} \times 4 = 80$, $w(4) = \binom{6}{4} \times 3 = 45$, $w(5) = \binom{6}{5} \times 2 = 12$ and $w(6) = \binom{6}{6} \times 1 = 1$. This implies that

$$E(0) = w(0) - w(1) + w(2) - w(3) + w(4) - w(5) + w(6)$$
$$= w(0) - 72 + 90 - 80 + 45 - 12 + 1$$
$$= w(0) - 28.$$

As we have $E(0) = 8$ by the assumptions given in the problem, it follows that the number of times that he has dined out is equal to $w(0) = E(0) + 28 = 36$.

11. Let S denote the set of all arrangements of 3 black balls, 4 red balls and 5 white balls in a row, and let B, R and W denote the set of arrangements of 3 black balls, 4 red balls and 5 white balls in a row for which the black, red and white balls form a single block in the arrangement respectively. Then we have

$$|S| = \frac{(3+4+5)!}{3!4!5!} = 27720,$$

$$|B| = \frac{(4+5+1)!}{4!5!} = 1260,$$

$$|R| = \frac{(3+5+1)!}{3!5!} = 504,$$

$$|W| = \frac{(3+4+1)!}{3!4!} = 280,$$

$$|B \cap R| = \frac{(5+1+1)!}{5!} = 42,$$

$$|B \cap W| = \frac{(4+1+1)!}{4!} = 30,$$

$$|R \cap W| = \frac{(3+1+1)!}{3!} = 20,$$

$$|B \cap R \cap W| = 3! = 6.$$

By (4.2.1), we have

$$|B \cup R \cup W|$$
$$= |B| + |R| + |W| - |B \cap R| - |B \cap W| - |R \cap W| + |B \cap R \cap W|$$
$$= 1260 + 504 + 280 - 42 - 30 - 20 + 6$$
$$= 1958.$$

Hence, the number of arrangements of 3 black balls, 4 red balls and 5 white balls in a row such that balls with the same colour do not form a single block is equal to

$$|S| - |B \cup R \cup W| = 27720 - 1958 = 25762.$$

12. (i) Let S denote the set of all arrangements of 3 a's, 3 b's and 3 c's in a row, and let X, Y and Z denote the set of arrangements of 3 a's, 3 b's and 3 c's in a row for which the a's, b's and c's form a single block in the arrangement respectively. Then we have

$$|S| = \frac{(3+3+3)!}{3!3!3!} = 1680,$$

$$|X| = |Y| = |Z| = \frac{(3+3+1)!}{3!3!} = 140,$$

$$|X \cap Y| = |X \cap Z| = |Y \cap Z| = \frac{(3+1+1)!}{3!} = 20,$$

$$|X \cap Y \cap Z| = 3! = 6.$$

By (4.2.1), we have

$$|X \cup Y \cup Z|$$
$$= |X| + |Y| + |Z| - |X \cap Y| - |X \cap Z| - |Y \cap Z| + |X \cap Y \cap Z|$$
$$= 3 \times 140 - 3 \times 20 + 6$$
$$= 366.$$

Hence, the number of arrangements of 3 a's, 3 b's and 3 c's in a row such that no three consecutive letters are the same is equal to

$$|S| - |X \cup Y \cup Z| = 1680 - 366 = 1314.$$

(ii) Let S denote the set of arrangements of 3 a's, 3 b's and 3 c's in a row such that no two consecutive letters are the same. For each $x \in S$, let $\bar{x}$ be what remains when we delete the c's from x (for instance, if $x = abcabcabc$, then $\bar{x} = ababab$). For $k = 1, 2, 3$, let A_k and B_k denote the set of elements $x \in S$ such that the a's (respectively b's) in $\bar{x}$ form k separate blocks. It is easy to see that $S = A_1 \cup A_2 \cup A_3 = B_1 \cup B_2 \cup B_3$, so that we have $|S| = \sum_{i=1}^{3} \sum_{j=1}^{3} |A_i \cap B_j|$. Let us compute $|A_i \cap B_j|$ for $i, j = 1, 2, 3$. Firstly, let us compute $|A_1 \cap B_1|$. Let x be an arrangement of 3 a's, 3 b's and 3 c's such that the a's form a single block and the b's form a single block in $\bar{x}$. This would then imply that there are either 2 consecutive a's or 2 consecutive b's in x. Consequently, this shows that $x \notin S$, so we must have $A_1 \cap B_1 = \emptyset$, or equivalently, $|A_1 \cap B_1| = 0$.

Next, let us compute $|A_1 \cap B_2|$. Let $x \in S$ be an arrangement of 3 a's, 3 b's and 3 c's such that $x \in A_1 \cap B_2$. Then it is easy to see that either $\bar{x} = bbaaab$ or $\bar{x} = baaabb$. In each of the 2 cases, a c must be placed in between the block of 2 b's, and the other 2 c's must be placed in between the leftmost a and the middle a, and in between the middle a and the rightmost a. So we have $|A_1 \cap B_2| = 2$. By symmetry, we have $|A_2 \cap B_1| = |A_1 \cap B_2| = 2$.

Next, let us compute $|A_1 \cap B_3|$. Since there are no arrangements of 3 a's and 3 b's for which the a's form a single block and the b's form three separate blocks, it follows that there are no arrangements $x \in S$ of 3 a's, 3 b's and 3 c's such that $x \in A_1 \cap B_3$. So $A_1 \cap B_3 = \emptyset$, and hence we must have $|A_1 \cap B_3| = 0$. By symmetry, we have $|A_3 \cap B_1| = |A_1 \cap B_3| = 0$.

Next, let us compute $|A_2 \cap B_2|$. Let $x \in S$ be an arrangement of 3 a's, 3 b's and 3 c's such that $\bar{x} \in A_2 \cap B_2$. Then it is easy to check that $\bar{x} = abaabb, abbaab, aababb, aabbab, babbaa, baabba, bbabaa$ or $bbaaba$. In each of the 8 cases, a c must be placed in between the block of 2 a's, and a c must be placed in between the block of 2 b's. Lastly, there are $\binom{7-2}{1} = 5$ choices to place the last c, such that no 2 c's are consecutive. So we have $|A_2 \cap B_2| = 8 \times 5 = 40$.

Next, let us compute $|A_2 \cap B_3|$. Let $x \in S$ be an arrangement of 3 a's, 3 b's and 3 c's such that $x \in A_2 \cap B_3$. Then it is easy to see that either $\bar{x} = baabab$ or $\bar{x} = babaab$. In each of the 2 cases, a c must be placed in between the block of 2 a's. Lastly, there are $\binom{7-1}{2} = 15$ choices to place the last 2 c's, such that no 2 c's are consecutive. So we have $|A_2 \cap B_3| = 2 \times 15 = 30$. By symmetry, we have $|A_3 \cap B_2| = |A_2 \cap B_3| = 30$.

Finally, let us compute $|A_3 \cap B_3|$. Let $x \in S$ be an arrangement of 3 a's, 3 b's and 3 c's such that $x \in A_3 \cap B_3$. Then it is easy to see that either $\bar{x} = ababab$ or $\bar{x} = bababa$. In each of the 2 cases, there are $\binom{7}{3} = 35$ choices left to place the 3 c's, such that no 2 c's are consecutive. So we have $|A_3 \cap B_3| = 2 \times 35 = 70$.

Therefore, we have

$$|S| = \sum_{i=1}^{3} \sum_{j=1}^{3} |A_i \cap B_j|$$
$$= 2|A_1 \cap B_2| + |A_2 \cap B_2| + 2|A_2 \cap B_3| + |A_3 \cap B_3|$$
$$= 2 \cdot 2 + 40 + 2 \cdot 30 + 70$$
$$= 174.$$

13. For any segment s in the rectangular grid, let P_s denote the property of the shortest route from corner X to corner Y passing through segment

s in the rectangular grid. By definition, we have $\omega(0)$ to be equal to the number of shortest routes from corner X to corner Y, which is then equal to $\binom{8+4}{4} = \binom{12}{4}$. Next, we have

$$\omega(P_{AB}) = \binom{2+1}{1}\binom{5+3}{3} = 3\binom{8}{3},$$

$$\omega(P_{BC}) = \binom{3+1}{1}\binom{5+2}{2} = 4\binom{7}{2},$$

$$\omega(P_{BD}) = \binom{3+1}{1}\binom{4+3}{3} = 4\binom{7}{3}.$$

This implies that $\omega(1) = 3\binom{8}{3} + 4\binom{7}{2} + 4\binom{7}{3}$. Next, we have

$$\omega(P_{AB}P_{BC}) = \binom{2+1}{1}\binom{5+2}{2} = 3\binom{7}{2},$$

$$\omega(P_{AB}P_{BD}) = \binom{2+1}{1}\binom{4+3}{3} = 3\binom{7}{3},$$

$$\omega(P_{BC}P_{BD}) = 0.$$

This implies that $\omega(2) = 3\binom{7}{2}+3\binom{7}{3}$. Finally, since there are no shortest routes from corner X to corner Y that passes through all of AB, BC and BD, we have

$$\omega(3) = \omega(P_{AB}P_{BC}P_{BD}) = 0.$$

Hence, the number of shortest routes from corner X to corner Y that does not pass through the segments AB, BC and BD is equal to

$$E(0) = \omega(0) - \omega(1) + \omega(2) - \omega(3)$$

$$= \binom{12}{4} - 3\binom{8}{3} - 4\binom{7}{2} - 4\binom{7}{3} + 3\binom{7}{2} + 3\binom{7}{3}$$

$$= 271.$$

14. Let $y_1 = x_1 - 3$, $y_2 = x_2$ and $y_3 = x_3 - 7$. The problem is equivalent to finding the number of non-negative integer solutions to the equation

$$y_1 + y_2 + y_3 = 18$$

satisfying $y_1 \leq 6$, $y_2 \leq 8$ and $y_3 \leq 10$. Let $n_1 = 7$, $n_2 = 9$, $n_3 = 11$, and for $i = 1, 2, 3$, let P_i denote the property of a non-negative integer solution (a_1, a_2, a_3) to the above equation satisfying $a_i \geq n_i$. By definition, we have $\omega(0) = \binom{18+2}{2} = 190$. Next, we have

$$\omega(P_1) = \binom{18-7+2}{2} = 78,$$

$$\omega(P_2) = \binom{18-9+2}{2} = 55,$$

$$\omega(P_3) = \binom{18-11+2}{2} = 36.$$

This implies that $\omega(1) = 78 + 55 + 36 = 169$. Next, we have

$$\omega(P_1 P_2) = \binom{18 - 7 - 9 + 2}{2} = 6,$$

$$\omega(P_1 P_3) = \binom{18 - 7 - 11 + 2}{2} = 1,$$

$$\omega(P_2 P_3) = 0.$$

This implies that $\omega(2) = 6 + 1 = 7$. Finally, we have $\omega(3) = \omega(P_1 P_2 P_3) = 0$. Hence, the number of integer solutions to the equation

$$x_1 + x_2 + x_3 = 28$$

satisfying $3 \leq x_1 \leq 9$, $0 \leq x_2 \leq 8$ and $7 \leq x_3 \leq 17$ is equal to

$$E(0) = \omega(0) - \omega(1) + \omega(2) - \omega(3)$$
$$= 190 - 169 + 7 - 0$$
$$= 28.$$

15. Let $y_1 = x_1 - 6$, $y_2 = x_2 - 5$ and $y_3 = x_3 - 10$. The problem is equivalent to finding the number of non-negative integer solutions to the equation

$$y_1 + y_2 + y_3 = 19$$

satisfying $y_1 \leq 9$, $y_2 \leq 15$ and $y_3 \leq 15$. Let $n_1 = 10$, $n_2 = n_3 = 16$, and for $i = 1, 2, 3$, let P_i denote the property of a non-negative integer solution (a_1, a_2, a_3) to the above equation satisfying $a_i \geq n_i$. By definition, we have $\omega(0) = \binom{19+2}{2} = \binom{21}{2}$. Next, we have

$$\omega(P_1) = \binom{19 - 10 + 2}{2} = \binom{11}{2},$$

$$\omega(P_2) = \binom{19 - 16 + 2}{2} = \binom{5}{2},$$

$$\omega(P_3) = \binom{19 - 16 + 2}{2} = \binom{5}{2}.$$

This implies that $\omega(1) = \binom{11}{2} + \binom{5}{2} + \binom{5}{2} = \binom{11}{2} + 2\binom{5}{2}$. Next, we have

$$\omega(P_1 P_2) = \omega(P_1 P_3) = \omega(P_2 P_3) = 0.$$

This implies that $\omega(2) = 0$. Finally, we have $\omega(3) = \omega(P_1 P_2 P_3) = 0$. Hence, the number of integer solutions to the equation

$$x_1 + x_2 + x_3 = 40$$

satisfying $6 \leq x_1 \leq 15$, $5 \leq x_2 \leq 20$ and $10 \leq x_3 \leq 25$ is equal to

$$E(0) = \omega(0) - \omega(1) + \omega(2) - \omega(3) = \binom{21}{2} - \binom{11}{2} - 2\binom{5}{2} = 30.$$

16. Let $y_1 = x_1 - 1$, $y_2 = x_2$, $y_3 = x_3 - 4$ and $y_4 = x_4 - 2$. The problem is equivalent to finding the number of non-negative integer solutions to the equation

$$y_1 + y_2 + y_3 + y_4 = 13$$

satisfying $y_1 \leq 4$, $y_2 \leq 7$, $y_3 \leq 4$ and $y_4 \leq 4$. Let $n_1 = n_3 = n_4 = 5$, $n_2 = 8$, and for $i = 1, 2, 3, 4$, let P_i denote the property of a non-negative integer solution (a_1, a_2, a_3, a_4) to the above equation satisfying $a_i \geq n_i$. By definition, we have $\omega(0) = \binom{13+3}{3} = \binom{16}{3}$. Next, we have

$$\omega(P_1) = \binom{13 - 5 + 3}{3} = \binom{11}{3},$$

$$\omega(P_2) = \binom{13 - 8 + 3}{3} = \binom{8}{3},$$

$$\omega(P_3) = \binom{13 - 5 + 3}{3} = \binom{11}{3},$$

$$\omega(P_4) = \binom{13 - 5 + 3}{3} = \binom{11}{3}.$$

This implies that $\omega(1) = \binom{11}{3} + \binom{8}{3} + \binom{11}{3} + \binom{11}{3} = 3\binom{11}{3} + \binom{8}{3}$. Next, we have

$$\omega(P_1 P_2) = \binom{13 - 5 - 8 + 3}{3} = 1,$$

$$\omega(P_1 P_3) = \binom{13 - 5 - 5 + 3}{3} = \binom{6}{3},$$

$$\omega(P_1 P_4) = \binom{13 - 5 - 5 + 3}{3} = \binom{6}{3},$$

$$\omega(P_2 P_3) = \binom{13 - 5 - 8 + 3}{3} = 1,$$

$$\omega(P_2 P_4) = \binom{13 - 5 - 8 + 3}{3} = 1,$$

$$\omega(P_3 P_4) = \binom{13 - 5 - 5 + 3}{3} = \binom{6}{3}.$$

This implies that $\omega(2) = 1 + \binom{6}{3} + \binom{6}{3} + 1 + 1 + \binom{6}{3} = 3\binom{6}{3} + 3$. Next, we have

$$\omega(3) = \omega(P_1 P_2 P_3) + \omega(P_1 P_2 P_4) + \omega(P_1 P_3 P_4) + \omega(P_2 P_3 P_4)$$

$$= 0 + 0 + 0 + 0$$

$$= 0, \text{ and}$$

$$\omega(4) = \omega(P_1 P_2 P_3 P_4)$$

$$= 0.$$

Hence, the number of integer solutions to the equation

$$x_1 + x_2 + x_3 + x_4 = 20$$

satisfying $1 \leq x_1 \leq 5$, $0 \leq x_2 \leq 7$, $4 \leq x_3 \leq 8$ and $2 \leq x_4 \leq 6$ is equal to

$$E(0) = \omega(0) - \omega(1) + \omega(2) - \omega(3) + \omega(4)$$

$$= \binom{16}{3} - 3\binom{11}{3} - \binom{8}{3} + 3\binom{6}{3} + 3$$

$$= 512.$$

17. For $i = 1, 2, \ldots, n$, let P_i denote the property of a non-negative integer solution $(a_1, a_2, \ldots, a_n)$ to the equation

$$x_1 + x_2 + \cdots + x_n = r$$

satisfying $a_i \geq k + 1$. By definition, we have $\omega(0) = \binom{r+n-1}{n-1}$. Next, let us compute $\omega(i)$ for all $i = 1, 2, \ldots, n$. To this end, let us fix $i \in \{1, 2, \ldots, n\}$, and take any i-tuple of integers $(n_1, n_2, \ldots, n_i)$ satisfying $1 \leq n_1 < n_2 < \cdots < n_i \leq n$. It is easy to see by definition that $\omega(P_{n_1} P_{n_2} \cdots P_{n_i}) = \binom{r-(k+1)i+n-1}{n-1}$. Since there are $\binom{n}{i}$ such i-tuples, it follows that we have

$$\omega(i) = \sum_{1 \leq n_1 < n_2 < \cdots < n_i \leq n} \omega(P_{n_1} P_{n_2} \cdots P_{n_i})$$

$$= \sum_{1 \leq n_1 < n_2 < \cdots < n_i \leq n} \binom{r - (k+1)i + n - 1}{n - 1}$$

$$= \binom{n}{i} \binom{r - (k+1)i + n - 1}{n - 1}.$$

Since $\omega(0) = \binom{r+n-1}{n-1} = \binom{n}{0}\binom{r-(k+1)\cdot 0 + n - 1}{n-1}$, it follows from (4.3.2) that the number of integer solutions to the equation

$$x_1 + x_2 + \cdots + x_n = r$$

satisfying $0 \leq x_i \leq k$ for each $i = 1, 2, \ldots, n$ is equal to

$$E(0) = \sum_{i=0}^{n} (-1)^i \omega(i) = \sum_{i=0}^{n} (-1)^i \binom{n}{i} \binom{r - (k+1)i + n - 1}{n - 1}.$$

18. For all $i = 1, 2, \ldots, n$, we let $y_i = x_i - 1$, so that the problem is equivalent to finding the number of non-negative integer solutions to the equation

$$y_1 + y_2 + \cdots + y_n = r - n.$$

satisfying $y_i \leq k - 1$ for all $i = 1, 2, \ldots, n$. For $i = 1, 2, \ldots, n$, let P_i denote the property of a non-negative integer solution $(a_1, a_2, \ldots, a_n)$ to the equation

$$y_1 + y_2 + \cdots + y_n = r - n$$

satisfying $a_i \geq k$. By definition, we have $\omega(0) = \binom{r-n+n-1}{n-1} = \binom{r-1}{n-1}$. Next, let us compute $\omega(i)$ for all $i = 1, 2, \ldots, n$. To this end, let us fix $i \in \{1, 2, \ldots, n\}$, and take any i-tuple of integers $(n_1, n_2, \ldots, n_i)$ satisfying $1 \leq n_1 < n_2 < \cdots < n_i \leq n$. It is easy to see by definition that $\omega(P_{n_1} P_{n_2} \cdots P_{n_i}) = \binom{r-n-ki+n-1}{n-1} = \binom{r-ki-1}{n-1}$. Since there are $\binom{n}{i}$ such i-tuples, it follows that we have

$$\omega(i) = \sum_{1 \leq n_1 < n_2 < \cdots < n_i \leq n} \omega(P_{n_1} P_{n_2} \cdots P_{n_i})$$

$$= \sum_{1 \leq n_1 < n_2 < \cdots < n_i \leq n} \binom{r - ki - 1}{n - 1}$$

$$= \binom{n}{i} \binom{r - ki - 1}{n - 1}.$$

Since $\omega(0) = \binom{r-1}{n-1} = \binom{n}{0} \binom{r-k \cdot 0 - 1}{n-1}$, it follows from (4.3.2) that the number of integer solutions to the equation

$$x_1 + x_2 + \cdots + x_n = r$$

satisfying $1 \leq x_i \leq k$ for each $i = 1, 2, \ldots, n$ is equal to

$$E(0) = \sum_{i=0}^{n} (-1)^i \omega(i) = \sum_{i=0}^{n} (-1)^i \binom{n}{i} \binom{r - ki - 1}{n - 1}.$$

19. For $i = 1, \ldots, n$, let P_i denote the property of an arrangement A that H_i is adjacent to W_i in A. By definition, we have $\omega(0) = (2n)!$. Next, let us compute $\omega(r)$ for all $r = 1, \ldots, n$. To this end, let us fix $r \in \{1, \ldots, n\}$, and take any r-tuple of integers $(n_1, \ldots, n_r)$ satisfying $1 \leq n_1 < \cdots < n_r \leq n$. For $i = 1, \ldots, r$, given that there are two ways to arrange H_{n_i} and W_{n_i} inside the same block when they are adjacent to each other. Then there are $(2n - 2r + r)! = (2n - r)!$ ways to arrange the r blocks consisting of the r couples $\{H_{n_1}, W_{n_1}\}, \{H_{n_2}, W_{n_2}\}, \ldots, \{H_{n_r}, W_{n_r}\}$, and the remaining $2n - 2r$ people. By (MP), we have $\omega(P_{n_1} P_{n_2} \cdots P_{n_r}) = 2^r (2n - r)!$. Since there

are $\binom{n}{r}$ such r-tuples, it follows that we have

$$
\begin{aligned}
\omega(r) &= \sum_{1 \le n_1 < n_2 < \cdots < n_r \le n} \omega(P_{n_1} P_{n_2} \cdots P_{n_r}) \\
&= \sum_{1 \le n_1 < n_2 < \cdots < n_r \le n} 2^r (2n - r)! \\
&= 2^r \binom{n}{r} (2n - r)!.
\end{aligned}
$$

Since $\omega(0) = (2n)! = \binom{n}{0} 2^0 (2n - 0)!$, it follows from (4.3.2) that the number of ways to arrange n couples $\{H_1, W_1\}, \ldots, \{H_n, W_n\}$ in a row, such that H_i is not adjacent to W_i for $i = 1, 2, \ldots, n$, is equal to

$$
E(0) = \sum_{r=0}^{n} (-1)^r \omega(r) = \sum_{r=0}^{n} (-1)^r 2^r \binom{n}{r} (2n - r)!.
$$

20. Let the colours be $c_1, c_2, \ldots, c_q$.

(i) Firstly, there are $q!$ ways to arrange the pq beads in a row, such that beads of the same colour must be in a single block. Next, we note that two arrangements of the pq beads in a string are identical if any one of them could be obtained from the other by flipping. Since p is odd and $q > 1$, it follows that any arrangement of the pq beads in a string gives rise to exactly two arrangements of the pq beads in a row, namely itself and its mirror image. So there are $\frac{q!}{2}$ ways to arrange the pq beads in a string, such that beads of the same colour must be in a single block.

For parts (ii) to (iv), we let S denote the set of arrangements of the $2q$ symbols $a_1, a_1, \ldots, a_q, a_q$ in a row, and for $i = 1, \ldots, q$, we let P_i denote the property of an arrangement a that the two symbols a_i form a single block in a. By definition, we have $\omega(0) = \frac{(2q)!}{2^q}$. Next, let us compute $\omega(j)$ for all $j = 1, \ldots, q$. Let us fix $j \in \{1, \ldots, q\}$, and take any j-tuple of integers $(n_1, \ldots, n_j)$ satisfying $1 \le n_1 < n_2 < \cdots < n_j \le q$. We note that there are $\frac{(2q - 2j + j)!}{2^{q-j}} = \frac{(2q - j)!}{2^{q-j}}$ ways to arrange $2q$ symbols in a row, such that the two symbols a_{n_i} form a single block in the arrangement for $i = 1, \ldots, j$. Since there are $\binom{q}{j}$ such j-tuples, it follows that we have

$$
\begin{aligned}
\omega(j) &= \sum_{1 \le n_1 < n_2 < \cdots < n_j \le q} \omega(P_{n_1} P_{n_2} \cdots P_{n_j}) \\
&= \sum_{1 \le n_1 < n_2 < \cdots < n_j \le q} \frac{(2q - j)!}{2^{q-j}}
\end{aligned}
$$

$$= \binom{q}{j} \frac{(2q-j)!}{2^{q-j}}.$$

Since $\omega(0) = \frac{(2q)!}{2^q} = \binom{q}{0} \frac{(2q-0)!}{2^{q-0}}$, it follows from (GPIE) that the number of ways to arrange the $2q$ symbols in a row, such that there are exactly j instances where two adjacent symbols are identical, is equal to

$$E(j) = \sum_{i=j}^{q} (-1)^{i-j} \binom{i}{j} \omega(i) = \sum_{i=j}^{q} (-1)^{i-j} \binom{i}{j} \binom{q}{i} \frac{(2q-i)!}{2^{q-i}}.$$

(ii) We first note that the number of ways to arrange the $2q$ symbols $a_1, a_1, \ldots, a_q, a_q$ in a row, such that no two adjacent symbols are identical, is equal to

$$E(0) = \sum_{i=0}^{q} (-1)^i \binom{i}{0} \binom{q}{i} \frac{(2q-i)!}{2^{q-i}} = \sum_{i=0}^{q} (-1)^i \binom{q}{i} \frac{(2q-i)!}{2^{q-i}}.$$

Next, for each colour c_i, $i = 1, 2, \ldots, q$ and each arrangement a of the $2q$ symbols in a row that contains no two identical adjacent symbols, there are $p - 1 = 2\lfloor \frac{p}{2} \rfloor$ ways to place the p beads of colour c_i into the two slots occupied by the two symbols a_i, such that each slot has at least 1 bead. By (MP), it follows that the number of ways to arrange the pq beads in a row, such that beads of the same colour must be in two separated blocks, is equal to

$$\left(2\left\lfloor \frac{p}{2} \right\rfloor \right)^q \sum_{i=0}^{q} (-1)^i \binom{q}{i} \frac{(2q-i)!}{2^{q-i}} = \left\lfloor \frac{p}{2} \right\rfloor^q \sum_{i=0}^{q} (-1)^i \binom{q}{i} 2^i (2q-i)!.$$

Consequently, the number of ways to arrange the pq beads in a string, such that beads of the same colour must be in two separated blocks, is equal to

$$\frac{1}{2} \left\lfloor \frac{p}{2} \right\rfloor^q \sum_{i=0}^{q} (-1)^i \binom{q}{i} 2^i (2q-i)!.$$

(iii) We first note that the number of ways to arrange the $2q$ symbols $a_1, a_1, a_2, a_2, \ldots, a_q, a_q$ in a row, such that there are exactly j instances where two adjacent symbols are identical, is equal to

$$E(j) = \sum_{i=j}^{q} (-1)^{i-j} \binom{i}{j} \omega(i) = \sum_{i=j}^{q} (-1)^{i-j} \binom{i}{j} \binom{q}{i} \frac{(2q-i)!}{2^{q-i}}.$$

Next, for each arrangement a of the $2q$ symbols in a row that has exactly j instances where two adjacent symbols are identical, and each integer i such that the two a_i's are not adjacent in a, there are $p - 1 = 2 \left\lfloor \frac{p}{2} \right\rfloor$ ways to place the p beads of colour c_i into the two slots occupied by the two symbols a_i, such that each slot has at least 1 bead. By (MP), it follows that the number of ways to arrange the pq beads in a row, such that there are exactly j instances where beads of the same colour form a single block, is equal to

$$\left(2 \left\lfloor \frac{p}{2} \right\rfloor\right)^{q-j} \sum_{i=j}^{q} (-1)^{i-j} \binom{i}{j} \binom{q}{i} \frac{(2q-i)!}{2^{q-i}}$$

$$= \frac{1}{2^j} \left\lfloor \frac{p}{2} \right\rfloor^{q-j} \sum_{i=j}^{q} (-1)^{i-j} \binom{i}{j} \binom{q}{i} 2^i (2q-i)!.$$

By (AP), the number of ways to arrange the pq beads in a row, such that beads of the same colour must be in at most two blocks, is equal to

$$\sum_{j=0}^{q} \frac{1}{2^j} \left\lfloor \frac{p}{2} \right\rfloor^{q-j} \sum_{i=j}^{q} (-1)^{i-j} \binom{i}{j} \binom{q}{i} 2^i (2q-i)!.$$

Consequently, we see that the number of ways to arrange the pq beads in a string, such that beads of the same colour must be in at most two blocks, is equal to

$$\frac{1}{2} \sum_{j=0}^{q} \frac{1}{2^j} \left\lfloor \frac{p}{2} \right\rfloor^{q-j} \sum_{i=j}^{q} (-1)^{i-j} \binom{i}{j} \binom{q}{i} 2^i (2q-i)!.$$

(iv) We first note that the number of ways to arrange the $2q$ symbols $a_1, a_1, a_2, a_2, \ldots, a_q, a_q$ in a row, such that there are exactly j instances where two adjacent symbols are identical, is equal to

$$E(j) = \sum_{i=j}^{q} (-1)^{i-j} \binom{i}{j} \omega(i) = \sum_{i=j}^{q} (-1)^{i-j} \binom{i}{j} \binom{q}{i} \frac{(2q-i)!}{2^{q-i}}.$$

Next, for each arrangement a of the $2q$ symbols in a row that has exactly j instances where two adjacent symbols are identical, and each integer i such that the two a_i's are not adjacent in a, there are $p - 3 = 2 \left\lfloor \frac{p-2}{2} \right\rfloor$ ways to place the p beads of colour c_i into the two slots occupied by the two symbols a_i, such that each slot has at least 2 beads. By (MP), it follows that the number of ways to arrange the pq beads in a row, such that there are exactly j

instances where beads of the same colour form a single block, is equal to

$$\left(2\left\lfloor\frac{p-2}{2}\right\rfloor\right)^{q-j}\sum_{i=j}^{q}(-1)^{i-j}\binom{i}{j}\binom{q}{i}\frac{(2q-i)!}{2^{q-i}}$$

$$=\frac{1}{2^j}\left\lfloor\frac{p-2}{2}\right\rfloor^{q-j}\sum_{i=j}^{q}(-1)^{i-j}\binom{i}{j}\binom{q}{i}2^i(2q-i)!.$$

By (AP), the number of ways to arrange the pq beads in a row, such that beads of the same colour must be in at most two blocks, and the size of each block is at least two, is equal to

$$\sum_{j=0}^{q}\frac{1}{2^j}\left\lfloor\frac{p-2}{2}\right\rfloor^{q-j}\sum_{i=j}^{q}(-1)^{i-j}\binom{i}{j}\binom{q}{i}2^i(2q-i)!.$$

Consequently, the number of ways to arrange the pq beads in a string, such that beads of the same colour must be in at most two blocks, is equal to

$$\frac{1}{2}\sum_{j=0}^{q}\frac{1}{2^j}\left\lfloor\frac{p-2}{2}\right\rfloor^{q-j}\sum_{i=j}^{q}(-1)^{i-j}\binom{i}{j}\binom{q}{i}2^i(2q-i)!.$$

21. (a) Firstly, there is exactly one way to place one object into each of the n boxes. Next, there are $H_{r-n}^n = \binom{r-n+n-1}{r-n} = \binom{r-1}{r-n}$ ways to distribute the remaining $r-n$ identical objects into the n distinct boxes. By (MP), the number of ways to distribute r identical objects into n distinct boxes, such that no boxes are empty, is equal to $1 \cdot \binom{r-1}{r-n} = \binom{r-1}{n-1}$.

(b) Let the n boxes be $b_1, b_2, \ldots, b_n$. For $i = 1, 2, \ldots, n$, let P_i denote the property that the box b_i is empty. By definition, we have $\omega(0) = H_r^n = \binom{r+n-1}{r}$. Next, since there are no distribution of r identical objects into n distinct boxes such that every box is empty, it follows that we have $\omega(n) = \omega(P_1 P_2 \cdots P_n) = 0$. It remains to compute $\omega(i)$ for all $i = 1, 2, \ldots, n-1$. To this end, let us fix $i \in \{1, 2, \ldots, n-1\}$, and take any i-tuple of integers $(n_1, n_2, \ldots, n_i)$ satisfying $1 \le n_1 < n_2 < \cdots < n_i \le n$. By definition, we have $\omega(P_{n_1} P_{n_2} \cdots P_{n_i}) = H_r^{n-i} = \binom{r+n-i-1}{r}$. Since there are $\binom{n}{i}$ such

i-tuples, it follows that we have

$$
\begin{aligned}
\omega(i) &= \sum_{1 \le n_1 < n_2 < \cdots < n_i \le n} \omega(P_{n_1} P_{n_2} \cdots P_{n_i}) \\
&= \sum_{1 \le n_1 < n_2 < \cdots < n_i \le n} \binom{r+n-i-1}{r} \\
&= \binom{n}{i}\binom{r+n-i-1}{r}.
\end{aligned}
$$

Since $\omega(0) = \binom{r+n-1}{r} = \binom{n}{0}\binom{r+n-0-1}{r}$, it follows from (4.3.2) that the number of ways to distribute r identical objects into n distinct boxes, such that no boxes are empty, is equal to

$$
\begin{aligned}
E(0) &= \sum_{i=0}^{n} (-1)^i \omega(i) \\
&= \sum_{i=0}^{n-1} (-1)^i \omega(i) \quad (\text{since } \omega(n) = 0) \\
&= \sum_{i=0}^{n-1} (-1)^i \binom{n}{i}\binom{r+n-i-1}{r}.
\end{aligned}
$$

By part (a), we have

$$
\sum_{i=0}^{n-1} (-1)^i \binom{n}{i}\binom{r+n-i-1}{r} = \binom{r-1}{n-1}
$$

as desired.

22. (a) Suppose that C is an r-element subset of A which contains B as a subset. Then we have $|C \setminus B| = r - m$. Since $C \setminus B \subseteq A \setminus B$ and $|A \setminus B| = n - m$, it follows that there are $\binom{n-m}{r-m} = \binom{n-m}{n-r}$ choices for $C \setminus B$. So there are $\binom{n-m}{n-r}$ r-element subsets of A that contain B as a subset.

(b) Let $B = \{b_1, \ldots, b_m\}$. For $i = 1, 2, \ldots, m$, let P_i denote the property of an r-element subset C of A that C does not contain the element b_i. By definition, we have $\omega(0) = \binom{n}{r}$. Next, let us compute $\omega(i)$ for all $i = 1, 2, \ldots, m$. To this end, let us fix $i \in \{1, 2, \ldots, m\}$, and take any i-tuple of integers $(n_1, n_2, \ldots, n_i)$ satisfying $1 \le n_1 < n_2 < \cdots < n_i \le m$. By definition, we have $\omega(P_{n_1} P_{n_2} \cdots P_{n_i}) = \binom{n-i}{r}$. Since there are $\binom{m}{i}$ such i-tuples, it follows that we have

$$\omega(i) = \sum_{1 \le n_1 < n_2 < \cdots < n_i \le n} \omega(P_{n_1} P_{n_2} \cdots P_{n_i})$$

$$= \sum_{1 \le n_1 < n_2 < \cdots < n_i \le n} \binom{n-i}{r}$$

$$= \binom{m}{i}\binom{n-i}{r}.$$

Since $\omega(0) = \binom{n}{r} = \binom{m}{0}\binom{n-0}{r}$, it follows from (4.3.2) that the number of r-element subsets of A that contain B as a subset, is equal to

$$E(0) = \sum_{i=0}^{m}(-1)^i \omega(i) = \sum_{i=0}^{m}(-1)^i \binom{m}{i}\binom{n-i}{r}.$$

By part (a), we have

$$\sum_{i=0}^{m}(-1)^i \binom{m}{i}\binom{n-i}{r} = \binom{n-m}{n-r}$$

as desired.

23. (a) Let a be a binary sequence of length n that does not contain "01" as a block. If the first digit in a is 0, then we must have $a = \underbrace{000 \cdots 00}_{n}$.

Else, a must start with exactly m 1's for some positive integer $m \le n$, in which case we must have $a = \underbrace{111 \cdots 11}_{m}\underbrace{000 \cdots 00}_{n-m}$. So there are $n+1$ binary sequences of length n that does not contain "01" as a block.

(b) For $i = 1, 2, \ldots, n-1$, let P_i denote the property of a binary sequence a of length n such that the i-th digit of a is equal to 0 and the $(i+1)$-th digit of a is equal to 1. By definition, we have $\omega(0) = 2^n$.

Next, let us compute $\omega(i)$ for all $i = 1, 2, \ldots, n-1$. To this end, let us fix $i \in \{1, 2, \ldots, n-1\}$, and take any i-tuple of integers $(n_1, n_2, \ldots, n_i)$ satisfying $1 \le n_1 < n_2 < \cdots < n_i \le n-1$. By the definition of the properties P_i's, it is easy to see that $\omega(P_{n_1} P_{n_2} \cdots P_{n_i}) > 0$ if and only if we have $n_j + 1 < n_{j+1}$ for all positive integers $j < i$. Let us denote the set of i-tuples $(n_1, n_2, \ldots, n_i)$ satisfying $1 \le n_1 < n_2 < \cdots < n_i \le n-1$ and $n_j + 1 < n_{j+1}$ for all positive integers $j < i$ by A_i, and let us set

$x_1 = n_1 - 1 \geq 0$ and $x_j = n_j - n_{j-1} \geq 2$ for all $j = 2, \ldots, i+1$. To determine $|A_i|$, it amounts to finding the number of non-negative integer solutions to the equation

$$x_1 + x_2 + \cdots + x_{i+1} = (n_1 - 1) + \sum_{j=2}^{i}(n_j - n_{j-1}) + (n+1-n_i) = n$$

satisfying $x_j \geq 2$ for all $j > 1$. By letting $y_1 = x_1$ and $y_j = x_j - 2$ for all $j > 1$, it is easy to see that the number of such i-tuples is equal to the number of non-negative integer solutions to the equation

$$y_1 + y_2 + \cdots + y_{i+1} = n - 2i,$$

which is then equal to $\binom{n-2i+i+1-1}{i+1-1} = \binom{n-i}{i}$. Consequently, we see that if $i > \lfloor \frac{n}{2} \rfloor$, then we have $n - i < i$, so that we have $A_i = \emptyset$. Consequently, we have $\omega(i) = 0$ for all $i > \lfloor \frac{n}{2} \rfloor$. Next, for all positive integers $i \leq \lfloor \frac{n}{2} \rfloor$, and any $(n_1, n_2, \ldots, n_i) \in A_i$, there are two choices each for each of the remaining $n - 2i$ digits in the binary sequence who has "01" blocks starting at the n_j-th places for $j = 1, 2, \ldots, i$. Consequently, we have

$$\omega(i) = \sum_{(n_1,n_2,\ldots,n_i)\in A_i} \omega(P_{n_1}P_{n_2}\cdots P_{n_i})$$

$$= \sum_{(n_1,n_2,\ldots,n_i)\in A_i} 2^{n-2i}$$

$$= \binom{n-i}{i}2^{n-2i}.$$

Since $\omega(0) = 2^n = \binom{n-0}{0}2^{n-2\cdot 0}$, it follows from (4.3.2) that the number of binary sequences of length n that does not contain "01" as a block is equal to

$$E(0) = \sum_{i=0}^{n}(-1)^i\omega(i)$$

$$= \sum_{i=0}^{\lfloor \frac{n}{2} \rfloor}(-1)^i\omega(i) \quad (\text{since } \omega(i) = 0 \text{ for all } i > \lfloor \frac{n}{2} \rfloor)$$

$$= \sum_{i=0}^{\lfloor \frac{n}{2} \rfloor}(-1)^i\binom{n-i}{i}2^{n-2i}.$$

By part (a), we have

$$n + 1 = \sum_{i=0}^{\lfloor \frac{n}{2} \rfloor} (-1)^i \binom{n-i}{i} 2^{n-2i}$$

as desired.

24. Let us label the configurations as follows:

(i) For $i = 1, 2, 3, 4$, let P_i denote the property of a coloured configuration that the two vertices incident to the edge e_i have the same colour. By definition, we have $\omega(0) = \lambda^4$. Next, we have $\omega(P_i) = \lambda^3$ for all $i = 1, 2, 3, 4$, which implies that $\omega(1) = 4\lambda^3$. Next, we have $\omega(P_i P_j) = \lambda^2$ for all positive integers i, j satisfying $1 \le i < j \le 4$, which implies that $\omega(2) = \binom{4}{2}\lambda^2 = 6\lambda^2$. Next, we have $\omega(P_1 P_2 P_3) = \lambda^2$ and $\omega(P_1 P_2 P_4) = \omega(P_1 P_3 P_4) = \omega(P_2 P_3 P_4) = \lambda$, which implies that $\omega(3) = \lambda^2 + 3\lambda$. Finally, we have $\omega(4) = \omega(P_1 P_2 P_3 P_4) = \lambda$.

Hence, the number of colour configurations that satisfies the property that any two vertices which are joined by a line segment must be coloured by different colours is equal to

$$\begin{aligned} E(0) &= \omega(0) - \omega(1) + \omega(2) - \omega(3) + \omega(4) \\ &= \lambda^4 - 4\lambda^3 + 6\lambda^2 - (\lambda^2 + 3\lambda) + \lambda \\ &= \lambda^4 - 4\lambda^3 + 5\lambda^2 - 2\lambda. \end{aligned}$$

(ii) For $i = 1, 2, 3, 4, 5$, let P_i denote the property of a coloured configuration that the two vertices incident to the edge e_i have the same colour. By definition, we have $\omega(0) = \lambda^4$. Next, we have $\omega(P_i) = \lambda^3$ for all $i = 1, 2, 3, 4, 5$, which implies that $\omega(1) = 5\lambda^3$. Next, we have $\omega(P_i P_j) = \lambda^2$ for all positive integers i, j satisfying $1 \le i < j \le 5$, which implies that $\omega(2) = \binom{5}{2}\lambda^2 = 10\lambda^2$. Next, we have

$$\omega(P_1 P_2 P_3) = \omega(P_3 P_4 P_5) = \lambda^2,$$
$$\omega(P_1 P_2 P_4) = \omega(P_1 P_2 P_5) = \omega(P_1 P_3 P_4) = \omega(P_1 P_3 P_5)$$
$$= \omega(P_1 P_4 P_5) = \omega(P_2 P_3 P_4) = \omega(P_2 P_3 P_5) = \omega(P_2 P_4 P_5) = \lambda.$$

This implies that $\omega(3) = 2\lambda^2 + 8\lambda$. Next, we have $\omega(P_i P_j P_k P_\ell) = \lambda$ for all positive integers i, j, k, ℓ satisfying $1 \le i < j < k < \ell \le 5$, which implies that $\omega(4) = \binom{5}{4}\lambda = 5\lambda$. Finally, we have $\omega(5) = \omega(P_1 P_2 P_3 P_4 P_5) = \lambda$.

Hence, the number of colour configurations that satisfies the property that any two vertices which are joined by a line segment must be coloured by different colours is equal to

$$E(0) = \omega(0) - \omega(1) + \omega(2) - \omega(3) + \omega(4) - \omega(5)$$
$$= \lambda^4 - 5\lambda^3 + 10\lambda^2 - (2\lambda^2 + 8\lambda) + 5\lambda - \lambda$$
$$= \lambda^4 - 5\lambda^3 + 8\lambda^2 - 4\lambda.$$

(iii) For $i = 1, 2, 3, 4, 5$, let P_i denote the property of a coloured configuration that the two vertices incident to the edge e_i have the same colour. By definition, we have $\omega(0) = \lambda^5$. Next, we have $\omega(P_i) = \lambda^4$ for all $i = 1, 2, 3, 4, 5$, which implies that $\omega(1) = 5\lambda^4$. Next, we have $\omega(P_i P_j) = \lambda^3$ for all positive integers i, j satisfying $1 \le i < j \le 5$, which implies that $\omega(2) = \binom{5}{2}\lambda^3 = 10\lambda^3$. Next, we have $\omega(P_i P_j P_k) = \lambda^2$ for all positive integers i, j, k satisfying $1 \le i < j < k \le 5$, which implies that $\omega(3) = \binom{5}{3}\lambda^2 = 10\lambda^2$. Next, we have $\omega(P_i P_j P_k P_\ell) = \lambda$ for all positive integers i, j, k, ℓ satisfying $1 \le i < j < k < \ell \le 5$, which implies that $\omega(4) = \binom{5}{4}\lambda = 5\lambda$. Finally, we have $\omega(5) = \omega(P_1 P_2 P_3 P_4 P_5) = \lambda$.

Hence, the number of colour configurations that satisfies the property that any two vertices which are joined by a line segment must be coloured by different colours is equal to

$$E(0) = \omega(0) - \omega(1) + \omega(2) - \omega(3) + \omega(4) - \omega(5)$$
$$= \lambda^5 - 5\lambda^4 + 10\lambda^3 - 10\lambda^2 + 5\lambda - \lambda$$
$$= \lambda^5 - 5\lambda^4 + 10\lambda^3 - 10\lambda^2 + 5\lambda - 1 - \lambda + 1$$
$$= (\lambda - 1)^5 - (\lambda - 1).$$

25. Let the rooms be $R_1, R_2, \ldots, R_q$. For $i = 1, 2, \ldots, q$, let P_i denote the property of an allocation that room R_i has exactly k persons. Let us compute $\omega(j)$ for $j = m, m+1, \ldots, q$. To this end, let us fix $j \in \{m, m+1, \ldots, q\}$, and take any j-tuple of integers $(n_1, n_2, \ldots, n_j)$ satisfying $1 \le n_1 < n_2 < \cdots < n_j \le q$. Firstly, there are $\binom{n}{kj}$ ways to choose kj persons out of n to be allocated into the rooms $R_{n_1}, R_{n_2}, \ldots, R_{n_j}$. Next, there are $\frac{(kj)!}{(k!)^j}$ ways to allocate the kj chosen persons into the rooms $R_{n_1}, R_{n_2}, \ldots, R_{n_j}$, such that each of the rooms $R_{n_1}, R_{n_2}, \ldots, R_{n_j}$

has exactly k persons. Finally, there are $(q - j)^{n-kj}$ ways to allocate the remaining $n - kj$ persons into the remaining $q - j$ rooms. By (MP), the number of ways to allocate n persons into the rooms $R_1, R_2, \ldots, R_q$ such that each of the rooms $R_{n_1}, R_{n_2}, \ldots, R_{n_j}$ has exactly k persons, is equal to

$$\binom{n}{kj} \cdot \frac{(kj)!}{(k!)^j} \cdot (q-j)^{n-kj} = \frac{n!}{(kj)!(n-kj)!} \cdot \frac{(kj)!}{(k!)^j} \cdot (q-j)^{n-kj}$$

$$= \frac{n!(q-j)^{n-kj}}{(k!)^j(n-kj)!}.$$

Since there are $\binom{q}{j} = \frac{q!}{j!(q-j)!}$ such j-tuples, it follows that we have

$$\omega(j) = \sum_{1 \le n_1 < n_2 < \cdots < n_j \le n} \omega(P_{n_1} P_{n_2} \cdots P_{n_j})$$

$$= \sum_{1 \le n_1 < n_2 < \cdots < n_j \le n} \frac{n!(q-j)^{n-kj}}{(k!)^j(n-kj)!}$$

$$= \frac{q!}{j!(q-j)!} \cdot \frac{n!(q-j)^{n-kj}}{(k!)^j(n-kj)!}$$

$$= \frac{q!n!(q-j)^{n-kj}}{(k!)^j(n-kj)!j!(q-j)!}.$$

By (4.3.2), the number of ways to allocate n persons into q distinct rooms, such that m of the q rooms have exactly k persons each, is equal to

$$E(m)$$

$$= \sum_{j=m}^{q} (-1)^{j-m} \binom{j}{m} \omega(j)$$

$$= \sum_{j=m}^{q} (-1)^j (-1)^{-m} \frac{j!}{m!(j-m)!} \cdot \frac{q!n!(q-j)^{n-kj}}{(k!)^j(n-kj)!j!(q-j)!}$$

$$= \sum_{j=m}^{q} (-1)^j (-1)^m \frac{q!n!(q-j)^{n-kj}}{m!(k!)^j(n-kj)!(j-m)!(q-j)!}$$

(since $(-1)^{-m} = (-1)^m$)

$$= (-1)^m \frac{q!n!}{m!} \sum_{j=m}^{q} (-1)^j \frac{(q-j)^{n-kj}}{(k!)^j(n-kj)!(j-m)!(q-j)!}.$$

26. For $i = 1, 2, \ldots, n$, let P_i denote the property of an arrangement a that the k x_i's form a single block in a. Let us compute $\omega(j)$ for all $j =$

$m, m+1, \ldots, n$. To this end, let us fix $i \in \{m, m+1, \ldots, n\}$, and take any i-tuple of integers $(n_1, n_2, \ldots, n_i)$ satisfying $1 \le n_1 < n_2 < \cdots < n_i \le n$. It is easy to see that there are $\frac{\{k(n-i)+i\}!}{(k!)^{n-i}} = \frac{(k!)^i \{kn-i(k-1)\}!}{(k!)^n}$ ways to arrange the members of A in a row, such that the k symbols x_{n_j} form a single block in the arrangement for $j = 1, \ldots, i$. Since there are $\binom{n}{i}$ such i-tuples, it follows that we have

$$
\omega(i) = \sum_{1 \le n_1 < n_2 < \cdots < n_i \le n} \omega(P_{n_1} P_{n_2} \cdots P_{n_i})
$$

$$
= \sum_{1 \le n_1 < n_2 < \cdots < n_i \le n} \frac{(k!)^i \{kn - i(k-1)\}!}{(k!)^n}
$$

$$
= \binom{n}{i} \frac{(k!)^i \{kn - i(k-1)\}!}{(k!)^n}.
$$

By (4.3.2), the number of ways to arrange the members of A in a row, such that the number of blocks containing all the k elements of the same type in the arrangement is exactly m, is equal to

$$
E(m)
$$

$$
= \sum_{i=m}^{n} (-1)^{i-m} \binom{i}{m} \omega(i)
$$

$$
= \sum_{i=m}^{n} (-1)^i (-1)^{-m} \binom{i}{m} \binom{n}{i} \frac{(k!)^i \{kn - i(k-1)\}!}{(k!)^n}
$$

$$
= \sum_{i=m}^{n} (-1)^i (-1)^m \binom{n}{i} \binom{i}{m} \frac{(k!)^i \{kn - i(k-1)\}!}{(k!)^n}
$$

$$
= \sum_{i=m}^{n} (-1)^i (-1)^m \binom{n}{m} \binom{n-m}{i-m} \frac{(k!)^i \{kn - i(k-1)\}!}{(k!)^n} \quad \text{(by (2.1.5))}
$$

$$
= \frac{(-1)^m}{(k!)^n} \binom{n}{m} \sum_{i=m}^{n} (-1)^i \binom{n-m}{i-m} (k!)^i \{kn - i(k-1)\}!.
$$

27. (4.6.2) Firstly, there are $\binom{r}{k}$ choices for the k fixed points. Next, for each k-element subset A of $\mathbf{N}_n$ corresponding to the k fixed points, we see that the remaining $r - k$ elements in the r-permutation form a $(r-k)$-permutation of the set $\mathbf{N}_n \setminus A$ with no fixed points, and there are $D(n-k, r-k, 0)$ such $(r-k)$-permutations of $\mathbf{N}_n \setminus A$ by definition. By (MP), the number of r-element permutations of $\mathbf{N}_n$ with exactly k fixed points is equal to $\binom{r}{k} D(n-k, r-k, 0)$, and hence by the definition

of $D(n, r, k)$, we have

$$D(n, r, k) = \binom{r}{k} D(n - k, r - k, 0)$$

as desired.

(4.6.3) Let $a = a_1 a_2 \cdots a_r$ be an r-element permutation of $\mathbf{N}_n$ with exactly k fixed points. Let us consider these cases:

Case 1: $a_1 = 1$. Then $a_2 a_3 \cdots a_r$ is a $(r - 1)$-element permutation of $\mathbf{N}_n \setminus \{1\}$ with exactly $k-1$ fixed points, and there are $D(n-1, r-1, k-1)$ such $(r - 1)$-permutations of $\mathbf{N}_n \setminus \{1\}$ by definition.

Case 2: $a_1 = m$ where $m > r$. For each $m > r$, we have $a_2 a_3 \cdots a_r$ to be a $(r - 1)$-element permutation of $\mathbf{N}_n \setminus \{m\}$ with exactly k fixed points, and there are $D(n-1, r-1, k)$ such $(r-1)$-permutations of $\mathbf{N}_n \setminus \{m\}$ by definition. Since there are $n-r$ choices for m, it follows from (MP) that there are $(n - r)D(n - 1, r - 1, k)$ r-element permutations $x_1 x_2 \cdots x_n$ of $\mathbf{N}_n$ with exactly k fixed points and $x_1 > r$.

Case 3: $a_1 = m$ where $2 \le m \le r$. Let us fix any $m \in \{2, 3, \ldots, r\}$. If $a_m = 1$, then we have $a_2 a_3 \cdots a_{m-1} a_{m+1} \cdots a_r$ is a $(r - 2)$-element permutation of $\mathbf{N}_n \setminus \{1, m\}$ with exactly k fixed points, and there are $D(n-2, r-2, k)$ such $(r-2)$-permutations of $\mathbf{N}_n \setminus \{1, m\}$ by definition. Else, we must have $a_m \ne 1$. For each $i = 2, 3, \ldots, n$, let us define b_i as follows:

$$b_i = \begin{cases} a_i & \text{if } a_i \ne 1, \\ m & \text{if } a_i = 1 \end{cases}$$

for all $i = 2, 3, \ldots, r$. Since $a_1 = m$ and $a_m \ne 1$, it follows that $b_m \ne m$, so it is easy to see that there is a bijection between the set of r-permutations $a_1 a_2 \cdots a_r$ of $\mathbf{N}_n$ with exactly k fixed points, $a_1 = m$ and $a_m \ne 1$, and the set of $(r - 1)$-permutations $b_2 b_3 \cdots b_r$ of $\mathbf{N}_n \setminus \{1\}$ with k fixed points and $b_m \ne m$.

Now, there are $D(n - 1, r - 1, k)$ $(r - 1)$-permutations $b_2 b_3 \cdots b_r$ of $\mathbf{N}_n \setminus \{1\}$ with k fixed points, and there are $D(n-2, r-2, k-1)$ $(r-1)$-permutations $b_2 b_3 \cdots b_r$ of $\mathbf{N}_n \setminus \{1\}$ with k fixed points and $b_m = m$. It follows that there are $D(n - 1, r - 1, k) - D(n - 2, r - 2, k - 1)$ $(r - 1)$-permutations $b_2 b_3 \cdots b_r$ of $\mathbf{N}_n \setminus \{1\}$ with k fixed points and $b_m \ne m$, and hence there are $D(n - 1, r - 1, k) - D(n - 2, r - 2, k - 1)$ r-permutations $a_1 a_2 \cdots a_r$ of $\mathbf{N}_n$ with exactly k fixed points, $a_1 = m$ and $a_m \ne 1$. Consequently, there are

$$D(n - 2, r - 2, k) + D(n - 1, r - 1, k) - D(n - 2, r - 2, k - 1)$$

r-permutations $a_1 a_2 \cdots a_r$ of $\mathbf{N}_n$ with exactly k fixed points and $a_1 = m$. Since there are $r - 1$ choices for m, it follows from (MP) that there are

$$(r - 1)[D(n - 2, r - 2, k) + D(n - 1, r - 1, k) - D(n - 2, r - 2, k - 1)]$$

r-permutations $a_1 a_2 \cdots a_r$ of $\mathbf{N}_n$ with exactly k fixed points and $2 \leq a_1 \leq r$.

By (AP), the number of r-element permutations of $\mathbf{N}_n$ with exactly k fixed points is equal to

$$
\begin{aligned}
& D(n - 1, r - 1, k - 1) + (n - r)D(n - 1, r - 1, k) \\
& \quad + (r - 1)[D(n - 2, r - 2, k) + D(n - 1, r - 1, k)] \\
& \quad - (r - 1)D(n - 2, r - 2, k - 1) \\
& = D(n - 1, r - 1, k - 1) + (n - 1)D(n - 1, r - 1, k) \\
& \quad + (r - 1)[D(n - 2, r - 2, k) - D(n - 2, r - 2, k - 1)].
\end{aligned}
$$

By the definition of $D(n, r, k)$, we have

$$
\begin{aligned}
D(n, r, k) = {} & D(n - 1, r - 1, k - 1) + (n - 1)D(n - 1, r - 1, k) \\
& + (r - 1)[D(n - 2, r - 2, k) - D(n - 2, r - 2, k - 1)]
\end{aligned}
$$

as desired.

(4.6.4) By setting $r = n$ in identity (4.6.3) and rearranging the terms in identity (4.6.3), we get

$$
\begin{aligned}
& D(n, n, k) - nD(n - 1, n - 1, k) \\
& = -[D(n - 1, n - 1, k) - (n - 1)D(n - 2, n - 2, k)] \\
& \quad + [D(n - 1, n - 1, k - 1) - (n - 1)D(n - 2, n - 2, k - 1)]. \quad (4.1)
\end{aligned}
$$

Let $f(n, k) = (-1)^{k-n}[D(n, n, k) - nD(n - 1, n - 1, k)]$ for all non-negative integers n and k. Then it is easy to see from the definition of $f(n, k)$ that $f(n, k) = 0$ whenever $n < k$. Also, we have

$$f(n, n) = (-1)^{n-n}[D(n, n, n) - nD(n - 1, n - 1, n)] = D(n, n, n) = 1$$

for all non-negative integers n. Next, we have

$$D(n, n, k) - nD(n - 1, n - 1, k) = (-1)^{n-k}f(n, k). \quad (4.2)$$

By combining equations (4.1) and (4.2), we get

$$(-1)^{n-k}f(n, k) = -(-1)^{n-k-1}f(n - 1, k) + (-1)^{n-k}f(n - 1, k - 1),$$

or equivalently,

$$f(n, k) = f(n - 1, k) + f(n - 1, k - 1) \quad (4.3)$$

for all positive integers n and k. Now, we recall that we have

$$\binom{n}{k} = \binom{n-1}{k} + \binom{n-1}{k-1}$$

for all positive integers n and k. This implies that the recursion formula (4.3) for the $f(n, k)$'s is the same as that for the binomial coefficients. Since we have $f(n, k) = 0 = \binom{n}{k}$ whenever $n < k$ and $f(n, n) = 1 = \binom{n}{n}$ for all non-negative integers n, it follows that we must have $f(n, k) = \binom{n}{k}$ for all non-negative integers n and k. Therefore, by equation (4.2), we have

$$D(n, n, k) = nD(n-1, n-1, k) + (-1)^{n-k} f(n, k)$$

$$= nD(n-1, n-1, k) + (-1)^{n-k} \binom{n}{k}$$

as desired.

(4.6.5) We have

$$\binom{k}{t} D(n, r, k) = \binom{k}{t}\binom{r}{k} D(n-k, r-k, 0) \quad \text{(by (4.6.2))}$$

$$= \binom{r}{t}\binom{r-t}{k-t} D(n-k, r-k, 0) \quad \text{(by (2.1.5))}$$

$$= \binom{r}{t} D(n-t, r-t, k-t) \quad \text{(by (4.6.2))}.$$

(4.6.6) Let $a_1 \cdots a_r$ be an r-permutation of $\mathbf{N}_n$ with exactly k fixed points. If $a_i \neq n$ for all $i = 1, 2, \ldots, r$ (which is possible since $r < n$), then $a_1 \cdots a_r$ is an r-permutation of $\mathbf{N}_n \setminus \{n\}$ with exactly k fixed points, and there are $D(n-1, r, k)$ such r-permutations of $\mathbf{N}_n \setminus \{n\}$ by definition. Else, we must have $a_i = n$ for some $i \in \{1, 2, \ldots, r\}$. Let us fix $i \in \{1, 2, \ldots, r\}$. In the case where $a_i = n$, the remaining $r - 1$ elements in the r-permutation $a_1 \cdots a_r$ form a $(r-1)$-permutation of the set $\mathbf{N}_n \setminus \{n\}$ with exactly k fixed points, and there are $D(n-1, r-1, k)$ such $(r-1)$-permutations of $\mathbf{N}_n \setminus \{n\}$ by definition. Since there are r choices for i, it follows from (MP) that there are $rD(n-1, r-1, k)$ such r-permutations of $\mathbf{N}_n$ in the case when $a_i = n$ for some $i \in \{1, 2, \ldots, r\}$. By (AP), the number of r-permutations of $\mathbf{N}_n$ with exactly k fixed points is equal to $rD(n-1, r-1, k) + D(n-1, r, k)$, and hence by the definition of $D(n, r, k)$, we have

$$D(n, r, k) = rD(n-1, r-1, k) + D(n-1, r, k)$$

as desired.

(4.6.7) Let us count the number of n-permutations $a_1 \cdots a_n$ of $\mathbf{N}_n$ that satisfies $a_j \neq j$ for all $j = 1, 2, \ldots, n - r$ in two different ways. To this end, let us first take any such n-permutation $a_1 \cdots a_n$ of $\mathbf{N}_n$. Since $a_j \neq j$ for all $j = 1, 2, \ldots, n - r$, it follows that $a_1 \cdots a_{n-r}$ is an $(n-r)$-permutation of $\mathbf{N}_n$ with no fixed points, and there are $D(n, n - r, 0)$ such $(n-r)$-permutations of $\mathbf{N}_n$ by definition. Next, since there are no restrictions on the r-permutation $a_{n-r+1} \cdots a_n$ of $\mathbf{N}_n \backslash \{a_1, \ldots, a_{n-r}\}$, it follows that there $r!$ possibilities for the r-permutation $a_{n-r+1} \cdots a_n$ of $\mathbf{N}_n \backslash \{a_1, \ldots, a_{n-r}\}$. By (MP), there are $r!D(n, n-r, 0)$ n-permutations $a_1 \cdots a_n$ of $\mathbf{N}_n$ that satisfies $a_j \neq j$ for all $j = 1, 2, \ldots, n - r$.

On the other hand, let the number of fixed points of $a_1 \cdots a_n$ be i. Since $a_j \neq j$ for all $j = 1, 2, \ldots, n - r$, it follows that the set of fixed points must be a subset of $\{n-r+1, \ldots, n\}$. Now, there are $\binom{r}{i}$ choices for the i fixed points. Next, for each i-element subset A of $\{n - r + 1, \ldots, n\}$ corresponding to the i fixed points, we see that the remaining $n - i$ elements in the n-permutation form a $(n - i)$-permutation of the set $\mathbf{N}_n \setminus A$ with no fixed points, and there are $D(n - i, n - i, 0)$ such $(n - i)$-permutations of $\mathbf{N}_n \setminus A$ by definition. By (MP), the number of n-element permutations $a_1 \cdots a_n$ of $\mathbf{N}_n$ with exactly i fixed points and $a_j \neq j$ for all $j = 1, 2, \ldots, n - r$ is equal to $\binom{r}{i}D(n - i, n - i, 0)$, and thus by (AP), the number of n-element permutations $a_1 \cdots a_n$ of $\mathbf{N}_n$ that satisfies $a_j \neq j$ for all $j = 1, 2, \ldots, n - r$ is equal to

$$\sum_{i=0}^{r} \binom{r}{i} D(n - i, n - i, 0).$$

Consequently, we have

$$r!D(n, n - r, 0) = \sum_{i=0}^{r} \binom{r}{i} D(n - i, n - i, 0),$$

as desired.

28. (i) For $k = 1, 2, \ldots, n - 1$, let P_k denote the property of a permutation a where k is followed immediately by $k + 1$ in a. By definition, we have $\omega(0) = n!$. Next, let us compute $\omega(i)$ for all $i = 1, 2, \ldots, n-1$. To this end, let us fix $i \in \{1, 2, \ldots, n-1\}$, and take any i-tuple of integers $(n_1, n_2, \ldots, n_i)$ satisfying $1 \leq n_1 < n_2 < \cdots < n_i \leq n-1$ and any permutation a that satisfies the properties $P_{n_1}, P_{n_2}, \ldots, P_{n_i}$. Furthermore, let us write $\{n_1, n_2, \ldots, n_i\} = \bigcup_{j=1}^{m} A_j$, where $A_j = \{p_j, p_j + 1, \ldots, q_j\}$ for some integers $p_j, q_j \in \{1, 2, \ldots, n - 1\}$, and

$q_j + 1 < p_{j+1}$ for all integers j satisfying $1 \leq i \leq m-1$. By setting $r_j = q_j + 1$ for all $j = 1, \ldots, m$, it is easy to see that the permutation a contains the m blocks $p_1 \cdots r_1, p_2 \cdots r_2, \ldots, p_m \cdots r_m$.

Now, there are $i + m$ distinct integers in the m blocks, which implies that there are $[n - (i+m) + m]! = (n-i)!$ ways to arrange the integers $1, 2, \ldots, n$, such that the permutation contains the m blocks $p_1 \cdots r_1, p_2 \cdots r_2, \ldots, p_m \cdots r_m$. Consequently, there are $(n-i)!$ permutations a that satisfy the properties $P_{n_1}, P_{n_2}, \ldots, P_{n_i}$. Since there are $\binom{n-1}{i}$ such i-tuples of integers $(n_1, n_2, \ldots, n_i)$ satisfying $1 \leq n_1 \cdots < n_i \leq n-1$, it follows that we have

$$\omega(i) = \sum_{1 \leq n_1 < \cdots < n_i \leq n-1} \omega(P_{n_1} \cdots P_{n_i})$$

$$= \sum_{1 \leq n_1 < \cdots < n_i \leq n-1} (n-i)!$$

$$= \binom{n-1}{i}(n-i)!.$$

Since $\omega(0) = n! = \binom{n-1}{0}(n-0)!$, it follows from (4.3.2) that the number of permutations of the set $\{1, 2, \ldots, n\}$, such that k is never followed immediately by $k+1$ for each $k = 1, 2, \ldots, n-1$, is equal to

$$C_n = E(0) = \sum_{i=0}^{n-1}(-1)^i \omega(i) = \sum_{i=0}^{n-1}(-1)^i \binom{n-1}{i}(n-i)!.$$

(ii) From part (i), we have

$$C_n = \sum_{i=0}^{n-1}(-1)^i \binom{n-1}{i}(n-i)!$$

$$= \sum_{i=0}^{n-1}(-1)^i \left(\frac{(n-1)!}{i!(n-i-1)!}(n-i)! \right)$$

$$= \sum_{i=0}^{n-1}(-1)^i \left(\frac{n!}{i!} - \frac{(n-1)!}{(i-1)!} \right)$$

$$= \sum_{i=0}^{n}(-1)^i \left(\frac{n!}{i!} - \frac{(n-1)!}{(i-1)!} \right)$$

$$= \sum_{i=0}^{n}(-1)^i \frac{n!}{i!} + \sum_{j=0}^{n-1}(-1)^j \frac{(n-1)!}{j!}$$

$$= D_n + D_{n-1} \quad \text{(by (4.6.8))}.$$

29. Since $\{a_1, a_2, \ldots, a_m\} = \{1, 2, \ldots, m\}$, it follows that we must have

$$\{a_{m+1}, a_{m+2}, \ldots, a_n\} = \{m+1, m+2, \ldots, n\}.$$

This implies that $a_1 a_2 \cdots a_m$ is a derangement of $\{1, 2, \ldots, m\}$ and $a_{m+1} a_{m+2} \cdots a_n$ is a derangement of $\{m+1, m+2, \ldots, n\}$. By definition, there are D_m possibilities for $a_1 a_2 \cdots a_m$ and D_{n-m} possibilities for $a_{m+1} a_{m+2} \cdots a_n$, so by (MP), the number of derangements $a_1 a_2 \cdots a_n$ of $\mathbf{N}_n$, such that $\{a_1, a_2, \ldots, a_m\} = \{1, 2, \ldots, m\}$, is equal to $D_m \cdot D_{n-m}$.

30. (i) Let $a_1 a_2 \cdots a_n$ be a permutation of $\mathbf{N}_n$ such that

$$\{a_1, a_2, \ldots, a_m\} = \{m+1, m+2, \ldots, 2m\}.$$

Then we must have $\{a_{m+1}, a_{m+2}, \ldots, a_n\} = \{1, 2, \ldots, m\}$, noting that $n = 2m$. This implies that any permutation $a_1 a_2 \cdots a_n$ of $\mathbf{N}_n$ such that $\{a_1, a_2, \ldots, a_m\} = \{m+1, m+2, \ldots, 2m\}$ must be a derangement of $\mathbf{N}_n$. Now, if $a_1 a_2 \cdots a_n$ is a permutation of $\mathbf{N}_n$ such that $\{a_1, a_2, \ldots, a_m\} = \{m+1, m+2, \ldots, 2m\}$, then there are $m!$ possibilities for $a_1 a_2 \cdots a_m$ and $m!$ possibilities for $a_{m+1} a_{m+2} \cdots a_n$, so by (MP), the number of derangements $a_1 a_2 \cdots a_n$ of $\mathbf{N}_n$, such that

$$\{a_1, a_2, \ldots, a_m\} = \{m+1, m+2, \ldots, 2m\},$$

is equal to $(m!)^2$.

(ii) Let S denote the set of permutations $a_1 a_2 \cdots a_n$ of $\mathbf{N}_n$ such that

$$\{a_1, a_2, \ldots, a_m\} = \{m+1, m+2, \ldots, 2m\}.$$

Then for each $a_1 a_2 \cdots a_n \in S$, we must have

$$\{a_{m+1}, a_{m+2}, \ldots, a_n\} = \{1, 2, \ldots, m, n\},$$

noting that $n = 2m + 1$. This implies that $a_1 a_2 \cdots a_n$ can have at most 1 fixed point, and that fixed point, if it exists, must be n. Let T denote the set of permutations $a_1 a_2 \cdots a_n$ of $\mathbf{N}_n$ such that $\{a_1, \ldots, a_m\} = \{m+1, \ldots, 2m\}$ and $a_n = n$. Then $T \subseteq S$, and $S \setminus T$ is the set of derangements $a_1 a_2 \cdots a_n$ of $\mathbf{N}_n$ such that $\{a_1, \ldots, a_m\} = \{m+1, \ldots, 2m\}$.

It remains to compute $|S \setminus T|$. Let us first take any $a_1 a_2 \cdots a_n \in S$. Then there are $m!$ possibilities for $a_1 a_2 \cdots a_m$ and $(n - m)! = (m + 1)!$ possibilities for $a_{m+1} a_{m+2} \cdots a_n$, so by (MP), we have $|S| = m!(m+1)!$. Next, let us take any $b_1 b_2 \cdots b_n \in T$. Then we must have $b_n = n$ and $\{b_{m+1}, b_{m+2}, \ldots, b_{2m}\} = \{1, 2, \ldots, m\}$.

Consequently, there are $m!$ possibilities for $b_1 b_2 \cdots b_m$ and $m!$ possibilities for $b_{m+1} b_{m+2} \cdots b_{2m}$, so by (MP), we have $|T| = (m!)^2$. Therefore, the number of derangements $a_1 a_2 \cdots a_n$ of $\mathbf{N}_n$, such that $\{a_1, a_2, \ldots, a_m\} = \{m+1, m+2, \ldots, 2m\}$, is equal to

$$|S \setminus T| = |S| - |T| = m!(m+1)! - (m!)^2 = m \cdot (m!)^2.$$

(iii) Let S denote the set of permutations $a_1 a_2 \cdots a_n$ of $\mathbf{N}_n$ such that

$$\{a_1, a_2, \ldots, a_m\} = \{m+1, m+2, \ldots, 2m\}.$$

Then for each $a_1 a_2 \cdots a_n \in S$, we must have

$$\{a_{m+1}, a_{m+2}, \ldots, a_{n-1}, a_n\} = \{1, 2, \ldots, m, 2m+1, 2m+2, \ldots, n\},$$

which implies that $1, 2, \ldots, 2m$ are not fixed points of any $a_1 a_2 \cdots a_n \in S$.

For $i = 1, 2, \ldots, r$, let P_i denote the property of an element $a_1 a_2 \cdots a_n \in S$ where $a_{2m+i} = 2m + i$. Then it is easy to see that $a_1 a_2 \cdots a_n$ is a derangement of $\mathbf{N}_n$ such that $\{a_1, a_2, \ldots, a_m\} = \{m+1, m+2, \ldots, 2m\}$ if and only if $a_1 a_2 \cdots a_n \in S$ and $a_1 a_2 \cdots a_n$ satisfies none of the properties $P_1, P_2, \ldots, P_r$. Let us first compute $\omega(0)$. Let us first take any $a_1 a_2 \cdots a_n \in S$. Then there are $m!$ possibilities for $a_1 a_2 \cdots a_m$ and $(n-m)! = (m+r)!$ possibilities for $a_{m+1} a_{m+2} \cdots a_n$, so by (MP), we have $\omega(0) = |S| = m!(m+r)!$.

Next, let us compute $\omega(i)$ for all $i = 1, 2, \ldots, r$. To this end, let us fix $i \in \{1, 2, \ldots, r\}$, and take any i-tuple of integers $(n_1, n_2, \ldots, n_i)$ satisfying $1 \le n_1 < n_2 < \cdots < n_i \le r$ and any permutation $a_1 a_2 \cdots a_n \in S$ that satisfies the properties $P_{n_1}, P_{n_2}, \ldots, P_{n_i}$. Since $\{a_1, a_2, \ldots, a_m\} = \{m+1, m+2, \ldots, 2m\}$, it follows that we have $m!$ possibilities for $a_1 a_2 \cdots a_m$. Next, since $a_{2m+n_j} = 2m + n_j$ for $j = 1, 2, \ldots, i$, it follows that we have

$$\{a_{m+1}, \ldots, a_n\} \setminus \{a_{2m+n_1}, \ldots, a_{2m+n_i}\}$$
$$= \{1, \ldots, m, 2m+1, \ldots, n\} \setminus \{2m + n_1, \ldots, 2m + n_i\},$$

so that there are $(n - m - i)! = (m + r - i)!$ possibilities for $a_{m+1} a_{m+2} \cdots a_n$ such that $a_{2m+n_j} = 2m + n_j$ for $j = 1, 2, \ldots, i$. By (MP), the number of permutations $a_1 a_2 \cdots a_n \in S$ that satisfies the properties $P_{n_1}, P_{n_2}, \ldots, P_{n_i}$ is equal to $m!(m + r - i)!$. Since there are $\binom{r}{i}$ such i-tuples of integers $(n_1, n_2, \ldots, n_i)$ satisfying

$1 \le n_1 \cdots < n_i \le r$, it follows that we have

$$
\begin{aligned}
\omega(i) &= \sum_{1 \le n_1 < \cdots < n_i \le r} \omega(P_{n_1} \cdots P_{n_i}) \\
&= \sum_{1 \le n_1 < \cdots < n_i \le r} m!(m + r - i)! \\
&= \binom{r}{i} m!(m + r - i)!.
\end{aligned}
$$

Since $\omega(0) = m!(m + r)! = \binom{r}{0} m!(m + r - 0)!$, it follows from (4.3.2) that the number of derangements $a_1 a_2 \cdots a_n$ of $\mathbf{N}_n$, such that $\{a_1, a_2, \ldots, a_m\} = \{m + 1, m + 2, \ldots, 2m\}$, is equal to

$$
\begin{aligned}
E(0) &= \sum_{i=0}^{r} (-1)^i \omega(i) \\
&= \sum_{i=0}^{r} (-1)^i \binom{r}{i} m!(m + r - i)! \\
&= m! \sum_{i=0}^{r} (-1)^i \binom{r}{i} (m + r - i)!.
\end{aligned}
$$

31. (4.6.10) We have

$$(n - 1)[D_{n-1} + D_{n-2}]$$

$$
= (n - 1) \left[(n - 1)! \sum_{i=0}^{n-1} \frac{(-1)^i}{i!} + (n - 2)! \sum_{i=0}^{n-2} \frac{(-1)^i}{i!} \right] \quad \text{(by (4.6.8))}
$$

$$
= (n - 1)! \left[(n - 1) \sum_{i=0}^{n-1} \frac{(-1)^i}{i!} + \sum_{i=0}^{n-2} \frac{(-1)^i}{i!} \right]
$$

$$
= (n - 1)! \left[(n - 1) \sum_{i=0}^{n-1} \frac{(-1)^i}{i!} + \sum_{i=0}^{n-1} \frac{(-1)^i}{i!} - \frac{(-1)^{n-1}}{(n - 1)!} \right]
$$

$$
= (n - 1)! \left[n \sum_{i=0}^{n-1} \frac{(-1)^i}{i!} + \frac{(-1)^n}{(n - 1)!} \right]
$$

$$
= n! \sum_{i=0}^{n-1} \frac{(-1)^i}{i!} + (-1)^n
$$

$$
= n! \sum_{i=0}^{n-1} \frac{(-1)^i}{i!} + n! \frac{(-1)^n}{n!}
$$

$$
= n! \sum_{i=0}^{n} \frac{(-1)^i}{i!}
$$

$= D_n$ (by (4.6.8)).

(4.6.11) We have

$$D_n = n! \sum_{i=0}^{n} \frac{(-1)^i}{i!} \quad \text{(by (4.6.8))}$$

$$= n! \sum_{i=0}^{n-1} \frac{(-1)^i}{i!} + n! \frac{(-1)^n}{n!}$$

$$= n(n-1)! \sum_{i=0}^{n-1} \frac{(-1)^i}{i!} + (-1)^n$$

$$= n D_{n-1} + (-1)^n \quad \text{(by (4.6.8))}.$$

32. By (4.6.11), we have $D_{2k} = 2k D_{2k-1} + (-1)^{2k}$ for all positive integers k. Since $2k$ is even, so is $2k D_{2k-1}$. As D_{2k} is the sum of an even integer $2k D_{2k-1}$ and an odd integer $(-1)^{2k}$, it follows that D_{2k} is odd for all integers k. Next, by (4.6.11) again, we have

$$D_{2k+1} = (2k+1) D_{2k} + (-1)^{2k+1}.$$

Since D_{2k} and $2k+1$ are both odd, so is $(2k+1)D_{2k}$. Since $(-1)^{2k+1}$ is odd and D_{2k+1} is the sum of two odd integers $(2k+1)D_{2k}$ and $(-1)^{2k+1}$, it follows that D_{2k+1} is even for all positive integers k. Since $D_1 = 0$ is even as well, this shows that D_n is even if and only if n is odd, and we are done.

33. (i) By (4.6.2), we have

$$D_n(k) = D(n, n, k) = \binom{n}{k} D(n-k, n-k, 0) = \binom{n}{k} D_{n-k}.$$

(ii) Let S denote the set of permutations of $\mathbf{N}_n$, and for each $k = 0, 1, \ldots, n$, let A_k denote the set of permutations of $\mathbf{N}_n$ with exactly k fixed points. Then we have $|S| = n!$, and $|A_k| = D_n(k) = \binom{n}{k} D_{n-k} = \binom{n}{n-k} D_{n-k}$ for all $k = 0, 1, \ldots, n$ by part (i). Since $S = \bigcup_{k=0}^{n} A_k$ and $A_i \cap A_j = \emptyset$ for all $i, j \in \{0, 1, \ldots, n\}$ with $i \neq j$, it follows that we have

$$n! = |S|$$

$$= \sum_{k=0}^{n} |A_k|$$

$$= \sum_{k=0}^{n} \binom{n}{n-k} D_{n-k}$$

$$= \sum_{j=0}^{n} \binom{n}{j} D_j \quad \text{(by setting } j = n - k)$$

$$= \binom{n}{0} D_0 + \binom{n}{1} D_1 + \cdots + \binom{n}{n} D_n.$$

(iii) Let S denote the set of permutations of $\mathbf{N}_{n+1}$ with $k+1$ fixed points. Let us count the number of pairs (a, j), where $a \in S$ and j is a fixed point of a in two different ways. Let us first fix $b \in S$. Since b has $k+1$ fixed points, it follows that there are $k+1$ such pairs (b, j), such that j is a fixed point of b. As there are $D_{n+1}(k+1)$ choices for b, it follows from (MP) that the number of such pairs (b, j), where $b \in S$ and j is a fixed point of b, is equal to $(k+1)D_{n+1}(k+1)$. Next, let us fix any $i \in \{1, 2, \ldots, n+1\}$, and let $c = c_1 c_2 \cdots c_{n+1} \in S$ be a permutation of $\mathbf{N}_{n+1}$ whose one of its fixed points is i. Then $c_1 c_2 \cdots c_{i-1} c_{i+1} \cdots c_{n+1}$ is a permutation of $\mathbf{N}_{n+1} \setminus \{i\}$ with exactly k fixed points, and there are $D_n(k)$ such permutations of $\mathbf{N}_{n+1} \setminus \{i\}$ by definition. Since there are $n + 1$ choices for i, it follows from (MP) that the number of such pairs (a, j), where $a \in S$ and j is a fixed point of a, is equal to $(n + 1)D_n(k)$. Consequently, we have

$$(k + 1)D_{n+1}(k + 1) = (n + 1)D_n(k)$$

as desired.

34. Let S denote the set of permutations of $\mathbf{N}_n$. Let us count the number of pairs (a, j), where $a \in S$ and j is a fixed point of a in two different ways. Let us first fix $b \in S$, and let the number of fixed points of b be k. Then we must have $k \in \{1, \ldots, n\}$. Since b has k fixed points, it follows that there are k such pairs (b, j), such that j is a fixed point of b. As there are $D_n(k)$ choices for b, it follows from (MP) that the number of such pairs (b, j), where b is a permutation of $\mathbf{N}_n$ with k fixed points and j is a fixed point of b, is equal to $k \cdot D_n(k)$. Since $k \in \{1, \ldots, n\}$, it follows from (AP) that the number of pairs (a, j), where $a \in S$ and j is a fixed point of a, is equal to $\sum_{k=1}^{n} k \cdot D_n(k)$. Next, let us fix any $i \in \{1, 2, \ldots, n\}$, and let $c = c_1 c_2 \cdots c_n \in S$ be a permutation of $\mathbf{N}_n$ where one of its fixed points is i. Then $c_1 c_2 \cdots c_{i-1} c_{i+1} \cdots c_n$ is a permutation of $\mathbf{N}_n \setminus \{i\}$, and there are $(n - 1)!$ such permutations of $\mathbf{N}_n \setminus \{i\}$ by definition. Since there are n choices for i, it follows from (MP) that the number of such pairs (a, j), where $a \in S$ and j is a fixed

point of a, is equal to $n \cdot (n-1)! = n!$. Consequently, we have

$$\sum_{k=1}^{n} k \cdot D_n(k) = n!,$$

or equivalently,

$$\sum_{k=0}^{n} k \cdot D_n(k) = n!$$

as desired.

35. We have

$$\begin{aligned}
D_n(0) - D_n(1) &= \binom{n}{0} D_{n-0} - \binom{n}{1} D_{n-1} \quad \text{(by Problem 4.33(i))} \\
&= D_n - n D_{n-1} \\
&= (-1)^n \quad \text{(by (4.6.11))}.
\end{aligned}$$

36. Let us first show that $\sum_{k=0}^{n} D_n(k) = n!$. Let S denote the set of permutations of $\mathbf{N}_n$, and for each $k = 0, 1, \ldots, n$, let A_k denote the set of permutations of $\mathbf{N}_n$ with exactly k fixed points. Then we have $|S| = n!$, and $|A_k| = D_n(k)$. Since $S = \bigcup_{k=0}^{n} A_k$ and $A_i \cap A_j = \emptyset$ for all $i, j \in \{0, 1, \ldots, n\}$ with $i \neq j$, it follows that we have

$$n! = |S| = \sum_{k=0}^{n} |A_k| = \sum_{k=0}^{n} D_n(k)$$

as claimed.

Next, let us show that $\sum_{k=0}^{n} k(k-1) D_n(k) = n!$. Let us count the number of pairs $(a, (i,j))$, where $a \in S$ and i, j are distinct fixed points of a in two different ways. Let us first fix $b \in S$, and let the number of fixed points of b be k. Then we must have $k \in \{2, \ldots, n\}$. Since b has k fixed points, it follows that there are $k(k-1)$ such pairs $(b, (i,j))$, such that i, j are distinct fixed points of b. As there are $D_n(k)$ choices for b, it follows from (MP) that the number of such pairs $(b, (i,j))$, where b is a permutation of $\mathbf{N}_n$ with k fixed points and i, j are distinct fixed points of b, is equal to $k(k-1) \cdot D_n(k)$. Since $k \in \{2, \ldots, n\}$, it follows from (AP) that the number of pairs $(a, (i,j))$, where $a \in S$ and i, j are distinct fixed points of a, is equal to $\sum_{k=2}^{n} k(k-1) \cdot D_n(k)$. Next, let us fix any distinct $i, j \in \{1, 2, \ldots, n\}$, and let $c \in S$ be a permutation of

$\mathbf{N}_n$ whose two of its fixed points are i and j. Then the remaining $n - 2$ elements of the n-permutation c is a permutation of $\mathbf{N}_n \setminus \{i, j\}$, and there are $(n - 2)!$ such permutations of $\mathbf{N}_n \setminus \{i, j\}$ by definition. Since there are $n(n - 1)$ choices for i and j, it follows from (MP) that the number of such pairs $(a, (i, j))$, where $a \in S$ and i, j are distinct fixed points of a, is equal to $n(n - 1) \cdot (n - 2)! = n!$. Consequently, we have

$$\sum_{k=2}^{n} k(k - 1) \cdot D_n(k) = n!,$$

or equivalently,

$$\sum_{k=0}^{n} k(k - 1) \cdot D_n(k) = n!$$

as claimed.

Therefore, we have

$$\sum_{k=0}^{n} (k - 1)^2 \cdot D_n(k) = \sum_{k=0}^{n} [k(k - 1) - (k - 1)] \cdot D_n(k)$$

$$= \sum_{k=0}^{n} k(k - 1) \cdot D_n(k) - \sum_{k=0}^{n} k \cdot D_n(k) + \sum_{k=0}^{n} D_n(k)$$

$$= n! - n! + n! \quad \text{(by Problem 4.34)}$$

$$= n!$$

as desired.

37. Let S denote the set of permutations of $\mathbf{N}_n$. Let us count the number of pairs $(a, (i_1, i_2, \ldots, i_r))$, where $a \in S$ and $i_1, i_2, \ldots, i_r$ are pairwise distinct fixed points of a in two different ways. Let us first fix $b \in S$, and let the number of fixed points of b be k. Then we must have $k \in \{0, 1, \ldots, n\}$. Since b has k fixed points, it follows that there are $k(k-1) \cdots (k-r+1)$ such pairs $(b, (i_1, i_2, \ldots, i_r))$, such that $i_1, i_2, \ldots, i_r$ are pairwise distinct fixed points of b. As there are $D_n(k)$ choices for b, it follows from (MP) that the number of such pairs $(b, (i_1, i_2, \ldots, i_r))$, where b is a permutation of $\mathbf{N}_n$ with k fixed points and $i_1, i_2, \ldots, i_r$ are distinct fixed points of b, is equal to $k(k - 1) \cdots (k - r + 1)D_n(k)$. Since $k \in \{0, 1, \ldots, n\}$, it follows from (AP) that the number of pairs $(a, (i_1, i_2, \ldots, i_r))$, where $a \in S$ and i, j are distinct fixed points of a, is equal to

$$\sum_{k=0}^{n} k(k - 1) \cdots (k - r + 1)D_n(k).$$

Next, let us fix any pairwise distinct $i_1, \ldots, i_r \in \{1, \ldots, n\}$, and let $c \in S$ be a permutation of $\mathbf{N}_n$ in which r of its fixed points are $i_1, \ldots, i_r$. Then the remaining $n - r$ elements of the n-permutation c is a permutation of $\mathbf{N}_n \setminus \{i_1, \ldots, i_r\}$, and there are $(n - r)!$ such permutations of $\mathbf{N}_n \setminus \{i_1, \ldots, i_r\}$ by definition. Since there are $n(n - 1) \cdots (n - r + 1)$ choices for $i_1, \ldots, i_r$, it follows from (MP) that the number of such pairs $(a, (i_1, \ldots, i_r))$, where $a \in S$ and $i_1, \ldots, i_r$ are pairwise distinct fixed points of a, is equal to $n(n-1) \cdots (n-r+1)(n-r)! = n!$. Consequently, we have

$$\sum_{k=0}^{n} k(k - 1) \cdots (k - r + 1) D_n(k) = n!.$$

Since $k(k - 1) \cdots (k - r + 1) = 0$ for all non-negative integers $k < r$, it follows that we have

$$\sum_{k=r}^{n} k(k-1) \cdots (k-r+1) D_n(k) = \sum_{k=0}^{n} k(k-1) \cdots (k-r+1) D_n(k) = n!$$

as desired.

38. (a) (i) We shall first make the following claims for any positive integers m, n, k:

Claim A: $(m + kn, n) = (m, n)$.

Claim B: $(m, k) = (n, k) = 1$ if and only if $(mn, k) = 1$.

Claim C: If $k | mn$ and $(m, k) = 1$, then $k | n$.

Claim D: If $(m, n) = 1$, then $\{k, k+m, k+2m, \ldots, k+(n-1)m\}$ form a complete set of residues modulo n.

The proofs of Claims A and B follow directly from the definition of the greatest common divisor, and shall be omitted. We will first show that Claim C holds. Suppose on the contrary that k does not divide n. Then we have $d = (n, k) < k$. Let $k' = \frac{k}{d} > 1$ and $n' = \frac{n}{d}$, so that we have $k' | mn'$ and $(n', k') = 1$. As $(m, k) = 1$ and $k' | k$, we must have $(m, k') = 1$. By Claim B, we must have $(mn', k') = 1$. On the other hand, since k' and mn' have a common divisor k', we must have $(mn', k') \geq k' > 1$, which contradicts the fact that $(mn', k') = 1$. So $k | n$.

Finally, we need to show that Claim D holds. Again, suppose on the contrary that the given statement is false. Then there exists integers i, j with $0 \leq i < j \leq n - 1$, such that $k + im \equiv k + jm \pmod{n}$, so we must have $n | (j - i)m$. As $(m, n) = 1$, it follows from Claim C that we must have $n \mid j - i$. As $0 < j - i < n - 1$, this is a contradiction, for both $j - i$ and n are integers. So $\{k, k+m, k+2m, \ldots, k+(n-1)m\}$ form a complete set of residues modulo n.

We shall now proceed to prove the given statement. Let m and n be positive integers with $(m, n) = 1$. Let

$$A_i = \{i, i+m, i+2m, \ldots, i+(n-1)m\}$$

for all $i = 1, \ldots, m$. Then it follows from Claim A that we have $(j, m) = (i, m)$ for all $i = 1, \ldots, m$ and $j \in A_i$. In particular, there are $\phi(m)$ elements i in $\{1, \ldots, m\}$ such that each element in A_i is coprime to m.

Next, let us pick any $i \in \{1, \ldots, m\}$ for which $(i, m) = 1$. By Claim D, A_i form a complete set of residues modulo n, so there are $\phi(n)$ elements in A_i that are coprime to n. By Claim B and (MP), there are $\phi(m)\phi(n)$ elements in $\{1, 2, \ldots, mn\}$ that are coprime to mn, and so by the definition of $\phi(mn)$, we must have $\phi(mn) = \phi(m)\phi(n)$ as desired.

(ii) Let us show that for all positive integers k and i, we have $(k, p^i) \neq 1$ if and only if p divides k. Clearly, if p divides k, then k and p^i have a common divisor p, which implies that $(k, p^i) \neq 1$. Conversely, if $(k, p^i) \neq 1$, then k is not coprime to p^i, so there exists some prime q that divides both p^i and k. This implies that q divides p. As p is a prime, we must have $p = q$, and hence p divides k, and this completes the claim. Now, there are $\left\lfloor \frac{p^i}{p} \right\rfloor = p^{i-1}$ elements in $\mathbf{N}_{p^i}$ that are divisible by p, which implies that $\phi(p^i) = p^i - p^{i-1}$ as desired.

(b) Let the prime factorization of n be $p_1^{m_1} p_2^{m_2} \cdots p_k^{m_k}$, where p_i is a prime number and m_i is a positive integer for all $i = 1, 2, \ldots, k$. We have

$$\begin{aligned}
\phi(n) &= \phi(p_1^{m_1} p_2^{m_2} \cdots p_k^{m_k}) \\
&= \phi(p_1^{m_1})\phi(p_2^{m_2}) \cdots \phi(p_k^{m_k}) \quad \text{(by part (i))} \\
&= (p_1^{m_1} - p_1^{m_1-1})(p_2^{m_2} - p_2^{m_2-1}) \cdots (p_k^{m_k} - p_k^{m_k-1}) \\
&\quad \text{(by part (ii))} \\
&= p_1^{m_1}\left(1 - \frac{1}{p_1}\right) p_2^{m_2}\left(1 - \frac{1}{p_2}\right) \cdots p_k^{m_k}\left(1 - \frac{1}{p_k}\right) \\
&= p_1^{m_1} p_2^{m_2} \cdots p_k^{m_k}\left(1 - \frac{1}{p_1}\right)\left(1 - \frac{1}{p_2}\right) \cdots \left(1 - \frac{1}{p_k}\right)
\end{aligned}$$

$$= n \left(1 - \frac{1}{p_1}\right) \left(1 - \frac{1}{p_2}\right) \cdots \left(1 - \frac{1}{p_k}\right)$$

$$= n \prod_{i=1}^{k} \left(1 - \frac{1}{p_i}\right).$$

39. (i) Since $100 = 2^2 \cdot 5^2$ and $300 = 2^2 \cdot 3 \cdot 5^2$, it follows from (4.7.1) that we have

$$\phi(100) = 100 \left(1 - \frac{1}{2}\right) \left(1 - \frac{1}{5}\right) = 100 \cdot \frac{1}{2} \cdot \frac{4}{5} = 40, \quad \text{and}$$

$$\phi(300) = 300 \left(1 - \frac{1}{2}\right) \left(1 - \frac{1}{3}\right) \left(1 - \frac{1}{5}\right) = 300 \cdot \frac{1}{2} \cdot \frac{2}{3} \cdot \frac{4}{5} = 80.$$

(ii) Let the prime factorization of m be $p_1^{a_1} p_2^{a_2} \cdots p_k^{a_k}$, where p_i is a prime number and a_i is a positive integer for all $i = 1, 2, \ldots, k$. Since m divides n, it follows that there exist positive integers $b_1, b_2, \ldots, b_k, \ell$, such that $b_i \geq a_i$ for all $i = 1, 2, \ldots, k$, $n = p_1^{b_1} p_2^{b_2} \cdots p_k^{b_k} \cdot \ell$, and $(p_1^{b_1} p_2^{b_2} \cdots p_k^{b_k}, \ell) = 1$. Now, we have

$$\frac{\phi(n)}{\phi(m)} = \frac{\phi(p_1^{b_1} p_2^{b_2} \cdots p_k^{b_k} \cdot \ell)}{\phi(p_1^{a_1} p_2^{a_2} \cdots p_k^{a_k})}$$

$$= \frac{\phi(p_1^{b_1} p_2^{b_2} \cdots p_k^{b_k}) \phi(\ell)}{\phi(p_1^{a_1} p_2^{a_2} \cdots p_k^{a_k})} \quad \text{(by Problem 4.38(a)(i))}$$

$$= \frac{p_1^{b_1} p_2^{b_2} \cdots p_k^{b_k} \left(1 - \frac{1}{p_1}\right) \left(1 - \frac{1}{p_2}\right) \cdots \left(1 - \frac{1}{p_k}\right) \phi(\ell)}{p_1^{a_1} p_2^{a_2} \cdots p_k^{a_k} \left(1 - \frac{1}{p_1}\right) \left(1 - \frac{1}{p_2}\right) \cdots \left(1 - \frac{1}{p_k}\right)}$$

(by (4.7.1))

$$= p_1^{b_1 - a_1} p_2^{b_2 - a_2} \cdots p_k^{b_k - a_k} \phi(\ell).$$

Since $b_i - a_i \geq 0$ for all $i = 1, 2, \ldots, k$, it follows that $p_1^{b_1 - a_1} p_2^{b_2 - a_2} \cdots p_k^{b_k - a_k}$ and hence $\frac{\phi(n)}{\phi(m)}$ are integers. Consequently, we have $\phi(m)$ to divide $\phi(n)$ whenever m divides n as desired.

40. Let $A = \left\{ \frac{k}{n} \mid 1 \leq k \leq n \right\}$, and for each positive divisor d of n, we let

$$A_d = \left\{ \frac{k}{d} \mid 1 \leq k \leq d, (k, d) = 1 \right\}.$$

Then it is clear that $\bigcup_{d \mid n} A_d \subseteq A$. Conversely, if $\frac{k}{n} \in A$, then we may let $d = (k, n)$, $d' = \frac{n}{d}$ and $k' = \frac{k}{d}$, so that d' is also a positive divisor of n, and $\frac{k}{n} = \frac{k/d}{n/d} = \frac{k'}{d'}$. Since $(k', d') = 1$, it follows that $\frac{k}{n} \in A_{d'}$, so we

have $\frac{k}{n} \in \bigcup_{d|n} A_d$. This shows that $A \subseteq \bigcup_{d|n} A_d$, so we have $A = \bigcup_{d|n} A_d$.
Since each positive rational number has a unique representation as an irreducible fraction with a positive denominator (that is, a representation of the form $\frac{a}{b}$, where a and b are coprime positive integers), it follows that we must have $A_d \cap A_{d'} = \emptyset$ for any distinct positive divisors d, d' of n. Since $|A| = n$ and $|A_d| = \varphi(d)$ for all positive divisors d of n, it follows that we have

$$\sum (\phi(d) \mid d \in \mathbf{N}, d|n) = \sum_{d \in \mathbf{N}, d|n} |A_d| = |A| = n$$

as desired.

41. If $h = 1$, then the equation holds trivially by Problem 4.38(a)(i). So let us assume that $h > 1$, and write $h = p_1^{a_1} p_2^{a_2} \cdots p_k^{a_k}$, where p_i is a prime number and a_i is a positive integer for all $i = 1, 2, \ldots, k$. As $(m, n) = h$, we may write $m = p_1^{b_1} p_2^{b_2} \cdots p_k^{b_k} \cdot r$ and $n = p_1^{c_1} p_2^{c_2} \cdots p_k^{c_k} \cdot s$, where $b_1, b_2, \ldots, b_k, c_1, c_2, \ldots, c_k, r, s$ are positive integers satisfying $b_i \geq a_i$ and $c_i \geq a_i$ for all $i = 1, 2, \ldots, k$, and $(p_1^{b_1} p_2^{b_2} \cdots p_k^{b_k}, r) = (p_1^{c_1} p_2^{c_2} \cdots p_k^{c_k}, s) = (r, s) = 1$. Then we have

$$\phi(mn) \cdot \phi(h)$$
$$= \phi(p_1^{b_1} \cdots p_k^{b_k} \cdot r \cdot p_1^{c_1} \cdots p_k^{c_k} \cdot s) \cdot \phi(p_1^{a_1} \cdots p_k^{a_k})$$
$$= \phi(p_1^{b_1+c_1} \cdots p_k^{b_k+c_k})\phi(r)\phi(s)\phi(p_1^{a_1} \cdots p_k^{a_k}) \quad \text{(by Problem 4.38(a)(i))}$$
$$= p_1^{b_1+c_1} \cdots p_k^{b_k+c_k} \left(1 - \frac{1}{p_1}\right) \cdots \left(1 - \frac{1}{p_k}\right) \phi(r)\phi(s)$$
$$\times p_1^{a_1} \cdots p_k^{a_k} \left(1 - \frac{1}{p_1}\right) \cdots \left(1 - \frac{1}{p_k}\right) \quad \text{(by (4.7.1))}$$
$$= p_1^{b_1+c_1} \cdots p_k^{b_k+c_k} \left(1 - \frac{1}{p_1}\right) \cdots \left(1 - \frac{1}{p_k}\right) \phi(r)\phi(s)$$
$$\times h \cdot \left(1 - \frac{1}{p_1}\right) \cdots \left(1 - \frac{1}{p_k}\right)$$
$$= p_1^{b_1} \cdots p_k^{b_k} \left(1 - \frac{1}{p_1}\right) \cdots \left(1 - \frac{1}{p_k}\right) \phi(r)$$
$$\times p_1^{c_1} \cdots p_k^{c_k} \left(1 - \frac{1}{p_1}\right) \cdots \left(1 - \frac{1}{p_k}\right) \phi(s) \cdot h$$
$$= \phi(p_1^{b_1} \cdots p_k^{b_k})\phi(r)\phi(p_1^{c_1} \cdots p_k^{c_k})\phi(s) \cdot h \quad \text{(by (4.7.1))}$$
$$= \phi(p_1^{b_1} \cdots p_k^{b_k} \cdot r)\phi(p_1^{c_1} \cdots p_k^{c_k} \cdot s) \cdot h \quad \text{(by Problem 4.38(a)(i))}$$
$$= \phi(m) \cdot \phi(n) \cdot h.$$

42. Let us take any positive integer $n \geq 3$, and write $n = 2^k \cdot m$, where k is a non-negative integer and m is a positive odd integer. Since $n \geq 3$, it follows that we must either have $k \geq 2$ or $m \geq 3$. If $k \geq 2$, then we have

$$
\begin{aligned}
\phi(n) &= \phi(2^k \cdot m) \\
&= \phi(2^k)\phi(m) \quad \text{(by Problem 4.38(a)(i))} \\
&= (2^k - 2^{k-1})\phi(m) \quad \text{(by Problem 4.38(a)(ii))} \\
&= 2^{k-1}\phi(m).
\end{aligned}
$$

Since we have $k - 1 \geq 1$, we have 2^{k-1} to be an even integer, and hence, $\phi(n) = 2^{k-1}\phi(m)$ is even. Else, we must have $m \geq 3$, and hence there exists some odd prime p that divides m. Let us write $m = p^\ell \cdot r$ where ℓ is a positive integer and r is a positive odd integer with $(p^\ell, r) = 1$. Then we have

$$
\begin{aligned}
\phi(n) &= \phi(2^k \cdot p^\ell \cdot r) \\
&= \phi(2^k)\phi(p^\ell)\phi(r) \quad \text{(by Problem 4.38(a)(i))} \\
&= \phi(2^k)(p^\ell - p^{\ell-1})\phi(r) \quad \text{(by Problem 4.38(a)(ii))}.
\end{aligned}
$$

Since p^ℓ and $p^{\ell-1}$ are both odd, we have $p^\ell - p^{\ell-1}$ to be an even integer, and hence $\phi(n) = \phi(2^k)(p^\ell - p^{\ell-1})\phi(r)$ is even as desired.

43. Let the k distinct prime factors of n be $p_1, p_2, \ldots, p_k$. Since $p_i \geq 2$ for all $i = 1, 2, \ldots, k$, it follows that we have $1 - \frac{1}{p_i} \geq 1 - \frac{1}{2} = \frac{1}{2}$ for all $i = 1, 2, \ldots, k$. Consequently, by (4.7.1), we have

$$
\phi(n) = n\left(1 - \frac{1}{p_1}\right)\left(1 - \frac{1}{p_2}\right)\cdots\left(1 - \frac{1}{p_k}\right) \geq n \cdot \underbrace{\frac{1}{2} \cdot \frac{1}{2} \cdots \frac{1}{2}}_{k} = n \cdot 2^{-k}
$$

as desired.

44. Let the k distinct odd prime factors of n be $p_1, p_2, \ldots, p_k$, and let us write $n = 2^m \cdot p_1^{a_1} p_2^{a_2} \cdots p_k^{a_k}$, where m is a non-negative integer and a_i is a positive integer for $i = 1, 2, \ldots, k$. Since $p_i^{a_i} - p_i^{a_i - 1}$ is even for all $i = 1, 2, \ldots, k$, we may write $p_i^{a_i} - p_i^{a_i - 1} = 2n_i$, where n_i is a positive integer for $i = 1, 2, \ldots, k$. Then we have

$$
\begin{aligned}
\phi(n) &= \phi(2^m \cdot p_1^{a_1} \cdots p_k^{a_k}) \\
&= \phi(2^m)\phi(p_1^{a_1}) \cdots \phi(p_k^{a_k}) \quad \text{(by Problem 4.38(a)(i))} \\
&= \phi(2^m)(p_1^{a_1} - p_1^{a_1 - 1}) \cdots (p_k^{a_k} - p_k^{a_k - 1}) \quad \text{(by Problem 4.38(a)(ii))} \\
&= \phi(2^m) \cdot 2n_1 \cdots 2n_k \\
&= 2^k \phi(2^m) \cdot n_1 \cdots n_k,
\end{aligned}
$$

which shows that 2^k divides $\phi(n)$ as desired.

45. Arguing by contradiction, suppose that there exists some positive integer n, such that $\phi(n) = 14$. By Problem 4.38(a)(ii), we have $\phi(2^k) = 2^k - 2^{k-1} = 2^{k-1}$ for all positive integers k. As 14 is not a power of 2, it follows that n is divisible by some odd prime p. Let us write $n = 2^r \cdot m$ where r is a non-negative integer and m is a positive odd integer with $m \geq 3$. Since p divides n and p does not divide 2, we must have p to divide m. By Problem 4.38(a)(i), we have $\phi(n) = \phi(2^r \cdot m) = \phi(2^r)\phi(m)$. Since $m \geq 3$, it follows from Problem 4.42 that $\phi(m)$ is even, so we may write $\phi(m) = 2m'$, where m' is a positive integer. This gives us

$$2m'\phi(2^r) = \phi(m)\phi(2^r) = \phi(n) = 14,$$

or equivalently, $m'\phi(2^r) = 7$. As $m'\phi(2^r)$ is odd and $\phi(2^k) = 2^{k-1}$ is even for all positive integers $k > 1$, this forces $r = 0$ or $r = 1$, and in both of these cases, we have $\phi(2^r) = 1$. This would then imply that $\phi(n) = \phi(2^r)\phi(m) = \phi(m)$. Now, let the number of distinct odd prime factors of m be ℓ. By Problem 4.44, we have 2^ℓ to divide $\phi(m)$. Since $\phi(m) = \phi(n) = 14$ and 2^k does not divide 14 for all positive integers $k > 1$, we must have $\ell = 1$, so m has exactly one odd prime factor. As p divides m, we must have $m = p^i$ for some positive integer i. By Problem 4.38(a)(ii), we have

$$\phi(m) = \phi(p^i) = p^i - p^{i-1} = p^{i-1}(p-1).$$

As $p - 1$ is even and $\phi(m) = 14 = 2 \cdot 7$, we must have $p - 1 = 2$ or $p - 1 = 14$, or equivalently, $p = 3$ or $p = 15$. As 15 is not a power of a prime, we must have $p = 3$. This would then imply that $p^{i-1} = \frac{\phi(m)}{p-1} = \frac{14}{3-1} = 7$. Since p is a prime, we must then have $p = 7$, which is a contradiction. So there are no positive integers n such that $\phi(n) = 14$.

46. If n is odd, then we have $(2, n) = 1$, so by Problem 4.38(a)(i), we have $\phi(2n) = \phi(2)\phi(n) = \phi(n)$. Else, if n is even, then we have $(2, n) = 2$. By setting $m = h = 2$ in Problem 4.41, we have

$$\phi(2n) = \phi(2n)\phi(2) = \phi(2)\phi(n) \cdot 2 = 2\phi(n).$$

47. Let S denote the universal set. Firstly, we observe that

$$E(m) - A(m, r) = \sum_{k=m}^{q}(-1)^{k-m}\binom{k}{m}\omega(k) - \sum_{k=m}^{r}(-1)^{k-m}\binom{k}{m}\omega(k)$$

$$= \sum_{k=r+1}^{q}(-1)^{k-m}\binom{k}{m}\omega(k).$$

For each $x \in S$, let us compute the contribution of the count by x to the sum $E(m) - A(m, r) = \sum_{k=r+1}^{q} (-1)^{k-m} \binom{k}{m} \omega(k)$. Let the number of properties that x satisfy be t. Let us consider these cases:

Case 1: $t \leq r$. In this case, x is counted 0 times in $\omega(k)$ for all $k = r+1, \ldots, q$. Thus, x contributes a count of 0 to the sum $E(m) - A(m, r) = \sum_{k=r+1}^{q} (-1)^{k-m} \binom{k}{m} \omega(k)$.

Case 2: $t > r$. In this case, x is counted $\binom{t}{k}$ times in $\omega(k)$ for $k = r+1, r+2, \ldots, t$ and is counted 0 times in $\omega(k)$ for $k = t+1, t+2, \ldots, q$. Thus, the count that x contributes to the sum $E(m) - A(m, r) = \sum_{k=r+1}^{q} (-1)^{k-m} \binom{k}{m} \omega(k)$ is equal to

$$\sum_{k=r+1}^{t} (-1)^{k-m} \binom{k}{m} \binom{t}{k}$$

$$= \sum_{k=r+1}^{t} (-1)^{k-m} \binom{t-m}{k-m} \binom{t}{m} \quad \text{(by (2.1.5))}$$

$$= \binom{t}{m} \sum_{k=r+1}^{t} (-1)^{k-m} \binom{t-m}{t-k}$$

$$= \binom{t}{m} \sum_{j=0}^{t-r-1} (-1)^{t-j-m} \binom{t-m}{j} \quad \text{(by setting } j = t-k\text{)}$$

$$= (-1)^{t-m} \binom{t}{m} \sum_{j=0}^{t-r-1} (-1)^{j} \binom{t-m}{j} \quad \text{(since } (-1)^{-j} = (-1)^{j}\text{)}$$

$$= (-1)^{t-m} \binom{t}{m} \cdot (-1)^{t-r-1} \binom{t-m-1}{t-r-1}$$

(by Problem 2.29, since $t - r - 1 < t - m$)

$$= (-1)^{r-m+1} \binom{t}{m} \binom{t-m-1}{t-r-1} \quad \text{(since } (-1)^{t-r-1} = (-1)^{-t+r+1}\text{)}.$$

Since there are $E(t)$ elements of S that has exactly t properties, it follows that we have

$$E(m) - A(m, r) = \sum_{t=r+1}^{q} E(t) \cdot (-1)^{r-m+1} \binom{t}{m} \binom{t-m-1}{t-r-1}$$

$$= (-1)^{r-m+1} \sum_{t=r+1}^{q} E(t) \binom{t}{m} \binom{t-m-1}{t-r-1}.$$

(i) If m and r have the same parity, then $r-m+1$ is odd, which implies that $(-1)^{r-m+1} = -1$. As $E(t)\binom{t}{m}\binom{t-m-1}{t-r-1}$ is non-negative for each $t = r+1, \ldots, q$, it follows that the sum $\sum\limits_{t=r+1}^{q} E(t)\binom{t}{m}\binom{t-m-1}{t-r-1}$ is non-negative, and therefore we have

$$E(m) - A(m,r) = (-1)^{r-m+1} \sum_{t=r+1}^{q} E(t)\binom{t}{m}\binom{t-m-1}{t-r-1}$$
$$= -\sum_{t=r+1}^{q} E(t)\binom{t}{m}\binom{t-m-1}{t-r-1}$$
$$\leq 0,$$

or equivalently, $E(m) \leq A(m,r)$.

(ii) If m and r have different parities, then $r - m + 1$ is even, which implies that $(-1)^{r-m+1} = 1$. As before in part (i), we have

$$E(m) - A(m,r) = (-1)^{r-m+1} \sum_{t=r+1}^{q} E(t)\binom{t}{m}\binom{t-m-1}{t-r-1}$$
$$= \sum_{t=r+1}^{q} E(t)\binom{t}{m}\binom{t-m-1}{t-r-1}$$
$$\geq 0,$$

or equivalently, $E(m) \geq A(m,r)$.

(iii) If $\omega(t) = 0$ for all t with $r < t \leq q$, then it follows that we must have $E(t) = 0$ for all t with $r < t \leq q$. Consequently, we have

$$E(m) - A(m,r) = (-1)^{r-m+1} \sum_{t=r+1}^{q} E(t)\binom{t}{m}\binom{t-m-1}{t-r-1}$$
$$= (-1)^{r-m+1} \sum_{t=r+1}^{q} 0 \cdot \binom{t}{m}\binom{t-m-1}{t-r-1}$$
$$= 0,$$

or equivalently, $E(m) = A(m,r)$. Conversely, if $\omega(t) > 0$ for some t with $r < t \leq q$, then there must exist some s, such that $r < s \leq q$ and $E(s) > 0$. As $s > r \geq m$ and $s - m - 1 \geq s - r - 1$, it follows that both $\binom{s}{m}$ and $\binom{s-m-1}{s-r-1}$ are positive. Furthermore, since $E(t)\binom{t}{m}\binom{t-m-1}{t-r-1}$ is non-negative for each $t = r+1, \ldots, q$, it follows that we have

$$\sum_{t=r+1}^{q} E(t)\binom{t}{m}\binom{t-m-1}{t-r-1} \geq E(s)\binom{s}{m}\binom{s-m-1}{s-r-1} > 0.$$

This implies that

$$E(m) - A(m,r) = (-1)^{r-m+1} \sum_{t=r+1}^{q} E(t) \binom{t}{m} \binom{t-m-1}{t-r-1}$$

is always non-zero regardless of the parities of r and m, so strict inequality in (i) and (ii) must hold.

48. If $j = 0$, then we have

$$\sum_{k=j}^{q} (-1)^{k-j} \omega(k) = \sum_{k=0}^{q} (-1)^{k} \omega(k) = E(0) \geq 0$$

by (4.3.2). So let us assume that $j > 0$. Let S denote the universal set. For each $x \in S$, let us compute the contribution of the count by x to the sum $\sum_{k=j}^{q} (-1)^{k-j} \omega(k)$. Let the number of properties that x satisfy be t. Let us consider these cases:

Case 1: $t < j$. In this case, x is counted 0 times in $\omega(k)$ for all $k = j, \ldots, q$. Thus, x contributes a count of 0 to the sum $\sum_{k=j}^{q} (-1)^{k-j} \omega(k)$.

Case 2: $t \geq j$. In this case, x is counted $\binom{t}{k}$ times in $\omega(k)$ for $k = j, j+1, \ldots, t$ and is counted 0 times in $\omega(k)$ for $k = t+1, t+2, \ldots, q$. Thus, the count that x contributes to the sum $\sum_{k=j}^{q} (-1)^{k-j} \omega(k)$ is equal to

$$\sum_{k=j}^{t} (-1)^{k-j} \binom{t}{k}$$

$$= \sum_{k=j}^{t} (-1)^{k-j} \binom{t}{t-k}$$

$$= \sum_{i=0}^{t-j} (-1)^{t-i-j} \binom{t}{i} \quad \text{(by setting } i = t - k\text{)}$$

$$= (-1)^{t-j} \sum_{i=0}^{t-j} (-1)^{i} \binom{t}{i} \quad \text{(since } (-1)^{-i} = (-1)^{i}\text{)}$$

$$= (-1)^{t-j} (-1)^{t-j} \binom{t-1}{t-j} \quad \text{(by Problem 2.29, since } t - j < t\text{)}$$

$$= \binom{t-1}{t-j}.$$

Since there are $E(t)$ elements of S that has exactly t properties, and $E(t)\binom{t-1}{t-j}$ is non-negative for each $t = j, j+1, \ldots, q$, it follows that we have

$$\sum_{k=j}^{q}(-1)^{k-j}\omega(k) = \sum_{t=j}^{q} E(t)\binom{t-1}{t-j} \geq 0$$

as desired.

49. (i) Let $A = \bigcup_{k=1}^{n} A_k$. Let $x \in A$. For each $i = 1, 2, \ldots, n$, let P_i denote the property that $x \in A_i$. For each $x \in A$, let us compute the contribution of the count by x to the number

$$\left|\bigcup_{k=1}^{n} A_k\right| - \sum_{k=1}^{n}|A_k| + \sum_{1\leq i<j\leq n} |A_i \cap A_j|$$

$$= |A| - \sum_{k=1}^{n}\omega(P_k) + \sum_{1\leq i<j\leq n} \omega(P_iP_j).$$

Let us take any $x \in A$, and let the properties that x satisfy be $P_{i_1}, P_{i_2}, \ldots, P_{i_m}$. Then by the definition of A, we have $m \geq 1$. We note that x contributes a count of 1 to $|A|$ and a count of m to $\sum_{k=1}^{n}\omega(P_k)$. If $m = 1$, then x contributes a count of 0 to $\omega(P_iP_j)$ for all $i, j \in \{1, 2, \ldots, n\}$ satisfying $i < j$, so that x contributes a count of 0 to $\sum_{1\leq i<j\leq n} \omega(P_iP_j)$.

Else, if $m > 1$, then x contributes a count of 1 to $\omega(P_iP_j)$ if we have $i = i_{n_1}$ and i_{n_2} for some $n_1, n_2 \in \{1, 2, \ldots, m\}$ satisfying $n_1 < n_2$, and a count of 0 otherwise. Thus x contributes a count of $\binom{m}{2} = \frac{m(m-1)}{2}$ to $\sum_{1\leq i<j\leq n} \omega(P_iP_j)$. Since $\frac{1(1-1)}{2} = 0$, it follows that x contributes a count of $\frac{m(m-1)}{2}$ to $\sum_{1\leq i<j\leq n} \omega(P_iP_j)$ in all cases.

Therefore, the count that x contributes to the number

$$|A| - \sum_{k=1}^{n}\omega(P_k) + \sum_{1\leq i<j\leq n} \omega(P_iP_j)$$

is equal to $1 - m + \frac{m(m-1)}{2} = \frac{(m-1)(m-2)}{2}$.

Now, it is easy to check that $\frac{(m-1)(m-2)}{2}$ is non-negative for all positive integers m. Since there are $E(m)$ elements of A that has exactly m properties, and $E(m)\frac{(m-1)(m-2)}{2}$ is non-negative for each

$m = 1, 2, \ldots, n$, it follows that we have

$$\left| \bigcup_{k=1}^{n} A_k \right| - \sum_{k=1}^{n} |A_k| + \sum_{1 \le i < j \le n} |A_i \cap A_j|$$

$$= |A| - \sum_{k=1}^{n} \omega(P_k) + \sum_{1 \le i < j \le n} \omega(P_i P_j)$$

$$= \sum_{m=1}^{n} E(m) \frac{(m-1)(m-2)}{2}$$

$$\ge 0,$$

or equivalently,

$$\left| \bigcup_{k=1}^{n} A_k \right| \ge \sum_{k=1}^{n} |A_k| - \sum_{1 \le i < j \le n} |A_i \cap A_j|$$

as desired.

(ii) The statement clearly holds for $n = 1$, so let us assume that $n > 1$. Let S denote the set of permutations of the n couples $H_1, W_1, \ldots, H_n, W_n$. For $k = 1, 2, \ldots, n$, let A_k denote the set of permutations $x \in S$ for which the couple H_k and W_k are adjacent in the row. Then it is easy to see that the set of permutations $x \in S$ such that x possesses P is the union $\bigcup_{k=1}^{n} A_k$. Firstly, let us fix $k \in \{1, 2, \ldots, n\}$, and compute $|A_k|$. There are two ways to arrange H_k and W_k, and there are $(2(n-1)+1)! = (2n-1)!$ ways to arrange the remaining $n - 1$ couples and the block containing the couple H_k and W_k. By (MP), we have $|A_k| = 2(2n-1)!$.

Next, let us fix $i, j \in \{1, 2, \ldots, n\}$ satisfying $i < j$, and compute $|A_i \cap A_j|$. There are two ways each to arrange H_i and W_i, and H_j and W_j. Next, there are $(2(n-2)+2)! = (2n-2)!$ ways to arrange the remaining $n - 2$ couples and the two blocks containing the couples H_i and W_i, and H_j and W_j. By (MP), we have $|A_i \cap A_j| = 4(2n-2)!$.

Thus, we have

$$\left| \bigcup_{k=1}^{n} A_k \right| \ge \sum_{k=1}^{n} |A_k| - \sum_{1 \le i < j \le n} |A_i \cap A_j|$$

$$= \sum_{k=1}^{n} 2(2n-1)! - \sum_{1 \le i < j \le n} 4(2n-2)!$$

$$= 2n(2n-1)! - \binom{n}{2} \cdot 4(2n-2)!$$

$$= \left[2n(2n-1) - 4\frac{n(n-1)}{2} \right] (2n-2)!$$

$$= 2n^2(2n-2)!$$

$$> (2n^2 - n)(2n-2)!$$

$$= \frac{(2n)!}{2}$$

$$= \frac{1}{2}|S|,$$

and consequently, there are more permutations with property P than without as desired.

50. (i) We have

$$S(r,k) = \frac{1}{k!} F(r,k)$$

$$= \frac{1}{k!} \sum_{i=0}^{k} (-1)^i \binom{k}{i} (k-i)^r \quad \text{(by (4.5.1))}$$

$$= \frac{1}{k!} \sum_{j=0}^{k} (-1)^{k-j} \binom{k}{k-j} j^r \quad \text{(by setting } j = k-i)$$

$$= \frac{1}{k!} \sum_{j=0}^{k} (-1)^{k-j} \binom{k}{j} j^r.$$

(ii) We have

$$B_r = \sum_{k=1}^{r} S(r,k)$$

$$= \sum_{k=0}^{\infty} S(r,k)$$

$$\text{(since } S(r,0) = 0 \text{ and } S(r,k) = 0 \text{ for all } k > r)$$

$$= \sum_{k=0}^{\infty} \frac{1}{k!} \sum_{j=0}^{k} (-1)^{k-j} \binom{k}{j} j^r \quad \text{(by part (i))}$$

$$= \sum_{0 \le j \le k < \infty} \frac{(-1)^{k-j}}{k!} \binom{k}{j} j^r$$

$$= \sum_{0 \le j, n < \infty} \frac{(-1)^n}{(n+j)!} \binom{n+j}{j} j^r \quad \text{(by setting } n = k - j)$$

$$= \sum_{n=0}^{\infty}\sum_{j=0}^{\infty} \frac{(-1)^n}{(n+j)!} \cdot \frac{(n+j)!}{n!j!} \cdot j^r$$

$$= \sum_{n=0}^{\infty} \frac{(-1)^n}{n!} \sum_{j=0}^{\infty} \frac{j^r}{j!}$$

$$= e^{-1} \sum_{j=0}^{\infty} \frac{j^r}{j!}.$$

51. For all positive integers n, we have

$$\frac{n}{n+r}a_{n-1} = \sum_{i=0}^{n-1}(-1)^i \frac{n}{n+r} \cdot \frac{r}{i+r}\binom{n-1}{i}$$

$$= \sum_{i=0}^{n-1}(-1)^i \frac{n}{n+r} \cdot \frac{r}{i+r} \cdot \frac{i+1}{n}\binom{n}{i+1} \quad \text{(by (2.1.2))}$$

$$= \sum_{i=0}^{n-1}(-1)^i \frac{r}{n+r} \cdot \frac{i+1}{i+r}\binom{n}{i+1}$$

$$= \sum_{i=0}^{n-1}(-1)^i \frac{r}{n+r} \cdot \frac{i+1}{i+r} \cdot \frac{n-i}{i+1}\binom{n}{i} \quad \text{(by (2.1.3))}$$

$$= \sum_{i=0}^{n-1}(-1)^i \frac{r}{n+r} \cdot \frac{n-i}{i+r}\binom{n}{i}$$

$$= \sum_{i=0}^{n-1}(-1)^i \frac{r}{i+r} \cdot \left(1 - \frac{i+r}{n+r}\right)\binom{n}{i}.$$

Next, we have

$$\sum_{i=0}^{n-1}(-1)^i \frac{r}{i+r} \cdot \left(1 - \frac{i+r}{n+r}\right)\binom{n}{i}$$

$$= \sum_{i=0}^{n-1}(-1)^i \frac{r}{i+r}\binom{n}{i} - \sum_{i=0}^{n-1}(-1)^i \frac{r}{i+r} \cdot \frac{i+r}{n+r}\binom{n}{i}$$

$$= \left[\sum_{i=0}^{n}(-1)^i \frac{r}{i+r}\binom{n}{i} - (-1)^n \frac{r}{n+r}\binom{n}{n}\right] - \frac{r}{n+r}\sum_{i=0}^{n-1}(-1)^i\binom{n}{i}$$

$$= a_n - (-1)^n \frac{r}{n+r} - \frac{r}{n+r}\left(\sum_{i=0}^{n}(-1)^i\binom{n}{i} - (-1)^n\binom{n}{n}\right)$$

$$= a_n - (-1)^n \frac{r}{n+r} + (-1)^n \frac{r}{n+r} \quad \text{(by (2.3.2), since } n \geq 1\text{)}$$

$$= a_n,$$

and consequently, we have

$$a_n = \frac{n}{n+r} a_{n-1}.$$

Now, we observe that

$$a_0 = \sum_{i=0}^{0} (-1)^i \frac{r}{i+r} \binom{0}{i} = (-1)^0 \frac{r}{0+r} \binom{0}{0} = 1 = \frac{1}{1} = \frac{1}{\binom{0+r}{r}}.$$

Also, for all positive integers n, it follows from (2.1.2) that

$$\binom{n+r}{r} = \binom{n+r}{n} = \frac{n+r}{n} \binom{n-1+r}{n-1} = \frac{n+r}{n} \binom{n-1+r}{r},$$

or equivalently,

$$\frac{1}{\binom{n+r}{r}} = \frac{n}{n+r} \cdot \frac{1}{\binom{n-1+r}{r}}.$$

This implies that the recursion formula for the a_n's is the same as that for the $\frac{1}{\binom{n+r}{r}}$'s. Consequently, we must have

$$a_n = \frac{1}{\binom{n+r}{r}}$$

for all non-negative integers n as desired.

52. Let us denote the universal set by S. For each $x \in S$, let us compute the contribution of the count by x to both $L(m)$ and $\sum_{k=m}^{q} (-1)^{k-m} \binom{k-1}{m-1} \omega(k)$. Let the number of properties that x satisfy be t. Let us consider these cases:

Case 1: $t < m$. Then x contributes a count of 0 to $L(m)$ and to $\omega(k)$ for all positive integers $k \geq m$, so that x contributes a total count of 0 to $\sum_{k=m}^{q} (-1)^{k-m} \binom{k-1}{m-1} \omega(k)$.

Case 2: $t \geq m$. Then x contributes a count of 1 to $L(m)$. Next, x is counted $\binom{t}{k}$ times in $\omega(k)$ for $k = 1, 2, \ldots, t$ and is counted 0 times in $\omega(k)$ for $k = t+1, t+2, \ldots, q$. Thus, the count that x contributes to the sum $\sum_{k=m}^{q} (-1)^{k-m} \binom{k-1}{m-1} \omega(k)$ is equal to

$$\sum_{k=m}^{t} (-1)^{k-m} \binom{k-1}{m-1} \binom{t}{k}$$

$$= \sum_{k=m}^{t} (-1)^{k-m} \cdot \frac{m}{k} \binom{k}{m} \binom{t}{k} \quad \text{(by (2.1.2), since } m \geq 1)$$

$$= \sum_{k=m}^{t} (-1)^{k-m} \cdot \frac{m}{k} \binom{t}{m} \binom{t-m}{k-m} \quad \text{(by (2.1.5))}$$

$$= \binom{t}{m} \sum_{i=0}^{t-m} (-1)^{i} \cdot \frac{m}{i+m} \binom{t-m}{i} \quad \text{(by setting } i = k - m)$$

$$= \binom{t}{m} \cdot \frac{1}{\binom{t-m+m}{m}} \quad \text{(by Problem 4.51)}$$

$$= 1.$$

In all cases, x contributes the same count to both $L(m)$ and $\sum_{k=m}^{q} (-1)^{k-m} \binom{k-1}{m-1} \omega(k)$, so we must have

$$L(m) = \sum_{k=m}^{q} (-1)^{k-m} \binom{k-1}{m-1} \omega(k)$$

as desired.

53. Let us first prove by induction on n that if $S_1, S_2, \ldots, S_n$ are any finite sets, then we have $\left| \bigcup_{k=1}^{n} S_k \right| \leq \sum_{k=1}^{n} |S_k|$, and equality holds if $S_1, S_2, \ldots, S_n$ are pairwise disjoint. The base case $n = 1$ is clear. For $n = 2$, we have

$$|S_1 \cup S_2| = |S_1| + |S_2| - |S_1 \cap S_2| \leq \sum_{k=1}^{2} |S_k|,$$

and equality holds if and only if $|S_1 \cap S_2| = 0$. Suppose that the assertion holds for any finite sets $S_1, S_2, \ldots, S_N$, where N is a fixed positive integer. Then by induction hypothesis, we have $\left| \bigcup_{k=1}^{N} S_k \right| \leq \sum_{k=1}^{N} |S_k|$, and equality holds if $S_1, S_2, \ldots, S_N$ are pairwise disjoint. Now, if S_{N+1} is another finite set, then by (4.1.1), we have

$$\left| \bigcup_{k=1}^{N+1} S_k \right| = \left| \bigcup_{k=1}^{N} S_k \right| + |S_{N+1}| - \left| \left(\bigcup_{k=1}^{N} S_k \right) \cap S_{N+1} \right|$$

$$\leq \sum_{k=1}^{N} |S_k| + |S_{N+1}|$$

$$= \sum_{k=1}^{N+1} |S_k|.$$

Furthermore, if $S_1, \ldots, S_{N+1}$ are pairwise disjoint, then we have $\left(\bigcup_{k=1}^{N} S_k \right) \cap S_{N+1} = \emptyset$, so by (4.1.1) and induction hypothesis again, we have

$$\left| \bigcup_{k=1}^{N+1} S_k \right| = \left| \bigcup_{k=1}^{N} S_k \right| + |S_{N+1}| - \left| \left(\bigcup_{k=1}^{N} S_k \right) \cap S_{N+1} \right|$$

$$= \sum_{k=1}^{N} |S_k| + |S_{N+1}|$$

$$= \sum_{k=1}^{N+1} |S_k|,$$

which completes the induction step, and we are done.

Now, let us proceed to find $\left| \bigcup_{k=1}^{1992} A_k \right|$. Let us first show that $\bigcap_{k=1}^{1992} A_k \neq \emptyset$.

Suppose on the contrary that $\bigcap_{k=1}^{1992} A_k = \emptyset$. Let the elements of A_1 be $a_1, a_2, \ldots, a_{44}$. For $i = 1, 2, \ldots, 44$, we let $N_i = \{j \mid 1 \leq j \leq 1992, a_i \in A_j\} \subseteq \{1, 2, \ldots, 1992\}$. We claim that $|N_i| \leq 44$ for all $i = 1, 2, \ldots, 44$. Suppose on the contrary that there exists some $i \in \{1, 2, \ldots, 44\}$, such that $|N_i| > 44$. Let $N_i = \{n_1, n_2, \ldots, n_m\}$. Since $\bigcap_{k=1}^{1992} A_k = \emptyset$, there exists some element $p \in \{1, 2, \ldots, 1992\}$, such that $a_i \notin A_p$. Let $A_p \cap A_{n_j} = \{b_{n_j}\}$ for all $j = 1, 2, \ldots, m$. Since $a_i \notin A_p$, it follows that we have $b_{n_j} \neq a_i$ for all $j = 1, 2, \ldots, m$, so that we have $b_{n_j} \in A_{n_j} \setminus \{a_i\}$ for all $j = 1, 2, \ldots, m$.

Next, since $a_i \in A_{n_j} \cap A_{n_k}$ for all $j, k \in \{1, \ldots, m\}$, and $|A_j \cap A_k| = 1$ for all distinct $j, k \in \{1, 2, \ldots, 1992\}$, it follows that we have $A_{n_j} \cap A_{n_k} = \{a_i\}$ for all distinct $j, k \in \{1, 2, \ldots, m\}$, so that we have $(A_{n_j} \setminus \{a_i\}) \cap (A_{n_k} \setminus \{a_i\}) = \emptyset$ for all distinct $j, k \in \{1, 2, \ldots, m\}$. Furthermore, since $b_{n_j} \neq a_i$ for all $j = 1, 2, \ldots, m$, it follows that we have $b_{n_j} \neq b_{n_k}$ for all distinct $j, k \in \{1, 2, \ldots, m\}$. We note that if $b_{n_j} = b_{n_k}$, then $b_{n_j} \in (A_{n_j} \setminus \{a_i\}) \cap (A_{n_k} \setminus \{a_i\}) \neq \emptyset$, a contradiction. As we have $b_{n_j} \in A_p$ for all $j = 1, 2, \ldots, m$, it follows that we have

$$|A_p| \geq |\{b_{n_1}, b_{n_2}, \ldots, b_{n_m}\}| = m = |N_i| > 44,$$

which contradicts the fact that $|A_p| = 44$. So we must have $|N_i| \leq 44$ for all $i = 1, 2, \ldots, 44$. Now, we have

$$\left| \bigcup_{i=1}^{44} N_i \right| \leq \sum_{i=1}^{44} |N_i| = \sum_{i=1}^{44} 44 = 1936 < 1992 = |\{1, 2, \ldots, 1992\}|,$$

so there must exist some $q \in \{1, 2, \ldots, 1992\}$, such that $q \notin \bigcup\limits_{i=1}^{44} N_i$. By the definition of N_i, it follows that we have $a_i \notin A_q$ for all $i = 1, 2, \ldots, 44$. As we have $A_1 = \{a_1, a_2, \ldots, a_{44}\}$, this would imply that $A_q \cap A_1 = \emptyset$, which contradicts the fact that $|A_q \cap A_1| = 1$. So we must have $\bigcap\limits_{k=1}^{1992} A_k \neq \emptyset$ as claimed. Now, we have

$$1 \le \left| \bigcap_{k=1}^{1992} A_k \right| \le |A_1 \cap A_2| = 1,$$

so we must have equality to hold throughout, that is, $\left| \bigcap\limits_{k=1}^{1992} A_k \right| = 1$. Without loss of generality, we may assume that $\bigcap\limits_{k=1}^{1992} A_k = \{a_1\}$. Next, since $a_1 \in A_j \cap A_k$ for all $j, k \in \{1, 2, \ldots, 1992\}$, and $|A_j \cap A_k| = 1$ for all distinct $j, k \in \{1, 2, \ldots, 1992\}$, it follows that we have $A_j \cap A_k = \{a_1\}$ for all distinct $j, k \in \{1, 2, \ldots, 1992\}$, so that we have $(A_j \setminus \{a_1\}) \cap (A_k \setminus \{a_1\}) = \emptyset$ for all distinct $j, k \in \{1, 2, \ldots, 1992\}$. As we have $\left(\bigcup\limits_{k=1}^{1992} A_k \right) \setminus \{a_1\} = \bigcup\limits_{k=1}^{1992} (A_k \setminus \{a_1\})$, it follows that we have

$$\left| \left(\bigcup_{k=1}^{1992} A_k \right) \setminus \{a_1\} \right| = \left| \bigcup_{k=1}^{1992} (A_k \setminus \{a_1\}) \right|$$

$$= \sum_{k=1}^{1992} |A_k \setminus \{a_1\}|$$

$$= \sum_{k=1}^{1992} 43$$

$$= 85656,$$

and therefore we have

$$\left| \bigcup_{k=1}^{1992} A_k \right| = \left| \left(\bigcup_{k=1}^{1992} A_k \right) \setminus \{a_1\} \right| + |\{a_1\}| = 85656 + 1 = 85657.$$

54. Let S denote the set of all sequences of the 28 random draws from the set

$$\{1, 2, 3, 4, 5, 6, 7, 8, 9, A, B, C, D, J, K, L, U, X, Y, Z\}.$$

For each $i = 1, 2, \ldots, 17$, we let P_i denote the property of an element $x \in S$ that the sequence $CUBAJULY1987$ occurs in that order in x

starting at the i-th draw. Since there are 20 choices for each of the 28 draws, we have $\omega(0) = |S| = 20^{28}$. Next, let us fix $i \in \{1, 2, \ldots, 17\}$, and pick $x \in S$ that satisfies P_i. Then there are 20 choices each for each of the remaining $28 - 12 = 16$ draws, so it follows that $\omega(P_i) = 20^{16}$, and hence we have

$$\omega(1) = \sum_{i=1}^{17} \omega(P_i) = \sum_{i=1}^{17} 20^{16} = 17 \cdot 20^{16}.$$

Next, let us take any pair (i, j) of integers satisfying $1 \le i < j \le 17$. By the definition of the properties P_i's, it is easy to see that $\omega(P_i P_j) > 0$ if and only if we have $j - i \ge 12$. Let us denote the set of pairs (i, j) satisfying $1 \le i < j \le 17$ and $j - i \ge 12$ by A, and find $|A|$. Equivalently, it amounts to finding the number of non-negative integer solutions to the equation

$$x_1 + x_2 + x_3 = 28$$

satisfying $x_2 \ge 12$ and $x_3 \ge 12$. By letting $y_1 = x_1$ and $y_j = x_j - 12$ for $j = 2, 3$, it is easy to see that the number of such pairs is equal to the number of non-negative integer solutions to the equation

$$y_1 + y_2 + y_3 = 4,$$

which is then equal to $\binom{4+3-1}{4} = 15$, and hence, we have $|A| = 15$. Next, for each $(i, j) \in A$, and each $x \in S$ satisfying properties P_i and P_j, there are 20 choices each for each of the remaining $28 - 2 \cdot 12 = 4$ draws, so it follows that $\omega(P_i P_j) = 20^4$. Consequently, we have

$$\omega(2) = \sum_{(i,j) \in A} \omega(P_i P_j) = \sum_{(i,j) \in A} 20^4 = 15 \cdot 20^4.$$

Thus, by (4.3.2), we have $E(0) = \sum_{i=0}^{17} (-1)^i \omega(i) = \omega(0) - \omega(1) + \omega(2)$, which implies that the number of sequences that contains the sequence $CUBAJULY1987$ in that order is equal to $\omega(0) - E(0) = \omega(1) - \omega(2)$. Hence, the probability that the sequence $CUBAJULY1987$ occurs in that order in any $x \in S$ is equal to $\frac{\omega(1)-\omega(2)}{\omega(0)} = \frac{17 \cdot 20^{16} - 15 \cdot 20^4}{20^{28}}$.

55. Let S denote the set of all sequences of the 35 random draws from the set

$$\{A, B, C, \ldots, X, Y, Z\}.$$

For each $i = 1, 2, \ldots, 22$, we let P_i denote the property of an element $x \in S$ that the sequence $MERRYCHRISTMAS$ occurs in that order in x starting at the i-th draw. Since there are 26 choices for each of the 35 draws, we have $\omega(0) = |S| = 26^{35}$. Next, let us fix $i \in \{1, 2, \ldots, 22\}$, and pick $x \in S$ that satisfies P_i. Then there are 26 choices each for each of the remaining $35 - 14 = 21$ draws, so it follows that $\omega(P_i) = 26^{21}$, and hence we have

$$\omega(1) = \sum_{i=1}^{22} \omega(P_i) = \sum_{i=1}^{22} 26^{21} = 22 \cdot 26^{21}.$$

Next, let us take any pair (i, j) of integers satisfying $1 \leq i < j \leq 22$. By the definition of the properties P_i's, it is easy to see that $\omega(P_i P_j) > 0$ if and only if we have $j - i \geq 14$. Let us denote the set of pairs (i, j) satisfying $1 \leq i < j \leq 22$ and $j - i \geq 14$ by A, and find $|A|$. Equivalently, it amounts to finding the number of non-negative integer solutions to the equation

$$x_1 + x_2 + x_3 = 35$$

satisfying $x_2 \geq 14$ and $x_3 \geq 14$. By letting $y_1 = x_1$ and $y_j = x_j - 14$ for $j = 2, 3$, it is easy to see that the number of such pairs is equal to the number of non-negative integer solutions to the equation

$$y_1 + y_2 + y_3 = 7,$$

which is then equal to $\binom{7+3-1}{7} = 36$, and hence, we have $|A| = 36$. Next, for each $(i, j) \in A$, and each $x \in S$ satisfying properties P_i and P_j, there are 26 choices each for each of the remaining $35 - 2 \cdot 14 = 7$ draws, so it follows that $\omega(P_i P_j) = 26^7$. Consequently, we have

$$\omega(2) = \sum_{(i,j) \in A} \omega(P_i P_j) = \sum_{(i,j) \in A} 26^7 = 36 \cdot 26^7.$$

Thus, by (4.3.2), we have $E(0) = \sum_{i=0}^{22} (-1)^i \omega(i) = \omega(0) - \omega(1) + \omega(2)$, which implies that the number of sequences that contains the sequence $MERRYCHRISTMAS$ in that order is equal to $\omega(0) - E(0) = \omega(1) - \omega(2)$. Hence, the probability that the sequence $MERRYCHRISTMAS$ occurs in that order in any $x \in S$ is equal to $\frac{\omega(1) - \omega(2)}{\omega(0)} = \frac{22 \cdot 26^{21} - 36 \cdot 26^7}{26^{35}}$.

56. Let S denote the set of 1990 people, and let us take any person $a \in S$ and any friend b of a. Let us denote the set of friends of a and b by F_a and F_b respectively, and the set of common friends of a and b by $F_{a,b}$.

Then we have $F_{a,b} = F_a \cap F_b$. We note that $a \notin F_{a,b}$ and $b \notin F_{a,b}$. It follows from (4.1.1) that

$$
\begin{aligned}
|F_{a,b}| &= |F_a \cap F_b| \\
&= |F_a| + |F_b| - |F_a \cup F_b| \\
&\geq |F_a| + |F_b| - |S| \\
&\geq 1327 + 1327 - 1990 \\
&= 664.
\end{aligned}
$$

Next, let us pick any $c \in F_{a,b}$. Let us denote the set of friends of c by F_c, and the set of common friends of a, b and c by $F_{a,b,c}$. Then we have $F_{a,b,c} = F_{a,b} \cap F_c$. It follows from (4.1.1) that

$$
\begin{aligned}
|F_{a,b,c}| &= |F_{a,b} \cap F_c| \\
&= |F_{a,b}| + |F_c| - |F_{a,b} \cup F_c| \\
&\geq |F_{a,b}| + |F_c| - |S| \\
&\geq 664 + 1327 - 1990 \\
&= 1.
\end{aligned}
$$

Consequently, the set $F_{a,b,c}$ of common friends of a, b and c is nonempty, so we can find some $d \in F_{a,b,c}$. Now, it is easy to check by the choice of b, c and d that every two of a, b, c and d are friends, and we are done.

57. For each $z \in S$, we define $f^0(z) = z$. We call $w \in S$ a finite-periodic point of f if there exists some positive integer n such that $f^n(w) = w$. For each positive integer k, we let $A_k = \{z \in S \mid f^k(z) = z\} = \{z \in S \mid z^{m^k} = z\}$, and let us denote the set of finite-periodic points of f by F. Then it is clear that $|A_k| = m^k$ for all positive integers k. Next, for each $w \in F$, we define $n(w)$ to be the smallest positive integer satisfying $w \in A_{n(w)}$ (in other words, w is a $n(w)$-periodic point of f).

Let us first show for any $w \in F$ and positive integer n that $w \in A_n$ if and only if $n(w)$ divides n. For the "if" direction, let us write $n = n(w) \cdot q$ where q is a positive integer. Then we have

$$
\begin{aligned}
f^{n(w) \cdot q}(w) &= \underbrace{f^{n(w)}(f^{n(w)}(\cdots(f^{n(w)}(w))))}_{q} \\
&= \underbrace{f^{n(w)}(f^{n(w)}(\cdots(f^{n(w)}(w))))}_{q-1} \\
&= \cdots \\
&= f^{n(w)}(w) \\
&= w,
\end{aligned}
$$

which shows that $w \in A_n$. Conversely, for the "only if" direction, let us write $n = n(w) \cdot q + r$, where q is a non-negative integer and $r \in \{0, 1, \ldots, n(w) - 1\}$. Then by our earlier argument, we have $f^{n(w) \cdot q}(w) = w$, and so we have

$$f^r(w) = f^r(f^{n(w) \cdot q}(w)) = f^{n(w) \cdot q + r}(w) = f^n(w) = w.$$

Since $n(w)$ is the smallest positive integer satisfying $w \in A_{n(w)}$, we have $f^i(w) \neq w$ for all positive integers $i < n(w)$. As $r < n(w) - 1$, this implies that we must have $r = 0$. So we must have $n = n(w) \cdot q$, that is, $n(w)$ divides n, and this completes the proof.

Next, let us show for any positive integers m and n that $A_m \cap A_n = A_{(m,n)}$. Indeed, let us take any $w \in A_m \cap A_n$. Then we have $n(w)|m$ and $n(w)|n$, so we have $n(w)|(m,n)$, and hence we have $w \in A_{(m,n)}$, which shows that $A_m \cap A_n \subseteq A_{(m,n)}$. Conversely, if $z \in A_{(m,n)}$, then we must have $n(z)|(m,n)$. Since $(m,n)|m$ and $(m,n)|n$, we must have $n(z)|m$ and $n(z)|n$, which would then imply that $z \in A_m$ and $z \in A_n$, or equivalently, $z \in A_m \cap A_n$. So this shows that $A_{(m,n)} \subseteq A_m \cap A_n$, and therefore we have $A_m \cap A_n = A_{(m,n)}$ as claimed.

Now, let us proceed to prove the main statement. We observe that the prime factors of 1989 are $3, 13$ and 17. Let P denote the set of 1989-periodic points of f. Then it is clear that $P \subseteq A_{1989}$. Let us show that $A_{1989} \setminus P = A_{117} \cup A_{153} \cup A_{663}$. Let us first take any $z \in A_{1989} \setminus P$. Then $n(z)|1989$. Since z is not 1989-periodic, we must have $n(z) < 1989$, and hence there exists some positive integer $q > 1$ for which $n(z) \cdot q = 1989$. Now $n(z) \in \{3, 9, 13, 17, 39, 51, 117, 153, 221, 663\}$. Using the result $A_m \cap A_n = A_{(m,n)}$, we see that $A_3 \cup A_9 \cup A_{13} \cup A_{17} \cup A_{39} \cup A_{51} \cup A_{221} \subseteq A_{117} \cup A_{153} \cup A_{663}$. It follows that we have $z \in A_{117} \cup A_{153} \cup A_{663}$. So this implies that $A_{1989} \setminus P \subseteq A_{117} \cup A_{153} \cup A_{663}$. Next, let us take any $w \in A_{117} \cup A_{153} \cup A_{663}$. Since $117, 153$ and 663 all divide 1989, it follows that we must have $n(w)|1989$, so that we have $w \in A_{1989}$. On the other hand, we have $n(w) \leq 663 < 1989$ by the definition of A_k, so we have $w \notin P$, and hence $w \in A_{1989} \setminus P$, which shows that $A_{117} \cup A_{153} \cup A_{663} \subseteq A_{1989} \setminus P$. Consequently, we have $A_{1989} \setminus P = A_{117} \cup A_{153} \cup A_{663}$ as claimed.

It remains to compute $|A_{117} \cup A_{153} \cup A_{663}|$. We observe that

$$|A_{117} \cap A_{153}| = |A_{(117,153)}| = |A_9| = m^9,$$
$$|A_{117} \cap A_{663}| = |A_{(117,663)}| = |A_{39}| = m^{39},$$
$$|A_{153} \cap A_{663}| = |A_{(153,663)}| = |A_{51}| = m^{51}, \text{ and}$$
$$|A_{117} \cap A_{153} \cap A_{663}| = |A_9 \cap A_{663}| = |A_{(9,663)}| = |A_3| = m^3.$$

By (4.2.1), we have

$$|A_{117} \cup A_{153} \cup A_{663}|$$
$$= |A_{117}| + |A_{153}| + |A_{163}|$$
$$-(|A_{117} \cap A_{153}| + |A_{117} \cap A_{663}| + |A_{153} \cap A_{663}|)$$
$$+|A_{117} \cap A_{153} \cap A_{663}|$$
$$= m^{117} + m^{153} + m^{663} - (m^9 + m^{39} + m^{51}) + m^3$$
$$= m^{663} + m^{153} + m^{117} - m^{51} - m^{39} - m^9 + m^3,$$

and hence the number of 1989-periodic points of f is equal to

$$|P| = |A_{1989}| - |A_{1989} \setminus P|$$
$$= |A_{1989}| - |A_{117} \cup A_{153} \cup A_{663}|$$
$$= m^{1989} - (m^{663} + m^{153} + m^{117} - m^{51} - m^{39} - m^9 + m^3)$$
$$= m^{1989} - m^{663} - m^{153} - m^{117} + m^{51} + m^{39} + m^9 - m^3.$$

58. For $j = 1, 2, \ldots, m$, we let P_j denote the property of a matix $M \in \mathcal{M}$ that the j-th row of M is a zero row, and for $i = 1, 2, \ldots, n$, we let P_{m+i} denote the property of a matix $M \in \mathcal{M}$ that the i-th column of M is a zero column. Furthermore, we let $E_r(j)$ denote the number of elements in $\mathcal{M}$ that has exactly j zero rows, $\omega_r(0) = |\mathcal{M}| = 2^{mn}$, and for $j = 1, 2, \ldots, m$, we let

$$\omega_r(j) = \sum_{1 \le k_1 < \cdots < k_j \le m} \omega(P_{k_1} \cdots P_{k_j}).$$

Firstly, let us compute $|\mathcal{M}_r| = |\mathcal{M}| - E_r(0)$. Let us compute $\omega_r(j)$ for $j = 1, 2, \ldots, m$. To this end, let us fix $j \in \{1, 2, \ldots, m\}$, and take any j-tuple of integers $(k_1, \ldots, k_j)$ satisfying $1 \le k_1 < \cdots < k_j \le m$. It is easy to see that there are 2 choices each for each of the entries in the remaining $(m - j) \times n$ submatrix, so that we have $\omega(P_{k_1} \cdots P_{k_j}) = 2^{n(m-j)}$. Since there are $\binom{m}{j}$ such j-tuples, we have

$$\omega_r(j) = \sum_{1 \le k_1 < \cdots < k_j \le m} \omega(P_{k_1} \cdots P_{k_j})$$
$$= \sum_{1 \le k_1 < \cdots < k_j \le m} 2^{n(m-j)}$$
$$= \binom{m}{j} 2^{n(m-j)}.$$

As we have $\omega_r(0) = 2^{nm} = \binom{m}{0}2^{n(m-0)}$, it follows from (4.3.2) that we have

$$E_r(0) = \sum_{j=0}^{m}(-1)^j\omega_r(j) = \sum_{j=0}^{m}(-1)^j\binom{m}{j}2^{n(m-j)} = (2^n - 1)^m,$$

where the last equality follows from the Binomial Theorem. Consequently, we have $|\mathcal{M}_r| = |\mathcal{M}| - E_r(0) = 2^{mn} - (2^n - 1)^m$.

Next, let us compute $|\mathcal{M}_r \cup \mathcal{M}_c| = |\mathcal{M}| - E(0)$ where $E(0)$ denotes the number of matrices having none of the properties $P_1, \ldots, P_m, P_{m+1}, \ldots, P_{m+n}$. Let us compute $\omega(\ell)$ for $\ell = 1, 2, \ldots, m+n$. To this end, let us fix $\ell \in \{1, 2, \ldots, m+n\}$ and $j \in \{0, 1, \ldots, \ell\}$, and take any ℓ-tuple of integers $(k_1, \ldots, k_\ell)$ satisfying $1 \le k_1 < \cdots < k_\ell \le m + n$ and $|\{p \mid 1 \le p \le \ell, 1 \le k_p \le m\}| = j$. Now, for each $M \in \mathcal{M}$ whose k_1-th, k_2-th, $\cdots$, k_j-th rows are zero rows and $(k_{j+1} - m)$-th, $(k_{j+2} - m)$-th, $\ldots$, $(k_\ell - m)$-th columns are zero columns, there are 2 choices each for each of the entries in the remaining $(m-j) \times (n-\ell+j)$ submatrix, so that we have $\omega(P_{k_1} \cdots P_{k_\ell}) = 2^{(m-j)(n-\ell+j)}$. As there are $\binom{m}{j}\binom{n}{\ell-j}$ such ℓ-tuples, it follows that we have

$$\omega(\ell) = \sum_{1 \le k_1 < \cdots < k_\ell \le m+n} \omega(P_{k_1} \cdots P_{k_\ell})$$

$$= \sum_{j=0}^{\ell} \sum_{\substack{1 \le k_1 < \cdots < k_\ell \le m+n, \\ |\{p \mid 1 \le p \le \ell, 1 \le k_p \le m\}|=j}} \omega(P_{k_1} \cdots P_{k_\ell})$$

$$= \sum_{j=0}^{\ell} \sum_{\substack{1 \le k_1 < \cdots < k_\ell \le m+n, \\ |\{p \mid 1 \le p \le \ell, 1 \le k_p \le m\}|=j}} 2^{(m-j)(n-\ell+j)}$$

$$= \sum_{j=0}^{\ell} \binom{m}{j}\binom{n}{\ell-j}2^{(m-j)(n-\ell+j)}.$$

Since $\omega(0) = 2^{mn} = \sum_{j=0}^{0}\binom{m}{j}\binom{n}{0-j}2^{(m-j)(n-0+j)}$, it follows from (4.3.2) that we have

$$E(0)$$

$$= \sum_{\ell=0}^{m+n}(-1)^\ell\omega(\ell)$$

$$= \sum_{\ell=0}^{m+n}(-1)^\ell\sum_{j=0}^{\ell}\binom{m}{j}\binom{n}{\ell-j}2^{(m-j)(n-\ell+j)}$$

$$= \sum_{i=0}^{m+n} \sum_{j=0}^{m+n} (-1)^{i+j} \binom{m}{j} \binom{n}{i} 2^{(m-j)(n-i)} \quad \text{(by setting } i = \ell - j\text{)}$$

$$= \sum_{i=0}^{n} (-1)^i \binom{n}{i} \sum_{j=0}^{m} (-1)^j \binom{m}{j} 2^{(m-j)(n-i)}$$

$$\text{(since } \binom{n}{i} = \binom{m}{j} = 0 \text{ for all } i > n \text{ and } j > m\text{)}$$

$$= \sum_{i=1}^{n} (-1)^i \binom{n}{i} \sum_{j=0}^{m} (-1)^j \binom{m}{j} (2^{n-i})^{m-j} + \sum_{j=0}^{m} (-1)^j \binom{m}{j} (2^n)^{m-j}$$

$$= \sum_{i=1}^{n} (-1)^i \binom{n}{i} (2^{n-i} - 1)^m + (2^n - 1)^m,$$

where the last equality follows from the Binomial Theorem. Consequently, we have

$$|\mathcal{M}_r \cup \mathcal{M}_c| = |\mathcal{M}| - E(0)$$

$$= 2^{mn} - \left(\sum_{i=1}^{n} (-1)^i \binom{n}{i} (2^{n-i} - 1)^m + (2^n - 1)^m \right)$$

$$= \sum_{i=1}^{n} (-1)^{i-1} \binom{n}{i} (2^{n-i} - 1)^m + 2^{mn} - (2^n - 1)^m.$$

Therefore, we have

$$|(\mathcal{M} \setminus \mathcal{M}_r) \cap \mathcal{M}_c|$$

$$= |\mathcal{M}_r \cup \mathcal{M}_c| - |\mathcal{M}_r|$$

$$= \left[\sum_{i=1}^{n} (-1)^{i-1} \binom{n}{i} (2^{n-i} - 1)^m + 2^{mn} - (2^n - 1)^m \right]$$

$$\quad - (2^{mn} - (2^n - 1)^m)$$

$$= \sum_{i=1}^{n} (-1)^{i-1} \binom{n}{i} (2^{n-i} - 1)^m$$

as desired.

59.   (i) Let S denote the set of permutations $a_1 a_2 \cdots a_n$ of $\{1, 2, \ldots, n\}$ that satisfies $a_m = m$. For $i = 1, 2, \ldots, m - 1$, we let P_i denote the property of a permutation $a_1 a_2 \cdots a_n \in S$ that $a_i = i$. By definition, we have $\omega(0) = |S| = (n - 1)!$. Next, let us compute $\omega(i)$ for $i = 1, 2, \ldots, m - 1$. To this end, let us fix $i \in \{1, 2, \ldots, m - 1\}$, and take any i-tuple of integers $(n_1, \ldots, n_i)$

satisfying $1 \leq n_1 < n_2 < \cdots < n_i \leq m - 1$. Then there are $(n - 1 - i)!$ ways to permute $\{1, 2, \ldots, n\} \setminus \{n_1, n_2, \ldots, n_i, m\}$, so that we have $\omega(P_{n_1} P_{n_2} \cdots P_{n_i}) = (n - 1 - i)!$. As there are $\binom{m-1}{i}$ such i-tuples, it follows that we have

$$\omega(i) = \sum_{1 \leq n_1 < n_2 < \cdots < n_i \leq m-1} \omega(P_{n_1} P_{n_2} \cdots P_{n_i})$$

$$= \sum_{1 \leq n_1 < n_2 < \cdots < n_i \leq m-1} (n - 1 - i)!$$

$$= \binom{m - 1}{i}(n - 1 - i)!.$$

As we have $\omega(0) = (n - 1)! = \binom{m-1}{0}(n - 1 - 0)!$, it follows from (4.3.2) that we have

$$P_n(m) = E(0) = \sum_{i=0}^{m-1} (-1)^i \omega(i) = \sum_{i=0}^{m-1} (-1)^i \binom{m - 1}{i}(n - 1 - i)!.$$

(ii) We have

$$P_n(m + 1)$$

$$= \sum_{i=0}^{m} (-1)^i \binom{m}{i}(n - 1 - i)! \quad \text{(by part (i))}$$

$$= (n - 1)! + \sum_{i=1}^{m} (-1)^i \binom{m}{i}(n - 1 - i)!$$

$$= (n - 1)! + \sum_{i=1}^{m} (-1)^i \frac{m}{i} \binom{m - 1}{i - 1}(n - 1 - i)! \quad \text{(by (2.1.2))}$$

$$= (n - 1)! + \sum_{i=1}^{m} (-1)^i \frac{m}{i} \cdot \frac{i}{m - i} \binom{m - 1}{i}(n - 1 - i)!$$

(by (2.1.3))

$$= (n - 1)! + \sum_{i=1}^{m} (-1)^i \frac{m}{m - i} \binom{m - 1}{i}(n - 1 - i)!$$

$$= \sum_{i=0}^{m} (-1)^i \frac{m}{m - i} \binom{m - 1}{i}(n - 1 - i)!$$

$$= \sum_{i=0}^{m} (-1)^i \left(1 + \frac{i}{m - i}\right) \binom{m - 1}{i}(n - 1 - i)!$$

$$= \sum_{i=0}^{m} (-1)^i \binom{m - 1}{i}(n - 1 - i)!$$

$$+ \sum_{i=0}^{m} (-1)^i \frac{i}{m-i} \binom{m-1}{i} (n-1-i)!$$

$$= \sum_{i=0}^{m-1} (-1)^i \binom{m-1}{i} (n-1-i)!$$

$$+ \sum_{i=1}^{m} (-1)^i \frac{i}{m-i} \binom{m-1}{i} (n-1-i)!$$

$$\left(\text{since } \binom{m-1}{m} = 0 \right)$$

$$= P_n(m) + \sum_{i=0}^{m-1} (-1)^{i+1} \frac{i+1}{m-i-1} \binom{m-1}{i+1} (n-2-i)!$$

(by part (i))

$$= P_n(m) - \sum_{i=0}^{m-1} (-1)^i \frac{i+1}{m-i-1} \cdot \frac{m-i-1}{i+1} \binom{m-1}{i} (n-2-i)!$$

(by (2.1.3))

$$= P_n(m) - \sum_{i=0}^{m-1} (-1)^i \binom{m-1}{i} (n-2-i)!$$

$$= P_n(m) - P_{n-1}(m) \quad \text{(by part (i))}.$$

60. Let us fix $k \in \{1, 2, \ldots, pq\}$, and let A_k denote the set of $p \times q$ $(0,1)$-matrices with exactly k entries equal to 1. For $j = 1, 2, \ldots, p$, we let P_j denote the property of a matix $M \in A_k$ that the j-th row of M is a zero row, and for $i = 1, 2, \ldots, q$, we let P_{p+i} denote the property of a matix $M \in A_k$ that the i-th column of M is a zero column. Let us compute $\omega(\ell)$ for $\ell = 1, 2, \ldots, p+q$. To this end, let us fix $\ell \in \{1, 2, \ldots, p+q\}$ and $j \in \{0, 1, \ldots, \ell\}$, and take any ℓ-tuple of integers $(k_1, \ldots, k_\ell)$ satisfying $1 \leq k_1 < \cdots < k_\ell \leq p+q$ and $|\{r \mid 1 \leq r \leq \ell, 1 \leq k_r \leq p\}| = j$.

Now, for each $M \in \mathcal{M}$ whose k_1-th, k_2-th, $\cdots$, k_j-th rows are zero rows and $(k_{j+1} - m)$-th, $(k_{j+2} - m)$-th, $\ldots$, $(k_\ell - m)$-th columns are zero columns, there are $\binom{(p-j)(q-\ell+j)}{k}$ ways to assign either 0 or 1 to each of the remaining entries in the $(p-j) \times (q-\ell+j)$ submatrix, so that there are exactly k entries equal to 1 in the $(p-j) \times (q-\ell+j)$ submatrix. Thus, we have $\omega(P_{k_1} \cdots P_{k_\ell}) = \binom{(p-j)(q-\ell+j)}{k}$. As there are

$\binom{p}{j}\binom{q}{\ell-j}$ such ℓ-tuples, it follows that we have

$$\omega(\ell) = \sum_{1\leq k_1 < \cdots < k_\ell \leq p+q} \omega(P_{k_1}\cdots P_{k_\ell})$$

$$= \sum_{j=0}^{\ell} \sum_{\substack{1\leq k_1<\cdots<k_\ell\leq p+q,\\ |\{p|1\leq p\leq\ell,1\leq k_p\leq p\}|=j}} \omega(P_{k_1}\cdots P_{k_\ell})$$

$$= \sum_{j=0}^{\ell} \sum_{\substack{1\leq k_1<\cdots<k_\ell\leq p+q,\\ |\{p|1\leq p\leq\ell,1\leq k_p\leq p\}|=j}} \binom{(p-j)(q-\ell+j)}{k}$$

$$= \sum_{j=0}^{\ell} \binom{p}{j}\binom{q}{\ell-j}\binom{(p-j)(q-\ell+j)}{k}.$$

Since $\omega(0) = \binom{pq}{k} = \sum\limits_{j=0}^{0} \binom{p}{j}\binom{q}{0-j}\binom{(p-j)(q-0+j)}{k}$, it follows from (4.3.2) that we have

$$N_k(p,q) = E(0)$$

$$= \sum_{\ell=0}^{p+q}(-1)^\ell \omega(\ell)$$

$$= \sum_{\ell=0}^{p+q}(-1)^\ell \sum_{j=0}^{\ell} \binom{p}{j}\binom{q}{\ell-j}\binom{(p-j)(q-\ell+j)}{k}$$

$$= \sum_{i=0}^{p+q}(-1)^{i+j} \binom{p}{j}\binom{q}{i}\binom{(p-j)(q-i)}{k}$$

(by setting $i = \ell - j$)

$$= \sum_{i=0}^{q}\sum_{j=0}^{p}(-1)^{i+j} \binom{p}{j}\binom{q}{i}\binom{(p-j)(q-i)}{k}$$

(since $\binom{q}{i} = \binom{p}{j} = 0$ for all $i > q$ and $j > p$)

$$= \sum_{i=0}^{q-1}\sum_{j=0}^{p-1}(-1)^{i+j} \binom{p}{j}\binom{q}{i}\binom{(p-j)(q-i)}{k}$$

(since $\binom{0}{k} = 0$, and $(p-j)(q-i)$ when $p = j$ or $q = i$).

Thus, we have

$$\sum_{k=1}^{pq}(-1)^{k-1}N_k(p,q)$$

$$= \sum_{k=1}^{pq}(-1)^{k-1}\sum_{i=0}^{q-1}\sum_{j=0}^{p-1}(-1)^{i+j}\binom{p}{j}\binom{q}{i}\binom{(p-j)(q-i)}{k}$$

$$= \sum_{i=0}^{q-1}\sum_{j=0}^{p-1}(-1)^{i+j}\binom{p}{j}\binom{q}{i}\sum_{k=1}^{pq}(-1)^{k-1}\binom{(p-j)(q-i)}{k}.$$

It remains to compute $\sum_{k=1}^{pq}(-1)^{k-1}\binom{(p-j)(q-i)}{k}$ for all $i \in \{0,1,\ldots,q-1\}$ and $j \in \{0,1,\ldots,p-1\}$. Let us fix $i \in \{0,1,\ldots,q-1\}$ and $j \in \{0,1,\ldots,p-1\}$. Then we have

$$\sum_{k=1}^{pq}(-1)^{k-1}\binom{(p-j)(q-i)}{k}$$

$$= \sum_{k=1}^{(p-j)(q-i)}(-1)^{k-1}\binom{(p-j)(q-i)}{k}$$

$$\left(\text{since } \binom{(p-j)(q-i)}{k} = 0 \text{ for all } k > (p-j)(q-i)\right)$$

$$= \sum_{k=0}^{(p-j)(q-i)}(-1)^{k-1}\binom{(p-j)(q-i)}{k} - (-1)^{0-1}\binom{(p-j)(q-i)}{0}$$

$$= 1,$$

where the last equality follows from (2.3.2), noting that $(p-j)(q-i) > 0$. Thus, we have

$$\sum_{k=1}^{pq}(-1)^{k-1}N_k(p,q)$$

$$= \sum_{i=0}^{q-1}\sum_{j=0}^{p-1}(-1)^{i+j}\binom{p}{j}\binom{q}{i}\sum_{k=1}^{pq}(-1)^{k-1}\binom{(p-j)(q-i)}{k}$$

$$= \sum_{i=0}^{q-1}\sum_{j=0}^{p-1}(-1)^{i+j}\binom{p}{j}\binom{q}{i}$$

$$= \sum_{i=0}^{q-1}(-1)^i\binom{q}{i}\sum_{j=0}^{p-1}(-1)^j\binom{p}{j}$$

$$= (-1)^{q-1} \binom{q-1}{q-1} \cdot (-1)^{p-1} \binom{p-1}{p-1}$$

(by Problem 2.29, since $p - 1 < p$ and $q - 1 < q$)

$$= (-1)^{p+q}.$$

61. Let $\{H_1, W_1\}, \{H_2, W_2\}, \ldots, \{H_n, W_n\}$ be n couples, where the H_i's denote the husbands and the W_i's denote the wives. Let $W_1, W_2, \ldots, W_n$ be seated around a table in that order in the clockwise direction. Let us first show that $M_n \geq n - 2$ for all $n \geq 2$. Clearly, the statement holds when $n = 2$, so let us assume that $n \geq 3$. Then for each $i = 2, 3, \ldots, n - 1$, we may seat H_1 in between W_i and W_{i+1}, and subsequently seat $H_1, H_2, H_3, \ldots, H_n$ around the table in that order in the clockwise direction, so that the resulting arrangement has no husband that is adjacent to his wife. Consequently, we must have $M_n \geq n - 2$ as claimed.

Next, let us show that $M_n \leq D_n$ for all $n \geq 2$. Indeed, let us take an arrangement s of the n husbands in the remaining n seats around the table such that no husband that is adjacent to his wife. Also, for $i = 1, 2, \ldots, n - 1$, we let a_i to be the unique integer such that H_{a_i} is seated between W_i and W_{i+1} in the arrangement s, and a_n to be the unique integer such that H_{a_n} is seated between W_n and W_1 in the arrangement s. As no husband is adjacent to his wife in the arrangement s, it follows that we must have $a_i \neq i$ for $i = 1, 2, \ldots, n$, so that $a_1 a_2 \cdots a_n$ is a derangement of $\mathbf{N}_n$. Consequently, this shows that there is an injection from the set of seating arrangements of the n husbands in the remaining n seats around the table with no husband adjacent to his wife to the set of derangements of $\mathbf{N}_n$, and hence we must have $M_n \leq D_n$ as claimed.

Now, let us proceed to show that the sequence $\left\{ \frac{M_n}{D_n} \right\}_{n=4}^{\infty}$ is (strictly) monotonically increasing. Equivalently, we would like to show that $\frac{M_n}{D_n} < \frac{M_{n+1}}{D_{n+1}}$ for all positive integers $n \geq 4$. By tables (4.6.2) and (4.8.1), we have

$$\frac{M_4}{D_4} = \frac{2}{9} = \frac{88}{396} < \frac{117}{396} = \frac{13}{44} = \frac{M_5}{D_5}.$$

Next, for all positive integers $n \geq 5$, we have

$$\frac{M_n}{D_n} + \frac{4}{n-1} \leq 1 + \frac{4}{5-1} = 2 \leq n - 3 \leq M_{n-1} < \frac{n+1}{n-1} M_{n-1}.$$

This implies that for all positive integers $n \geq 5$, we have

$$(-1)^{n+1}\left(\frac{M_n}{D_n} + \frac{4}{n-1}\right) \leq \frac{M_n}{D_n} + \frac{4}{n-1} < \frac{n+1}{n-1}M_{n-1}$$

$$\Rightarrow (-1)^{n+1}\frac{M_n}{D_n} < \frac{n+1}{n-1}M_{n-1} + (-1)^{n+2}\frac{4}{n-1}$$

$$\Rightarrow (-1)^{n+1}M_n < \frac{n+1}{n-1}M_{n-1}D_n + (-1)^{n+2}\frac{4}{n-1}D_n$$

$$\Rightarrow M_n[(n+1)D_n + (-1)^{n+1}]$$
$$< D_n\left[(n+1)M_n + \frac{n+1}{n-1}M_{n-1} + (-1)^{n+2}\frac{4}{n-1}\right]$$

$$\Rightarrow M_n D_{n+1} < D_n M_{n+1} \quad \text{(by (4.6.11) and (4.8.4))}$$

$$\Rightarrow \frac{M_n}{D_n} < \frac{M_{n+1}}{D_{n+1}}.$$

Consequently, this shows that the sequence $\left\{\frac{M_n}{D_n}\right\}_{n=4}^{\infty}$ is (strictly) monotonically increasing. Next, we have

$$\lim_{n\to\infty}\frac{M_n}{D_n} = \lim_{n\to\infty}\frac{M_n}{n!} \cdot \lim_{n\to\infty}\frac{n!}{D_n}$$

$$= \frac{1}{e^2}\cdot e \quad \text{(by (4.6.9) and (4.8.6))}$$

$$= \frac{1}{e}.$$

62. Let $(x)_0 = 1$ and $(x)_m = x(x-1)\cdots(x-m+1)$ for all positive integers n. Then we have

$$\sum_{k=0}^{n}k^r\binom{n}{k}D_{n-k}$$

$$= \sum_{k=0}^{n}k^r D_n(k) \quad \text{(by Problem 4.33)}$$

$$= \sum_{k=0}^{n}\left(\sum_{m=0}^{r}S(r,m)(k)_m\right)D_n(k) \quad \text{(by Problem 1.85)}$$

$$= \sum_{k=0}^{n}(k)_m D_n(k)\sum_{m=0}^{r}S(r,m)$$

$$= \sum_{k=m}^{n}(k)_m D_n(k)\sum_{m=0}^{r}S(r,m)$$

(since $(k)_m = 0$ for all non-negative integers $k < m$)

$$= \sum_{k=m}^{n} (k)_m D_n(k) \sum_{m=0}^{\min\{r,n\}} S(r,m) \quad \text{(since } m \le r \text{ and } m \le n\text{)}$$

$$= n! \sum_{m=0}^{\min\{r,n\}} S(r,m) \quad \text{(by Problem 4.37)}.$$

When $n \ge r$, we have $\min\{r,n\} = r$, so that we have

$$\sum_{k=0}^{n} k^r \binom{n}{k} D_{n-k} = n! \sum_{m=0}^{\min\{r,n\}} S(r,m) = n! \sum_{m=0}^{r} S(r,m) = B_r \cdot n!.$$

63. Let $S = \{1, 2, 3, \ldots, 280\}$. For all $k = 2, 3, 5, 7$ and $x \in S$, we say that x has property P_k if x is divisible by k. Then we have

$$\omega(P_2) = \left\lfloor \frac{280}{2} \right\rfloor = 140,$$

$$\omega(P_3) = \left\lfloor \frac{280}{3} \right\rfloor = 93,$$

$$\omega(P_5) = \left\lfloor \frac{280}{5} \right\rfloor = 56,$$

$$\omega(P_7) = \left\lfloor \frac{280}{7} \right\rfloor = 40.$$

This implies that $\omega(1) = 140 + 93 + 56 + 40 = 329$. Next, we have

$$\omega(P_2 P_3) = \left\lfloor \frac{280}{2 \cdot 3} \right\rfloor = 46,$$

$$\omega(P_2 P_5) = \left\lfloor \frac{280}{2 \cdot 5} \right\rfloor = 28,$$

$$\omega(P_2 P_7) = \left\lfloor \frac{280}{2 \cdot 7} \right\rfloor = 20,$$

$$\omega(P_3 P_5) = \left\lfloor \frac{280}{3 \cdot 5} \right\rfloor = 18,$$

$$\omega(P_3 P_7) = \left\lfloor \frac{280}{3 \cdot 7} \right\rfloor = 13,$$

$$\omega(P_5 P_7) = \left\lfloor \frac{280}{5 \cdot 7} \right\rfloor = 8.$$

This implies that $\omega(2) = 46 + 28 + 20 + 18 + 13 + 8 = 133$. Next, we

have

$$\omega(P_2 P_3 P_5) = \left\lfloor \frac{280}{2 \cdot 3 \cdot 5} \right\rfloor = 9,$$

$$\omega(P_2 P_3 P_7) = \left\lfloor \frac{280}{2 \cdot 3 \cdot 7} \right\rfloor = 6,$$

$$\omega(P_2 P_5 P_7) = \left\lfloor \frac{280}{2 \cdot 5 \cdot 7} \right\rfloor = 4,$$

$$\omega(P_3 P_5 P_7) = \left\lfloor \frac{280}{3 \cdot 5 \cdot 7} \right\rfloor = 2.$$

This implies that $\omega(3) = 9 + 6 + 4 + 2 = 21$. Finally, we have

$$\omega(4) = \omega(P_2 P_3 P_5 P_7) = \left\lfloor \frac{280}{2 \cdot 3 \cdot 5 \cdot 7} \right\rfloor = 1.$$

Hence, we have

$$\begin{aligned}
E(0) &= \omega(0) - \omega(1) + \omega(2) - \omega(3) + \omega(4) \\
&= 280 - 329 + 133 - 21 + 1 \\
&= 64.
\end{aligned}$$

Thus, the number of elements of S that is divisible by at least one of $2, 3, 5$ and 7 is equal to $|S| - E(0) = 280 - 64 = 216$. Let us denote the set of elements of S that is divisible by at least one of $2, 3, 5$ and 7 by A. As each element in A has a smallest prime factor $p \in \{2, 3, 5, 7\}$, it follows from (PP) that any 5-element subset of A contains two numbers that has a common prime factor $q \in \{2, 3, 5, 7\}$. Consequently, we must have $n \geq 217$. It remains to show that $n = 217$. To this end, we let

$$\begin{aligned}
B_1 &= \{1\} \cup \{m \in S \mid m \text{ is a prime number}\}, \\
B_2 &= \{11 \cdot 11, 7 \cdot 13, 5 \cdot 17, 3 \cdot 19, 2 \cdot 23\}, \\
B_3 &= \{11 \cdot 13, 7 \cdot 17, 5 \cdot 19, 3 \cdot 23, 2 \cdot 29\}, \\
B_4 &= \{13 \cdot 13, 11 \cdot 17, 7 \cdot 19, 5 \cdot 23, 3 \cdot 29, 2 \cdot 31\}, \\
B_5 &= \{13 \cdot 17, 11 \cdot 19, 7 \cdot 23, 5 \cdot 29, 3 \cdot 31, 2 \cdot 37\}, \\
B_6 &= \{13 \cdot 19, 11 \cdot 23, 7 \cdot 29, 5 \cdot 31, 3 \cdot 37, 2 \cdot 41\},
\end{aligned}$$

and $B = \bigcup_{i=1}^{6} B_i$. Then it is clear that $B \subseteq S$ and $B_i \cap B_j = \emptyset$ for all distinct $i, j \in \{1, 2, \ldots, 6\}$. Also, it is easy to see that any two distinct numbers in B_i are relatively prime for all $i = 1, 2, \ldots, 6$. Let us compute $|B|$. To this end, it suffices to compute $|B_1|$. We observe that

$$S \setminus B_1$$
$$= \{m \in S \mid m \text{ is composite}\}$$
$$= (A \setminus \{2, 3, 5, 7\})$$
$$\cup \{m \in S \mid m \text{ is composite and is not divisible by } 2, 3, 5, 7\}$$
$$= (A \setminus \{2, 3, 5, 7\})$$
$$\cup \{11 \cdot 11, 11 \cdot 13, 11 \cdot 17, 11 \cdot 19, 11 \cdot 23, 13 \cdot 13, 13 \cdot 17, 13 \cdot 19\},$$

where the last equality follows from the fact that each composite number in S must have a prime factor $p \leq \sqrt{280} < 17$. This implies that

$$|S \setminus B_1| = |A \setminus \{2, 3, 5, 7\}| + 8 = 216 - 4 + 8 = 220,$$

so that we have $|B_1| = |S| - |S \setminus B_1| = 280 - 220 = 60$. Hence, we have

$$|B| = \left| \bigcup_{i=1}^{6} B_i \right| = \sum_{i=1}^{6} |B_i| = 60 + 5 + 5 + 6 + 6 + 6 = 88.$$

Now, let us take any subset C of S with $|C| = 217$, and let $D = B \cap C$. Then by (4.1.1), we have

$$|D| = |B \cap C| = |B| + |C| - |B \cup C| \geq 88 + 217 - 280 = 25.$$

Since $D \subseteq B$ and $25 > 6 \cdot 4$, it follows from (PP) that there exists some $k \in \{1, 2, \ldots, 6\}$, such that $|D \cap B_k| \geq 5$. As $D \cap B_k = (B \cap C) \cap B_k = C \cap B_k$, and any two distinct numbers in B_i are relatively prime for all $i = 1, 2, \ldots, 6$, it follows that C contains at least 5 numbers (from B_k) that are relatively prime. So this shows that $n = 217$, and we are done.

Remark. The idea behind the second part of the solution is courtesy of John Scholes. We refer the reader to http://www.cs.cornell.edu/~asdas/imo/imo/isoln/isoln913.html for the original solution.

Solutions to Exercise 5

1. We have
$$(x^3 + x^4 + x^5 + \cdots)^3 = [x^3(1 + x + x^2 + \cdots)]^3$$
$$= x^9(1-x)^{-3} \quad \text{(by (5.1.3))}$$
$$= x^9 \sum_{k=0}^{\infty} \binom{k+3-1}{k} x^k \quad \text{(by (5.1.6))}$$
$$= x^9 \sum_{k=0}^{\infty} \binom{k+2}{2} x^k.$$

Thus, the coefficient of x^{20} in the expansion of $(x^3 + x^4 + x^5 + \cdots)^3$ is equal to the coefficient of $x^{20-9} = x^{11}$ in $\sum_{k=0}^{\infty} \binom{k+2}{2} x^k$, which is then equal to $\binom{11+2}{2} = \binom{13}{2}$.

2. We have
$$(1 + x + x^2 + \cdots + x^5)^4$$
$$= (1 - x^6)^4(1-x)^{-4} \quad \text{(by (5.1.3))}$$
$$= (1 - x^6)^4 \sum_{k=0}^{\infty} \binom{k+4-1}{k} x^k \quad \text{(by (5.1.6))}$$
$$= (1 - 4x^6 + 6x^{12} - 4x^{18} + x^{24}) \sum_{k=0}^{\infty} \binom{k+3}{3} x^k.$$

Thus, the coefficient of x^9 in the expansion of $(1 + x + x^2 + \cdots + x^5)^4$ is equal to $\binom{9+3}{3} - 4\binom{3+3}{3} = \binom{12}{3} - 4\binom{6}{3}$, and the coefficient of x^{14} in the expansion of $(1 + x + x^2 + \cdots + x^5)^4$ is equal to $\binom{14+3}{3} - 4\binom{8+3}{3} + 6\binom{2+3}{3} = \binom{17}{3} - 4\binom{11}{3} + 6\binom{5}{3}$.

3. Theorem 5.1.1 (iv): We have
$$x^m A(x) = x^m(a_0 + a_1 x + a_2 x^2 + \cdots) = a_0 x^m + a_1 x^{m+1} + a_2 x^{m+2} + \cdots.$$

Hence, it follows that $x^m A(x)$ is the generating function for the sequence (c_r), where

$$c_r = \begin{cases} 0 & \text{if } 0 \leq r \leq m - 1, \\ a_{r-m} & \text{if } r \geq m \end{cases}$$

Theorem 5.1.1 (vi): We have

$$\begin{aligned}
(1 - x)A(x) &= (1 - x)(a_0 + a_1 x + a_2 x^2 + \cdots) \\
&= (a_0 + a_1 x + a_2 x^2 + \cdots) - (a_0 x + a_1 x^2 + a_2 x^3 + \cdots) \\
&= a_0 + (a_1 - a_0)x + (a_2 - a_1)x^2 + \cdots.
\end{aligned}$$

Hence, it follows that $(1 - x)A(x)$ is the generating function for the sequence (c_r), where

$$c_0 = a_0 \quad \text{and} \quad c_r = a_r - a_{r-1} \quad \text{for all } r \geq 1.$$

Theorem 5.1.1 (viii): Since $A'(x) = a_1 + 2a_2 x + 3a_3 x^2 + \cdots$, it follows that $A'(x)$ is the generating function for the sequence (c_r), where

$$c_r = (r + 1)a_{r+1} \quad \text{for all } r.$$

Theorem 5.1.1 (ix): We have

$$x A'(x) = x(a_1 + 2a_2 x + 3a_3 x^2 + \cdots) = a_1 x + 2a_2 x^2 + 3a_3 x^3 + \cdots.$$

Hence, it follows that $x A'(x)$ is the generating function for the sequence (c_r), where

$$c_r = r a_r \quad \text{for all } r.$$

Theorem 5.1.1 (x): We have

$$\begin{aligned}
\int_0^x A(t)\, dt &= \int_0^x (a_0 + a_1 t + a_2 t^2 + \cdots)\, dt \\
&= \left[a_0 t + \frac{a_1}{2} t^2 + \frac{a_2}{3} t^3 + \cdots \right]_0^x \\
&= a_0 x + \frac{a_1}{2} x^2 + \frac{a_2}{3} x^3 + \cdots.
\end{aligned}$$

Hence, it follows that $\int_0^x A(t)\, dt$ is the generating function for the sequence (c_r), where

$$c_0 = 0 \quad \text{and} \quad c_r = \frac{a_{r-1}}{r} \quad \text{for all } r \geq 1.$$

4. Let $a_i = i$ and $b_i = i^2$ for all non-negative integers r. By (5.1.5), the generating function for the sequence (a_r) is $A(x) = \frac{x}{(1-x)^2}$. Since we have $b_r = ra_r$ for all non-negative integers r, it follows from Theorem 5.1.1 (ix) that the generating function $B(x)$ for the sequence (b_r) is

$$xA'(x) = x \cdot \frac{(1-x)^2 + x \cdot 2(1-x)}{(1-x)^4} = \frac{x+x^2}{(1-x)^3}.$$

Next, since we have $c_0 = 0 = b_0$ $c_r = \sum_{i=1}^{r} i^2 = \sum_{i=0}^{r} i^2 = \sum_{i=0}^{r} b_i$ for all positive integers r, it follows from Theorem 5.1.1 (vii) that the generating function $C(x)$ for the sequence (c_r) is

$$\frac{B(x)}{1-x} = \frac{x+x^2}{(1-x)^4}.$$

Now, we observe that

$$C(x) = \frac{x+x^2}{(1-x)^4}$$

$$= (x+x^2) \sum_{r=0}^{\infty} \binom{r+4-1}{r} x^r \quad \text{(by (5.1.6))}$$

$$= \sum_{r=0}^{\infty} \binom{r+3}{3} x^{r+1} + \sum_{r=0}^{\infty} \binom{r+3}{3} x^{r+2}$$

$$= \sum_{r=1}^{\infty} \binom{r+2}{3} x^r + \sum_{r=2}^{\infty} \binom{r+1}{3} x^r$$

$$= x + \sum_{r=2}^{\infty} \left(\binom{r+1}{3} + \binom{r+2}{3} \right) x^r.$$

Since the equation $c_r = \binom{r+1}{3} + \binom{r+2}{3}$ also holds when $r = 1$, it follows that we must have

$$\sum_{i=1}^{r} i^2 = c_r = \binom{r+1}{3} + \binom{r+2}{3}$$

for all positive integers r as desired.

5. Let $a_i = 2^i$ and $b_i = i2^i$ for all non-negative integers i. By (5.1.4), the generating function for the sequence (a_i) is $A(x) = \frac{1}{1-2x}$. Since we have $b_i = ia_i$ for all non-negative integers i, it follows from Theorem 5.1.1 (ix) that the generating function $B(x)$ for the sequence (b_i) is

$$xA'(x) = x \cdot \frac{2}{(1-2x)^2} = \frac{2x}{(1-2x)^2}.$$

Next, since we have $c_r = \sum\limits_{i=0}^{r} i2^i = \sum\limits_{i=0}^{r} b_i$ for all non-negative integers r, it follows from Theorem 5.1.1 (vii) that the generating function $C(x)$ for the sequence (c_r) is

$$\frac{B(x)}{1-x} = \frac{2x}{(1-2x)^2(1-x)}.$$

Now, we observe that

$$C(x) = \frac{2x}{(1-2x)^2(1-x)} = \frac{2}{1-x} - \frac{4}{1-2x} + \frac{2}{(1-2x)^2}.$$

Since $A(x) = \frac{1}{1-2x} = \sum\limits_{r=0}^{\infty} 2^r x^r$, and $A'(x) = \frac{2}{(1-2x)^2}$, it follows that we have

$$\frac{1}{(1-2x)^2} = \frac{1}{2}A'(x) = \frac{1}{2}\sum_{r=1}^{\infty} r2^r x^{r-1} = \sum_{r=0}^{\infty}(r+1)2^r x^r.$$

This implies that

$$C(x) = \frac{2}{1-x} - \frac{4}{1-2x} + \frac{2}{(1-2x)^2}$$

$$= 2\sum_{r=0}^{\infty} x^r - 4\sum_{r=0}^{\infty} 2^r x^r + 2\sum_{r=0}^{\infty}(r+1)2^r x^r$$

$$= \sum_{r=1}^{\infty}(2 - 4\cdot 2^r + 2(r+1)2^r)x^r$$

$$= \sum_{r=1}^{\infty}(2 + (r-1)2^{r+1})x^r.$$

Since the equation $c_r = 2 + (r-1)2^{r+1}$ also holds when $r = 0$, it follows that we must have

$$\sum_{i=0}^{r} i2^i = c_r = 2 + (r-1)2^{r+1}$$

for all non-negative integers r as desired.

6. (i) Let the generating function for the sequence (a_r) be $A(x)$. Then

we have

$$A(x) = \sum_{r=0}^{\infty} a_r x^r$$

$$= \sum_{r=0}^{\infty} \frac{1}{4^r} \binom{2r}{r} x^r$$

$$= \sum_{r=0}^{\infty} \frac{(2r)!}{(2^r r!)^2} x^r$$

$$= 1 + \sum_{r=1}^{\infty} \frac{1}{2^r r!} \cdot \frac{1 \cdot 2 \cdot 3 \cdots 2r}{2^r \cdot 1 \cdot 2 \cdots r} x^r$$

$$= 1 + \sum_{r=1}^{\infty} \frac{1}{2^r r!} \cdot \frac{1 \cdot 2 \cdot 3 \cdots 2r}{2 \cdot 4 \cdots 2r} x^r$$

$$= 1 + \sum_{r=1}^{\infty} \frac{1 \cdot 3 \cdot 5 \cdots (2r-1)}{2^r r!} x^r$$

$$= 1 + \sum_{r=1}^{\infty} (-1)^{2r} \frac{\frac{1}{2} \cdot \frac{3}{2} \cdot \frac{5}{2} \cdots \frac{2r-1}{2}}{r!} x^r$$

$$= 1 + \sum_{r=1}^{\infty} \frac{\left(-\frac{1}{2}\right) \cdot \left(-\frac{3}{2}\right) \cdot \left(-\frac{5}{2}\right) \cdots \left(-\frac{2r-1}{2}\right)}{r!} (-x)^r$$

$$= 1 + \sum_{r=1}^{\infty} \frac{\left(-\frac{1}{2}\right) \cdot \left(-\frac{1}{2}-1\right) \cdot \left(-\frac{1}{2}-2\right) \cdots \left(-\frac{1}{2}-r+1\right)}{r!} (-x)^r$$

$$= 1 + \sum_{r=1}^{\infty} \binom{-\frac{1}{2}}{r} (-x)^r$$

$$= \sum_{r=0}^{\infty} \binom{-\frac{1}{2}}{r} (-x)^r$$

$$= (1-x)^{-\frac{1}{2}},$$

where the last equality follows from (5.1.1).

(ii) For each non-negative integer n, we let

$$c_n = \sum_{k=0}^{n} a_k a_{n-k}$$

$$= \sum_{k=0}^{n} \frac{1}{4^k} \binom{2k}{k} \cdot \frac{1}{4^{n-k}} \binom{2(n-k)}{n-k}$$

$$= \frac{1}{4^n} \sum_{k=0}^{n} \binom{2k}{k} \binom{2(n-k)}{n-k}.$$

Then it follows from Theorem 5.1.1 (iii) that $A^2(x) = (1-x)^{-1}$ is the generating function for the sequence (c_r). Since we have $\sum_{n=0}^{\infty} c_n x^n = (1-x)^{-1} = \sum_{n=0}^{\infty} x^n$, it follows by comparing the coefficients of x^n on both sides of the equation that we have $c_n = 1$ for all non-negative integers n, or equivalently,

$$\sum_{k=0}^{n} \binom{2k}{k} \binom{2(n-k)}{n-k} = 4^n$$

for all non-negative integers n as desired.

7. By (2.1.2), we have $\frac{r}{n} \binom{n}{r} = \binom{n-1}{r-1}$ for all positive integers r. This implies that we have

$$\frac{1}{n} \sum_{r=1}^{n} r \binom{n}{r} \binom{m}{r} = \sum_{r=1}^{n} \binom{n-1}{r-1} \binom{m}{r}$$

$$= \sum_{r=1}^{n} \binom{n-1}{n-r} \binom{m}{r}$$

$$= \sum_{r=0}^{n} \binom{n-1}{n-r} \binom{m}{r},$$

where the last equality follows from the fact that $\binom{n-1}{n} = 0$. Now, for each non-negative integer r, we let $a_r = \binom{m}{r}$ and $b_r = \binom{n-1}{r}$. Then the generating functions for the sequences (a_r) and (b_r) are $A(x) = (1+x)^m$ and $B(x) = (1+x)^{n-1}$ respectively. Let $c_r = \sum_{k=0}^{r} a_k b_{r-k} = \sum_{k=0}^{r} \binom{n-1}{r-k} \binom{m}{k}$ for all non-negative integers r. Then it follows from Theorem 5.1.1 (ii) that the generating function for (c_r) is

$$C(x) = A(x)B(x) = (1+x)^{n+m-1} = \sum_{r=0}^{\infty} \binom{n+m-1}{r} x^r.$$

Consequently, we have

$$\frac{1}{n} \sum_{r=1}^{n} r \binom{n}{r} \binom{m}{r} = \sum_{r=0}^{n} \binom{n-1}{n-r} \binom{m}{r} = c_n = \binom{n+m-1}{n},$$

or equivalently,

$$\sum_{r=1}^{n} r \binom{n}{r} \binom{m}{r} = n \binom{n+m-1}{n}$$

as desired.

8. For each non-negative integer r, we let a_r denote the number of ways to distribute r identical pieces of candy to 3 children so that no child gets more than 4 pieces. Then it is easy to see that the generating function for (a_r) is

$$(1 + x + \cdots + x^4)^3$$
$$= (1 - x^5)^3 (1 - x)^{-3} \quad \text{(by (5.1.3))}$$
$$= (1 - 3x^5 + 3x^{10} - x^{15}) \sum_{r=0}^{\infty} \binom{r + 3 - 1}{r} x^r \quad \text{(by (5.1.6))}$$
$$= (1 - 3x^5 + 3x^{10} - x^{15}) \sum_{r=0}^{\infty} \binom{r + 2}{2} x^r.$$

Thus, the number of ways to distribute 10 identical pieces of candy to 3 children so that no child gets more than 4 pieces is equal to

$$a_{10} = \binom{10 + 2}{2} - 3\binom{10 - 5 + 2}{2} + 3\binom{10 - 10 + 2}{2}$$
$$= \binom{12}{2} - 3\binom{7}{2} + 3$$
$$= 6.$$

9. For each non-negative integer r, we let a_r denote the number of ways to distribute r identical balls to 7 distinct boxes, such that box 1 must hold at least 3, and at most 10, of the balls. Then it is easy to see that the generating function for (a_r) is

$$(x^3 + x^4 + \cdots + x^{10})(1 + x + x^2 \cdots)^6$$
$$= x^3 (1 + x + \cdots + x^7)(1 + x + x^2 \cdots)^6$$
$$= x^3 (1 - x^8)(1 - x)^{-7} \quad \text{(by (5.1.3))}$$
$$= x^3 (1 - x^8) \sum_{r=0}^{\infty} \binom{r + 7 - 1}{r} x^r \quad \text{(by (5.1.6))}$$
$$= x^3 (1 - x^8) \sum_{r=0}^{\infty} \binom{r + 6}{6} x^r.$$

Thus, the number of ways to distribute 40 identical balls to 7 distinct boxes, such that box 1 must hold at least 3, and at most 10, of the balls, is equal to

$$a_{40} = \binom{40 - 3 + 6}{6} - \binom{40 - 11 + 6}{6} = \binom{43}{6} - \binom{35}{6}.$$

10. For each non-negative integer r, we let a_r denote the number of ways to select r balls from n identical blue balls, n identical red balls and n identical white balls. Then it is easy to see that the generating function for (a_r) is

$$(1 + x + \cdots + x^n)^3$$
$$= (1 - x^{n+1})^3(1 - x)^{-3} \quad \text{(by (5.1.3))}$$
$$= (1 - 3x^{n+1} + 3x^{2n+2} - x^{3n+3}) \sum_{r=0}^{\infty} \binom{r + 3 - 1}{r} x^r \quad \text{(by (5.1.6))}$$
$$= (1 - 3x^{n+1} + 3x^{2n+2} - x^{3n+3}) \sum_{r=0}^{\infty} \binom{r + 2}{2} x^r.$$

Thus, the number of ways to select $2n$ balls from n identical blue balls, n identical red balls and n identical white balls, is equal to

$$a_{2n} = \binom{2n + 2}{2} - 3\binom{2n - (n + 1) + 2}{2} = \binom{2n + 2}{2} - 3\binom{n + 1}{2}.$$

11. The problem is equivalent to finding the number of ways to divide 10 identical chairs into 4 different rooms so that each room will have $1, 2, 3, 4$ or 5 chairs. For each non-negative integer r, we let a_r denote the number of ways to divide r identical chairs into 4 different rooms such that each room will have $1, 2, 3, 4$ or 5 chairs. Then it is easy to see that the generating function for (a_r) is

$$(x + x^2 + \cdots + x^5)^4$$
$$= x^4(1 + x + \cdots + x^4)^4$$
$$= x^4(1 - x^5)^4(1 - x)^{-4} \quad \text{(by (5.1.3))}$$
$$= x^4(1 - 4x^5 + 6x^{10} - 4x^{15} + x^{20}) \sum_{r=0}^{\infty} \binom{r + 4 - 1}{r} x^r \quad \text{(by (5.1.6))}$$
$$= x^4(1 - 4x^5 + 6x^{10} - 4x^{15} + x^{20}) \sum_{r=0}^{\infty} \binom{r + 3}{3} x^r.$$

Thus, the number of ways to divide 10 identical chairs into 4 different rooms such that each room will have $1, 2, 3, 4$ or 5 chairs is equal to

$$a_{10} = \binom{10 - 4 + 3}{3} - 4\binom{10 - 9 + 3}{3} = \binom{9}{3} - 4\binom{4}{3} = 68.$$

12. (i) By (5.1.3), the generating function $A(x)$ for (a_r) is

$$(x + x^2 + x^3 + \cdots)^3(1 + x + x^2 + \cdots)^2 = x^3(1 + x + x^2 + \cdots)^5$$
$$= x^3(1 - x)^{-5}.$$

(ii) By (5.1.3), the generating function $B(x)$ for (b_r) is

$$(x^2 + x^3 + x^4 \cdots)^2 (1 + x + x^2 + \cdots)^3 = x^4(1 + x + x^2 + \cdots)^5$$
$$= x^4(1 - x)^{-5}.$$

(iii) As $B(x) = xA(x)$, it follows from Theorem 5.1.1 (iv) that we have $b_r = a_{r-1}$ for each $r = 1, 2, \ldots$, or equivalently, $a_r = b_{r+1}$ for each $r = 1, 2, \ldots$ as desired.

13. The generating function for (a_r) is

$$(x^3 + x^4 + \cdots + x^9)(1 + x + \cdots + x^8)(x^7 + x^8 + \cdots + x^{17})$$
$$= x^{10}(1 + x + \cdots + x^6)(1 + x + \cdots + x^8)(1 + x + \cdots + x^{10})$$
$$= x^{10}(1 - x^7)(1 - x^9)(1 - x^{11})(1 - x)^{-3}.$$

By (5.1.6), we have

$$x^{10}(1 - x^7)(1 - x^9)(1 - x^{11})(1 - x)^{-3}$$
$$= x^{10}(1 - x^7 - x^9 - x^{11} + x^{16} + x^{18} + x^{20} - x^{27}) \sum_{r=0}^{\infty} \binom{r + 3 - 1}{r} x^r$$
$$= x^{10}(1 - x^7 - x^9 - x^{11} + x^{16} + x^{18} + x^{20} - x^{27}) \sum_{r=0}^{\infty} \binom{r + 2}{2} x^r.$$

Thus, we have

$$a_{28} = \binom{20}{2} - \binom{13}{2} - \binom{11}{2} - \binom{9}{2} + \binom{4}{2} + 1.$$

14. The problem is equivalent to finding the number of ways to divide $\frac{3000}{25} = 120$ identical pencils amongst 4 student groups such that each group gets at least $\frac{150}{25} = 6$, but not more than $\frac{1000}{25} = 40$ of the pencils. For each non-negative integer r, we let a_r denote the number of ways to divide k identical pencils amongst 4 student groups such that each group gets at least 6, but not more than 40 of the pencils. By (5.1.3) and (5.1.6), we see that the generating function for (a_r) is

$$(x^6 + x^7 + \cdots + x^{40})^4$$
$$= x^{24}(1 + x + \cdots + x^{34})^4$$
$$= x^{24}(1 - x^{35})^4(1 - x)^{-4}$$
$$= x^{24}(1 - 4x^{35} + 6x^{70} - 4x^{105} + x^{140}) \sum_{r=0}^{\infty} \binom{r + 4 - 1}{r} x^r$$
$$= (x^{24} - 4x^{59} + 6x^{94} - 4x^{129} + x^{164}) \sum_{r=0}^{\infty} \binom{r + 3}{3} x^r.$$

Thus, the number of ways to divide 120 identical pencils amongst 4 student groups such that each group gets at least 6, but not more than 40 of the pencils, is equal to

$$a_{120} = \binom{120 - 24 + 3}{3} - 4\binom{120 - 59 + 3}{3} + 6\binom{120 - 94 + 3}{3}$$

$$= \binom{99}{3} - 4\binom{64}{3} + 6\binom{29}{3}.$$

15. For each non-negative integer r, we let a_r denote the number of ways to select r letters from "F, U, N, C, T, I, O" that contain at most 3 U's and at least 1 O. Then it is easy to see that the generating function for (a_r) is

$$(1 + x + x^2 + x^3)(x + x^2 + x^3 + \cdots)(1 + x + x^2 + \cdots)^5$$
$$= x(1 - x^4)(1 - x)^{-1}(1 + x + x^2 + \cdots)^6$$
$$= x(1 - x^4)(1 - x)^{-7} \quad \text{(by (5.1.3))}$$
$$= x(1 - x^4) \sum_{r=0}^{\infty} \binom{r + 7 - 1}{r} x^r \quad \text{(by (5.1.6))}$$
$$= x(1 - x^4) \sum_{r=0}^{\infty} \binom{r + 6}{6} x^r.$$

Thus, the number of ways to select 10 letters from "F, U, N, C, T, I, O" that contain at most 3 U's and at least 1 O, is equal to

$$a_{10} = \binom{10 - 1 + 6}{6} - \binom{10 - 5 + 6}{6} = \binom{15}{6} - \binom{11}{6}.$$

16. (i) By (5.1.3), the generating function for (a_r) is

$$(1 + x + x^2 + x^3)(x^2 + x^3 + x^4 + \cdots)(1 + x + x^2 + \cdots)^5$$
$$= x^2(1 + x + x^2 + x^3)(1 + x + x^2 + \cdots)^6$$
$$= x^2(1 + x + x^2 + x^3)(1 - x)^{-6}.$$

(ii) A similar argument as in Example 5.3.2 shows that the generating function for (a_r) is

$$[(1 - x)(1 - x^2)(1 - x^3)(1 - x^5)(1 - x^8)]^{-1}.$$

(iii) A similar argument as in Example 5.3.3 shows that the generating function for (a_r) is

$$(1 + x^5)(1 + x^{10})(1 + x^{15}).$$

(iv) A similar argument as in Example 5.3.4 shows that the generating function for (a_r) is

$$\prod_{k=0}^{\infty}(1+x^{2k+1}).$$

(v) A similar argument as in Example 5.3.4 shows that the generating function for (a_r) is

$$\prod_{k=1}^{\infty}(1+x^{2k}).$$

(vi) The number of integer solutions to the inequality

$$x_1 + x_2 + x_3 + x_4 + x_5 \leq r$$

with $1 \leq x_i \leq 6$ for each $i = 1, 2, 3, 4, 5$ is equal to the number of integer solutions to the equation

$$x_1 + x_2 + x_3 + x_4 + x_5 + x_6 = r$$

with $1 \leq x_i \leq 6$ for each $i = 1, 2, 3, 4, 5$ and $x_6 \geq 0$. Thus, it follows from (5.1.3) that the generating function for (a_r) is

$$(x + x^2 + \cdots + x^6)^5(1 + x + x^2 + \cdots)$$
$$= x^5(1 + x + \cdots + x^5)^5(1-x)^{-1}$$
$$= x^5(1 - x^6)^5(1-x)^{-6}.$$

17. For each non-negative integer r, we let a_r denote the number of r-element multi-subsets of the multi-set $\{(3n) \cdot x, (3n) \cdot y, (3n) \cdot z\}$. Then it is easy to see that the generating function for (a_r) is

$$(1 + t + \cdots + t^{3n})^3$$
$$= (1 - t^{3n+1})^3(1 - t)^3$$
$$= (1 - 3t^{3n+1} + 3t^{6n+2} - t^{9n+3}) \sum_{r=0}^{\infty}\binom{r+3-1}{r}t^r \quad \text{(by (5.1.6))}$$
$$= (1 - 3t^{3n+1} + 3t^{6n+2} - t^{9n+3}) \sum_{r=0}^{\infty}\binom{r+2}{2}t^r.$$

Thus, the number of $4n$-element multi-subsets of the multi-set

$$\{(3n) \cdot x, (3n) \cdot y, (3n) \cdot z\}$$

is equal to

$$a_{4n} = \binom{4n+2}{2} - 3\binom{4n - (3n+1) + 2}{2} = \binom{4n+2}{2} - 3\binom{n+1}{2}.$$

18. For each non-negative integer r, we let a_r denote the number of r-element multi-subsets of the multi-set $\{n \cdot z_1, n \cdot z_2, \ldots, n \cdot z_m\}$. Then it is easy to see that the generating function for (a_r) is

$$(1 + x + \cdots + x^n)^m$$
$$= (1 - x^{n+1})^m (1 - x)^{-m}$$
$$= \sum_{k=0}^{m} \binom{m}{k} (-1)^k x^{k(n+1)} \sum_{r=0}^{\infty} \binom{r + m - 1}{r} x^r \quad \text{(by (5.1.6))}$$
$$= \sum_{k=0}^{m} \sum_{r=0}^{\infty} (-1)^k \binom{m}{k} \binom{r + m - 1}{m - 1} x^{k(n+1)+r}.$$

Thus, the number of $3n$-element multi-subsets of the multi-set

$$\{n \cdot z_1, n \cdot z_2, \ldots, n \cdot z_m\}$$

is equal to

$$a_{3n} = \binom{m}{0}\binom{3n + m - 1}{m - 1} - \binom{m}{1}\binom{2n - 1 + m - 1}{m - 1}$$
$$+ \binom{m}{2}\binom{n - 2 + m - 1}{m - 1}$$
$$= \binom{3n + m - 1}{m - 1} - m\binom{2n + m - 2}{m - 1} + \binom{m}{2}\binom{n + m - 3}{m - 1}.$$

19. For each non-negative integer r, we let a_r denote the number of ways to yield a sum of r from rolling 5 distinct dice. Then it is easy to see that the generating function for (a_r) is

$$(x + x^2 + \cdots + x^6)^5$$
$$= x^5 (1 + x + \cdots + x^5)^5$$
$$= x^5 (1 - x^6)^5 (1 - x)^{-5}$$
$$= x^5 \sum_{k=0}^{5} \binom{5}{k} (-1)^k x^{6k} \sum_{r=0}^{\infty} \binom{r + 5 - 1}{r} x^r \quad \text{(by (5.1.6))}$$
$$= \sum_{k=0}^{5} \sum_{r=0}^{\infty} (-1)^k \binom{5}{k} \binom{r + 4}{4} x^{6k+5+r}.$$

Thus, the number of ways to yield a sum of 17 from rolling 5 distinct dice is equal to

$$a_{17} = \binom{5}{0}\binom{12 + 4}{4} - \binom{5}{1}\binom{6 + 4}{4} + \binom{5}{2}\binom{0 + 4}{4} = 780.$$

As there are a total of 6^5 outcomes from rolling 5 dice, it follows that the probability that a roll of 5 distinct dice yields a sum of 17 is equal to $\frac{780}{6^5} = \frac{65}{648}$.

20. For each non-negative integer r and n, we let $a_{r,n}$ denote the number of ways to yield a sum of r from rolling n distinct dice. Then it is easy to see that the generating $A_n(x)$ for $(a_{r,n})_{r=0}^{\infty}$ is $(x + x^2 + \cdots + x^6)^n$. Since we have $a_r = \sum_{n=0}^{\infty} a_{r,n}$ for all non-negative integers r, it follows that the generating function for (a_r) is

$$\sum_{n=0}^{\infty} A_n(x) = \sum_{n=0}^{\infty} (x + x^2 + \cdots + x^6)^n = [1 - (x + x^2 + \cdots + x^6)]^{-1}.$$

21. (i) By (5.1.3), the generating function $A(x)$ for (a_r) is

$$(x + x^2 + x^3 + \cdots)^{k+1}(1 + x + x^2 + \cdots)^k$$
$$= x^{k+1}(1 + x + x^2 + \cdots)^{2k+1}$$
$$= x^{k+1}(1 - x)^{-(2k+1)}.$$

(ii) By (5.1.3), the generating function $B(x)$ for (b_r) is

$$(x^m + x^{m+1} + x^{m+2}\cdots)^k(1 + x + x^2 + \cdots)^{k+1}$$
$$= x^{mk}(1 + x + x^2 + \cdots)^{2k+1}$$
$$= x^{mk}(1 - x)^{-(2k+1)}.$$

(iii) As $B(x) = x^{(m-1)k-1}A(x)$, it follows from Theorem 5.1.1 (iv) that we have $b_r = a_{r-(m-1)k-1}$ for each $r = (m-1)k - 1, (m-1)k, \ldots$, or equivalently, $a_r = b_{r+(m-1)k-1}$ as desired.

22. By (5.1.3), the generating function for (a_r) is

$$(1 + x + x^2 + \cdots)(1 + x^2 + x^4 + \cdots)(1 + x^3 + \cdots)(1 + x^4 + \cdots)$$
$$= [(1 - x)(1 - x^2)(1 - x^3)(1 - x^4)]^{-1}.$$

23. We first note that a_r is equal to the number of ways to arrange $r - 4$ 0's and 4 1's, such that no two 1's are consecutive, which is then equal to the number of integer solutions to the equation

$$x_1 + x_2 + x_3 + x_4 + x_5 = r - 4$$

that satisfy $x_1, x_5 \geq 0$ and $x_2, x_3, x_4 \geq 1$. Now, for each integer r, we let b_r denote the number of integer solutions to the equation

$$x_1 + x_2 + x_3 + x_4 + x_5 = r - 4$$

that satisfy $x_1, x_5 \geq 0$ and $x_2, x_3, x_4 \geq 1$. By (5.1.3), it follows that the generating function for (b_r) is given by

$$(x+x^2+x^3+\cdots)^3(1+x+x^2+\cdots)^2 = x^3(1+x+x^2+\cdots)^5 = x^3(1-x)^{-5}.$$

Since $a_r = b_{r-4}$, it follows from Theorem 5.1.1 (iv) that the generating function for (a_r) is $x^7(1-x)^{-5}$. Now, by (5.1.6), we have

$$
\begin{aligned}
x^7(1-x)^{-5} &= x^7 \sum_{t=0}^{\infty} \binom{t+5-1}{t} x^t \\
&= \sum_{t=0}^{\infty} \binom{t+4}{4} x^{t+7} \\
&= \sum_{r=7}^{\infty} \binom{r-3}{4} x^r \\
&= \sum_{r=0}^{\infty} \binom{r-3}{4} x^r,
\end{aligned}
$$

where the last equality follows from the fact that $\binom{k}{4} = 0$ for all integers $k < 4$. Thus, we have $a_r = \binom{r-3}{4}$.

24. We first note that a_r is equal to the number of ways to arrange $r - m$ 0's and m 1's, such that there are at least $t - 1$ 0's between any two 1's. This is then equal to the number of integer solutions to the equation

$$x_1 + x_2 + \cdots + x_{m+1} = r - m$$

that satisfy $x_1, x_{m+1} \geq 0$ and $x_2, x_3, \ldots, x_m \geq t - 1$. Now, for each non-negative integer r, we let b_r denote the number of integer solutions to the equation

$$x_1 + x_2 + \cdots + x_{m+1} = r$$

that satisfy $x_1, x_{m+1} \geq 0$ and $x_2, x_3, \ldots, x_m \geq t - 1$. By (5.1.3), it follows that the generating function for (b_r) is given by

$$
\begin{aligned}
(x^{t-1} &+ x^t + x^{t+1} + \cdots)^{m-1}(1 + x + x^2 + \cdots)^2 \\
&= x^{(t-1)(m-1)}(1 + x + x^2 + \cdots)^{m+1} \\
&= x^{(t-1)(m-1)}(1 - x)^{-(m+1)}.
\end{aligned}
$$

Since $a_r = b_{r-m}$, it follows from Theorem 5.1.1 (iv) that the generating function for (a_r) is

$$x^{(t-1)(m-1)+m}(1-x)^{-(m+1)} = x^{t(m-1)+1}(1-x)^{-(m+1)}.$$

Now, by (5.1.6), we have

$$x^{t(m-1)+1}(1-x)^{-(m+1)} = x^{t(m-1)+1} \sum_{r=0}^{\infty} \binom{r+m+1-1}{r} x^r$$

$$= \sum_{r=0}^{\infty} \binom{r+m}{m} x^{r+t(m-1)+1}$$

$$= \sum_{r=t(m-1)+1}^{\infty} \binom{r-t(m-1)-1+m}{m} x^r$$

$$= \sum_{r=0}^{\infty} \binom{r-(m-1)(t-1)}{m} x^r,$$

where the last equality follows from the fact that $\binom{k}{m} = 0$ for all integers $k < m$. Thus, we have $a_r = \binom{r-(m-1)(t-1)}{m}$.

25. The number of integer solutions to the inequality

$$x_1 + x_2 + x_3 + x_4 \le r$$

with $3 \le x_1 \le 9$, $1 \le x_2 \le 10$, $x_3 \ge 2$ and $x_4 \ge 0$, is equal to the number of integer solutions to the equation

$$x_1 + x_2 + x_3 + x_4 + x_5 = r$$

with $3 \le x_1 \le 9$, $1 \le x_2 \le 10$, $x_3 \ge 2$ and $x_4, x_5 \ge 0$. Thus, it follows from (5.1.3) that the generating function for (a_r) is

$$(x^3 + x^4 + \cdots + x^9)(x + \cdots + x^{10})(x^2 + x^3 + \cdots)(1 + x + \cdots)^2$$
$$= x^6(1 + x + \cdots + x^6)(1 + x + \cdots + x^9)(1 + x + x^2 + \cdots)^3$$
$$= x^6(1 - x^7)(1 - x^{10})(1 - x)^{-5}.$$

By (5.1.6), we have

$$x^6(1 - x^7)(1 - x^{10})(1 - x)^{-5}$$

$$= (x^6 - x^{13} - x^{16} + x^{23}) \sum_{r=0}^{\infty} \binom{r+5-1}{r} x^r$$

$$= (x^6 - x^{13} - x^{16} + x^{23}) \sum_{r=0}^{\infty} \binom{r+4}{4} x^r.$$

Therefore, we have

$$a_{20} = \binom{20-6+4}{4} - \binom{20-13+4}{4} - \binom{20-16+4}{4}$$

$$= \binom{18}{4} - \binom{11}{4} - \binom{8}{4}.$$

26. It is clear that the inequalities hold when $n = 0$, so let us assume that $n > 0$. We first observe that $\binom{2\alpha}{2n}$ is the coefficient of x^{2n} in the expansion of $(1 + x)^{2\alpha}$. On the other hand, it follows from (5.1.1) that we have

$$(1 + x)^{2\alpha} = (1 + x)^\alpha (1 + x)^\alpha$$

$$= \sum_{i=0}^{\infty} \binom{\alpha}{i} x^i \sum_{j=0}^{\infty} \binom{\alpha}{j} x^j$$

$$= \sum_{i=0}^{\infty} \binom{\alpha}{i} x^i \sum_{k=i}^{\infty} \binom{\alpha}{k-i} x^{k-i} \quad \text{(by setting } k = i + j\text{)}$$

$$= \sum_{i=0}^{\infty} \sum_{k=i}^{\infty} \binom{\alpha}{i} \binom{\alpha}{k-i} x^k$$

$$= \sum_{k=0}^{\infty} \sum_{i=0}^{k} \binom{\alpha}{i} \binom{\alpha}{k-i} x^k.$$

This implies that the coefficient of x^{2n} in the expansion of $(1 + x)^{2\alpha}$ is equal to $\sum_{i=0}^{2n} \binom{\alpha}{i} \binom{\alpha}{2n-i}$, so we have

$$\binom{2\alpha}{2n} = \sum_{i=0}^{2n} \binom{\alpha}{i} \binom{\alpha}{2n-i} = \binom{\alpha}{n}^2 + 2 \sum_{i=n+1}^{2n} \binom{\alpha}{i} \binom{\alpha}{2n-i}.$$

Next, for each positive integer r, we let $a_r = \frac{r-1-\alpha}{r}$. Then it follows that we have $\binom{\alpha}{r} = (-1)^r \prod_{i=1}^{r} a_i$ for all non-negative integers r. Now, let us consider these cases:

Case 1: $-1 < \alpha < 0$. Then we have $a_r = \frac{r-1-\alpha}{r} > \frac{r-1}{r} > 0$ for all positive integers r. Moreover, we have

$$(r - 1 - \alpha)(r + 1) = r^2 - \alpha r - 1 - \alpha < r^2 - \alpha r = r(r - \alpha),$$

or equivalently,

$$a_r = \frac{r - 1 - \alpha}{r} < \frac{r - \alpha}{r + 1} = a_{r+1}$$

for all positive integers r. In particular, we have $a_i < a_j$ for all positive integers i, j satisfying $i < j$. Now, for all $i = n + 1, \ldots, 2n$ and $j = n + 1, n + 2, \ldots, i$, we have $j \geq n - i + j$, which implies that we have

$a_j \geq a_{n-i+j} > 0$. Thus, for all $i = n+1, \ldots, 2n$, we have

$$\binom{\alpha}{i}\binom{\alpha}{2n-i}$$

$$= (-1)^i \prod_{j=1}^{i} a_j \cdot (-1)^{2n-i} \prod_{j=1}^{2n-i} a_j$$

$$= \prod_{j=1}^{n} a_j \prod_{j=n+1}^{i} a_j \prod_{j=1}^{2n-i} a_j$$

$$\geq \prod_{j=1}^{n} a_j \prod_{j=n+1}^{i} a_{n-i+j} \prod_{j=1}^{2n-i} a_j$$

(since $a_j \geq a_{n-i+j} > 0$ for all $j = n+1, n+2, \ldots, i$)

$$= \prod_{j=1}^{n} a_j \prod_{k=2n+1-i}^{n} a_k \prod_{j=1}^{2n-i} a_j \quad \text{(by setting } k = n-i+j\text{)}$$

$$= \prod_{j=1}^{n} a_j \prod_{j=1}^{n} a_j$$

$$= (-1)^n \prod_{j=1}^{n} a_j (-1)^n \prod_{j=1}^{n} a_j$$

$$= \binom{\alpha}{n}^2.$$

Consequently, we have

$$\binom{2\alpha}{2n} = \binom{\alpha}{n}^2 + 2 \sum_{i=n+1}^{2n} \binom{\alpha}{i}\binom{\alpha}{2n-i}$$

$$\geq \binom{\alpha}{n}^2 + 2 \sum_{i=n+1}^{2n} \binom{\alpha}{n}^2$$

$$= (2n+1)\binom{\alpha}{n}^2.$$

Case 2: $\alpha < -1$. Then we have $a_r = \frac{r-1-\alpha}{r} > 1$ for all positive integers r. Moreover, we have

$$(r-1-\alpha)(r+1) = r^2 - \alpha r - 1 - \alpha > r^2 - \alpha r = r(r-\alpha),$$

or equivalently,

$$a_r = \frac{r-1-\alpha}{r} > \frac{r-\alpha}{r+1} = a_{r+1}.$$

for all positive integers r. In particular, we have $a_i > a_j$ for all positive integers i, j satisfying $i < j$. A similar calculation as in Case 1 would now show that $\binom{\alpha}{i}\binom{\alpha}{2n-i} \leq \binom{\alpha}{n}^2$ for all $i = n+1, \ldots, 2n$, and hence we have

$$\binom{2\alpha}{2n} = \binom{\alpha}{n}^2 + 2\sum_{i=n+1}^{2n}\binom{\alpha}{i}\binom{\alpha}{2n-i}$$

$$\leq \binom{\alpha}{n}^2 + 2\sum_{i=n+1}^{2n}\binom{\alpha}{n}^2$$

$$= (2n+1)\binom{\alpha}{n}^2.$$

27. (i) We have

$$B(x) = \sum_{k=0}^{\infty}\sum_{i=0}^{k}\binom{n}{i}x^k$$

$$= \sum_{i=0}^{\infty}\sum_{k=i}^{\infty}\binom{n}{i}x^k$$

$$= \sum_{i=0}^{\infty}\binom{n}{i}x^i\sum_{k=i}^{\infty}x^{k-i}$$

$$= \sum_{i=0}^{\infty}\binom{n}{i}x^i\sum_{j=0}^{\infty}x^j \quad \text{(by setting } j = k - i)$$

$$= \sum_{i=0}^{n}\binom{n}{i}x^i\sum_{j=0}^{\infty}x^j \quad \left(\text{since } \binom{n}{i} = 0 \text{ for all } i > n\right)$$

$$= (1+x)^n(1-x)^{-1} \quad \text{(by (5.1.3)).}$$

(ii) We first observe that for all $k = 0, 1, \ldots, n-1$, we have

$$\binom{n}{k+1} + \cdots + \binom{n}{n} = \binom{n}{n-k-1} + \cdots + \binom{n}{0} = b_{n-1-k}.$$

This implies that for all positive integers n, we have

$$a_{n-1}$$

$$= \sum_{k=0}^{n-1}\left[\binom{n}{0} + \cdots + \binom{n}{k}\right]\left[\binom{n}{k+1} + \cdots + \binom{n}{n}\right]$$

$$= \sum_{k=0}^{n-1}b_k b_{n-1-k}.$$

By Theorem 5.1.1 (iii), it follows that $B(x)^2 = (1+x)^{2n}(1-x)^{-2}$ is the generating function for (a_n). Now, by (5.1.5), we have

$$B(x)^2$$

$$= (1+x)^{2n}(1-x)^{-2}$$

$$= \sum_{r=0}^{2n} \binom{2n}{r} x^r \sum_{j=0}^{\infty} (j+1)x^j$$

$$= \sum_{r=0}^{\infty} \binom{2n}{r} x^r \sum_{j=0}^{\infty} (j+1)x^j \quad \text{(since } \binom{2n}{i} = 0 \text{ for all } i > 2n\text{)}$$

$$= \sum_{r=0}^{\infty} \sum_{j=0}^{\infty} \binom{2n}{r} (j+1)x^{r+j}$$

$$= \sum_{r=0}^{\infty} \sum_{k=r}^{\infty} \binom{2n}{r} (k-r+1)x^k \quad \text{(by setting } k = r+j\text{)}$$

$$= \sum_{k=0}^{\infty} \sum_{r=0}^{k} \binom{2n}{r} (k-r+1)x^k.$$

Thus, we have

$$a_{n-1} = \sum_{r=0}^{n-1} \binom{2n}{r} (n-1-r+1) = \sum_{r=0}^{n-1} \binom{2n}{r} (n-r).$$

(iii) We have

$$a_{n-1}$$

$$= \sum_{r=0}^{n-1} \binom{2n}{r} (n-r)$$

$$= n \sum_{r=0}^{n-1} \binom{2n}{r} - \sum_{r=0}^{n-1} r\binom{2n}{r}$$

$$= n \left(\sum_{r=0}^{n} \binom{2n}{r} - \binom{2n}{n} \right) - \left(\sum_{r=0}^{n} r\binom{2n}{r} - n\binom{2n}{n} \right)$$

$$= n \sum_{r=0}^{n} \binom{2n}{r} - \sum_{r=0}^{n} r\binom{2n}{r}$$

$$= n \left(2^{2n-1} + \frac{1}{2}\binom{2n}{n} \right) - n2^{2n-1} \quad \text{(by Problems 2.33 and 2.34)}$$

$$= \frac{n}{2}\binom{2n}{n}.$$

28. (i) Let us prove by induction on n that $\binom{n}{r}_m$ is equal to the number of integer solutions to the equation

$$x_1 + x_2 + \cdots + x_n = r$$

that satisfy $0 \le x_i \le m - 1$ for each $i = 1, 2, \ldots, n$, with the base case $n = 1$ being trivial. Suppose that the assertion holds for some positive integer $n = k \ge 1$. Let $(n_1, n_2, \ldots, n_{k+1})$ be an integer solution to the equation

$$x_1 + x_2 + \cdots + x_{k+1} = r$$

that satisfy $0 \le x_i \le m - 1$ for each $i = 1, 2, \ldots, k + 1$. Then $(n_1, n_2, \ldots, n_k)$ is an integer solution to the equation

$$x_1 + x_2 + \cdots + x_k = r - n_{k+1}$$

that satisfy $0 \le x_i \le m - 1$ for each $i = 1, 2, \ldots, k$. Now, by induction hypothesis, there are $\binom{k}{r-n_{k+1}}_m$ integer solutions to the equation

$$x_1 + x_2 + \cdots + x_k = r - n_{k+1}$$

that satisfy $0 \le x_i \le m - 1$ for each $i = 1, \ldots, k$. Since $n_{k+1} \in \{0, 1, \ldots, m - 1\}$, it follows from (AP) that the number of integer solutions to the equation to the equation

$$x_1 + x_2 + \cdots + x_{k+1} = r$$

that satisfy $0 \le x_i \le m - 1$ for each $i = 1, 2, \ldots, k + 1$, is equal to $\sum_{i=0}^{m-1} \binom{k}{r-i}_m = \binom{k+1}{r}_m$. This completes the induction step, and we are done.

(ii) By part (i), we have $\binom{n}{0}_m$ to be equal the number of integer solutions to the equation

$$x_1 + x_2 + \cdots + x_n = 0$$

that satisfy $0 \le x_i \le m - 1$ for each $i = 1, 2, \ldots, n$. Since the only integer solution to the equation

$$x_1 + x_2 + \cdots + x_n = 0$$

that satisfy $0 \le x_i \le m - 1$ for each $i = 1, 2, \ldots, n$ is $(0, 0, \ldots, 0)$, it follows that $\binom{n}{0}_m = 1$.

(iii) By part (i), we have $\binom{n}{1}_m$ to be equal the number of integer solutions to the equation

$$x_1 + x_2 + \cdots + x_n = 1$$

that satisfy $0 \leq x_i \leq m - 1$ for each $i = 1, 2, \ldots, n$. Since $m \geq 2$, it follows that $\binom{n}{1}_m$ is equal the number of non-negative integer solutions to the equation

$$x_1 + x_2 + \cdots + x_n = 1,$$

which is then equal to $\binom{n+1-1}{1} = n$. So we have $\binom{n}{1}_m = n$.

(iv) Let A denote the set of integer solutions to the equation

$$x_1 + x_2 + \cdots + x_n = r$$

that satisfy $0 \leq x_i \leq m - 1$ for each $i = 1, 2, \ldots, n$, and let B denote the set of integer solutions to the equation

$$x_1 + x_2 + \cdots + x_n = s$$

that satisfy $0 \leq x_i \leq m - 1$ for each $i = 1, 2, \ldots, n$. Let us define $f : A \to B$ by $f((k_1, k_2, \ldots, k_n)) = (m - 1 - k_1, m - 1 - k_2, \ldots, m - 1 - k_n)$ for all $(k_1, k_2, \ldots, k_n) \in A$. We claim that f is a well-defined map. Indeed, since $(k_1, k_2, \ldots, k_n) \in A$, it follows that we have $0 \leq k_i \leq m - 1$ for each $i = 1, 2, \ldots, n$, which would then imply that we have $0 \leq m - 1 - k_i \leq m - 1$ for each $i = 1, 2, \ldots, n$. Also, we have

$$\sum_{i=1}^{n}(m - 1 - k_i) = \sum_{i=1}^{n}(m - 1) - \sum_{i=1}^{n} k_i = n(m - 1) - r = s,$$

so this implies that $(m-1-k_1, m-1-k_2, \ldots, m-1-k_n) \in B$, and hence f is well-defined. Now, it is easy to see from the definition of f that f is a bijection. By (BP) and part (i), we have

$$\binom{n}{r}_m = |A| = |B| = \binom{n}{s}_m$$

as desired.

(v) By part (i), it follows that $\sum_{r=0}^{n(m-1)} \binom{n}{r}_m$ is equal to the number of integer solutions to the inequality

$$x_1 + x_2 + \cdots + x_n \leq n(m - 1)$$

that satisfy $0 \leq x_i \leq m - 1$ for each $i = 1, 2, \ldots, n$. However, any n-tuple $(k_1, k_2, \ldots, k_n)$ that satisfies $0 \leq k_i \leq m - 1$ for each $i = 1, 2, \ldots, n$ will also satisfy the inequality

$$0 \leq k_1 + k_2 + \cdots + k_n \leq n(m - 1),$$

so it follows that $\sum_{r=0}^{n(m-1)} \binom{n}{r}_m$ is equal to the number of n-tuples $(k_1, k_2, \ldots, k_n)$ that satisfy $0 \leq k_i \leq m - 1$ for each $i = 1, 2, \ldots, n$. As there are m^n n-tuples $(k_1, k_2, \ldots, k_n)$ that satisfy $0 \leq k_i \leq m - 1$, it follows that we must have $\sum_{r=0}^{n(m-1)} \binom{n}{r}_m = m^n$ as desired.

(vi) By part (i), we have $\binom{n}{r}_m$ to be equal to the number of integer solutions to the equation

$$x_1 + x_2 + \cdots + x_n = r$$

that satisfy $0 \leq x_i \leq m - 1$ for each $i = 1, 2, \ldots, n$, so it follows that the generating function for $\left(\binom{n}{r}_m\right)_{r=0}^{\infty}$ is $(1 + x + \cdots + x^{m-1})^n$.

(vii) By part (vi), we have

$$\sum_{r=0}^{\infty} \binom{n}{r}_m x^r = (1 + x + \cdots + x^{m-1})^n. \tag{5.1}$$

As $(1 + x + \cdots + x^{m-1})^n$ is a polynomial in x of degree $n(m-1)$, we must have $\sum_{r=0}^{\infty} \binom{n}{r}_m x^r$ to be a polynomial in x of degree $n(m - 1)$ as well, so it follows that we have $\binom{n}{r}_m = 0$ for all $r > n(m - 1)$, that is,

$$\sum_{r=0}^{\infty} \binom{n}{r}_m x^r = \sum_{r=0}^{n(m-1)} \binom{n}{r}_m x^r.$$

Now, by setting $x = -1$ on both sides of equation (5.1), we deduce that

$$\sum_{r=0}^{n(m-1)} \binom{n}{r}_m (-1)^r = \sum_{r=0}^{\infty} \binom{n}{r}_m (-1)^r$$

$$= (1 + -1 + \cdots + (-1)^{m-1})^n$$

$$= \begin{cases} 0 & \text{if } m \text{ is even} \\ 1 & \text{if } m \text{ is odd.} \end{cases}$$

(viii) By differentiating both sides of equation (5.1) with respect to x, we get

$$\sum_{r=1}^{n(m-1)} r \binom{n}{r}_m x^{r-1}$$

$$= \sum_{r=1}^{\infty} r \binom{n}{r}_m x^{r-1}$$

$$= n(1 + x + \cdots + x^{m-1})^{n-1}(1 + 2x + \cdots + (m-1)x^{m-2}) \quad (5.2)$$

when $m \geq 2$, and

$$\sum_{r=1}^{n(m-1)} r \binom{n}{r}_m x^{r-1} = \sum_{r=1}^{\infty} r \binom{n}{r}_m x^{r-1} = 0 \quad (5.3)$$

when $m = 1$. By setting $x = 1$ on both sides of equation (5.2), we deduce that

$$\sum_{r=1}^{n(m-1)} r \binom{n}{r}_m = n(\underbrace{1 + 1 + \cdots + 1}_{m})^{n-1}(1 + 2 + \cdots + (m-1))$$

$$= nm^{n-1} \cdot \frac{m(m-1)}{2}$$

$$= \frac{n(m-1)m^n}{2}$$

when $m \geq 2$. Next, by setting $x = 1$ on both sides of equation (5.3), we deduce that

$$\sum_{r=1}^{n(m-1)} r \binom{n}{r}_m = 0 = \frac{n(m-1)m^n}{2}$$

when $m = 1$. So we have

$$\sum_{r=1}^{n(m-1)} r \binom{n}{r}_m x^{r-1} = \frac{n(m-1)m^n}{2}$$

as desired.

(ix) By setting $x = -1$ on both sides of equation (5.2), we deduce that

$$\sum_{r=1}^{n(m-1)} (-1)^{r-1} r \binom{n}{r}_m$$

$$= n(1 - 1 + \cdots + (-1)^{m-1})^{n-1}(1 - 2 + \cdots + (-1)^{m-2}(m-1))$$

$$(5.4)$$

when $m \geq 2$, and by setting $x = -1$ on both sides of equation (5.3), we deduce that

$$\sum_{r=1}^{n(m-1)} r \binom{n}{r}_m (-1)^{r-1} = 0 = \frac{n(1-m)}{2}$$

when $m = 1$. Now, if m is even, then we must have $m \geq 2$ and

$$(1 - 1 + \cdots + (-1)^{m-1})^{n-1} = 0,$$

so it follows from equation (5.4) that

$$\sum_{r=1}^{n(m-1)} (-1)^{r-1} r \binom{n}{r}_m$$
$$= n(1 - 1 + \cdots + (-1)^{m-1})^{n-1}(1 - 2 + \cdots + (-1)^{m-2}(m-1))$$
$$= 0$$

when m is even. Else, if m is odd and $m \geq 2$, that is, $m = 2k+1$ for some positive integer k, then we have $(1-1+\cdots+(-1)^{m-1})^{n-1} = 1$ and

$$1 - 2 + \cdots + (-1)^{m-2}(m-1)$$
$$= \sum_{i=1}^{m-1} (-1)^i i$$
$$= \sum_{i=1}^{2k} (-1)^i i$$
$$= \sum_{i=1}^{k} (-1)^{2i-1-1}(2i-1) + \sum_{i=1}^{k} (-1)^{2i-1} 2i$$
$$= \sum_{i=1}^{k} (2i-1) - \sum_{i=1}^{k} 2i$$
$$= -k$$
$$= \frac{1-m}{2}.$$

Thus, it follows from equation (8) that

$$\sum_{r=1}^{n(m-1)} (-1)^{r-1} r \binom{n}{r}_m$$
$$= n(1 - 1 + \cdots + (-1)^{m-1})^{n-1}(1 - 2 + \cdots + (-1)^{m-2}(m-1))$$
$$= \frac{n(1-m)}{2}$$

when m is odd and $m \geq 2$. So we conclude that

$$\sum_{r=1}^{n(m-1)} (-1)^{r-1} r \binom{n}{r}_m = \begin{cases} 0 & \text{if } m \text{ is even} \\ \frac{n(1-m)}{2} & \text{if } m \text{ is odd.} \end{cases}$$

(x) For each non-negative integer r, we let $c_r = \sum_{i=0}^{r} \binom{p}{i}_m \binom{q}{r-i}_m$. By part (vi), the generating functions for $\left(\binom{p}{r}_m\right)_{r=0}^{\infty}$ and $\left(\binom{q}{r}_m\right)_{r=0}^{\infty}$ are $(1 + x + \cdots + x^{m-1})^p$ and $(1 + x + \cdots + x^{m-1})^q$ respectively. Hence, it follows from Theorem 5.1.1 (ii) that the generating function for (c_r) is $(1 + x + \cdots + x^{m-1})^p (1 + x + \cdots + x^{m-1})^q = (1 + x + \cdots + x^{m-1})^{p+q}$, and therefore by part (vi) again, we have

$$\sum_{i=0}^{r} \binom{p}{i}_m \binom{q}{r-i}_m = c_r = \binom{p+q}{r}_m.$$

(xi) We recall from part (vi) that the generating function for $\left(\binom{n}{r}_m\right)_{r=0}^{\infty}$ is

$$(1 + x + \cdots + x^{m-1})^n.$$

By (5.1.6), we have

$$(1 + x + \cdots + x^{m-1})^n$$
$$= (1 - x^m)^n (1 - x)^{-n}$$
$$= \sum_{i=0}^{n} \binom{n}{i} (-x^m)^i \sum_{r=0}^{\infty} \binom{r+n-1}{r} x^r$$
$$= \sum_{i=0}^{n} \binom{n}{i} (-x^m)^i \sum_{r=0}^{\infty} \binom{r+n-1}{r} x^r$$
$$= \sum_{i=0}^{n} \sum_{r=0}^{\infty} (-1)^i \binom{n}{i} \binom{r+n-1}{n-1} x^{r+mi}$$
$$= \sum_{i=0}^{n} \sum_{k=mi}^{\infty} (-1)^i \binom{n}{i} \binom{n-1+k-mi}{n-1} x^k$$
(by setting $k = r + mi$)
$$= \sum_{i=0}^{n} \sum_{k=0}^{\infty} (-1)^i \binom{n}{i} \binom{n-1+k-mi}{n-1} x^k$$
(since $\binom{i}{n-1} = 0$ for all $i < n-1$)
$$= \sum_{k=0}^{\infty} \sum_{i=0}^{n} (-1)^i \binom{n}{i} \binom{n-1+k-mi}{n-1} x^k.$$

Hence, we conclude that

$$\binom{n}{r}_m = \sum_{i=0}^{n}(-1)^i \binom{n}{i}\binom{n-1+r-mi}{n-1}.$$

29. We first observe that $2^{r-2n}(1+x)^{2n-r} = \left(\frac{1+x}{2}\right)^{2n-r}$ is the generating function for the sequence $\left(2^{r-2n}\binom{2n-r}{m}\right)_{m=0}^{\infty}$. By Theorem 5.1.1 (i), it follows that $\sum_{r=0}^{n}\left(\frac{1+x}{2}\right)^{2n-r}$ is the generating function for the sequence $\left(\sum_{r=0}^{n}2^{r-2n}\binom{2n-r}{m}\right)_{m=0}^{\infty}$. Now, we have

$$\sum_{r=0}^{n}\left(\frac{1+x}{2}\right)^{2n-r}$$

$$= \sum_{k=n}^{2n}\left(\frac{1+x}{2}\right)^{k} \quad \text{(by setting } k = 2n-r\text{)}$$

$$= \left(\frac{1+x}{2}\right)^{n}\sum_{k=0}^{n}\left(\frac{1+x}{2}\right)^{k}$$

$$= \left(\frac{1+x}{2}\right)^{n}\left[1-\left(\frac{1+x}{2}\right)^{n+1}\right]\left[1-\left(\frac{1+x}{2}\right)\right]^{-1}$$

$$= \left[\left(\frac{1+x}{2}\right)^{n}-\left(\frac{1+x}{2}\right)^{2n+1}\right]\left(\frac{1}{2}-\frac{x}{2}\right)^{-1}$$

$$= 2(2^{-n}(1+x)^{n} - 2^{-2n-1}(1+x)^{2n+1})(1-x)^{-1}$$

$$= 2\left(2^{-n}\sum_{i=0}^{\infty}\binom{n}{i}x^i - 2^{-2n-1}\sum_{i=0}^{\infty}\binom{2n+1}{i}x^i\right)\sum_{j=0}^{\infty}x^j \quad \text{(by (5.1.3))}$$

$$= 2\sum_{i=0}^{\infty}\left(2^{-n}\binom{n}{i} - 2^{-2n-1}\binom{2n+1}{i}\right)x^i\sum_{j=0}^{\infty}x^j$$

$$= \sum_{i=0}^{\infty}\sum_{j=0}^{\infty}\left(2^{-n+1}\binom{n}{i} - 2^{-2n}\binom{2n+1}{i}\right)x^{i+j}$$

$$= \sum_{i=0}^{\infty}\sum_{t=i}^{\infty}\left(2^{-n+1}\binom{n}{i} - 2^{-2n}\binom{2n+1}{i}\right)x^{t} \quad \text{(by setting } t = i+j\text{)}$$

$$= \sum_{t=0}^{\infty}\sum_{i=0}^{t}\left(2^{-n+1}\binom{n}{i} - 2^{-2n}\binom{2n+1}{i}\right)x^{t}.$$

Hence, we have

$$S_n = \sum_{r=0}^{n} 2^{r-2n} \binom{2n-r}{n}$$

$$= \sum_{i=0}^{n} \left(2^{-n+1} \binom{n}{i} - 2^{-2n} \binom{2n+1}{i} \right)$$

$$= 2^{-n+1} \sum_{i=0}^{n} \binom{n}{i} - 2^{-2n} \sum_{i=0}^{n} \binom{2n+1}{i}$$

$$= 2^{-n+1} 2^n - 2^{-2n} 2^{2n} \quad \text{(by (2.3.1) and Problem 2.32)}$$

$$= 1.$$

30. The exponential generating function for (a_r) is

$$\sum_{r=0}^{\infty} \frac{a_r}{r!} x^r$$

$$= 1 + \sum_{r=1}^{\infty} \frac{4 \cdot 7 \cdots (3r+1)}{r!} x^r$$

$$= 1 + \sum_{r=1}^{\infty} (-1)^{2r} \frac{\frac{4}{3} \cdot \frac{7}{3} \cdot \frac{10}{3} \cdots \cdot \frac{(3r+1)}{3}}{r!} (3x)^r$$

$$= 1 + \sum_{r=1}^{\infty} \frac{\left(-\frac{4}{3}\right) \cdot \left(-\frac{7}{3}\right) \cdot \left(-\frac{10}{3}\right) \cdots \cdot \left(-\frac{3r+1}{3}\right)}{r!} (-3x)^r$$

$$= 1 + \sum_{r=1}^{\infty} \frac{\left(-\frac{4}{3}\right) \cdot \left(-\frac{4}{3}-1\right) \cdot \left(-\frac{4}{3}-2\right) \cdots \cdot \left(-\frac{4}{3}-r+1\right)}{r!} (-3x)^r$$

$$= 1 + \sum_{r=1}^{\infty} \binom{-\frac{4}{3}}{r} (-3x)^r$$

$$= \sum_{r=0}^{\infty} \binom{-\frac{4}{3}}{r} (-3x)^r$$

$$= (1-3x)^{-\frac{4}{3}},$$

where the last equality follows from (5.1.1).

31. For each positive integer n, we let a_n denote the number of ways to colour the n squares of a $1 \times n$ chessboards using the colours blue, red and white, such that each square is coloured by a colour and an even number of squares are to be coloured red. Then the exponential generating function for (a_n) is given by

$$(e^x)^2 \left(\frac{e^x + e^{-x}}{2} \right) = \frac{e^{3x} + e^x}{2} = \sum_{n=0}^{\infty} \frac{3^n + 1}{2n!} x^n.$$

Therefore, the desired answer is equal to $a_n = \frac{3^n+1}{2}$.

32. For each positive integer n, we let a_n denote the number of n-digit quaternary sequences that contain an odd number of 0's, an even number of 1's and at least one 3. Then the exponential generating function for (a_n) is given by

$$\left(\frac{e^x + e^{-x}}{2}\right)\left(\frac{e^x - e^{-x}}{2}\right)e^x(e^x - 1) = \frac{e^{4x} - e^{3x} - 1 + e^{-x}}{4}$$

$$= -\frac{1}{4} + \sum_{n=0}^{\infty} \frac{4^n - 3^n + (-1)^n}{4n!}x^n.$$

Therefore, the desired answer is equal to $a_n = \frac{4^n - 3^n + (-1)^n}{4}$.

33. (i) For each positive integer n, we let a_n denote the number of words of length n formed by the symbols $\alpha, \beta, \gamma, \delta, \epsilon, \lambda$, in which the total number of α's and β's is even. Then either the number of α's and the number of β's are both even, or the number of α's and the number of β's are both odd. Thus, it follows that the exponential generating function for (a_n) is given by

$$(e^x)^4\left[\left(\frac{e^x + e^{-x}}{2}\right)^2 + \left(\frac{e^x - e^{-x}}{2}\right)^2\right] = \frac{e^{6x} + e^{2x}}{2}$$

$$= \sum_{n=0}^{\infty} \frac{6^n + 2^n}{2n!}x^n.$$

Therefore, the desired answer is equal to $a_n = \frac{6^n + 2^n}{2}$.

(ii) For each positive integer n, we let a_n denote the number of words of length n formed by the symbols $\alpha, \beta, \gamma, \delta, \epsilon, \lambda$, in which the total number of α's and β's is odd. Then either the number of α's is even and the number of β's is odd, or the number of α's is odd and the number of β's is even. Thus, it follows that the exponential generating function for (a_n) is given by

$$2(e^x)^4\left(\frac{e^x + e^{-x}}{2}\right)\left(\frac{e^x - e^{-x}}{2}\right) = \frac{e^{6x} - e^{2x}}{2} = \sum_{n=0}^{\infty} \frac{6^n - 2^n}{2n!}x^n.$$

Therefore, the desired answer is equal to $a_n = \frac{6^n - 2^n}{2}$.

34. For each non-negative integer r, we let a_r denote the number of ways of distributing r distinct objects into 5 distinct boxes such that each of the boxes $1, 3$ and 5 must hold an odd number of objects, while each of the remaining boxes must hold an even number of objects. By a similar

argument as in Example 5.4.9, we see that the exponential generating function for (a_r) is given by

$$\left(\frac{e^x - e^{-x}}{2}\right)^3 \left(\frac{e^x + e^{-x}}{2}\right)^2$$

$$= \frac{e^{5x} - e^{3x} - 2e^x + 2e^{-x} + e^{-3x} - e^{-5x}}{32}$$

$$= \sum_{r=0}^{\infty} \frac{5^r - 3^r - 2 + 2(-1)^r + (-3)^r - (-5)^r}{32r!} x^r.$$

Therefore, the desired answer is equal to

$$a_r = \frac{5^r - 3^r - 2 + 2(-1)^r + (-3)^r - (-5)^r}{32}.$$

35. (i) Firstly, we observe that for all non-negative integers n, k satisfying $k \le n$, we have

$$\binom{z}{2k}\binom{z-2k}{n-k} = \prod_{i=1}^{2k} \frac{z-i+1}{i} \prod_{j=1}^{n-k} \frac{z-2k-j+1}{j}$$

$$= \prod_{i=1}^{n+k} \frac{z-i+1}{i} \frac{(n+k)!}{(2k)!(n-k)!}$$

$$= \binom{z}{n+k}\binom{n+k}{n-k}.$$

Next, we observe that $(1+x)^{2z} = [1 + (x^2 + 2x)]^z$ is the generating function for $\left(\binom{2z}{r}\right)$. By (5.1.1), we have

$$[1 + (x^2 + 2x)]^z$$

$$= \sum_{j=0}^{\infty} \binom{z}{j}(x^2 + 2x)^j$$

$$= \sum_{j=0}^{\infty} \binom{z}{j} x^j (x+2)^j$$

$$= \sum_{j=0}^{\infty} \binom{z}{j} x^j \sum_{i=0}^{j} \binom{j}{i} 2^{j-i} x^i$$

$$= \sum_{j=0}^{\infty} \sum_{i=0}^{\infty} \binom{z}{j}\binom{j}{i} 2^{j-i} x^{j+i} \quad \left(\text{since } \binom{j}{i} = 0 \text{ for all } i > j\right)$$

$$= \sum_{j=0}^{\infty} \sum_{r=j}^{\infty} \binom{z}{j}\binom{j}{r-j} 2^{2j-r} x^r \quad (\text{by setting } r = j+i)$$

$$= \sum_{r=0}^{\infty} \sum_{j=0}^{r} \binom{z}{j}\binom{j}{r-j} 2^{2j-r} x^r.$$

Thus, we have

$$\binom{2z}{2n} = \sum_{j=0}^{2n} \binom{z}{j}\binom{j}{2n-j} 2^{2j-2n}.$$

Finally, we have

$$\sum_{j=0}^{2n} \binom{z}{j}\binom{j}{2n-j} 2^{2j-2n}$$

$$= \sum_{j=n}^{2n} \binom{z}{j}\binom{j}{2n-j} 2^{2j-2n} \quad \text{(since } j < 2n-j \text{ for all } j < n)$$

$$= \sum_{k=0}^{n} \binom{z}{n+k}\binom{n+k}{n-k} 2^{2k} \quad \text{(by setting } k = j-n)$$

$$= \sum_{k=0}^{n} \binom{z}{2k}\binom{z-2k}{n-k} 2^{2k}.$$

Therefore, we have

$$\sum_{k=0}^{n} \binom{z}{2k}\binom{z-2k}{n-k} 2^{2k} = \sum_{j=0}^{2n} \binom{z}{j}\binom{j}{2n-j} 2^{2j-2n} = \binom{2z}{2n}$$

as desired.

(ii) Firstly, we observe that for all non-negative integers n, k satisfying $k \le n$, we have

$$\binom{z+1}{2k+1}\binom{z-2k}{n-k} = \prod_{i=1}^{2k+1} \frac{z-i+2}{i} \prod_{j=1}^{n-k} \frac{z-2k-j+1}{j}$$

$$= \prod_{i=1}^{n+k+1} \frac{z-i+2}{i} \frac{(n+k+1)!}{(2k+1)!(n-k)!}$$

$$= \binom{z+1}{n+k+1}\binom{n+k+1}{n-k}.$$

Next, we observe that $(1+x)^{2z+2} = [1+(x^2+2x)]^{z+1}$ is the generating function for $\left(\binom{2z+2}{r}\right)$. By (5.1.1), we have

$$[1 + (x^2 + 2x)]^{z+1}$$

$$= \sum_{j=0}^{\infty} \binom{z+1}{j} x^j (x+2)^j$$

$$= \sum_{j=0}^{\infty} \binom{z+1}{j} x^j \sum_{i=0}^{j} \binom{j}{i} 2^{j-i} x^i$$

$$= \sum_{j=0}^{\infty} \sum_{i=0}^{\infty} \binom{z+1}{j} \binom{j}{i} 2^{j-i} x^{j+i} \quad \left(\text{since } \binom{j}{i} = 0 \text{ for all } i > j\right)$$

$$= \sum_{j=0}^{\infty} \sum_{r=j}^{\infty} \binom{z+1}{j} \binom{j}{r-j} 2^{2j-r} x^r \quad (\text{by setting } r = j+i)$$

$$= \sum_{r=0}^{\infty} \sum_{j=0}^{r} \binom{z+1}{j} \binom{j}{r-j} 2^{2j-r} x^r.$$

Thus, we have

$$\binom{2z+2}{2n+1} = \sum_{j=0}^{2n+1} \binom{z+1}{j} \binom{j}{2n+1-j} 2^{2j-2n-1}.$$

Finally, we have

$$\sum_{j=0}^{2n+1} \binom{z+1}{j} \binom{j}{2n+1-j} 2^{2j-2n-1}$$

$$= \sum_{j=n+1}^{2n+1} \binom{z+1}{j} \binom{j}{2n+1-j} 2^{2j-2n-1}$$

$$(\text{since } j < 2n+1-j \text{ for all } j < n+1)$$

$$= \sum_{k=0}^{n} \binom{z+1}{n+k+1} \binom{n+k+1}{n-k} 2^{2k+1}$$

$$(\text{by setting } k = j - n - 1)$$

$$= \sum_{k=0}^{n} \binom{z+1}{2k+1} \binom{z-2k}{n-k} 2^{2k+1}.$$

Therefore, we have

$$\sum_{k=0}^{n} \binom{z+1}{2k+1} \binom{z-2k}{n-k} 2^{2k+1}$$

$$= \sum_{j=0}^{2n+1} \binom{z+1}{j}\binom{j}{2n+1-j}2^{2j-2n-1}$$

$$= \binom{2z+2}{2n+1}$$

as desired.

36. We have

$$\sum_{r=1}^{n}\sum_{k=0}^{r}(-1)^{k+1}\frac{k}{r}\binom{r}{k}k^{n-1}$$

$$= \sum_{r=1}^{n}\sum_{k=1}^{r}(-1)^{k+1}\frac{k}{r}\binom{r}{k}k^{n-1}$$

$$= \sum_{r=1}^{n}\sum_{k=1}^{r}(-1)^{k+1}\binom{r-1}{k-1}k^{n-1} \quad \text{(by (2.1.2))}$$

$$= \sum_{k=1}^{n}\sum_{r=k}^{n}(-1)^{k+1}\binom{r-1}{k-1}k^{n-1}$$

$$= \sum_{k=1}^{n}(-1)^{k+1}k^{n-1}\sum_{r=k}^{n}\binom{r-1}{k-1}$$

$$= \sum_{k=1}^{n}(-1)^{k+1}k^{n-1}\binom{n}{k} \quad \text{(by (2.5.1))}$$

$$= \sum_{k=0}^{n}(-1)^{-k-1}\binom{n}{k}k^{n-1}$$

$$= (-1)^{-n-1}\sum_{k=0}^{n}(-1)^{n-k}\binom{n}{k}k^{n-1}$$

$$= (-1)^{-n-1}n!S(n-1,n) \quad \text{(by Problem 4.50(i))}$$

$$= 0.$$

37. Let T denote the set of all n-tuples $(k_1,\ldots,k_n)$ of non-negative integers such that $\sum_{i=1}^{n} k_i = r$. By the Multinomial Theorem, we have

$$(x_1 + x_2 + \cdots + x_n)^r = \sum_{(k_1,\ldots,k_n)\in T}\binom{r}{k_1,\ldots,k_n}x_1^{k_1}x_2^{k_2}\cdots x_n^{k_n}$$

$$= r!\sum_{(k_1,\ldots,k_n)\in T}\frac{1}{k_1!k_2!\cdots k_n!}x_1^{k_1}x_2^{k_2}\cdots x_n^{k_n}.$$

$$(5.5)$$

By setting $x_i = x^i$ for each $i = 1, 2, \ldots, n$, it follows from equation (5.5) that we have

$$(x + x^2 + \cdots + x^n)^r = r! \sum_{(k_1, \ldots, k_n) \in T} \frac{1}{k_1! k_2! \cdots k_n!} x^{k_1 + 2k_2 + \cdots + nk_n}.$$

Now, by (5.1.6), we have

$$
\begin{aligned}
(x_1 + x_2 + \cdots + x_n)^r \\
= x^r (1 + x + \cdots + x^{n-1})^r \\
= x^r (1 - x^n)^r (1 - x)^{-r} \\
= x^r \left[\sum_{i=0}^{r} (-1)^i \binom{r}{i} x^{ni} \right] \left[\sum_{j=0}^{\infty} \binom{r+j-1}{j} x^j \right] \\
= \sum_{i=0}^{r} \sum_{j=0}^{\infty} (-1)^i \binom{r}{i} \binom{r+j-1}{j} x^{ni+j+r}.
\end{aligned}
$$

Thus, we have

$$
\begin{aligned}
r! \sum_{(k_1, \ldots, k_n) \in T} \frac{1}{k_1! k_2! \cdots k_n!} x^{k_1 + 2k_2 + \cdots + nk_n} \\
= \sum_{i=0}^{r} \sum_{j=0}^{\infty} (-1)^i \binom{r}{i} \binom{r+j-1}{j} x^{ni+j+r}.
\end{aligned}
\tag{5.6}
$$

Now, let us compare the coefficient of x^n on both sides of equation (5.6). Since $r > 0$, it follows that the only pair (i, j) of non-negative integers that satisfy $ni + j + r = n$ is $(0, n - r)$. We observe that in this case, we have

$$(-1)^i \binom{r}{i} \binom{r+j-1}{j} = (-1)^0 \binom{r}{0} \binom{n-1}{n-r} = \binom{n-1}{r-1}.$$

Therefore, we have

$$\sum_{\substack{(k_1, \ldots, k_n) \in T \\ k_1 + 2k_2 + \cdots + nk_n = n}} \frac{1}{k_1! k_2! \cdots k_n!} = \frac{1}{r!} \binom{n-1}{r-1}$$

as desired.

38. (i) The problem is equivalent to finding the number of 18-digit quaternary sequences that contain at least one 0, one 1, one 2 and one 3. For each non-negative integer r, we let a_r denote the number of r-digit quaternary sequences that contain at least one 0, one

1, one 2 and one 3. Then the exponential generating function for (a_r) is given by

$$(e^x - 1)^4 = e^{4x} - 4e^{3x} + 6e^{2x} - 4e^x + 1$$

$$= 1 + \sum_{r=0}^{\infty} \frac{4^r - 4 \cdot 3^r + 6 \cdot 2^r - 4}{r!} x^r.$$

Therefore, the number of ways to assign ten female workers and eight male workers to work in one of four different departments of a company, such that each department gets at least one worker, is equal to

$$a_{18} = 4^{18} - 4 \cdot 3^{18} + 6 \cdot 2^{18} - 4.$$

(ii) The problem is equivalent to finding the number of pairs (a, b), where a is a 10-digit quaternary sequence and 8-digit quaternary sequence, such that a contains at least one 0, one 1, one 2 and one 3. By (MP), we have 4^8 choices for b, and by part (i), the number of choices for a is equal to

$$a_{10} = 4^{10} - 4 \cdot 3^{10} + 6 \cdot 2^{10} - 4.$$

By (MP), the number of ways to assign ten female workers and eight male workers to work in one of four different departments of a company, such that each department gets at least one female worker, is equal to

$$4^8(4^{10} - 4 \cdot 3^{10} + 6 \cdot 2^{10} - 4).$$

(iii) The problem is equivalent to finding the number of pairs (a, b), where a is a 10-digit quaternary sequence and 8-digit quaternary sequence, such that a and b both contain at least one 0, one 1, one 2 and one 3. By part (i), the number of choices for a is equal to

$$a_{10} = 4^{10} - 4 \cdot 3^{10} + 6 \cdot 2^{10} - 4,$$

and the number of choices for b is equal to

$$a_8 = 4^8 - 4 \cdot 3^8 + 6 \cdot 2^8 - 4.$$

By (MP), the number of ways to assign ten female workers and eight male workers to work in one of four different departments of a company, such that each department gets at least one female worker and at least one male worker, is equal to

$$(4^{10} - 4 \cdot 3^{10} + 6 \cdot 2^{10} - 4) \cdot (4^8 - 4 \cdot 3^8 + 6 \cdot 2^8 - 4).$$

39. For each non-negative integer r, we let a_r denote the number of r-permutations of multi-set
$$\{\infty \cdot \alpha, \infty \cdot \beta, \infty \cdot \gamma, \infty \cdot \lambda\}$$
in which the number of α's is odd and the number of λ's is even. The exponential generating function for (a_r) is given by
$$(e^x)^2 \left(\frac{e^x - e^{-x}}{2}\right)\left(\frac{e^x + e^{-x}}{2}\right) = \frac{e^{4x} - 1}{4}$$
$$= -\frac{1}{4} + \sum_{r=0} \frac{4^r}{4 r!} x^r$$
$$= \sum_{r=1} \frac{4^{r-1}}{r!} x^r.$$

Therefore, the desired answer is given by
$$a_r = \begin{cases} 0 & \text{if } r = 0, \\ 4^{r-1} & \text{if } r \geq 1. \end{cases}$$

40. We first note that a_r is equal to the number of r-digit n-ary sequences that contains all of the digits $1, 2, \ldots, n$. Thus it follows that the exponential generating function for (a_r) is $(e^x - 1)^n$, that is,
$$\sum_{r=0}^{\infty} n! S(r, n) \frac{x^r}{r!} = (e^x - 1)^n.$$

Now, we have
$$\sum_{r=0}^{\infty} \left(\sum_{m=0}^{\infty} (-1)^m m! S(r, m+1)\right) \frac{x^r}{r!}$$
$$= \sum_{m=0}^{\infty} \frac{(-1)^m}{m+1}\left(\sum_{r=0}^{\infty} (m+1)! S(r, m+1)\frac{x^r}{r!}\right)$$
$$= \sum_{m=0}^{\infty} \frac{(-1)^m}{m+1}(e^x - 1)^{m+1}$$
$$= \sum_{m=1}^{\infty} \frac{(-1)^{m-1}}{m}(e^x - 1)^m$$
$$= \sum_{m=1}^{\infty} \frac{(-1)^{m+1}}{m}(e^x - 1)^m$$
$$= \log(1 + (e^x - 1))$$
$$= \log e^x$$
$$= x. \tag{5.7}$$

By comparing the coefficient of x^r on both sides of equation (5.7), we deduce that

$$\sum_{m=0}^{\infty} (-1)^m m! S(r, m+1) = 0$$

for all non-negative integers $r \geq 2$ as desired.

41. By Problem 5.40, we have

$$A_n(x) = \sum_{r=0}^{\infty} S(r, n) \frac{x^r}{r!} = \sum_{r=0}^{\infty} \frac{1}{n!} F(r, n) \frac{x^r}{r!} = \frac{1}{n!} (e^x - 1)^n.$$

Thus, for all positive integers $n \geq 2$, we have

$$\frac{d}{dx} A_n(x) = \frac{1}{(n-1)!} e^x (e^x - 1)^{n-1}$$

$$= \frac{1}{(n-1)!} (e^x - 1)^n + \frac{1}{(n-1)!} (e^x - 1)^{n-1}$$

$$= n A_n(x) + A_{n-1}(x).$$

42. The exponential generating function for (B_r) is given by

$$\sum_{r=0}^{\infty} B_r \frac{x^r}{r!} = 1 + \sum_{r=1}^{\infty} B_r \frac{x^r}{r!}$$

$$= 1 + \sum_{r=1}^{\infty} \sum_{k=1}^{r} S(r, k) \frac{x^r}{r!}$$

$$= 1 + \sum_{k=1}^{\infty} \sum_{r=k}^{\infty} S(r, k) \frac{x^r}{r!}$$

$$= 1 + \sum_{k=1}^{\infty} \sum_{r=0}^{\infty} S(r, k) \frac{x^r}{r!} \quad \text{(since } S(r, k) = 0 \text{ for all } r < k)$$

$$= 1 + \sum_{k=1}^{\infty} \frac{1}{k!} (e^x - 1)^k \quad \text{(by Problem 5.41)}$$

$$= \sum_{k=0}^{\infty} \frac{1}{k!} (e^x - 1)^k$$

$$= e^{e^x - 1}.$$

43. (a) Let us fix a positive integer n, and for each non-negative integer r, we let b_r denote the number of ways of distributing r distinct objects into n distinct boxes such that the objects in each box are ordered. Then b_r is equal to the number of ways of arranging $n - 1$

1's and $2, 3, \ldots, r + 1$ in a row. Firstly, there are $\binom{n+r-1}{r}$ ways to arrange r 0's and $n - 1$ 1's in a row. Next, there are $r!$ ways to place the r integers $2, 3, \ldots, r + 1$ into the slots occupied by the 0's. By (MP), we have $b_r = r!\binom{n+r-1}{r} = n(n+1) \cdots (n+r-1)$.

(b) (i) By part (a), we have $b_r = n^{(r)}$ for all non-negative integers r. Now, for each non-negative integer $i \leq r$, there are $\binom{r}{i}$ ways to choose i objects out of r distinct objects, and there are b_r ways to distribute these i objects into n distinct boxes such that the objects in each box are ordered. By (AP) and (MP), we have

$$a_r = \sum_{i=0}^{r} \binom{r}{i} b_i = \sum_{i=0}^{r} \binom{r}{i} n^{(i)}.$$

(ii) Since $n^{(i)} = b_i = i!\binom{n+i-1}{i}$, it follows that the exponential generating function for (a_r) is

$$\sum_{r=0}^{\infty} \frac{a_r}{r!} x^r = \sum_{r=0}^{\infty} \sum_{i=0}^{r} \frac{1}{r!} \binom{r}{i} n^{(i)} x^r$$

$$= \sum_{r=0}^{\infty} \sum_{i=0}^{r} \frac{i!}{r!} \binom{r}{i} \binom{n+i-1}{i} x^r$$

$$= \sum_{i=0}^{\infty} \sum_{r=i}^{\infty} \frac{1}{(r-i)!} \binom{n+i-1}{i} x^r$$

$$= \sum_{i=0}^{\infty} \binom{n+i-1}{i} x^i \sum_{r=i}^{\infty} \frac{1}{(r-i)!} x^{r-i}$$

$$= \sum_{i=0}^{\infty} \binom{n+i-1}{i} x^i \sum_{k=0}^{\infty} \frac{1}{k!} x^k$$

(by setting $k = r - i$)

$$= e^x \sum_{i=0}^{\infty} \binom{n+i-1}{i} x^i$$

$$= e^x (1 - x)^{-n} \quad \text{(by (5.1.6))}.$$

44. (i) By (5.1.3), the generating function for (a_r) is

$$(1 + x + x^2 + \cdots)^4 = (1 - x)^{-4}.$$

(ii) By (5.1.3), the generating function for (a_r) is

$$(x + x^2 + x^3 + \cdots)^4 = x^4 (1 + x + x^2 + \cdots)^4 = x^4 (1 - x)^{-4}.$$

(iii) We first note that a_r is equal to the number of partitions of r into 4 parts. By Theorem 5.3.5, a_r is equal to the number of partitions of r into parts the largest size of which is 4. By (5.1.3), the generating function for (a_r) is

$$(1 + x + x^2 + \cdots)(1 + x^2 + \cdots)(1 + x^3 + \cdots)(x^4 + x^8 + \cdots)$$
$$= x^4[(1 - x)(1 - x^2)(1 - x^3)(1 - x^4)]^{-1}.$$

(iv) We first note that a_r is equal to the number of partitions of r into at most 4 parts. By the Corollary to Theorem 5.3.5, a_r is equal to the number of partitions of r into parts with sizes not exceeding 4. Hence the generating function for (a_r) is

$$\prod_{i=1}^{4}(1 - x^i)^{-1} = [(1 - x)(1 - x^2)(1 - x^3)(1 - x^4)]^{-1}.$$

45. For each positive integer n, we let a_n denote the number of partitions of n into parts where no even part occurs more than once, and b_n denote the number of partitions of n in which parts of each size occur at most three times. Then it follows from (5.1.3) that the generating function $A(x)$ for (a_n) is

$$\prod_{k=1}^{\infty}(1 + x^{2k-1} + x^{2(2k-1)} + \cdots)(1 + x^{2k}) = \prod_{k=1}^{\infty}(1 + x^{2k})(1 - x^{2k-1})^{-1},$$

and the generating function $B(x)$ for (b_n) is

$$\prod_{n=1}^{\infty}(1 + x^n + x^{2n} + x^{3n}) = \prod_{n=1}^{\infty}(1 - x^{4n})(1 - x^n)^{-1}.$$

Now, we have

$$B(x)$$
$$= \prod_{n=1}^{\infty}(1 - x^{4n})(1 - x^n)^{-1}$$
$$= \frac{1 - x^4}{1 - x} \cdot \frac{1 - x^8}{1 - x^2} \cdot \frac{1 - x^{12}}{1 - x^3} \cdot \frac{1 - x^{16}}{1 - x^4} \cdots$$
$$= \frac{(1 + x^2)(1 - x^2)}{1 - x} \cdot \frac{(1 + x^4)(1 - x^4)}{1 - x^2} \cdot \frac{(1 + x^6)(1 - x^6)}{1 - x^3} \cdots$$
$$= \frac{1 + x^2}{1 - x} \cdot \frac{1 + x^4}{1 - x^3} \cdot \frac{1 + x^6}{1 - x^5} \cdot \frac{1 + x^8}{1 - x^7} \cdots$$
$$= \prod_{k=1}^{\infty}(1 + x^{2k})(1 - x^{2k-1})^{-1}$$
$$= A(x).$$

Consequently, this shows that $a_n = b_n$ for all positive integers n as desired.

46. We note that a_r is equal to the number of partitions of r into exactly n parts. By the Corollary to Theorem 5.3.5, the generating function for (a_r) is $x^n \prod_{i=1}^{n} (1 - x^i)^{-1}$.

47. We note that a_r is equal to the number of partitions of r into at most n parts. By the Corollary to Theorem 5.3.5, the generating function for (a_r) is $\prod_{i=1}^{n} (1 - x^i)^{-1}$.

48. It is easy to see that the generating function for (b_r) is $\prod_{i=1}^{n} (1-x^i)^{-1}$. On the other hand, we see that a_r is equal to the number of partitions of r into at most n parts. By the Corollary to Theorem 5.3.5, the generating function for (a_r) is $\prod_{i=1}^{n} (1 - x^i)^{-1}$. Since the generating functions for (a_r) and (b_r) are the same, it follows that we must have $a_r = b_r$ for all non-negative integers r.

49. (i) The generating function for (a_r) is

$$(1 + x)(1 + x^2)(1 + x^4) \cdots$$
$$= \frac{(1 + x)(1 - x)}{(1 - x)} \cdot \frac{(1 + x^2)(1 - x^2)}{(1 - x^2)} \frac{(1 + x^4)(1 - x^4)}{(1 - x^4)} \cdots$$
$$= \frac{1 - x^2}{1 - x} \cdot \frac{1 - x^4}{1 - x^2} \cdot \frac{1 - x^8}{1 - x^4} \cdots$$
$$= (1 - x)^{-1}.$$

(ii) By part (i) and (5.1.3), we have $a_r = 1$ for all positive integers r.

(iii) Every natural number can be expressed uniquely as a sum of some distinct powers of 2.

50. By Theorem 5.3.2, it suffices to show that the number of partitions of $2n$ into distinct even parts is equal to the number of partitions of n into distinct parts. Let $\mathcal{P}$ denote the set of partitions of n into distinct parts, and let $\mathcal{Q}$ denote the set of partitions of $2n$ into distinct even parts. Let us define $f : \mathcal{P} \to \mathcal{Q}$ as follows: for each $P \in \mathcal{P}$ with Ferrers diagram $\mathcal{F}(P)$, we let n_P denote the number of rows in $\mathcal{F}(P)$, and we let $\mathcal{F}_{f,P}$ denote the Ferrers diagram with n_P rows, and whose number of asterisks in the i-th row is twice the number of asterisks in the i-th row of $\mathcal{F}(P)$ for $i = 1, 2, \ldots, n_P$. We then put $f(P)$ to be the partition of $2n$ whose Ferrers diagram is $\mathcal{F}_{f,P}$. Then it is easy to see that f is a well-defined bijection between $\mathcal{P}$ and $\mathcal{Q}$, and so we must have $|\mathcal{P}| = |\mathcal{Q}|$

by (BP) as required.

51. By Theorem 5.3.2, it suffices to show that the number of partitions of kn into distinct parts whose sizes are multiples of k is equal to the number of partitions of n into distinct parts. Let $\mathcal{P}$ denote the set of partitions of n into distinct parts, and let $\mathcal{Q}$ denote the set of partitions of kn into distinct parts whose sizes are multiples of k. Let us define $f \colon \mathcal{P} \to \mathcal{Q}$ as follows: for each $P \in \mathcal{P}$ with Ferrers diagram $\mathcal{F}(P)$, we let n_P denote the number of rows in $\mathcal{F}(P)$, and we let $\mathcal{F}_{f,P}$ denote the Ferrers diagram with n_P rows, and whose number of asterisks in the i-th row is k times the number of asterisks in the i-th row of $\mathcal{F}(P)$ for $i = 1, 2, \ldots, n_P$. We then put $f(P)$ to be the partition of kn whose Ferrers diagram is $\mathcal{F}_{f,P}$. Then it is easy to see that f is a well-defined bijection between $\mathcal{P}$ and $\mathcal{Q}$, and so we must have $|\mathcal{P}| = |\mathcal{Q}|$ by (BP) as required.

52. Let $\hat{p}(n)$ denote the number of partitions of n in which each part has size at least 2. We make the following two observations:

Observation 1: $p(n+1) - p(n) = \hat{p}(n+1)$. To see that this is indeed the case, let us take any partition Q of $n+1$ in which the smallest part is 1. Removing the part of size 1 from Q, we have a partition of n. We notice that this is a one-to-one correspondence, and the observation follows.

Observation 2: $\hat{p}(n) \le \hat{p}(n+1)$. To see that this is indeed the case, let us take any partition Q of n where each part has size at least 2. If we add one to the largest part of Q, we have a partition of $n+1$ where each part has size at least 2. This correspondence is injective and the observation follows.

Using Observations 1 and 2, we see that

$$p(n) - p(n-1) = \hat{p}(n) \le \hat{p}(n+1) = p(n+1) - p(n).$$

The desired inequality now follows by rearranging the terms in the inequality.

53. (i) There is only one partition of 5 into exactly 1 part, namely 5, so $p(5,1) = 1$.

The partitions of 5 into exactly 2 parts are $4+1$ and $3+2$. So we have $p(5,2) = 2$.

The partitions of 5 into exactly 3 parts are $3+1+1$ and $2+2+1$. So we have $p(5,3) = 2$.

The partitions of 8 into exactly 3 parts are $6+1+1, 5+2+1, 4+3+1, 4+2+2$ and $3+3+2$, so we have $p(8,3) = 5$.

(ii) Let $\mathcal{P}$ denote the set of partitions of n with at most m parts, and let $\mathcal{Q}$ denote the set of partitions of $n+m$ with m parts. Then it is easy to see from the definition of $p(n,k)$ that $|\mathcal{P}| = \sum\limits_{k=1}^{m} p(n,k)$ and $|\mathcal{Q}| = p(n+m,m)$. Thus it remains to show that $|\mathcal{P}| = |\mathcal{Q}|$. Let us define $f: \mathcal{P} \to \mathcal{Q}$ as follows: for each $P \in \mathcal{P}$ with Ferrers diagram $\mathcal{F}(P)$, we let n_P denote the number of rows in $\mathcal{F}(P)$. Then we have $n_P \leq m$. Next, we let $\mathcal{F}_{f,P}$ denote the Ferrers diagram with m rows whose number of asterisks in the i-th row is 1 more than the number of asterisks in the i-th row of $\mathcal{F}(P)$ for $i = 1, 2, \ldots, n_P$, and whose number of asterisks in the j-th row is equal to 1 for $j = n_P + 1, n_P + 2, \ldots, m$. We then put $f(P)$ to be the partition of $n+m$ whose Ferrers diagram is $\mathcal{F}_{f,P}$. It is now easy to see that f is a bijection between $\mathcal{P}$ and $\mathcal{Q}$, and so we must have $|\mathcal{P}| = |\mathcal{Q}|$ by (BP), or equivalently,

$$\sum_{k=1}^{m} p(n,k) = p(n+m,m)$$

as desired.

54. (i) The partitions of 5 into exactly 3 parts are $3+1+1$ and $2+2+1$. So we have $p(5,3) = 2$.

The partitions of 7 into exactly 2 parts are $6+1, 5+2$ and $4+3$, so we have $p(7,2) = 3$.

The partitions of 8 into exactly 3 parts are $6+1+1, 5+2+1, 4+3+1, 4+2+2$ and $3+3+2$, so we have $p(8,3) = 5$.

(ii) Let $\mathcal{P}$ denote the set of partitions of n into k parts, $\mathcal{P}_1$ denote the set of partitions of n into k parts the smallest of which is 1, and $\mathcal{P}_2$ denote the set of partitions of n into k parts that does not contain a part of size 1. Then it is easy to see that $\mathcal{P}_1 \cap \mathcal{P}_2 = \emptyset$ and $\mathcal{P}_1 \cup \mathcal{P}_2 = \mathcal{P}$, so that we have $|\mathcal{P}_1| + |\mathcal{P}_2| = |\mathcal{P}| = p(n,k)$. Also, we let $\mathcal{Q}_1$ denote the set of partitions of $n-1$ into $k-1$ parts, and $\mathcal{Q}_2$ denote the set of partitions of $n-k$ into k parts.

Let us define $f: \mathcal{P}_1 \to \mathcal{Q}_1$ as follows: for each $P_1 \in \mathcal{P}_1$ with Ferrers diagram $\mathcal{F}(P_1)$, we let $f(P_1)$ denote the partition of $n-1$ whose Ferrers diagram is obtained from $\mathcal{F}(P_1)$ by removing the bottommost asterisk in the leftmost column of $\mathcal{F}(P_1)$. It is easy to see that f is a well-defined bijection between $\mathcal{P}_1$ and $\mathcal{Q}_1$, and so we must have $|\mathcal{P}_1| = |\mathcal{Q}_1| = p(n-1,k-1)$ by (BP).

Next, let us define $g: \mathcal{P}_2 \to \mathcal{Q}_2$ as follows: for each $P_2 \in \mathcal{P}_2$ with Ferrers diagram $\mathcal{F}(P_2)$, we let $g(P_2)$ denote the partition of $n-k$

whose Ferrers diagram is obtained from $\mathcal{F}(P_2)$ by removing the rightmost asterisk in each row of $\mathcal{F}(P_2)$. It is now easy to see that g is a well-defined bijection between $\mathcal{P}_2$ and $\mathcal{Q}_2$, and so we must have $|\mathcal{P}_2| = |\mathcal{Q}_2| = p(n - k, k)$ by (BP). Thus, we have

$$p(n - 1, k - 1) + p(n - k, k) = |\mathcal{P}_1| + |\mathcal{P}_2| = |\mathcal{P}| = p(n, k)$$

as desired.

55. Let us first show that $p(n + k, k) = p(n)$ for all positive integers n, k satisfying $n \leq k$. Let $\mathcal{P}$ denote the set of partitions of n, and $\mathcal{Q}$ denote the set of partitions of $n + k$ into k parts. Let us define $f: \mathcal{P} \to \mathcal{Q}$ as follows: for each $P \in \mathcal{P}$ with Ferrers diagram $\mathcal{F}(P)$, we let n_P denote the number of rows in $\mathcal{F}(P)$. Then we have $n_P \leq n \leq k$. Next, we let $\mathcal{F}_{f,P}$ denote the Ferrers diagram with k rows whose number of asterisks in the i-th row is 1 more than the number of asterisks in the i-th row of $\mathcal{F}(P)$ for $i = 1, 2, \ldots, n_P$, and whose number of asterisks in the j-th row is equal to 1 for $j = n_P + 1, n_P + 2, \ldots, k$. We then put $f(P)$ to be the partition of $n + k$ whose Ferrers diagram is $\mathcal{F}_{f,P}$. It is now easy to see that f is a well-defined bijection between $\mathcal{P}$ and $\mathcal{Q}$, and so we must have $p(n) = |\mathcal{P}| = |\mathcal{Q}| = p(n + k, k)$ by (BP). Next, by setting $k = n$ in the equation $p(n + k, k) = p(n)$, we get $p(2n, n) = p(n)$. So we have

$$p(n + k, k) = p(2n, n) = p(n)$$

for all positive integers n, k satisfying $n \leq k$ as desired.

56. Let $\mathcal{P}$ denote the collection of finite multi-sets $\{r_1 \cdot 1, r_2 \cdot 2, \ldots, r_n \cdot n\}$ that satisfy $\sum_{i=1}^{n} r_i = k$ and $\sum_{i=1}^{n} i r_i = n$. Then it follows that $|\mathcal{P}| = p(n, k)$. Next, we let S denote the set of positive integer solutions to the equation

$$x_1 + x_2 + \cdots + x_k = n.$$

Let us define $f: S \to \mathcal{P}$ by $f((x_1, \ldots, x_k)) = \{x_1, \ldots, x_k\}$ for all $(x_1, \ldots, x_k) \in S$. Now, it is easy to see that for all $P = \{r_1 \cdot 1, r_2 \cdot 2, \ldots, r_n \cdot n\} \in \mathcal{P}$, we have $|f^{-1}(P)| = \binom{k}{r_1, r_2, \ldots, r_n} \leq k!$. Since $S = \bigcup_{P \in \mathcal{P}} f^{-1}(P)$ and $f^{-1}(P) \cap f^{-1}(Q) = \emptyset$ for all distinct $P, Q \in \mathcal{P}$, it follows that we have

$$|S| = \left| \bigcup_{P \in \mathcal{P}} f^{-1}(P) \right| = \sum_{P \in \mathcal{P}} |f^{-1}(P)| \leq |\mathcal{P}| \cdot k! = k! p(n, k).$$

Now, $|S|$ is equal to the number of non-negative integer solutions to the equation

$$x_1 + x_2 + \cdots + x_k = n - k,$$

which is then equal to $H_{n-k}^k = \binom{n-1}{k-1}$. Therefore, we have

$$\binom{n-1}{k-1} = |S| \le k! p(n, k),$$

or equivalently,

$$p(n, k) \ge \frac{1}{k!} \binom{n-1}{k-1}$$

as desired.

57. Let $\mathcal{P}$ denote the set of partitions of $n - \binom{k}{2}$ into k parts, and let $\mathcal{Q}$ denote the set of partitions of n into k distinct parts. Let us define $f : \mathcal{P} \to \mathcal{Q}$ as follows: for each $P \in \mathcal{P}$ with Ferrers diagram $\mathcal{F}(P)$, we let $\mathcal{F}_{f,P}$ denote the Ferrers diagram with k rows whose number of asterisks in the i-th row is $i - 1$ more than the number of asterisks in the i-th row of $\mathcal{F}(P)$ for $i = 1, 2, \ldots, k$. We then put $f(P)$ to be the partition of $n - \binom{k}{2} + \sum_{i=1}^{k}(i - 1) = n - \binom{k}{2} + \frac{k(k-1)}{2} = n$ whose Ferrers diagram is $\mathcal{F}_{f,P}$. It is now easy to see from the above construction that f is a well-defined bijection between $\mathcal{P}$ and $\mathcal{Q}$, and so we must have

$$|\mathcal{Q}| = |\mathcal{P}| = p\left(n - \binom{k}{2}, k\right)$$

by (BP) as desired.

58. Let $\mathcal{P}$ denote the set of partitions of n into at most m parts and $\mathcal{Q}$ denote the set of partitions of n into parts with sizes not exceeding m. Also, for each $i = 1, 2, \ldots, m$, we let $\mathcal{P}_i$ denote the set of partitions of n into i parts, and $\mathcal{Q}_i$ denote the set of partitions into parts the largest of which is i. Since $\mathcal{P} = \bigcup_{i=1}^{m} \mathcal{P}_i$, $\mathcal{Q} = \bigcup_{i=1}^{m} \mathcal{Q}_i$, and $\mathcal{P}_i \cap \mathcal{P}_j = \emptyset = \mathcal{Q}_i \cap \mathcal{Q}_j$ for all distinct i, j, it follows from (AP) that $|\mathcal{P}| = \sum_{i=1}^{m} |\mathcal{P}_i|$ and $|\mathcal{Q}| = \sum_{i=1}^{m} |\mathcal{Q}_i|$. Next, it follows from Theorem 5.3.5 that $|\mathcal{P}_i| = |\mathcal{Q}_i|$. So we have

$$|\mathcal{P}| = \sum_{i=1}^{m} |\mathcal{P}_i| = \sum_{i=1}^{m} |\mathcal{Q}_i| = |\mathcal{Q}|$$

as desired.

59. (i) Let us fix a positive integer k. For each positive integer n, we let a_n denote the number of partitions of n into parts, each of which appears at most k times, and we let b_n denote the number

of partitions of n into parts the sizes of which are not divisible by $k+1$. Then it follows that the generating function $A(x)$ for (a_n) is $\prod_{r=1}^{\infty}(1+x^r+x^{2r}+\cdots+x^{kr})$, and by (5.1.3), the generating function $B(x)$ for (b_n) is $\prod_{\substack{r\in\mathbf{N}\\k+1\nmid r}}(1+x^r+x^{2r}+\cdots) = \prod_{\substack{r\in\mathbf{N}\\k+1\nmid r}}(1-x^r)^{-1}$. Now, we have

$$
\begin{aligned}
A(x) &= \prod_{r=1}^{\infty}(1+x^r+x^{2r}+\cdots+x^{kr}) \\
&= \prod_{r=1}^{\infty}(1+x^r+x^{2r}+\cdots+x^{kr})(1-x^r)(1-x^r)^{-1} \\
&= \prod_{r=1}^{\infty}(1-x^{(k+1)r})(1-x^r)^{-1} \\
&= \prod_{\substack{r\in\mathbf{N}\\k+1\nmid r}}(1-x^r)^{-1} \\
&= B(x).
\end{aligned}
$$

Consequently, this shows that $a_n = b_n$ for all positive integers n as desired.

(ii) Firstly, we see that n is not congruent to 1 or -1 (mod 6) if and only if the greatest common divisor $\gcd(n,6)$ of n and 6 is greater than 1. Equivalently, n is not congruent to 1 or -1 (mod 6) if and only if either n is even or n is a odd multiple of 3. For each positive integer n, we let a_n denote the number of partitions of n into parts each of which appears at least twice, and we let b_n denote the number of partitions of n into parts the sizes of which are not congruent to 1 or -1 (mod 6). By (5.1.3), it follows that the generating function $A(x)$ for (a_n) is

$$
\begin{aligned}
\prod_{r=1}^{\infty}(1+x^{2r}+x^{3r}+x^{4r}\cdots) &= \prod_{r=1}^{\infty}[(1+x^r+x^{2r}+\cdots)-x^r] \\
&= \prod_{r=1}^{\infty}[(1-x^r)^{-1}-x^r].
\end{aligned}
$$

and the generating function $B(x)$ for (b_n) is

$$
\prod_{\substack{r\in\mathbf{N}\\\gcd(r,6)>1}}(1+x^r+x^{2r}+\cdots) = \prod_{\substack{r\in\mathbf{N}\\\gcd(r,6)>1}}(1-x^r)^{-1}.
$$

Now, we have

$$A(x) = \prod_{r=1}^{\infty}[(1-x^r)^{-1} - x^r]$$

$$= \prod_{r=1}^{\infty}(1 - x^r + x^{2r})(1-x^r)^{-1}$$

$$= \prod_{r=1}^{\infty}(1 - x^r + x^{2r})(1+x^r)(1+x^r)^{-1}(1-x^r)^{-1}$$

$$= \prod_{r=1}^{\infty}(1 + x^{3r})(1 - x^{2r})^{-1}$$

$$= \prod_{r=1}^{\infty}(1 + x^{3r})(1 - x^{3r})(1 - x^{3r})^{-1}(1 - x^{2r})^{-1}$$

$$= \prod_{r=1}^{\infty}(1 - x^{6r})(1 - x^{3r})^{-1}(1 - x^{2r})^{-1}$$

$$= \prod_{r=1}^{\infty}(1 - x^{2r})^{-1} \prod_{r=1}^{\infty}(1 - x^{6r})(1 - x^{3r})^{-1}$$

$$= \prod_{r=1}^{\infty}(1 - x^{2r})^{-1} \prod_{r=1}^{\infty}(1 - x^{3(2r-1)})^{-1}$$

$$= \prod_{\substack{r \in \mathbf{N} \\ \gcd(r,6)>1}} (1 - x^r)^{-1}$$

$$= B(x).$$

Consequently, this shows that $a_n = b_n$ for all positive integers n as desired.

60. We shall show that there exists a polynomial $Q(x)$, such that $C(n) = \lfloor Q(n) \rfloor$ for all positive integers n. To this end, we first observe that the generating function for $(C(n))$ is given by

$$\prod_{r=0}^{\infty}(1 + x^{2^r} + x^{2 \cdot 2^r} + x^{3 \cdot 2^r}) = \prod_{r=0}^{\infty}\frac{1 - x^{4 \cdot 2^r}}{1 - x^{2^r}}$$

$$= \frac{1-x^4}{1-x} \cdot \frac{1-x^8}{1-x^2} \cdot \frac{1-x^{16}}{1-x^4} \cdot \frac{1-x^{32}}{1-x^8} \cdots$$

$$= \frac{1}{(1-x)(1-x^2)}$$

$$= \frac{1}{(1-x)(1-x)(1+x)}$$

$$= \frac{1}{(1+x)(1-x)^2}.$$

By (5.1.6), we have

$$\frac{1}{(1+x)(1-x)^2} = \frac{1+x}{(1+x)^2(1-x)^2}$$

$$= \frac{1+x}{(1-x^2)^2}$$

$$= (1+x)\sum_{i=0}^{\infty} \binom{i+2-1}{i}(x^2)^i$$

$$= \sum_{i=0}^{\infty}(i+1)(x^{2i}+x^{2i+1}).$$

Since $\lfloor \frac{2i+2}{2} \rfloor = i+1 = \lfloor \frac{2i+3}{2} \rfloor$ for all integers i, it follows that we have

$$\frac{1}{(1+x)(1-x)^2} = \sum_{i=0}^{\infty}(i+1)(x^{2i}+x^{2i+1}) = \sum_{n=0}^{\infty}\left\lfloor \frac{n+2}{2} \right\rfloor x^n.$$

As $\frac{1}{(1+x)(1-x)^2}$ is the generating function for $(C(n))$, it follows that we must have $C(n) = \lfloor \frac{n+2}{2} \rfloor$ for all positive integers n. The desired statement now follows by taking $Q(x) = \frac{x+2}{2}$.

61. The solution for this problem is exactly the same as that of the solution for the previous problem, and shall be omitted.

62. (a) (i) The partitions of 8 into 3 parts are $6+1+1$, $5+2+1$, $4+3+1$, $4+2+2$ and $3+2+2$.

    (ii) The noncongruent triangles whose sides are of integer length a, b, c such that $a+b+c = 16$ have sides $(7,7,2)$, $(7,6,3)$, $(7,5,4)$, $(6,6,4)$ and $(6,5,5)$.

    (iii) The number of partitions obtained in (i) and the number of noncongruent triangles obtained in (ii) are both equal to 5.

    (b) (i) For each non-negative integer r, we let T denote the set of noncongruent triangles whose sides are of integer length a, b, c such that $a+b+c = 2r$, and we let $\mathcal{P}$ denote the set of partitions of r into 3 parts. Moreover, let us denote an element $\Delta \in T$ by a 3-element multi-set $\{a, b, c\}$, where a, b, c denotes the lengths of the sides of Δ, and denote an element $P \in \mathcal{P}$ by a 3-element multi-set $\{m, n, p\}$, where m, n and p are the parts of P. Let us define $f : T \to \mathcal{P}$ by $f(\{a, b, c\}) = \{r-a, r-b, r-c\}$. We claim that f is well-defined. Indeed, we have

$$(r - a) + (r - b) + (r - c) = 3r - (a + b + c) = r,$$
$$r - a = \frac{a + b + c}{2} - a = \frac{b + c - a}{2} > 0,$$
$$r - b = \frac{a + b + c}{2} - b = \frac{a + c - b}{2} > 0,$$
$$r - c = \frac{a + b + c}{2} - c = \frac{a + b - c}{2} > 0,$$

where the last three inequalities follow from the fact that the sum of the lengths of any two sides of a (non-degenerate) triangle is strictly greater than the length of the third side. So f is well-defined. Moreover, it is clear from the definition of f that f is injective. It remains to show that f is surjective. To this end, let us take any partition $P \in \mathcal{P}$ with parts m, n, p. Then we have $m + n + p = r$, and the 3-element multi-set $\{m + n, m + p, n + p\}$ clearly defines a (non-degenerate) triangle whose sides have integer length and sum up to $2r$. Since

$$f(\{m + n, m + p, n + p\})$$
$$= \{r - (m + n), r - (m + p), r - (n + p)\}$$
$$= \{p, n, m\}$$
$$= P,$$

this shows that f is surjective. So f is a bijection from T to $\mathcal{P}$, and by (BP), we have

$$a_r = |T| = |\mathcal{P}| = b_r$$

for all non-negative integers r as desired.

(ii) By Theorem 5.3.5, b_r is equal to the number of partitions of r into parts the largest size of which is 3. By (5.1.3), the generating function for (b_r) is

$$(1 + x + x^2 + \cdots)(1 + x^2 + x^4 + \cdots)(x^3 + x^6 + \cdots)$$
$$= x^3[(1 - x)(1 - x^2)(1 - x^3)]^{-1}.$$

Since $a_r = b_r$ for all non-negative integers r, it follows that the generating function for (a_r) is the same as that of the generating function for (b_r), so the generating function for (a_r) is $x^3[(1 - x)(1 - x^2)(1 - x^3)]^{-1}$.

63. (i) The self-conjugate partitions of 15 are $6 + 3 + 3 + 1 + 1 + 1$, $5 + 4 + 3 + 2 + 1$, $4 + 4 + 4 + 3$, $8 + 1 + 1 + 1 + 1 + 1 + 1 + 1$.

(ii) The partitions of 15 into distinct odd parts are 15, $11 + 3 + 1$, $9 + 5 + 1$ and $7 + 5 + 3$.

(iii) Let us fix a positive integer n, and let $\mathcal{P}$ denote the set of self-conjugate partitions of n, and $\mathcal{Q}$ denote the set of partitions of n into distinct odd parts. Also, for any Ferrers diagram $\mathcal{F}$ with n total asterisks and k asterisks on the leading diagonal of $\mathcal{F}$, we let r_i denote the number of asterisks to the right of the asterisk located at the i-th row and the i-th column of $\mathcal{F}$, and b_i denote the number of asterisks below the asterisk located at the i-th row and the i-th column of $\mathcal{F}$ for all $i = 1, 2, \ldots, k$. Then it is clear that we have $\sum_{i=1}^{k} (r_i + b_i + 1) = n$ and $r_1 + b_1 > r_2 + b_2 > \cdots > r_k + b_k$. Moreover, we see that $\mathcal{F}$ is self-conjugate if and only if we have $r_i = b_i$ for all $i = 1, 2, \ldots, k$.

Henceforth, let us define a map $f \colon \mathcal{P} \to \mathcal{Q}$ as follows: for each $P \in \mathcal{P}$ with Ferrers diagram $\mathcal{F}(P)$, we let k_P denote the number of asterisks on the leading diagonal of $\mathcal{F}(P)$. Next, we let $\mathcal{F}_{f,P}$ denote the Ferrers diagram with k_P rows whose number of asterisks in the i-th row is equal to $r_i + b_i + 1 = 2r_i + 1$ for $i = 1, 2, \ldots, k_P$. We then put $f(P)$ to be the partition of n whose Ferrers diagram is $\mathcal{F}_{f,P}$. Then it is easy to see from the above construction that f is a well-defined bijection between $\mathcal{P}$ and $\mathcal{Q}$, and so we must have $|\mathcal{P}| = |\mathcal{Q}|$ by (BP) as required.

64. Let $\mathcal{P}$ denote the set of self-conjugate partitions of n with largest size equal to m, and let $\mathcal{Q}$ denote the set of self-conjugate partitions of $n - 2m + 1$ with largest size not exceeding $m - 1$. Given any self-conjugate Ferrers diagram $\mathcal{F}$ with k columns (that is, the partition corresponding to $\mathcal{F}$ has largest size k), we see that the number of asterisks that lie either in the leftmost column or the topmost row of $\mathcal{F}$ is equal to $2k - 1$, and the Ferrers diagram obtained by removing the leftmost column and the topmost row of $\mathcal{F}$ is self-conjugate as well.

Henceforth, let us define $f \colon \mathcal{P} \to \mathcal{Q}$ as follows: for each $P \in \mathcal{P}$ with Ferrers diagram $\mathcal{F}(P)$, we let $f(P)$ denote the partition of $n - 2m + 1$ whose Ferrers diagram is obtained from $\mathcal{F}(P)$ by removing the leftmost column and the topmost row of $\mathcal{F}(P)$. It is then easy to see that f establishes a bijection between $\mathcal{P}$ and $\mathcal{Q}$, and so we must have $|\mathcal{P}| = |\mathcal{Q}|$ by (BP) as desired.

65. (i) For each non-negative integer r, we let $\mathcal{P}_{r,m}$ denote the set of self-conjugate partitions of r, such that the Durfee square of the Ferrers

diagram of the partition is an $m \times m$ square and $\mathcal{Q}_{r,m}$ denote the set of partitions of r into even parts with the largest size not exceeding $2m$. We seek to establish a bijection between $\mathcal{P}_{r,m}$ and $\mathcal{Q}_{r-m^2,m}$. Given any self-conjugate Ferrers diagram $\mathcal{F}$ whose Durfee square is an $m \times m$ square, we let $\mathcal{F}_b$ (respectively $\mathcal{F}_r$) denote the Ferrers diagram obtained by removing the topmost (respectively leftmost) m rows (respectively columns) of $\mathcal{F}$. Then the partition corresponding to the Ferrers diagram $\mathcal{F}_b$ has largest size not exceeding m, and the Ferrers diagrams $\mathcal{F}_b$ and $\mathcal{F}_r$ are conjugates of each other. Henceforth, let us define $f: \mathcal{P}_{r,m} \to \mathcal{Q}_{r-m^2,m}$ as follows: for each $P \in \mathcal{P}_{r,m}$ with Ferrers diagram $\mathcal{F}(P)$, we let n_P denote the number of rows of $\mathcal{F}(P)_b$. Next, we let $\mathcal{F}_{f,P}$ denote the Ferrers diagram with n_P rows whose number of asterisks in the i-th row is twice the number of asterisks in the i-th row of $\mathcal{F}(P)_b$ for $i = 1, 2, \ldots, n_P$. We then put $f(P)$ to be the partition of $r - m^2$ whose Ferrers diagram is $\mathcal{F}_{f,P}$. Then it is easy to see from the above construction that f is a well-defined bijection between $\mathcal{P}_{r,m}$ and $\mathcal{Q}_{r-m^2,m}$, and hence we have $|\mathcal{P}_{r,m}| = |\mathcal{Q}_{r-m^2,m}|$ by (BP).

It remains to compute the generating function for $(|\mathcal{P}_{r,m}|)_{r=0}^{\infty} = (|\mathcal{Q}_{r,m}|)_{r=0}^{\infty}$ for all non-negative integers r. For each non-negative integer r and positive integer m, we let $a_{r,m}$ denote the number of partitions of r into even parts with the largest size not exceeding $2m$, and $b_{r,m}$ denote the number of partitions of r into even parts the largest size of which is $2m$. By (5.1.3), it is easy to see that the generating function for $(b_{r,m})_{r=0}^{\infty}$ is

$$(1 + x^2 + x^4 + \cdots)(1 + x^4 + x^8 + \cdots) \cdots (x^{2m} + x^{4m} + \cdots)$$
$$= x^{2m} \prod_{i=1}^{m} (1 - x^{2i})^{-1}.$$

We shall now prove by induction on m that the generating function for $(a_{r,m})_{r=0}^{\infty}$ is $\prod_{i=1}^{m} (1 - x^{2i})^{-1}$, with the base case $m = 1$ being trivial. Suppose that the assertion holds for some positive integer $m = k \geq 1$. By induction hypothesis, the generating function for $(a_{r,k})$ is $\prod_{i=1}^{k} (1 - x^{2i})^{-1}$. As we have

$$a_{r,k+1} = a_{r,k} + b_{r,k+1}$$

for all non-negative integers r, it follows from Theorem 5.1.1 (i) that the generating function for $(a_{r,k+1})_{r=0}^{\infty}$ is

$$\prod_{i=1}^{k}(1-x^{2i})^{-1} + x^{2(k+1)}\prod_{i=1}^{k+1}(1-x^{2i})^{-1}$$

$$= (1 - x^{2(k+1)})\prod_{i=1}^{k+1}(1-x^{2i})^{-1} + x^{2(k+1)}\prod_{i=1}^{k+1}(1-x^{2i})^{-1}$$

$$= \prod_{i=1}^{k+1}(1-x^{2i})^{-1},$$

thereby completing the inductive step. So the assertion follows. Since $|\mathcal{P}_{r,m}| = |\mathcal{Q}_{r-m^2,m}| = a_{r-m^2,m}$, it follows from Theorem 5.1.1 (iv) that the generating function for $(|\mathcal{P}_{r,m}|)$, the number of self-conjugate partitions of r whose Durfee square is an $m \times m$ square, is equal to

$$\frac{x^{m^2}}{\prod_{k=1}^{m}(1-x^{2k})}.$$

(ii) For each non-negative integer r, we let c_r denote the number of self-conjugate partitions of r. By Problem 5.63(iii), c_r is equal to the number of partitions of r into distinct odd parts. Thus, it follows that the generating function for (c_r) is equal to $\prod_{k=0}^{\infty}(1 + x^{2k+1})$. Next, for each positive integer r and positive integer m, we let $c_{r,m}$ denote the number of self-conjugate partitions of r whose Durfee square is an $m \times m$ square. By part (i), the generating function $C_m(x)$ for $(c_{r,m})_{r=0}^{\infty}$ is $x^{m^2}\prod_{k=1}^{m}(1-x^{2k})^{-1}$ for all positive integers m. Since we have $c_r = \sum_{m=1}^{\infty} c_{r,m}$ for all positive integers r, it follows that the generating function for (c_r) is

$$\sum_{m=1}^{\infty} C_m(x) = \sum_{m=1}^{\infty} \frac{x^{m^2}}{\prod_{k=1}^{m}(1-x^{2k})}.$$

Therefore, we have

$$\prod_{k=0}^{\infty}(1 + x^{2k+1}) = 1 + \sum_{m=1}^{\infty} \frac{x^{m^2}}{\prod_{k=1}^{m}(1-x^{2k})}$$

as desired.

66. (i) By (5.1.3), we have

$$A(x) = \prod_{k=1}^{\infty}(1+x^k+x^{2k}+\cdots) = \prod_{k=1}^{\infty}(1-x^k)^{-1} = \left[\prod_{k=1}^{\infty}(1-x^k)\right]^{-1}.$$

(ii) For each non-negative integer r and positive integer m, we let $\mathcal{P}_{r,m}$ denote the set of partitions of r, such that the Durfee square of the Ferrers diagram of the partition is an $m \times m$ square and $\mathcal{Q}_{r,m}$ denote the set of pairs of partitions (Q_1, Q_2), such that Q_1 is a partition with largest size not exceeding m, Q_2 is a partition into at most m parts, and the total number of asterisks in the Ferrers diagram of Q_1 and Q_2 is equal to r. We seek to establish a bijection between $\mathcal{P}_{r,m}$ and $\mathcal{Q}_{r-m^2,m}$. Given any Ferrers diagram $\mathcal{F}$ whose Durfee square is an $m \times m$ square, we let $\mathcal{F}_b$ (respectively $\mathcal{F}_r$) denote the Ferrers diagram obtained by removing the topmost (respectively leftmost) m rows (respectively columns) of $\mathcal{F}$. Then the partition corresponding to the Ferrers diagram $\mathcal{F}_b$ has largest size not exceeding m, and the partition corresponding to the Ferrers diagram $\mathcal{F}_r$ has at most m parts.

Henceforth, let us define $f: \mathcal{P}_{r,m} \to \mathcal{Q}_{r-m^2,m}$ as follows: for each $P \in \mathcal{P}_{r,m}$ with Ferrers diagram $\mathcal{F}(P)$, we let Q_1 and Q_2 denote the partitions whose Ferrers diagrams are $\mathcal{F}(P)_r$ and $\mathcal{F}(P)_b$ respectively, and we let $f(P) = (Q_1, Q_2)$. Then it is easy to see from the above construction that f is a well-defined bijection between $\mathcal{P}_{r,m}$ and $\mathcal{Q}_{r-m^2,m}$, and hence we have $|\mathcal{P}_{r,m}| = |\mathcal{Q}_{r,m}|$ by (BP).

It remains to compute the generating function for $(|\mathcal{P}_{r,m}|)_{r=0}^{\infty} = (|\mathcal{Q}_{r-m^2,m}|)_{r=0}^{\infty}$ for all non-negative integers r. For each non-negative integer r and positive integer m, we let $a_{r,m}$ denote the number of partitions of r into parts with the largest size not exceeding m. By a similar argument as in Problem 5.65(ii), we see that the generating function for the sequences $(a_{r,m})_{r=0}^{\infty}$ is $\prod\limits_{k=1}^{m} (1-x^k)^{-1}$.

By Theorem 5.3.5, we have $|\mathcal{Q}_{r,m}| = \sum\limits_{k=0}^{r-m^2} a_{k,m} a_{r-m^2-k,m}$, which implies that the generating function $C_m(x)$ for $(|\mathcal{P}_{r,m}|)_{r=0}^{\infty}$ is $x^{m^2} \prod\limits_{i=1}^{m} (1 - x^i)^{-2}$ by Theorem 5.1.1 (iii) and (iv). Finally, since $p(r) = \sum\limits_{m=1}^{\infty} |\mathcal{P}_{r,m}|$ for all non-negative integers r, it follows that the generating function for $(p(r))$ is

$$\sum_{m=1}^{\infty} C_m(x) = \sum_{m=1}^{\infty} \frac{x^{m^2}}{\prod_{k=1}^{m} (1 - x^k)^2}.$$

Therefore, we have

$$\left[\prod_{k=1}^{\infty}(1-x^k)\right]^{-1} = 1 + \sum_{m=1}^{\infty}\frac{x^{m^2}}{\prod_{k=1}^{m}(1-x^k)^2}.$$

67. For each non-negative integer r and positive integer m, we let $\mathcal{P}_{r,m}$ denote the set of partitions of r into distinct parts such that the largest isosceles right triangle of asterisks in the upper left-hand corner of the Ferrers diagram has m asterisks in the top horizon row. Also, we let $\mathcal{Q}_{r,m}$ denote the set of partitions of r into at most m parts. We seek to establish a bijection between $\mathcal{P}_{r,m}$ and $\mathcal{Q}_{r-\frac{m(m+1)}{2},m}$. Given any partition of r into distinct parts such that the largest isosceles right triangle of asterisks in the upper left-hand corner of the Ferrers diagram has m asterisks in the top horizon row, we let $\mathcal{F}_m$ denote the Ferrers diagram obtained by removing $m+1-i$ asterisks from the i-th row of $\mathcal{F}$ for $i = 1, \ldots, m$. Then the partition corresponding to the Ferrers diagram $\mathcal{F}_m$ has at most m parts.

Henceforth, let us define $f \colon \mathcal{P}_{r,m} \to \mathcal{Q}_{r-\frac{m(m+1)}{2},m}$ as follows: for each $P \in \mathcal{P}_{r,m}$ with Ferrers diagram $\mathcal{F}(P)$, we let $f(P)$ be the partition whose Ferrers diagram is $\mathcal{F}(P)_m$. Then it is easy to see from the above construction that f is a well-defined bijection between $\mathcal{P}_{r,m}$ and $\mathcal{Q}_{r-\frac{m(m+1)}{2},m}$, and hence we have $|\mathcal{P}_{r,m}| = \left|\mathcal{Q}_{r-\frac{m(m+1)}{2},m}\right|$ by (BP).

It remains to compute the generating function for $(|\mathcal{P}_{r,m}|)_{r=0}^{\infty} = \left(\left|\mathcal{Q}_{r-\frac{m(m+1)}{2},m}\right|\right)_{r=0}^{\infty}$ for all non-negative integers r. For each non-negative integer r and positive integer m, we let $a_{r,m}$ denote the number of partitions of r into at most m parts, and c_r denote the number of partitions of r into distinct parts.

By a similar argument as in Problem 5.65(ii), we see that the generating function for the sequences $(a_{r,m})_{r=0}^{\infty}$ and (c_r) are $\prod_{i=1}^{m}(1-x^i)^{-1}$ and $\prod_{k=1}^{\infty}(1+x^k)$ respectively, which implies that the generating function $C_m(x)$ for $(|\mathcal{P}_{r,m}|)_{r=0}^{\infty}$ is $x^{\frac{m(m+1)}{2}}\prod_{i=1}^{m}(1-x^i)^{-1}$ by Theorem 5.1.1 (iv).

Finally, since $c_r = \sum_{m=1}^{\infty}|\mathcal{P}_{r,m}|$ for all non-negative integers r, it follows that the generating function for (c_r) is

$$\sum_{m=1}^{\infty}C_m(x) = \sum_{m=1}^{\infty}\frac{x^{\frac{m(m+1)}{2}}}{\prod_{k=1}^{m}(1-x^k)}.$$

Therefore, we have

$$\prod_{k=1}^{\infty}(1 + x^k) = \sum_{m=1}^{\infty} \frac{x^{\frac{m(m+1)}{2}}}{\prod_{k=1}^{m}(1 - x^k)}.$$

The desired identity now follows by replacing x by x^2 in the above equation.

68. Let $\mathcal{P}$ denote the set of partitions of $r - p$ into $q - 1$ parts with sizes not exceeding p, and let $\mathcal{Q}$ denote the set of partitions of $r - q$ into $p - 1$ parts with sizes not exceeding q. We seek to establish a bijection between $\mathcal{P}$ and $\mathcal{Q}$. To this end, let us take any $P \in \mathcal{P}$ with Ferrers diagram $\mathcal{F}(P)$. We let $\mathcal{F}(P)_1$ be the Ferrers diagram obtained from $\mathcal{F}(P)$ by deleting an asterisk from each of the $q - 1$ rows of $\mathcal{F}(P)$. Next, we let $n_{P,C}$ (respectively $n_{P,R}$) denote the number of columns (respectively rows) in $\mathcal{F}(P)_1$. By the definition of $\mathcal{P}$, we must have $n_{P,C} \leq p - 1$ and $n_{P,R} \leq q - 1$. Also, there are $r - p - q + 1$ asterisks in $\mathcal{F}(P)_1$. Henceforth, we let $\mathcal{F}(P)_2$ be the Ferrers diagram whose number of asterisks in the i-th column is 1 more than the number of asterisks in the i-th column of $\mathcal{F}(P)_1$ for $i = 1, \ldots, n_{P,C}$, and whose number of asterisks in the j-th column is equal to 1 for $j = n_{P,C} + 1, \ldots, p - 1$. Then the number of columns in $\mathcal{F}(P)_2$ is exactly $p - 1$, the number of rows in $\mathcal{F}(P)_2$ is at most q, and there are $r - q$ asterisks in $\mathcal{F}(P)_2$. Finally, we let $f(P)$ be the partition whose Ferrers diagram is $\mathcal{F}(P)_2^t$. Then it is clear that $f(P)$ is a partition of $r - q$ into exactly $p - 1$ parts with sizes not exceeding q, that is, $f(P) \in \mathcal{Q}$. Now, it is easy to see from the above construction that the map $f \colon \mathcal{P} \to \mathcal{Q}$ defined by $P \mapsto f(P)$ is a bijection. Consequently, we have $|\mathcal{P}| = |\mathcal{Q}|$ by (BP) as desired.

69. For any Ferrers diagram $\mathcal{F}$, we let $d_{\mathcal{F}}$ denote the size of the Durfee square of $\mathcal{F}$. We let $\mathcal{F}_b$ (respectively $\mathcal{F}_r$) be the Ferrers diagram obtained from $\mathcal{F}$ by removing the topmost (respectively leftmost) $d_{\mathcal{F}}$ rows (respectively columns) of $\mathcal{F}$. Then $\mathcal{F}_b$ and $\mathcal{F}_r$ define partitions into distinct parts. Let $r_{\mathcal{F}}$ denote the number of rows of $\mathcal{F}_r$. Next, we define $\mathcal{F}_t$ to the Ferrers diagram obtained from $\mathcal{F}_r$ by removing $r_{\mathcal{F}} + 1 - i$ asterisks from the i-th row of $\mathcal{F}_r$ for all $i = 1, \ldots, r_{\mathcal{F}}$. Finally, we let $\ell_{\mathcal{F}}$ denote the number of asterisks in the last row of $\mathcal{F}$, and $t_{\mathcal{F}}$ denote the smallest index for which the number of asterisks in the $t_{\mathcal{F}}$-th row of $\mathcal{F}_r$ is strictly greater than the number of asterisks in the $(t_{\mathcal{F}} + 1)$-th row of $\mathcal{F}_r$ (if no such index exists, then we set $t_{\mathcal{F}} = d_{\mathcal{F}}$). For example, if P_1 and P_2 are the partitions $9 + 8 + 6 + 5$ and $5 + 4 + 3 + 1$ respectively,

then $\ell_{\mathcal{F}(P_1)} = 5$, $\ell_{\mathcal{F}(P_2)} = 1$, $t_{\mathcal{F}(P_1)} = 2$ and $t_{\mathcal{F}(P_2)} = 3$.

Now, let us fix any $n, j \in \mathbf{N}$, and let $\mathcal{P}_{n,j}^d$ denote the set of partitions of n into exactly j distinct parts. Let $\mathcal{P}_{n,j}^{ne}$ denote the subset of $\mathcal{P}_{n,j}^d$ consisting of partitions P such that either of the Ferrers diagrams $\mathcal{F}(P)_b$ or $\mathcal{F}(P)_t$ is non-empty. Also, we let $\mathcal{E}_n^d = \bigcup_{j \text{ even}} \mathcal{P}_{n,j}^d$, $\mathcal{O}_n^d = \bigcup_{j \text{ odd}} \mathcal{P}_{n,j}^d$, $\mathcal{E}_n^{ne} = \bigcup_{j \text{ even}} \mathcal{P}_{n,j}^{ne}$ and $\mathcal{O}_n^{ne} = \bigcup_{j \text{ odd}} \mathcal{P}_{n,j}^{ne}$.

Given any partition $P \in \mathcal{P}_{n,j}^{ne}$, we shall define a partition $f(P) \in \mathcal{P}_{n,j-1}^{ne} \cup \mathcal{P}_{n,j+1}^{ne}$ as follows:

Case 1: $\ell_{\mathcal{F}(P)} \leq t_{\mathcal{F}(P)}$. Then it is easy to see in this case that the Ferrers diagram $\mathcal{F}_b$ must be non-empty. We let $f(\mathcal{F}(P))$ denote the Ferrers diagram obtained from $\mathcal{F}(P)$ by removing the last row of $\mathcal{F}(P)$, and subsequently adding 1 asterisk to each of the first $\ell_{\mathcal{F}(P)}$ rows of the resulting Ferrers diagram. We then let $f(P)$ be the partition whose Ferrers diagram is $f(\mathcal{F}(P))$. Then it is clear that $f(P) \in \mathcal{P}_{n,j-1}^{ne}$.

Case 2: $\ell_{\mathcal{F}(P)} > t_{\mathcal{F}(P)}$. Then it is easy to see in this case that the Ferrers diagram $\mathcal{F}_t$ must be non-empty. We let $f(\mathcal{F}(P))$ denote the Ferrers diagram obtained from $\mathcal{F}(P)$ by first removing 1 asterisk from each of the first $t_{\mathcal{F}(P)}$ rows of $\mathcal{F}(P)$, and subsequently adding a row of $t_{\mathcal{F}(P)}$ asterisks below the last row of the resulting Ferrers diagram. Then it is clear that $f(P) \in \mathcal{P}_{n,j+1}^{ne}$.

Now, we see from our above construction that if $\ell_{\mathcal{F}(P)} \leq t_{\mathcal{F}(P)}$, then we have $\ell_{\mathcal{F}(f(P))} > t_{\mathcal{F}(f(P))}$, and if $\ell_{\mathcal{F}(P)} > t_{\mathcal{F}(P)}$, then we have $\ell_{\mathcal{F}(f(P))} \leq t_{\mathcal{F}(f(P))}$. This immediately implies that we have $f(f(P)) = P$ for all $P \in \mathcal{P}_{n,j}^{ne}$. Henceforth, let us define $g \colon \mathcal{E}_n^{ne} \to \mathcal{O}_n^{ne}$ and $h \colon \mathcal{O}_n^{ne} \to \mathcal{E}_n^{ne}$ by $g(P) = f(P)$ and $h(Q) = f(Q)$ for all $P \in \mathcal{E}_n^{ne}$ and $Q \in \mathcal{O}_n^{ne}$. Then it is clear from our above construction and observations about the partition $f(P)$ that g and h are well-defined mappings, and $h \circ g$ (respectively $g \circ h$) is the identity mapping on $\mathcal{E}_n^{ne}$ (respectively $\mathcal{O}_n^{ne}$). By (BP), it follows that $|\mathcal{E}_n^{ne}| = |\mathcal{O}_n^{ne}|$.

It remains to show that $p_e(n) - p_o(n) = (-1)^k$ if $n = \frac{k(3k \pm 1)}{2}$ for some $k \geq 0$, and $p_e(n) - p_o(n) = 0$ otherwise. Indeed, we see that for all $n, j \in \mathbf{N}$, we have $\mathcal{P}_{n,j}^{ne}$ to be a proper subset of $\mathcal{P}_{n,j}^d$ if and only if there exists a partition $P \in \mathcal{P}_{n,j}^d$ such that both Ferrers diagrams $\mathcal{F}(P)_b$ and $\mathcal{F}(P)_t$ are empty. We note that the Ferrers diagram $\mathcal{F}(P)_b$ is empty if $j = d_{\mathcal{F}(P)}$, while $\mathcal{F}(P)_t$ is empty if $\mathcal{F}(P)_t$ is the Ferrers diagram corresponding to the partition

$$d_{\mathcal{F}(P)} + (d_{\mathcal{F}(P)} - 1) + \cdots + 1 = j + (j-1) + \cdots + 1$$

or the partition

$$(d_{\mathcal{F}(P)} - 1) + (d_{\mathcal{F}(P)} - 2) + \cdots + 1 = (j - 1) + (j - 2) + \cdots + 1.$$

In the former case, we have

$$n = j^2 + \frac{j(j+1)}{2} = \frac{j(3j+1)}{2},$$

and in the latter case, we have

$$n = j^2 + \frac{j(j-1)}{2} = \frac{j(3j-1)}{2}.$$

This implies that $|\mathcal{P}_{n,j}^d| = |\mathcal{P}_{n,j}^{ne}| + 1$ if $n = \frac{j(3j\pm1)}{2}$, and $|\mathcal{P}_{n,j}^d| = |\mathcal{P}_{n,j}^{ne}|$ otherwise. As $\frac{j(3j-1)}{2} = \frac{(-j)(3(-j)+1)}{2}$, and $\frac{j(3j+1)}{2} \neq \frac{k(3k+1)}{2}$ for all distinct $j, k \in \mathbf{Z}$, it follows that we have

$$p_e(n) = |\mathcal{E}_n^d|$$
$$= \sum_{j \text{ even}} |\mathcal{P}_{n,j}^d|$$
$$= \sum_{j \text{ even}} |\mathcal{P}_{n,j}^{ne}| + 1$$
$$= |\mathcal{E}_n^{ne}| + 1$$
$$= |\mathcal{O}_n^{ne}| + 1$$
$$= \sum_{j \text{ odd}} |\mathcal{P}_{n,j}^{ne}| + 1$$
$$= \sum_{j \text{ odd}} |\mathcal{P}_{n,j}^d| + 1$$
$$= |\mathcal{O}_n^d| + 1$$
$$= p_o(n) + 1$$

if $n = \frac{k(3k\pm1)}{2}$ for some even $k \geq 0$. Likewise, we have $p_o(n) = p_e(n) + 1$ if $n = \frac{k(3k\pm1)}{2}$ for some odd $k \geq 0$, and $p_e(n) = p_o(n)$ if $n \neq \frac{k(3k\pm1)}{2}$ for all $k \geq 0$. The desired statement now follows.

70. Let $a_r = p_e(r) - p_o(r)$ for all non-negative integers r. Then it is clear that the generating function for (a_r) is $\prod_{k=1}^{\infty} (1 - x^k)$. Also, by Problem 5.69, the generating function for (a_r) is given by

$$1 + \sum_{k=1}^{\infty} (-1)^k \left(x^{\frac{1}{2}k(3k-1)} + x^{\frac{1}{2}k(3k+1)} \right)$$

$$= 1 + \sum_{k=1}^{\infty} (-1)^k x^{\frac{1}{2}k(3k-1)} + \sum_{j=-\infty}^{-1} (-1)^{-j} x^{-\frac{1}{2}j(-3j+1)}$$

$$= 1 + \sum_{k=1}^{\infty} (-1)^k x^{\frac{1}{2}k(3k-1)} + \sum_{j=-\infty}^{-1} (-1)^j x^{\frac{1}{2}j(3j-1)}$$

$$= \sum_{m=-\infty}^{\infty} (-1)^m x^{\frac{1}{2}m(3m-1)}.$$

Thus, we have

$$\prod_{k=1}^{\infty} (1 - x^k) = \sum_{m=-\infty}^{\infty} (-1)^m x^{\frac{1}{2}m(3m-1)}$$

as desired.

71. By Problem 5.70, it follows that we have

$$\prod_{k=1}^{\infty} (1 - x^k)^{-1} \sum_{m=-\infty}^{\infty} (-1)^m x^{\frac{1}{2}m(3m-1)} = 1.$$

Moreover, by Problem 5.66(i), we have

$$\prod_{k=1}^{\infty} (1 - x^k)^{-1} = \sum_{i=0}^{\infty} p(i) x^i,$$

and the proof of Problem 5.70 tells us that we have

$$\sum_{m=-\infty}^{\infty} (-1)^m x^{\frac{1}{2}m(3m-1)}$$

$$= 1 + \sum_{k=1}^{\infty} (-1)^k \left(x^{\frac{1}{2}k(3k-1)} + x^{\frac{1}{2}k(3k+1)} \right)$$

$$= 1 - (x + x^2) + (x^5 + x^7) + \cdots$$

$$+ (-1)^m \left(x^{\frac{1}{2}m(3m-1)} + x^{\frac{1}{2}m(3m+1)} \right) + \cdots.$$

By comparing the coefficients of x^n for all $n \geq 1$, it follows that we have

$$p(n) - p(n - 1) - p(n - 2) + p(n - 5) + p(n - 7)$$

$$+ \cdots + (-1)^m p \left(n - \frac{1}{2}m(3m - 1) \right) + \cdots$$

$$+ (-1)^m p \left(n - \frac{1}{2}m(3m + 1) \right) + \cdots = 0$$

as desired.

72. We first observe that $\beta(j) = \beta(j+1) - (3j+2)$ and $\beta(j) = \beta(j-1) + (3j-1)$. For any partition P, we let the largest part and number of parts of P be m_P and n_P respectively (here, if P is the empty partition, then $m_P = n_P = 0$). Next, we let $\mathcal{P}_j$ denote the set of partitions of $n - \beta(j)$ (note that $\mathcal{P}_j = \emptyset$ if $\beta(j) > n$), and we let $\mathcal{P}_E = \bigcup_{j \text{ even}} \mathcal{P}_j$ and $\mathcal{P}_O = \bigcup_{j \text{ odd}} \mathcal{P}_j$.

Given any partition $P \in \mathcal{P}_j$ with the form $n - \beta(j) = k_1 + \cdots + k_{n_P}$, where $k_1 \geq k_2 \geq \cdots \geq k_{n_P} \geq 1$, we shall define a partition $f(P) \in \mathcal{P}_{j-1} \cup \mathcal{P}_{j+1}$ as follows:

Case 1: $n_P + 3j \geq m_P$. If P is the empty partition, then we let $f(P)$ be the partition with a single part of size $3j - 1$. So let us assume that P is not the empty partition. Then we have $n_P + 3j - 1 \geq m_P - 1 = k_1 - 1 \geq k_2 - 1 \geq \cdots \geq k_{n_P} - 1 \geq 0$. Also, we have

$$n_P + 3j - 1 + \sum_{i=1}^{n_P}(k_i - 1) = 3j - 1 + \sum_{i=1}^{n_P} k_i$$
$$= 3j - 1 + n - \beta(j)$$
$$= n - \beta(j-1).$$

Now, we let $f(P)$ be the partition of $n - \beta(j-1)$ with the form

$$n - \beta(j-1) = (n_P + 3j - 1) + (k_1 - 1) + \cdots + (k_{n_P} - 1).$$

Case 2: $n_P + 3j < m_P$. As $j \geq 0$, let us assume that P is not the empty partition. Then we have $k_i + 1 > 1$ for all $i = 2, \ldots, n_P$ and $m_P - 3j - n_P - 1 \geq 0$. Also, we have

$$\sum_{i=2}^{n_P}(k_i + 1) + m_P - 3j - n_P - 1 = \sum_{i=1}^{n_P} k_i + (n_P - 1) - 3j - n_P - 1$$
$$= n - \beta(j) - 3j - 2$$
$$= n - \beta(j+1).$$

Now, we let $f(P)$ be the partition of $n - \beta(j+1)$ with the form

$$n - \beta(j+1) = (k_2 + 1) + \cdots + (k_{n_P} + 1) + \underbrace{1 + \cdots + 1}_{m_P - 3j - n_P - 1}.$$

Next, we claim that $f(f(P)) = P$ for all partitions P. To this end, we will first make some observations about the partition $f(P)$. As before, we shall consider these cases:

Case 1: $n_P + 3j \geq m_P$. Then we must have

$$n_{f(P)} + 3(j-1) \leq n_P + 1 + 3(j-1) < n_P + 3j - 1 = m_{f(P)}.$$

If either P or $f(P)$ is the empty partition, then it is clear that we must have $f(f(P)) = P$, so let us assume that both P and $f(P)$ are not the empty partition. Let the number of parts of P greater than 1 be equal to s. Let $\ell_1 = n_P + 3j - 1$ and $\ell_{i+1} = k_i - 1$ for all $i = 1, \ldots, s$. Then we have $m_{f(P)} = n_P + 3j - 1$ and $n_{f(P)} = s + 1$, so the partition $f(f(P))$ has the form

$$n - \beta(j) = (\ell_2 + 1) + \cdots + (\ell_{s+1} + 1) + \underbrace{1 + 1 + 1 + \cdots + 1 + 1}_{m_{f(P)} - 3(j-1) - n_{f(P)} - 1}.$$

As $\ell_{i+1} + 1 = k_i$ for all $i = 1, \ldots, s$ and

$$m_{f(P)} - 3(j-1) - n_{f(P)} - 1 = n_P + 3j - 1 - 3(j-1) - s - 1 - 1 = n_P - s,$$

it follows that $f(f(P))$ must be the partition P.

Case 2: $n_P + 3j < m_P$. Then we must have

$$n_{f(P)} + 3(j+1) = n_P - 1 + m_P - 3j - n_P - 1 + 3j + 3 = m_P + 1 \geq m_{f(P)}.$$

Again, if either P or $f(P)$ is the empty partition, then it is clear that we must have $f(f(P)) = P$, so let us assume that both P and $f(P)$ are not the empty partition. Let $\ell_i = k_{i+1} + 1$ for all $i = 1, \ldots, n_P - 1$ and $\ell_j = 1$ for all $j = n_P, \ldots, m_P - 3j - 2$. Then we have $m_{f(P)} = k_2 + 1$ and $n_{f(P)} = m_P - 3j - 2$, so the partition $f(f(P))$ has the form

$$n - \beta(j) = (n_{f(P)} + 3j + 2) + (\ell_1 - 1) + \cdots + (\ell_{n_P - 1} - 1).$$

As $n_{f(P)} + 3j + 2 = m_P = k_1$ and $\ell_{i-1} - 1 = k_i$ for all $i = 2, \ldots, n_P$, it follows that $f(f(P))$ must be the partition P.

Henceforth, let us define $g \colon \mathcal{P}_E \to \mathcal{P}_O$ and $h \colon \mathcal{P}_O \to \mathcal{P}_E$ by $g(P) = f(P)$ and $h(Q) = f(Q)$ for all $P \in \mathcal{P}_E$ and $Q \in \mathcal{P}_O$. Then it is clear from our above construction and observations about the partition $f(P)$ that g and h are well-defined mappings, and $h \circ g$ (respectively $g \circ h$) is the identity mapping on $\mathcal{P}_E$ (respectively $\mathcal{P}_O$). By (BP), it follows that $|\mathcal{P}_E| = |\mathcal{P}_O|$.

Now, we see that for all distinct j, k, we have $\beta(j) \neq \beta(k)$, which implies that $\mathcal{P}_j \cap \mathcal{P}_k = \emptyset$ for all distinct j, k. Moreover, since $\lim_{j \to \infty} \beta(j) = \infty = \lim_{j \to \infty} \beta(-j)$, it follows that there exists some positive integer N, such that $\mathcal{P}_j = \emptyset = \mathcal{P}_{-j}$ for all $j > N$. This implies that the sums $\sum_{j \text{ even}} |\mathcal{P}_j| = \sum_{j \text{ even}} p(n - \beta(j))$ and $\sum_{j \text{ odd}} |\mathcal{P}_j| = \sum_{j \text{ odd}} p(n - \beta(j))$ are finite. By (AP), it follows that we have $|\mathcal{P}_E| = \sum_{j \text{ even}} |\mathcal{P}_j| = \sum_{j \text{ even}} p(n - \beta(j))$ and $|\mathcal{P}_O| = \sum_{j \text{ odd}} |\mathcal{P}_j| = \sum_{j \text{ odd}} p(n - \beta(j))$. Thus, by (BP), we have

$$\sum_{j \text{ even}} p(n - \beta(j)) = |\mathcal{P}_E| = |\mathcal{P}_O| = \sum_{j \text{ odd}} p(n - \beta(j)).$$

73. Let $a_k = 1 + r + \cdots + r^k$ for all $k \geq 0$. Let $\mathcal{A}$ denote the set of partitions of n of the form $n = n_1 + \cdots + n_s$ (with $n_s > 0$) that satisfy $n_i \geq rn_{i+1}$ for all $i = 1, \ldots, s-1$, and let $\mathcal{B}$ denote the set of partitions of n, which each part is equal to a_k for some $k \geq 0$. We seek to establish a bijection between $\mathcal{A}$ and $\mathcal{B}$.

Let us take a partition $P \in \mathcal{A}$ of the form $n = n_1 + \cdots + n_s$, and we let $m_i = n_i - rn_{i+1}$ for all $i = 1, \ldots, s-1$ and $m_s = n_s$. Then it follows that we have $m_i \geq 0$ for all $i = 1, \ldots, s$. We then let $f(P)$ be the partition with exactly m_i parts of size a_{i-1} for all $i = 1, \ldots, s$. We claim that $f(P) \in \mathcal{B}$. Indeed, we see that

$$
\begin{aligned}
\sum_{i=1}^{s} m_i a_{i-1} &= \sum_{i=1}^{s-1} \sum_{k=0}^{i-1} (n_i - rn_{i+1}) r^k + n_s \sum_{k=0}^{s-1} r^k \\
&= \sum_{i=1}^{s-1} \sum_{k=0}^{i-1} n_i r^k - \sum_{i=1}^{s-1} \sum_{k=0}^{i-1} n_{i+1} r^{k+1} + n_s \sum_{k=0}^{s-1} r^k \\
&= \sum_{i=1}^{s} \sum_{k=0}^{i-1} n_i r^k - \sum_{i=2}^{s} \sum_{k=1}^{i-1} n_i r^k \\
&= n_1 + \sum_{i=2}^{s} \sum_{k=0}^{i-1} n_i r^k - \sum_{i=2}^{s} \sum_{k=1}^{i-1} n_i r^k \\
&= n_1 + \sum_{i=2}^{s} n_i \\
&= \sum_{i=1}^{s} n_i \\
&= n,
\end{aligned}
$$

which shows that $f(P)$ is a partition of n, and hence $f(P) \in \mathcal{B}$. Consequently, it follows that the map $f \colon \mathcal{A} \to \mathcal{B}$, defined by $P \mapsto f(P)$ for all $P \in \mathcal{A}$, is well-defined.

Next, we will show that f is surjective. Let us take a partition $Q \in \mathcal{B}$, and let the number of parts of size a_{i-1} of Q be m_i for all $i \geq 1$, and let the largest part of Q be a_{k-1}, where $k \geq 1$. Then we have $m_k > 0$, and $m_i = 0$ for all $i > k$. We let $n_i = \sum_{j=0}^{k-i} r^j m_{i+j}$ for all $i = 1, \ldots, k$. We first observe that $n_k = m_k > 0$, and

$$
n_i - rn_{i+1} = \sum_{j=0}^{k-i} r^j m_{i+j} - \sum_{j=0}^{k-i-1} r^{j+1} m_{i+1+j}
$$

$$= \sum_{j=0}^{k-i} r^j m_{i+j} - \sum_{j=1}^{k-i} r^j m_{i+j}$$

$$= m_i$$

$$\geq 0$$

for all $i = 1, \ldots, k - 1$. Finally, we see that

$$\sum_{i=1}^{k} n_i = \sum_{i=1}^{k} \sum_{j=0}^{k-i} r^j m_{i+j}$$

$$= \sum_{i=1}^{k} \sum_{\ell=i}^{k} r^{\ell-i} m_\ell \quad \text{(by setting } \ell = i + j)$$

$$= \sum_{\ell=1}^{k} m_\ell \sum_{i=1}^{\ell} r^{\ell-i}$$

$$= \sum_{\ell=1}^{k} m_\ell \sum_{p=0}^{\ell-1} r^p \quad \text{(by setting } p = \ell - i)$$

$$= \sum_{\ell=1}^{k} m_\ell a_{\ell-1}$$

$$= n.$$

Consequently, we see that the partition P of n of the form $n = n_1 + \cdots + n_k$ is in $\mathcal{P}$, and $f(P) = Q$. So this shows that f is surjective.

Finally, we will show that f is injective. Suppose $P, P' \in \mathcal{A}$ satisfy $f(P) = f(P')$. Write P and P' in the form $n = n_1 + \cdots + n_s$ and $n = n'_1 + \cdots + n'_t$ respectively. Let the number of parts of size a_{i-1} of the partition $f(P) = f(P')$ be m_i for all $i \geq 1$, and let the largest part of $f(P) = f(P')$ be a_{k-1}, where $k \geq 1$. Then we have $m_k > 0$, and $m_i = 0$ for all $i > k$. Moreover, it follows from our above construction that k is equal to the number of parts of P and P', so we have $s = k = t$. Thus, it follows from our above argument that we have $n_i = \sum_{j=0}^{k-i} r^j m_{i+j} = n'_i$ for all $i = 1, \ldots, k$. So this shows that $P = P'$, and hence f is injective. So f is a bijection, and hence we have $f(r, n) = |A| = |B| = g(r, n)$ by (BP) as desired.

Solutions to Exercise 6

1. The given recurrence relation may be written as $a_n - 3a_{n-1} + 2a_{n-2} = 0$. Its characteristic equation is $x^2 - 3x + 2 = 0$, and its characteristic roots are $\alpha_1 = 1$ and $\alpha_2 = 2$. Thus, by result (I), the general solution of the recurrence relation is given by $a_n = A + B \cdot 2^n$, where A and B are constants to be determined. The initial conditions $a_0 = 2$ and $a_1 = 3$ imply that

$$A + B = 2, \quad \text{and}$$
$$A + 2B = 3.$$

Solving the system for A and B gives us $A = B = 1$. So we have $a_n = 1 + 2^n$ for all $n \geq 0$.

2. The characteristic equation of the given recurrence relation is $x^2 - 6x + 9 = 0$, and its characteristic root is $\alpha = 3$ of multiplicity 2. Thus, by result (II), the general solution of the recurrence relation is given by $a_n = (A + Bn) \cdot 3^n$, where A and B are constants to be determined. The initial conditions $a_0 = 2$ and $a_1 = 3$ imply that

$$A = 2, \quad \text{and}$$
$$3(A + B) = 3.$$

Solving the system for A and B gives us $A = 2$ and $B = -1$. So we have $a_n = (2 - n)3^n$ for all $n \geq 0$.

3. The given recurrence relation may be written as $a_n - \frac{1}{2}a_{n-1} - \frac{1}{2}a_{n-2} = 0$. Its characteristic equation is $x^2 - \frac{1}{2}x - \frac{1}{2} = 0$, and its characteristic roots are $\alpha_1 = 1$ and $\alpha_2 = -\frac{1}{2}$. Thus, by result (I), the general solution of the recurrence relation is given by $a_n = A + B\left(-\frac{1}{2}\right)^n$, where A and B are constants to be determined. The initial conditions $a_0 = 0$ and

$a_1 = 1$ imply that
$$A + B = 0, \quad \text{and}$$
$$A - \frac{1}{2}B = 1.$$

Solving the system for A and B gives us $A = \frac{2}{3}$ and $B = -\frac{2}{3}$. So we have $a_n = \frac{2}{3}\left(1 - \left(-\frac{1}{2}\right)^n\right)$ for all $n \geq 0$.

4. The characteristic equation of the given recurrence relation is $x^2 - 4x + 4 = 0$, and its characteristic root is $\alpha = 2$ of multiplicity 2. Thus, by result (II), the general solution of the recurrence relation is given by $a_n = (A + Bn)2^n$, where A and B are constants to be determined. The initial conditions $a_0 = -\frac{1}{4}$ and $a_1 = 1$ imply that
$$A = -\frac{1}{4}, \quad \text{and}$$
$$2(A + B) = 1.$$

Solving the system for A and B gives us $A = -\frac{1}{4}$ and $B = \frac{3}{4}$. So we have $a_n = \left(-\frac{1}{4} + \frac{3n}{4}\right)2^n = (3n - 1)2^{n-2}$ for all $n \geq 0$.

5. The given recurrence relation may be written as $2a_n - a_{n-1} - 2a_{n-2} + a_{n-3} = 0$. Its characteristic equation is $2x^3 - x^2 - 2x + 1 = 0$, and its characteristic roots are $\alpha_1 = 1$, $\alpha_2 = -1$ and $\alpha_3 = \frac{1}{2}$. Thus, by result (I), the general solution of the recurrence relation is given by $a_n = A + B(-1)^n + C\left(\frac{1}{2}\right)^n$, where A, B and C are constants to be determined. The initial conditions $a_0 = 0$, $a_1 = 1$ and $a_2 = 2$ imply that
$$A + B + C = 0,$$
$$A - B + \frac{1}{2}C = 1, \quad \text{and}$$
$$A + B + \frac{1}{4}C = 2.$$

Solving the system for A, B and C gives us $A = \frac{5}{2}$, $B = \frac{1}{6}$ and $C = -\frac{8}{3}$. So we have $a_n = \frac{5}{2} + \frac{1}{6}(-1)^n - \frac{8}{3}\left(\frac{1}{2}\right)^n$ for all $n \geq 0$.

6. The characteristic equation of the given recurrence relation is $x^3 - 6x^2 + 11x - 6 = 0$, and its characteristic roots are $\alpha_1 = 1$, $\alpha_2 = 2$ and $\alpha_3 = 3$. Thus, by result (I), the general solution of the recurrence relation is given by $a_n = A + B \cdot 2^n + C \cdot 3^n$, where A, B and C are constants to be determined. The initial conditions $a_0 = \frac{1}{3}$, $a_1 = 1$ and $a_2 = 2$ imply that
$$A + B + C = \frac{1}{3},$$
$$A + 2B + 3C = 1, \quad \text{and}$$
$$A + 4B + 9C = 2.$$

Solving the system for A, B and C gives us $A = -\frac{1}{2}$, $B = 1$ and $C = -\frac{1}{6}$. So we have $a_n = -\frac{1}{2} + 2^n - \frac{1}{6} \cdot 3^n$ for all $n \geq 0$.

7. The given recurrence relation may be written as $a_n + a_{n-1} - 16a_{n-2} + 20a_{n-3} = 0$. Its characteristic equation is $x^3 + x^2 - 16x + 20 = 0$, and its characteristic roots are $\alpha_1 = 2$ of multiplicity 2 and $\alpha_2 = -5$. Thus, by result (II), the general solution of the recurrence relation is given by $a_n = (A + Bn)2^n + C(-5)^n$, where A, B and C are constants to be determined. The initial conditions $a_0 = 0$, $a_1 = 1$ and $a_2 = -1$ imply that

$$A + C = 0,$$
$$2A + 2B - 5C = 1, \quad \text{and}$$
$$4A + 8B + 25C = -1.$$

Solving the system for A, B and C gives us $A = \frac{5}{49}$, $B = \frac{1}{7}$ and $C = -\frac{5}{49}$. So we have

$$\begin{aligned}
a_n &= \left(\frac{5}{49} + \frac{n}{7}\right)2^n - \frac{5}{49}(-5)^n \\
&= \left(\frac{1}{7} + \frac{n}{7}\right)2^n - \frac{2}{49} \cdot 2^n - \frac{5}{49}(-1)^n 5^n \\
&= \frac{1}{7}(n+1)2^n - \frac{1}{49}(2^{n+1} + (-1)^n 5^{n+1})
\end{aligned}$$

for all $n \geq 0$.

8. The characteristic equation of the given recurrence relation is $x^4 + x^3 - 3x^2 - 5x - 2 = 0$, and its characteristic roots are $\alpha_1 = -1$ of multiplicity 3 and $\alpha_2 = 2$. By result (II), the general solution of the recurrence relation is given by

$$a_n = (A + Bn + Cn^2)(-1)^n + D \cdot 2^n$$

for all $n \geq 0$.

9. Let us first find $a_n^{(h)}$. The characteristic equation of $a_n - \frac{1}{2}a_{n-1} = 0$ is $x - \frac{1}{2} = 0$, and its characteristic root is $\alpha = \frac{1}{2}$. Thus, we have $a_n^{(h)} = A\left(\frac{1}{2}\right)^n$, where A is a constant.

Next, let us find $a_n^{(p)}$. Since $f(n) = -3$ is a polynomial in n of degree 0, and 1 is not a characteristic root, we let $a_n^{(p)} = B$, where B is a constant. Since $a_n^{(p)}$ satisfies the equation $a_n = \frac{1}{2}a_{n-1} - 3$, we must have $B = \frac{1}{2}B - 3$, or equivalently, $B = -6$, and hence $a_n^{(p)} = -6$. By (6.4.2), the general solution of the equation $a_n = \frac{1}{2}a_{n-1} - 3$ is given by

$a_n = a_n^{(h)} + a_n^{(p)} = A\left(\frac{1}{2}\right)^n - 6$. The initial condition $a_0 = 2(3 + \sqrt{3})$ implies that $A - 6 = 2(3 + \sqrt{3})$, or equivalently, $A = 2(\sqrt{3} + 6)$. Therefore, the required solution is $a_n = 2(\sqrt{3} + 6)\left(\frac{1}{2}\right)^n - 6 = \left(\frac{1}{2}\right)^{n-1}(\sqrt{3} + 6) - 6$ for all $n \geq 0$.

10. Let us first find $a_n^{(h)}$. The characteristic equation of $a_n - 3a_{n-1} = 0$ is $x - 3 = 0$, and its characteristic root is $\alpha = 3$. Thus, we have $a_n^{(h)} = A \cdot 3^n$, where A is a constant.

Next, let us find $a_n^{(p)}$. Since $f(n) = 3 \cdot 2^n - 4n$ is a sum of the exponential function $3 \cdot 2^n$ and the polynomial $-4n$ in n of degree 1, and 1 and 2 are not characteristic roots, we let $a_n^{(p)} = B \cdot 2^n + Cn + D$, where B, C and D are constants. Since $a_n^{(p)}$ satisfies the equation $a_n - 3a_{n-1} = 3 \cdot 2^n - 4n$, we must have

$$(B \cdot 2^n + Cn + D) - 3(B \cdot 2^{n-1} + C(n - 1) + D) = 3 \cdot 2^n - 4n.$$

By equating the coefficients of 2^n, n and the constant terms, we get

$$-\frac{1}{2}B = 3,$$
$$-2C = -4, \quad \text{and}$$
$$3C - 2D = 0.$$

Solving the system for B, C and D gives us $B = -6$, $C = 2$ and $D = 3$, and hence we have $a_n^{(p)} = -6 \cdot 2^n + 2n + 3$. By (6.4.2), the general solution of the equation $a_n - 3a_{n-1} = 3 \cdot 2^n - 4n$ is given by $a_n = a_n^{(h)} + a_n^{(p)} = A \cdot 3^n - 6 \cdot 2^n + 2n + 3$. The initial condition $a_1 = 2$ implies that $3A - 7 = 2$, or equivalently, $A = 3$.

Therefore, the required solution is $a_n = 3 \cdot 3^n - 6 \cdot 2^n + 2n + 3 = 3^{n+1} - 6 \cdot 2^n + 2n + 3$ for all $n \geq 1$.

11. Let us first find $a_n^{(h)}$. The characteristic equation of $a_n - a_{n-1} = 0$ is $x - 1 = 0$, and its characteristic root is $\alpha = 1$. Thus, we have $a_n^{(h)} = A$, where A is a constant.

Next, let us find $a_n^{(p)}$. Since $f(n) = 4n - 1$ is a polynomial in n of degree 1, and 1 is a characteristic root of multiplicity 1, we let $a_n^{(p)} = Bn + Cn^2$, where B and C are constants. Since $a_n^{(p)}$ satisfies the equation $a_n - a_{n-1} = 4n - 1$, we must have

$$(Bn + Cn^2) - (B(n - 1) + C(n - 1)^2) = 4n - 1.$$

By equating the coefficients of n and the constant terms, we get

$$2C = 4, \quad \text{and}$$
$$B - C = -1.$$

Solving the system for B and C gives us $B = 1$ and $C = 2$, and hence we have $a_n^{(p)} = n + 2n^2$. By (6.4.2), the general solution of the equation $a_n - a_{n-1} = 4n - 1$ is given by $a_n = a_n^{(h)} + a_n^{(p)} = A + n + 2n^2$. The initial condition $a_0 = 1$ implies that $A = 1$.

Therefore, the required solution is $a_n = 1 + n + 2n^2$ for all $n \geq 0$.

12. Let us first find $a_n^{(h)}$. The characteristic equation of $a_n - pa_{n-1} = 0$ is $x - p = 0$, and its characteristic root is $\alpha = p$. Thus, we have $a_n^{(h)} = Ap^n$, where A is a constant. Next, let us find $a_n^{(p)}$. Let us consider the following cases:

Case 1: $p \neq 1$. Since $f(n) = q$ is a polynomial in n of degree 0, and 1 is not a characteristic root, we let $a_n^{(p)} = B$, where B is a constant. Since $a_n^{(p)}$ satisfies the equation $a_n = pa_{n-1} + q$, we must have $B = pB + q$, or equivalently, $B = \frac{q}{1-p}$, and hence $a_n^{(p)} = \frac{q}{1-p}$. By (6.4.2), the general solution of the equation $a_n = pa_{n-1} + q$ is given by $a_n = a_n^{(h)} + a_n^{(p)} = Ap^n + \frac{q}{1-p}$. The initial condition $a_0 = r$ implies that $A + \frac{q}{1-p} = r$, or equivalently, $A = r - \frac{q}{1-p}$.

Therefore, the required solution for this case is

$$a_n = \left(r - \frac{q}{1-p} \right) p^n + \frac{q}{1-p} = rp^n + \frac{(1-p^n)q}{1-p}$$

for all $n \geq 0$.

Case 2: $p = 1$. Since $f(n) = q$ is a polynomial in n of degree 0, and 1 is a characteristic root of multiplicity 1, we let $a_n^{(p)} = Bn$, where B is a constant. Since $a_n^{(p)}$ satisfies the equation $a_n = pa_{n-1} + q = a_{n-1} + q$, we must have $Bn = B(n-1) + q$, or equivalently, $B = q$, and hence $a_n^{(p)} = qn$. By (6.4.2), the general solution of the equation $a_n = pa_{n-1} + q$ is given by $a_n = a_n^{(h)} + a_n^{(p)} = Ap^n + qn = A + qn$. The initial condition $a_0 = r$ implies that $A = r$.

Therefore, the required solution for this case is $a_n = r + qn$ for all $n \geq 0$.

We conclude that

$$a_n = \begin{cases} rp^n + \frac{(1-p^n)q}{1-p} & \text{if } p \neq 1, \\ r + qn & \text{if } p = 1 \end{cases}$$

for all $n \geq 0$.

13. As the sequence $\left(a_n - \frac{1}{10}a_{n-1} \right)$ is a geometric progression with common ratio $\frac{1}{2}$, it follows that we have $a_n - \frac{1}{10}a_{n-1} = C \left(\frac{1}{2} \right)^n$ for all $n \geq 1$ for some constant C. As $a_1 - \frac{1}{10}a_0 = \frac{3}{5} - \frac{1}{10} = \frac{1}{2}$, it follows that we must have $C = 1$, so that we have $a_n - \frac{1}{10}a_{n-1} = \left(\frac{1}{2} \right)^n$ for all $n \geq 1$. Now,

the characteristic equation of $a_n - \frac{1}{10}a_{n-1} = 0$ is $x - \frac{1}{10} = 0$, and its characteristic root is $\alpha = \frac{1}{10}$. Thus, we have $a_n^{(h)} = A\left(\frac{1}{10}\right)^n$, where A is a constant.

Next, let us find $a_n^{(p)}$. Since $f(n) = \left(\frac{1}{2}\right)^n$ is an exponential function and $\frac{1}{2}$ is not a characteristic root, we let $a_n^{(p)} = B\left(\frac{1}{2}\right)^n$, where B is a constant. Since $a_n^{(p)}$ satisfies the equation $a_n - \frac{1}{10}a_{n-1} = \left(\frac{1}{2}\right)^n$, we must have $B\left(\frac{1}{2}\right)^n - \frac{B}{10}\left(\frac{1}{2}\right)^{n-1} = \left(\frac{1}{2}\right)^n$, or equivalently, $B = \frac{5}{4}$. By (6.4.2), the general solution of the equation $a_n - \frac{1}{10}a_{n-1} = \left(\frac{1}{2}\right)^n$ is given by $a_n = a_n^{(h)} + a_n^{(p)} = A\left(\frac{1}{10}\right)^n + \frac{5}{4}\left(\frac{1}{2}\right)^n$. The initial condition $a_0 = 1$ implies that $A + \frac{5}{4} = 1$, or equivalently, $A = -\frac{1}{4}$.

Therefore, we have $a_n = -\frac{1}{4}\left(\frac{1}{10}\right)^n + \frac{5}{4}\left(\frac{1}{2}\right)^n = \frac{1}{4}\left[5\left(\frac{1}{2}\right)^n - \left(\frac{1}{10}\right)^n\right]$ for all $n \geq 0$.

14. By taking logarithm to the base 10 on both sides of the recurrence relation, we get

$$4\log a_n + \log a_{n-1} = 10.$$

We let $b_n = \log a_n$ for all $n \geq 0$, so that we have $4b_n + b_{n-1} = 10$ and $b_0 = \log a_0 = 0$. The characteristic equation of $4b_n + b_{n-1} = 0$ is $4x + 1 = 0$, and its characteristic root is $\alpha = -\frac{1}{4}$. Thus, we have $b_n^{(h)} = A\left(-\frac{1}{4}\right)^n = A(-1)^n 2^{-2n}$, where A is a constant.

Next, let us find $b_n^{(p)}$. Since $f(n) = 10$ is a polynomial in n of degree 0, and 1 is not a characteristic root, we let $b_n^{(p)} = B$, where B is a constant. Since $b_n^{(p)}$ satisfies the equation $4b_n + b_{n-1} = 10$, we must have $5B = 10$, or equivalently, $B = 2$, and hence $b_n^{(p)} = 2$. By (6.4.2), the general solution of the equation $4b_n + b_{n-1} = 10$ is given by $b_n = b_n^{(h)} + b_n^{(p)} = 2 + A(-1)^n 2^{-2n}$. The initial condition $b_0 = 0$ implies that $A + 2 = 0$, or equivalently, $A = -2$.

Therefore, we have $b_n = 2 - 2(-1)^n 2^{-2n} = 2(1 + (-1)^{n+1} 2^{-2n})$ for all $n \geq 0$, and consequently we have $a_n = 10^{b_n} = 10^{2(1+(-1)^{n+1}2^{-2n})}$ for all $n \geq 0$.

15. By taking logarithm on both sides of the recurrence relation, we get

$$\log a_n = \log 2 + \frac{1}{2}\log a_{n-1}.$$

We let $b_n = \log a_n$ for all $n \geq 0$, so that we have $b_n - \frac{1}{2}b_{n-1} = \log 2$ and $b_0 = \log a_0 = \log 25$. The characteristic equation of $b_n - \frac{1}{2}b_{n-1} = 0$ is $x - \frac{1}{2} = 0$, and its characteristic root is $\alpha = \frac{1}{2}$. Thus, we have $b_n^{(h)} = A\left(\frac{1}{2}\right)^n = A2^{-n}$, where A is a constant.

Next, let us find $b_n^{(p)}$. Since $f(n) = \log 2$ is a polynomial in n of degree 0, and 1 is not a characteristic root, we let $b_n^{(p)} = B$, where B is a constant. Since $b_n^{(p)}$ satisfies the equation $b_n - \frac{1}{2}b_{n-1} = \log 2$, we must have $\frac{1}{2}B = \log 2$, or equivalently, $B = \log 4$, and hence $b_n^{(p)} = \log 4$. By (6.4.2), the general solution of the equation $b_n - \frac{1}{2}b_{n-1} = \log 2$ is given by $b_n = b_n^{(h)} + b_n^{(p)} = \log 4 + A2^{-n}$. The initial condition $b_0 = \log 25$ implies that $A + \log 4 = \log 25$, or equivalently, $A = \log \frac{25}{4}$.

Therefore, we have $b_n = \log 4 + 2^{-n} \log \frac{25}{4}$ for all $n \geq 0$, and consequently we have $a_n = 10^{b_n} = 10^{(\log 4 + 2^{-n} \log \frac{25}{4})} = 4 \cdot 10^{2^{-n} \log \frac{25}{4}}$ for all $n \geq 0$. As $\lim\limits_{n \to \infty} 2^{-n} \log \frac{25}{4} = 0$, it follows that we have $\lim\limits_{n \to \infty} 10^{2^{-n} \log \frac{25}{4}} = 1$, and therefore we have

$$\lim_{n \to \infty} a_n = \lim_{n \to \infty} 4 \cdot 10^{2^{-n} \log \frac{25}{4}} = 4.$$

16. We first observe that we have $\frac{a_{n-1}}{a_{n-2}} = \left(\frac{a_n}{a_{n-1}}\right)^2 \geq 0$ for all $n \geq 2$. As $a_{n-1} \neq 0$ for all $n \geq 2$, it follows that we must have $\frac{a_{n-1}}{a_{n-2}} > 0$ for all $n \geq 2$. Finally, we shall prove by induction on n that we have $a_n > 0$ for all $n \geq 0$, with the base case $n = 0$ being trivial. Suppose that the statement holds for some non-negative integer $n = k \geq 0$. By induction hypothesis, we have $a_k > 0$. As we have $\frac{a_{n-1}}{a_{n-2}} > 0$ for all $n \geq 2$, it follows in particular that we have $\frac{a_{k+1}}{a_k} > 0$, and therefore we have $a_{k+1} = a_k \cdot \frac{a_{k+1}}{a_k} > 0$, thereby completing the induction step. So we have $a_n > 0$ for all $n \geq 0$.

Now, by taking logarithms base 2 on both sides of the recurrence relation, we get

$$2 \log_2 a_n - 2 \log_2 a_{n-1} = \log_2 a_{n-1} - \log_2 a_{n-2},$$

or equivalently,

$$2 \log_2 a_n - 3 \log_2 a_{n-1} + \log_2 a_{n-2} = 0.$$

We let $b_n = \log_2 a_n$ for all $n \geq 0$, so that we have $2b_n - 3b_{n-1} + b_{n-2} = 0$, $b_0 = \log_2 \frac{1}{4} = -2$ and $b_1 = \log_2 1 = 0$. The characteristic equation of $2b_n - 3b_{n-1} + b_{n-2} = 0$ is $2x^2 - 3x + 1 = 0$, and its characteristic roots are $\alpha_1 = 1$ and $\alpha_2 = \frac{1}{2}$. Thus, by result (I), the general solution of the recurrence relation $2b_n - 3b_{n-1} + b_{n-2} = 0$ is given by $b_n = A + B\left(\frac{1}{2}\right)^n = A + B2^{-n}$, where A and B are constants to be determined. The initial conditions $b_0 = -2$ and $b_1 = 0$ imply that

$$A + B = -2, \quad \text{and}$$

$$A + \frac{1}{2}B = 0.$$

Solving the system for A and B gives us $A = 2$ and $B = -4$. So we have $b_n = 2 - 4 \cdot 2^{-n} = 2(1 - 2^{1-n})$, and therefore we have $a_n = 2^{b_n} = 2^{2(1-2^{1-n})}$ for all $n \geq 0$.

17. Let us first find $a_n^{(h)}$. The characteristic equation of $a_n + 3a_{n-1} = 0$ is $x + 3 = 0$, and its characteristic root is $\alpha = -3$. Thus, we have $a_n^{(h)} = A(-3)^n$, where A is a constant.

Next, let us find $a_n^{(p)}$. Since $f(n) = 4n^2 - 2n + 2^n$ is a sum of the exponential function 2^n and the polynomial $4n^2 - 2n$ in n of degree 2, and 0, $\frac{1}{2}$ and 2 are not characteristic roots, we let $a_n^{(p)} = B \cdot 2^n + Cn^2 + Dn + E$, where B, C, D and E are constants. Since $a_n^{(p)}$ satisfies the equation $a_n + 3a_{n-1} = 4n^2 - 2n + 2^n$, we must have

$$(B \cdot 2^n + Cn^2 + Dn + E)$$
$$+3(B \cdot 2^{n-1} + C(n-1)^2 + D(n-1) + E)$$
$$= 4n^2 - 2n + 2^n.$$

By equating the coefficients of 2^n, n^2, n and the constant terms, we get

$$\frac{5}{2}B = 1,$$
$$4C = 4,$$
$$4D - 6C = -2, \quad \text{and}$$
$$4E + 3C - 3D = 0.$$

Solving the system for B, C, D and E gives us $B = \frac{2}{5}$, $C = 1$, $D = 1$ and $E = 0$, and hence we have $a_n^{(p)} = \frac{2}{5} \cdot 2^n + n^2 + n = n^2 + n + \frac{1}{5} \cdot 2^{n+1}$. By (6.4.2), the general solution of the equation $a_n + 3a_{n-1} = 4n^2 - 2n + 2^n$ is given by $a_n = a_n^{(h)} + a_n^{(p)} = A \cdot (-3)^n + n^2 + n + \frac{1}{5} \cdot 2^{n+1}$. The initial condition $a_0 = 1$ implies that $A + \frac{2}{5} = 1$, or equivalently, $A = \frac{3}{5}$. Therefore, the required solution is $a_n = \frac{3}{5}(-3)^n + n^2 + n + \frac{1}{5} \cdot 2^{n+1}$ for all $n \geq 0$.

18. The characteristic equation of the given recurrence relation is $x^2 - 2x + 2 = 0$, and its characteristic roots are $\alpha = 1 + i$ and $\bar{\alpha} = 1 - i$. By expressing α and $\bar{\alpha}$ in trigonometric form, we have $\alpha = \sqrt{2}\left(\cos\frac{\pi}{4}i + \sin\frac{\pi}{4}\right)$ and $\bar{\alpha} = \sqrt{2}\left(\cos\frac{\pi}{4} - i\sin\frac{\pi}{4}\right)$. Thus, by result (I), the general solution of the recurrence relation is given by

$$a_n = A\alpha^n + B\bar{\alpha}^n$$
$$= (\sqrt{2})^n \left[A\left(\cos\frac{n\pi}{4} + i\sin\frac{n\pi}{4}\right) + B\left(\cos\frac{n\pi}{4} - i\sin\frac{n\pi}{4}\right)\right]$$
$$= (\sqrt{2})^n \left(C\cos\frac{n\pi}{4} + D\sin\frac{n\pi}{4}\right),$$

where $C = A + B$ and $D = i(A - B)$ are constants to be determined. The initial conditions $a_0 = 1$ and $a_1 = 2$ imply that

$$C = 1, \quad \text{and}$$

$$\sqrt{2}\left(\frac{\sqrt{2}}{2}C + \frac{\sqrt{2}}{2}D\right) = 2.$$

Solving the system for C and D gives us $C = D = 1$. So we have

$$a_n = (\sqrt{2})^n \left(\cos\frac{n\pi}{4} + \sin\frac{n\pi}{4}\right)$$

for all $n \geq 0$.

19. Let us first find $a_n^{(h)}$. The characteristic equation of $a_n - 4a_{n-1} + 4a_{n-2} = 0$ is $x^2 - 4x + 4 = 0$, and its characteristic root is $\alpha = 2$ of multiplicity 2. Thus, we have $a_n^{(h)} = (A + Bn)2^n$, where A and B are constants.

Next, let us find $a_n^{(p)}$. Since $f(n) = 2^n$ is an exponential function, and 2 is a characteristic root of multiplicity 2, we let $a_n^{(p)} = Cn^2 2^n$, where C is a constant. Since $a_n^{(p)}$ satisfies the equation $a_n - 4a_{n-1} + 4a_{n-2} = 2^n$, we must have

$$Cn^2 2^n - 4C(n-1)^2 2^{n-1} + 4C(n-2)^2 2^{n-2} = 2^n.$$

By equating the coefficients of 2^n, we get $2C = 1$, or equivalently, $C = \frac{1}{2}$, and hence we have $a_n^{(p)} = \frac{1}{2}n^2 2^n$. By (6.4.2), the general solution of the equation $a_n - 4a_{n-1} + 4a_{n-2} = 2^n$ is given by $a_n = a_n^{(h)} + a_n^{(p)} = \left(A + Bn + \frac{1}{2}n^2\right)2^n$. The initial conditions $a_0 = 0$ and $a_1 = 3$ imply that

$$A = 0, \quad \text{and}$$

$$2(A + B + \frac{1}{2}) = 3.$$

Solving the system for A and B gives us $A = 0$ and $B = 1$.

Therefore, the required solution is $a_n = \left(n + \frac{1}{2}n^2\right)2^n$ for all $n \geq 0$.

20. Let us first find $a_n^{(h)}$. The characteristic equation of $a_n - a_{n-1} - 2a_{n-2} = 0$ is $x^2 - x - 2 = 0$, and its characteristic roots are $\alpha_1 = 2$ and $\alpha_2 = -1$. Thus, we have $a_n^{(h)} = A \cdot 2^n + B(-1)^n$, where A and B are constants.

Next, let us find $a_n^{(p)}$. Since $f(n) = 4n$ is a polynomial in n of degree 1, and 1 is not a characteristic root, we let $a_n^{(p)} = Cn + D$, where C and D are constants. Since $a_n^{(p)}$ satisfies the equation $a_n - a_{n-1} - 2a_{n-2} = 4n$, we must have

$$(Cn + D) - (C(n-1) + D) - 2(C(n-2) + D) = 4n.$$

By equating the coefficients of n and the constant terms, we get

$$-2C = 4, \quad \text{and}$$
$$5C - 2D = 0.$$

Solving the system for C and D gives us $C = -2$ and $D = -5$, and hence we have $a_n^{(p)} = -2n - 5$. By (6.4.2), the general solution of the equation $a_n - a_{n-1} - 2a_{n-2} = 4n$ is given by $a_n = a_n^{(h)} + a_n^{(p)} = A \cdot 2^n + B(-1)^n - 2n - 5$. The initial conditions $a_0 = -4$ and $a_1 = -5$ imply that

$$A + B - 5 = -4, \quad \text{and}$$
$$2A - B - 7 = -5.$$

Solving the system for A and B gives us $A = 1$ and $B = 0$.
Therefore, the required solution is $a_n = 2^n - 2n - 5$ for all $n \geq 0$.

21. Let us first find $a_n^{(h)}$. The characteristic equation of $a_n + a_{n-1} - 2a_{n-2} = 0$ is $x^2 + x - 2 = 0$, and its characteristic roots are $\alpha_1 = 1$ and $\alpha_2 = -2$. Thus, we have $a_n^{(h)} = A + B(-2)^n$, where A and B are constants.
Next, let us find $a_n^{(p)}$. Since $f(n) = 2^{n-2}$ is an exponential function, and 2 is not a characteristic root, we let $a_n^{(p)} = C2^n$, where C is a constant. Since $a_n^{(p)}$ satisfies the equation $a_n + a_{n-1} - 2a_{n-2} = 2^{n-2}$, we must have

$$C2^n + C2^{n-1} - 2C2^{n-2} = 2^{n-2}.$$

By equating the coefficients of 2^n, we get $C = \frac{1}{4}$, and hence we have $a_n^{(p)} = \frac{1}{4} \cdot 2^n = 2^{n-2}$. By (6.4.2), the general solution of the equation $a_n + a_{n-1} - 2a_{n-2} = 2^{n-2}$ is given by $a_n = a_n^{(h)} + a_n^{(p)} = A + B(-2)^n + 2^{n-2}$. The initial conditions $a_0 = 0$ and $a_1 = 0$ imply that

$$A + B + \frac{1}{4} = 0, \quad \text{and}$$
$$A - 2B + \frac{1}{2} = 0.$$

Solving the system for A and B gives us $A = -\frac{1}{3}$ and $B = \frac{1}{12}$.
Therefore, we have $a_n = -\frac{1}{3} + \frac{1}{12}(-2)^n + 2^{n-2} = -\frac{1}{3} + \frac{1}{3}(-2)^{n-2} + 2^{n-2}$ for all $n \geq 0$. Now, let us consider these cases:
Case 1: n is odd. Then we have $(-2)^{n-2} = -2^{n-2}$, and hence we have

$$a_n = -\frac{1}{3} + \frac{1}{3}(-2)^{n-2} + 2^{n-2}$$
$$= -\frac{1}{3} - \frac{1}{3} \cdot 2^{n-2} + 2^{n-2}$$

$$= -\frac{1}{3} + \frac{2}{3} \cdot 2^{n-2}$$

$$= -\frac{1}{3} + \frac{1}{3} \cdot 2^{n-1}$$

$$= \frac{1}{3}(2^{n-1} - 1).$$

Case 2: n is even. Then we have $(-2)^{n-2} = 2^{n-2}$, and hence we have

$$a_n = -\frac{1}{3} + \frac{1}{3}(-2)^{n-2} + 2^{n-2}$$

$$= -\frac{1}{3} + \frac{1}{3} \cdot 2^{n-2} + 2^{n-2}$$

$$= -\frac{1}{3} + \frac{4}{3} \cdot 2^{n-2}$$

$$= -\frac{1}{3} + \frac{1}{3} \cdot 2^n$$

$$= \frac{1}{3}(2^n - 1).$$

Hence, we conclude that

$$a_n = \begin{cases} \frac{1}{3}(2^{n-1} - 1) & \text{if } n \text{ is odd,} \\ \frac{1}{3}(2^n - 1) & \text{if } n \text{ is even} \end{cases}$$

for all $n \geq 0$.

22. Let us first find $a_n^{(h)}$. The characteristic equation of $a_n - 3a_{n-1} + 2a_{n-2} = 0$ is $x^2 - 3x + 2 = 0$, and its characteristic roots are $\alpha_1 = 1$ and $\alpha_2 = 2$. Thus, we have $a_n^{(h)} = A + B \cdot 2^n$, where A and B are constants.

Next, let us find $a_n^{(p)}$. Since $f(n) = 2^n$ is an exponential function, and 2 is a characteristic root of multiplicity 1, we let $a_n^{(p)} = Cn2^n$, where C is a constant. Since $a_n^{(p)}$ satisfies the equation $a_n - 3a_{n-1} + 2a_{n-2} = 2^n$, we must have

$$Cn2^n - 3C(n-1)2^{n-1} + 2C(n-2)2^{n-2} = 2^n.$$

By equating the coefficients of 2^n, we get $\frac{1}{2}C = 1$, or equivalently, $C = 2$, and hence we have $a_n^{(p)} = 2n \cdot 2^n = n2^{n+1}$. By (6.4.2), the general solution of the equation $a_n - 3a_{n-1} + 2a_{n-2} = 2^n$ is given by $a_n = a_n^{(h)} + a_n^{(p)} = A + B \cdot 2^n + n2^{n+1}$. The initial conditions $a_0 = 0$ and $a_1 = 5$ imply that

$$A + B = 0, \quad \text{and}$$

$$A + 2B + 4 = 5.$$

Solving the system for A and B gives us $A = -1$ and $B = 1$.
Therefore, the required solution is $a_n = -1 + 2^n + n2^{n+1}$ for all $n \geq 0$.

23. Let us first find $a_n^{(h)}$. The characteristic equation of $a_n + 5a_{n-1} + 6a_{n-2} = 0$ is $x^2 + 5x + 6 = 0$, and its characteristic roots are $\alpha_1 = -2$ and $\alpha_2 = -3$. Thus, we have $a_n^{(h)} = A(-2)^n + B(-3)^n$, where A and B are constants.

Next, let us find $a_n^{(p)}$. Since $f(n) = 3n^2$ is a polynomial in n of degree 2, and 1 is not a characteristic root, we let $a_n^{(p)} = Cn^2 + Dn + E$, where C, D and E are constants. Since $a_n^{(p)}$ satisfies the equation $a_n + 5a_{n-1} + 6a_{n-2} = 3n^2$, we must have

$$(Cn^2 + Dn + E) + 5(C(n-1)^2 + D(n-1) + E)$$
$$+ 6(C(n-2)^2 + D(n-2) + E)$$
$$= 3n^2.$$

By equating the coefficients of n^2, n and the constant terms, we get

$$12C = 3,$$
$$12D - 34C = 0, \quad \text{and}$$
$$12E - 17D + 29C = 0.$$

Solving the system for C, D and E gives us $C = \frac{1}{4}$, $D = \frac{17}{24}$ and $E = \frac{115}{288}$, and hence we have $a_n^{(p)} = \frac{1}{4}n^2 + \frac{17}{24}n + \frac{115}{288}$. By (6.4.2), the general solution of the equation $a_n + 5a_{n-1} + 6a_{n-2} = 3n^2$ is given by $a_n = a_n^{(h)} + a_n^{(p)} = A(-2)^n + B(-3)^n + \frac{1}{4}n^2 + \frac{17}{24}n + \frac{115}{288}$. The initial conditions $a_0 = 0$ and $a_1 = 1$ imply that

$$A + B + \frac{115}{288} = 0, \quad \text{and}$$
$$-2A - 3B + \frac{391}{288} = 1.$$

Solving the system for A and B gives us $A = -\frac{14}{9}$ and $B = \frac{37}{32}$. Therefore, the required solution is $a_n = -\frac{14}{9}(-2)^n + \frac{37}{32}(-3)^n + \frac{1}{4}n^2 + \frac{17}{24}n + \frac{115}{288}$ for all $n \geq 0$.

24. The characteristic equation of the given recurrence relation is $px^2 + qx + r = 0$, and its discriminant D is equal to

$$q^2 - 4pr = (-p-r)^2 - 4pr = p^2 + 2pr + r^2 - 4pr = p^2 - 2pr + r^2 = (p-r)^2.$$

Let us consider cases:

Case 1: $p \neq r$. Then the characteristic roots of the characteristic equation are $\alpha_1 = \frac{-q+p-r}{2p} = 1$ and $\alpha_2 = \frac{-q+r-p}{2p} = \frac{r}{p}$. Thus, by result (I), the general solution of the recurrence relation is given by

$a_n = A + B \left(\frac{r}{p}\right)^n$, where A and B are constants to be determined. The initial conditions $a_0 = s$ and $a_1 = t$ imply that

$$A + B = s, \quad \text{and}$$

$$A + \frac{rB}{p} = t.$$

Solving the system for A and B gives us $A = s + \frac{p(t-s)}{p-r}$ and $B = -\frac{p(t-s)}{p-r}$. So we have $a_n = s + \frac{p(t-s)}{p-r} \left(1 - \left(\frac{r}{p}\right)^n\right)$ for all $n \geq 0$.

Case 2: $p = r$. Then the characteristic root of the characteristic equation is $\alpha = \frac{-q}{2p} = 1$ of multiplicity 2. Thus, by result (II), the general solution of the recurrence relation is given by $a_n = A + Bn$, where A and B are constants to be determined. The initial conditions $a_0 = s$ and $a_1 = t$ imply that

$$A = s, \quad \text{and}$$

$$A + B = t.$$

Solving the system for A and B gives us $A = s$ and $B = t - s$. So we have $a_n = s + (t - s)n$ for all $n \geq 0$.

Hence, we conclude that

$$a_n = \begin{cases} s + \frac{p(t-s)}{p-r} \left(1 - \left(\frac{r}{p}\right)^n\right) & \text{if } p \neq r, \\ s + (t - s)n & \text{if } p = r, \end{cases}$$

for all $n \geq 0$.

25. (i) We have

$$ra_n + s = \frac{rpa_{n-1} + rq}{ra_{n-1} + s} + s$$

$$= \frac{rpa_{n-1} + ps - ps + rq}{ra_{n-1} + s} + s$$

$$= \frac{p(ra_{n-1} + s)}{ra_{n-1} + s} + \frac{rq - ps}{ra_{n-1} + s} + s$$

$$= p + s + \frac{rq - ps}{ra_{n-1} + s}.$$

(ii) By making the substitution $ra_n + s = \frac{b_{n+1}}{b_n}$ into (1), we get

$$\frac{b_{n+1}}{b_n} = p + s + \frac{rq - ps}{\frac{b_n}{b_{n-1}}} = p + s + \frac{(rq - ps)b_{n-1}}{b_n}. \tag{6.1}$$

By multiplying b_n on both sides of equation (6.1), we get

$$b_{n+1} = (p + s)b_n + (rq - ps)b_{n-1},$$

or equivalently,

$$b_{n+1} - (p+s)b_n + (ps - rq)b_{n-1} = 0$$

as desired.

26. Let us prove by induction on n that we have $a_n > 0$ for all $n \geq 0$, with the base case $n = 0$ being trivial. Suppose that the statement holds for some non-negative integer $n = k \geq 0$. By induction hypothesis, we have $a_k > 0$, and hence we have $a_{k+1} = \frac{3a_k}{2a_k+1} > 0$, thereby completing the induction step. So we have $a_n > 0$ for all $n \geq 0$.

Now, we let $b_0 = 1$, and $b_{n+1} = b_n(2a_n + 1)$ for all $n \geq 0$. Let us prove by induction on n that we have $b_n > 0$ for all $n \geq 0$, with the base case $n = 0$ being trivial. Suppose that the statement holds for some non-negative integer $n = k \geq 0$. By induction hypothesis, we have $b_k > 0$. Since $a_k > 0$, it follows that we have $2a_k + 1 > 0$, and hence we have $b_{k+1} = b_k(2a_k + 1) > 0$, thereby completing the induction step. So we have $b_n > 0$ for all $n \geq 0$. Consequently, we have $2a_n + 1 = \frac{b_{n+1}}{b_n}$ for all $n \geq 0$.

. By making the substitution $2a_n + 1 = \frac{b_{n+1}}{b_n}$ into the recurrence relation, the recurrence relation reduces to the following recurrence relation by Problem 6.25(ii):

$$b_{n+1} - (3 + 1)b_n + (3 \cdot 1 - 0 \cdot 2)b_{n-1} = 0,$$

or equivalently,

$$b_{n+1} - 4b_n + 3b_{n-1} = 0.$$

Its characteristic equation is $x^2 - 4x + 3 = 0$, and its characteristic roots are $\alpha_1 = 1$ and $\alpha_2 = 3$. Thus, by result (I), the general solution of the recurrence relation is given by $b_n = A + B \cdot 3^n$, where A and B are constants to be determined. The initial conditions $b_0 = 1$ and $b_1 = b_0(2a_0 + 1) = \frac{3}{2}$ imply that

$$A + B = 1, \quad \text{and}$$

$$A + 3B = \frac{3}{2}.$$

Solving the system for A and B gives us $A = \frac{3}{4}$ and $B = \frac{1}{4}$. So we have $b_n = \frac{3}{4} + \frac{1}{4} \cdot 3^n = \frac{3}{4}(3^{n-1} + 1)$ for all $n \geq 0$. Hence, we have

$$a_n = \frac{1}{2}\left(\frac{b_{n+1}}{b_n} - 1\right)$$

$$= \frac{1}{2}\left(\frac{\frac{3}{4}(3^n + 1)}{\frac{3}{4}(3^{n-1} + 1)} - 1\right)$$

$$= \frac{1}{2} \left(\frac{3^n + 1}{3^{n-1} + 1} - 1 \right)$$

$$= \frac{1}{2} \cdot \frac{(3^n + 1) - (3^{n-1} + 1)}{3^{n-1} + 1}$$

$$= \frac{3^{n-1}}{3^{n-1} + 1}$$

for all $n \geq 0$.

27. Let us prove by induction on n that we have $a_n > 0$ for all $n \geq 0$, with the base case $n = 0$ being trivial. Suppose that the statement holds for some non-negative integer $n = k \geq 0$. By induction hypothesis, we have $a_k > 0$, and hence we have $a_{k+1} = \frac{3a_k + 1}{a_k + 3} > 0$, thereby completing the induction step. So we have $a_n > 0$ for all $n \geq 0$.

Now, we let $b_0 = 1$, and $b_{n+1} = b_n(a_n + 3)$ for all $n \geq 0$. Let us prove by induction on n that we have $b_n > 0$ for all $n \geq 0$, with the base case $n = 0$ being trivial. Suppose that the statement holds for some non-negative integer $n = k \geq 0$. By induction hypothesis, we have $b_k > 0$. Since $a_k > 0$, it follows that we have $a_k + 3 > 0$, and hence we have $b_{k+1} = b_k(a_k + 3) > 0$, thereby completing the induction step. So we have $b_n > 0$ for all $n \geq 0$. Consequently, we can write $a_n + 3 = \frac{b_{n+1}}{b_n}$ for all $n \geq 0$.

By making the substitution $a_n + 3 = \frac{b_{n+1}}{b_n}$ into the recurrence relation, the recurrence relation reduces to the following recurrence relation by Problem 6.25(ii):

$$b_{n+1} - (3 + 3)b_n + (3 \cdot 3 - 1 \cdot 1)b_{n-1} = 0,$$

or equivalently,

$$b_{n+1} - 6b_n + 8b_{n-1} = 0.$$

Its characteristic equation is $x^2 - 6x + 8 = 0$, and its characteristic roots are $\alpha_1 = 2$ and $\alpha_2 = 4$. Thus, by result (I), the general solution of the recurrence relation is given by $b_n = A \cdot 2^n + B \cdot 4^n$, where A and B are constants to be determined. The initial conditions $b_0 = 1$ and $b_1 = b_0(a_0 + 3) = 8$ imply that

$$A + B = 1, \quad \text{and}$$
$$2A + 4B = 8.$$

Solving the system for A and B gives us $A = -2$ and $B = 3$. So we have $b_n = -2 \cdot 2^n + 3 \cdot 4^n = 2^{n+1}(3 \cdot 2^{n-1} - 1)$ for all $n \geq 0$. Hence, we

have

$$a_n = \frac{b_{n+1}}{b_n} - 3$$

$$= \frac{2^{n+2}(3 \cdot 2^n - 1)}{2^{n+1}(3 \cdot 2^{n-1} - 1)} - 3$$

$$= \frac{3 \cdot 2^{n+1} - 2}{3 \cdot 2^{n-1} - 1} - 3$$

$$= \frac{3 \cdot 2^{n+1} - 2 - 3(3 \cdot 2^{n-1} - 1)}{3 \cdot 2^{n-1} - 1}$$

$$= \frac{12 \cdot 2^{n-1} - 2 - 9 \cdot 2^{n-1} + 3}{3 \cdot 2^{n-1} - 1}$$

$$= \frac{3 \cdot 2^{n-1} + 1}{3 \cdot 2^{n-1} - 1}$$

for all $n \geq 0$.

28. Let $a_1 = a$, $b_1 = 1$, and $b_{n+1} = b_n(2 - a_n)$ for all $n \geq 1$. Let us prove by induction on n that we have $b_n \neq 0$ for all $n \geq 1$, with the base case $n = 0$ being trivial. Suppose that the statement holds for some non-negative integer $n = k \geq 1$. By induction hypothesis, we have $b_k > 0$. Since $(2 - a_k)a_{k+1} = 1$, it follows that we have $2 - a_k \neq 0$, and hence we have $b_{k+1} = b_k(2 - a_k) \neq 0$, thereby completing the induction step. So we have $b_n \neq 0$ for all $n \geq 1$. Consequently, we can write $2 - a_n = \frac{b_{n+1}}{b_n}$ for all $n \geq 1$.

By making the substitution $2 - a_n = \frac{b_{n+1}}{b_n}$ into the recurrence relation, the recurrence relation reduces to the following recurrence relation by Problem 6.25(ii):

$$b_{n+1} - (0 + 2)b_n + (0 \cdot 2 - 1 \cdot (-1))b_{n-1} = 0,$$

or equivalently,

$$b_{n+1} - 2b_n + b_{n-1} = 0.$$

Its characteristic equation is $x^2 - 2x + 1 = 0$, and its characteristic root is $\alpha = 1$ of multiplicity 2. Thus, by result (II), the general solution of the recurrence relation is given by $b_n = A + Bn$, where A and B are constants to be determined. The initial conditions $b_1 = 1$ and $b_2 = b_1(2 - a_1) = 2 - a$ imply that

$$A + B = 1, \quad \text{and}$$
$$A + 2B = 2 - a.$$

Solving the system for A and B gives us $A = a$ and $B = 1 - a$. So we have $b_n = a + (1 - a)n$ for all $n \geq 1$. Hence, we have

$$a_n = 2 - \frac{b_{n+1}}{b_n}$$

$$= 2 - \frac{a + (1 - a)(n + 1)}{a + (1 - a)n}$$

$$= \frac{2(a + (1 - a)n) - (a + (1 - a)(n + 1))}{a + (1 - a)n}$$

$$= \frac{2a + 2(1 - a)n - a - (1 - a)n - 1 + a}{a + (1 - a)n}$$

$$= \frac{(2a - 1) + (1 - a)n}{a + (1 - a)n}$$

$$= 1 + \frac{a - 1}{a + (1 - a)n}$$

for all $n \geq 1$. If $a = 1$, then $a_n = 1$ for all $n \geq 1$ and thus we have $\lim_{n \to \infty} a_n = 1$. Else, if $a_n \neq 1$, then

$$\lim_{n \to \infty} a_n = \lim_{n \to \infty} \left[1 + \frac{a - 1}{a + (1 - a)n} \right] = 1.$$

In both cases, we conclude that $\lim_{n \to \infty} a_n = 1$.

29. Let $a = a_1 a_2 \cdots a_n$ be a derangement of $\mathbf{N}_n$, and let $a_1 = m$. Then we must have $2 \leq m \leq n$. If $a_m = 1$, then $a_2 a_3 \cdots a_{m-1} a_{m+1} \cdots a_n$ is a derangement of $\mathbf{N}_n \setminus \{1, m\}$, and there are D_{n-2} such derangements of $\mathbf{N}_n \setminus \{1, m\}$ by definition. Else, we must have $a_m \neq 1$. For each $i = 2, 3, \ldots, n$, let us define b_i as follows:

$$b_i = \begin{cases} a_i & \text{if } a_i \neq 1, \\ m & \text{if } a_i = 1. \end{cases}$$

Since $a_1 = m$ and $a_m \neq 1$, it follows that $b_m \neq m$, so it is easy to see that there is a bijection between the set of derangements $a_1 a_2 \cdots a_n$ of $\mathbf{N}_n$ with $a_1 = m$ and $a_m \neq 1$, and the set of derangements $b_2 b_3 \cdots b_n$ of $\mathbf{N}_n \setminus \{1\}$. Now, by definition, there are D_{n-1} derangements $b_2 b_3 \cdots b_r$ of $\mathbf{N}_n \setminus \{1\}$, and thus there are D_{n-1} derangements $a_1 a_2 \cdots a_n$ of $\mathbf{N}_n$ with $a_1 = m$ and $a_m \neq 1$. By (AP), there are $D_{n-1} + D_{n-2}$ derangements $a_1 a_2 \cdots a_n$ of $\mathbf{N}_n$ with $a_1 = m$. As there are $n - 1$ choices for m, it follows from (MP) that there are $(n - 1)(D_{n-1} + D_{n-2})$ derangements $a_1 a_2 \cdots a_n$ of $\mathbf{N}_n$. As there are D_n derangements $a_1 a_2 \cdots a_n$ of $\mathbf{N}_n$ by definition, it follows that we must have $D_n = (n - 1)(D_{n-1} + D_{n-2})$ as desired.

30. Let $d = d_1 d_2 \cdots d_n$ be a ternary sequence of length n in which no two 0's are adjacent. Let us consider the following cases:

Case 1: $d_1 \neq 0$. Then $d_2 d_3 \cdots d_n$ is a ternary sequence of length $n-1$ in which no two 0's are adjacent, and there are no restrictions on d_2. Thus, for a fixed $i \in \{1, 2\}$, there is a bijection between the set of ternary sequences $d_1 d_2 \cdots d_n$ of length n in which no two 0's are adjacent and $d_1 = i$, and the set of ternary sequences of length $n-1$ in which no two 0's are adjacent. Since there are a_{n-1} ternary sequences of length $n-1$ in which no two 0's are adjacent, it follows that there are a_{n-1} ternary sequences $d_1 d_2 \cdots d_n$ of length n in which no two 0's are adjacent and $d_1 = i$. As there are 2 choices for i, it follows from (MP) that there are $2a_{n-1}$ ternary sequences $d_1 d_2 \cdots d_n$ of length n in which no two 0's are adjacent and $d_1 \neq 0$.

Case 2: $d_1 = 0$. Then $d_2 d_3 \cdots d_n$ is a ternary sequence of length $n-1$ in which no two 0's are adjacent, and $d_2 \neq 0$. A similar argument as in Case 1 would show that there are $2a_{n-2}$ ternary sequences $d_1 d_2 \cdots d_n$ of length n in which no two 0's are adjacent and $d_1 = 0$.

By (AP), we must have $a_n = 2a_{n-1} + 2a_{n-2}$. The given recurrence relation may be written as $a_n - 2a_{n-1} - 2a_{n-2} = 0$. Its characteristic equation is $x^2 - 2x - 2 = 0$, and its characteristic roots are $\alpha_1 = 1 + \sqrt{3}$ and $\alpha_2 = 1 - \sqrt{3}$. Thus, by result (I), the general solution of the recurrence relation is given by $a_n = A(1 + \sqrt{3})^n + B(1 - \sqrt{3})^n$, where A and B are constants to be determined. The initial conditions $a_1 = 3$ and $a_2 = 3^2 - 1 = 8$ imply that

$$A(1 + \sqrt{3}) + B(1 - \sqrt{3}) = 3, \quad \text{and}$$
$$A(1 + \sqrt{3})^2 + B(1 - \sqrt{3})^2 = 8.$$

Solving the system for A and B gives us $A = \frac{3 + 2\sqrt{3}}{6}$ and $B = \frac{3 - 2\sqrt{3}}{6}$. So we have $a_n = \frac{3 + 2\sqrt{3}}{6}(1 + \sqrt{3})^n + \frac{3 - 2\sqrt{3}}{6}(1 - \sqrt{3})^n$ for all $n \geq 1$.

31. (i) Suppose that $n \geq 1$. As the hyperbola is symmetric about the y-axis, it follows that the center of the circle C_n must lie on the y-axis as well. Let the center of C_n be $(0, y_n)$. Then it follows that the equation of C_n is $x^2 + (y - y_n)^2 = a_n^2$. As C_n is tangent to C_{n-1}, we must have $y_n - y_{n-1} = a_n + a_{n-1}$.

Next, we let (p, q) be a point of tangency between the circle C_n and the hyperbola $x^2 - y^2 = 1$. Then the equation of the tangent lines to C_n and the hyperbola $x^2 - y^2 = 1$ at (p, q) are $px + (q - y_n)y = a_n^2 + y_n(q - y_n)$ and $px - qy = 1$ respectively. As (p, q) is a point of tangency between C_n and the hyperbola $x^2 - y^2 = 1$, the two tangent lines must be the same, so we must have $q - y_n = -q$, or equivalently, $q = \frac{y_n}{2}$. Also, we must have $a_n^2 + y_n(q - y_n) = 1$, or equivalently, $y_n^2 = 2a_n^2 - 2$. This implies that we have $y_n^2 - y_{n-1}^2 = 2(a_n^2 - a_{n-1}^2)$, where $y_0 = 0$. That is, $(y_n + y_{n-1})(y_n - y_{n-1}) = 2(a_n + a_{n-1})(a_n - a_{n-1})$. As $y_n - y_{n-1} = a_n + a_{n-1}$, we must have $y_n + y_{n-1} = 2(a_n - a_{n-1})$, which would then imply that $2y_n = 3a_n - a_{n-1}$. We now substitute $y_n = \frac{1}{2}(3a_n - a_{n-1})$ into the equation $y_n - y_{n-1} = a_n + a_{n-1}$ to deduce that $a_n = 6a_{n-1} - a_{n-2}$ for all $n \geq 2$.

(ii) From the equations $y_1 = y_0 + a_1 + a_0 = a_1 + 1$ and $2y_1 = 3a_1 - a_0 = 3a_1 - 1$, we get $a_1 = 3$. By the recurrence relation $a_n = 6a_{n-1} - a_{n-2}$ and the initial conditions $a_0 = 1$ and $a_1 = 3$, it follows that a_n must be an integer for all $n \geq 0$. The characteristic equation of the recurrence relation $a_n = 6a_{n-1} - a_{n-2}$ is $x^2 - 6x + 1 = 0$, and its characteristic roots are $\alpha_1 = 3 + 2\sqrt{2}$ and $\alpha_2 = 3 - 2\sqrt{2}$. Thus, by result (I), the general solution of the recurrence relation is given by $a_n = A(3 + 2\sqrt{2})^n + B(3 - 2\sqrt{2})^n$, where A and B are constants to be determined. The initial conditions $a_0 = 1$ and $a_1 = 3$ imply that

$$A + B = 1, \quad \text{and}$$
$$(3 + 2\sqrt{2})A + (3 - 2\sqrt{2})B = 3.$$

Solving the system for A and B gives us $A = B = \frac{1}{2}$. So we have $a_n = \frac{1}{2}[(3 + 2\sqrt{2})^n + (3 - 2\sqrt{2})^n]$ for all $n \geq 0$.

32. By applying the cofactor expansion of a_n along the first row, it follows that we have $a_n = (p + q)a_{n-1} - pqb_{n-1}$, where

$$b_{n-1} = \left.\begin{vmatrix} 1 & pq & 0 & 0 & \cdots & 0 & 0 \\ 0 & p+q & pq & 0 & \cdots & 0 & 0 \\ 0 & 1 & p+q & pq & \cdots & 0 & 0 \\ 0 & 0 & 1 & p+q & \cdots & 0 & 0 \\ \vdots & \vdots & \vdots & \vdots & \ddots & \vdots & \vdots \\ 0 & 0 & 0 & 0 & \cdots & p+q & pq \\ 0 & 0 & 0 & 0 & \cdots & 1 & p+q \end{vmatrix}\right\} n - 1.$$

$$\underbrace{\qquad\qquad\qquad\qquad}_{n-1}$$

Next, by applying the cofactor expansion of b_{n-1} along the first column, we see that $b_{n-1} = a_{n-2}$. It follows from the above equation that we have $a_n = (p + q)a_{n-1} - pqa_{n-2}$. Its characteristic equation is

$x^2 - (p+q)x + pq = 0$, and its characteristic roots are p and q. Let us consider the following cases:

Case 1: $p \neq q$. By result (I), the general solution of the recurrence relation is given by $a_n = Ap^n + Bq^n$, where A and B are constants to be determined. The initial conditions $a_1 = p+q$ and $a_2 = (p+q)^2 - pq = p^2 + pq + q^2$ imply that

$$Ap + Bq = p + q, \quad \text{and}$$
$$Ap^2 + Bq^2 = p^2 + pq + q^2.$$

Solving the system for A and B gives us $A = \frac{p}{p-q}$ and $B = -\frac{q}{p-q}$. Thus, we have $a_n = \frac{p^{n+1} - q^{n+1}}{p-q}$ in this case.

Case 2: $p = q$. By result (II), the general solution of the recurrence relation is given by $a_n = (A + Bn)p^n$, where A and B are constants to be determined. The initial conditions $a_1 = 2p$ and $a_2 = 3p^2$ imply that

$$(A + B)p = 2p, \quad \text{and}$$
$$(A + 2B)p^2 = 3p^2.$$

Solving the system for A and B gives us $A = B = 1$. Thus, we have $a_n = (1 + n)p^n$ in this case.

We conclude that

$$a_n = \begin{cases} \frac{p^{n+1} - q^{n+1}}{p-q} & \text{if } p \neq q, \\ (1 + n)p^n & \text{if } p = q. \end{cases}$$

33. By applying the cofactor expansion of a_n along the first row, it follows that we have $a_n = (pq + 1)a_{n-1} - qb_{n-1}$, where

$$b_{n-1} = \left.\begin{vmatrix} p & q & 0 & 0 & \cdots & 0 \\ 0 & pq+1 & q & 0 & \cdots & 0 \\ 0 & p & pq+1 & q & \cdots & 0 \\ 0 & 0 & p & pq+1 & \cdots & 0 \\ \vdots & \vdots & \vdots & \vdots & \ddots & \vdots \\ 0 & 0 & 0 & 0 & \cdots & q \\ 0 & 0 & 0 & 0 & \cdots & pq+1 \end{vmatrix}\right\} n-1.$$

$$\underbrace{}_{n-1}$$

Next, by applying the cofactor expansion of b_{n-1} along the first column we see that $b_{n-1} = pa_{n-2}$. It follows from the above equation that we have $a_n = (pq + 1)a_{n-1} - pqa_{n-2}$. Its characteristic equation is $x^2 - (pq + 1)x + pq = 0$, and its characteristic roots are pq and 1.

Let us consider the following cases:

Case 1: $pq \neq 1$. By result (I), the general solution of the recurrence relation is given by $a_n = A + B(pq)^n$, where A and B are constants to be determined. The initial conditions $a_1 = pq + 1$ and $a_2 = (pq+1)^2 - pq = p^2q^2 + pq + 1$ imply that

$$A + Bpq = pq + 1, \quad \text{and}$$
$$A + Bp^2q^2 = p^2q^2 + pq + 1.$$

Solving the system for A and B gives us $A = \frac{1}{1-pq}$ and $B = -\frac{pq}{1-pq}$. Thus, we have $a_n = \frac{1-(pq)^{n+1}}{1-pq}$ in this case.

Case 2: $pq = 1$. By result (II), the general solution of the recurrence relation is given by $a_n = A + Bn$, where A and B are constants to be determined. The initial conditions $a_1 = 2$ and $a_2 = 3$ imply that

$$A + B = 2, \quad \text{and}$$
$$A + 2B = 3.$$

Solving the system for A and B gives us $A = B = 1$. Thus, we have $a_n = 1 + n$ in this case.

We conclude that

$$a_n = \begin{cases} \frac{1-(pq)^{n+1}}{1-pq} & \text{if } pq \neq 1, \\ 1 + n & \text{if } pq = 1. \end{cases}$$

34. For each $n \in \mathbf{N}$, we let a_n be the number of sequences $d_1 d_2 \cdots d_n$ of length n with $d_i \in \{1, 2, 3, 4\}$ for all $i = 1, \ldots, n$ in which 1 and 2 are not adjacent. Let $d_1 d_2 \cdots d_n$ be such a sequence. Let us consider the following cases:

Case 1: $d_1 \in \{3, 4\}$. Then $d_2 d_3 \cdots d_n$ is a sequence of length $n - 1$ with $d_i \in \{1, 2, 3, 4\}$ for all $i = 2, \ldots, n$ in which 1 and 2 are not adjacent, and there are no restrictions on d_2. Thus, for a fixed $j \in \{3, 4\}$, there is a bijection between the set of sequences $d_1 d_2 \cdots d_n$ of length n with $d_i \in \{1, 2, 3, 4\}$ for all $i = 1, \ldots, n$ in which 1 and 2 are not adjacent and $d_1 = j$, and the set of sequences $d_2 \cdots d_n$ of length $n - 1$ with $d_i \in \{1, 2, 3, 4\}$ for all $i = 2, \ldots, n$ in which 1 and 2 are not adjacent. Since there are a_{n-1} sequences $d_2 \cdots d_n$ of length $n - 1$ with $d_i \in \{1, 2, 3, 4\}$ for all $i = 2, \ldots, n$ in which 1 and 2 are not adjacent, it follows that there are a_{n-1} sequences $d_1 d_2 \cdots d_n$ of length n with $d_i \in \{1, 2, 3, 4\}$ for all $i = 1, \ldots, n$ in which 1 and 2 are not adjacent and $d_1 = j$. As there are 2 choices for j, it follows from (MP) that there are $2a_{n-1}$ sequences

$d_1 d_2 \cdots d_n$ of length n with $d_i \in \{1, 2, 3, 4\}$ for all $i = 1, \ldots, n$ in which 1 and 2 are not adjacent and $d_1 \in \{3, 4\}$.

Case 2: $d_1 = 1$ and $d_2 \in \{3, 4\}$. Then $d_3 \cdots d_n$ is a sequence of length $n - 2$ with $d_i \in \{1, 2, 3, 4\}$ for all $i = 3, \ldots, n$ in which 1 and 2 are not adjacent, and there are no restrictions on d_3. A similar argument as in Case 1 would now show that there are $2a_{n-2}$ sequences $d_1 d_2 \cdots d_n$ of length n with $d_i \in \{1, 2, 3, 4\}$ for all $i = 1, \ldots, n$ in which 1 and 2 are not adjacent, $d_1 = 1$ and $d_2 \in \{3, 4\}$.

Case 3: $(d_1, d_2) \in \{(1, 1), (2, 2), (2, 3), (2, 4)\}$. Then $d_2 \cdots d_n$ is a sequence of length $n - 1$ with $d_i \in \{1, 2, 3, 4\}$ for all $i = 2, \ldots, n$ in which 1 and 2 are not adjacent. Moreover, d_1 is uniquely determined by d_2 in this case. Thus, there is a bijection between the set of sequences $d_1 d_2 \cdots d_n$ of length n with $d_i \in \{1, 2, 3, 4\}$ for all $i = 1, \ldots, n$ in which 1 and 2 are not adjacent and $(d_1, d_2) = (1, 1), (2, 2), (2, 3), (2, 4)$, and the set of sequences $d_2 \cdots d_n$ of length $n - 1$ with $d_i \in \{1, 2, 3, 4\}$ for all $i = 2, \ldots, n$ in which 1 and 2 are not adjacent. A similar argument as in Case 1 would now show that there are a_{n-1} sequences $d_1 d_2 \cdots d_n$ of length n with $d_i \in \{1, 2, 3, 4\}$ for all $i = 1, \ldots, n$ in which 1 and 2 are not adjacent and $(d_1, d_2) = (1, 1), (2, 2), (2, 3), (2, 4)$.

By (AP), we must have $a_n = 3a_{n-1} + 2a_{n-2}$. The given recurrence relation may be written as $a_n - 3a_{n-1} - 2a_{n-2} = 0$. Its characteristic equation is $x^2 - 3x - 2 = 0$, and its characteristic roots are $\alpha_1 = \frac{3+\sqrt{17}}{2}$ and $\alpha_2 = \frac{3-\sqrt{17}}{2}$. Thus, by result (I), the general solution of the recurrence relation is given by $a_n = A\left(\frac{3+\sqrt{17}}{2}\right)^n + B\left(\frac{3-\sqrt{17}}{2}\right)^n$, where A and B are constants to be determined. The initial conditions $a_1 = 4$ and $a_2 = 4^2 - 2 = 14$ imply that

$$A\left(\frac{3+\sqrt{17}}{2}\right) + B\left(\frac{3-\sqrt{17}}{2}\right) = 4, \quad \text{and}$$

$$A\left(\frac{3+\sqrt{17}}{2}\right)^2 + B\left(\frac{3-\sqrt{17}}{2}\right)^2 = 14.$$

Solving the system for A and B gives us $A = \frac{17+5\sqrt{17}}{34}$ and $B = \frac{17-5\sqrt{17}}{34}$. Therefore, we have $a_n = \frac{17+5\sqrt{17}}{34}\left(\frac{3+\sqrt{17}}{2}\right)^n + \frac{17-5\sqrt{17}}{34}\left(\frac{3-\sqrt{17}}{2}\right)^n$ for all $n \geq 1$.

35. Let us take any pavement of the $2 \times n$ rectangle with 1×2 identical blocks and 2×2 identical blocks. Let us consider the following cases:

Case 1: The top leftmost square of the $2 \times n$ rectangle is covered by a vertical 1×2 block. Then there are no restrictions on how the top leftmost square of the remaining $2 \times (n-1)$ rectangle can be covered, and so there are a_{n-1} such pavements in this case.

Case 2: The top leftmost square of the $2 \times n$ rectangle is covered by a horizontal 1×2 block. Then the bottom leftmost square of the $2 \times n$ rectangle must be covered by a horizontal 1×2 block as well, and there are no restrictions on how the top leftmost square of the remaining $2 \times (n-2)$ rectangle can be covered. Thus, there are a_{n-2} such pavements in this case.

Case 3: The top leftmost square of the $2 \times n$ rectangle is covered by a 2×2 block. A similar argument as in Case 1 shows that there are a_{n-2} such pavements in this case.

By (AP), we must have $a_n = a_{n-1} + 2a_{n-2}$. The given recurrence relation may be written as $a_n - a_{n-1} - 2a_{n-2} = 0$. Its characteristic equation is $x^2 - x - 2 = 0$, and its characteristic roots are $\alpha_1 = 2$ and $\alpha_2 = -1$. Thus, by result (I), the general solution of the recurrence relation is given by $a_n = A \cdot 2^n + B(-1)^n$, where A and B are constants to be determined. The initial conditions $a_1 = 1$ and $a_2 = 3$ imply that

$$2A - B = 1, \quad \text{and}$$
$$4A + B = 3.$$

Solving the system for A and B gives us $A = \frac{2}{3}$ and $B = \frac{1}{3}$. Therefore, we have $a_n = \frac{1}{3}(2^{n+1} + (-1)^n)$ for all $n \geq 1$.

36. For each $n \geq 1$, we let b_n (respectively b'_n) denote the number of pavements of a $3 \times 2n$ rectangle such that the top (respectively bottom) leftmost square of the $3 \times 2n$ rectangle is covered by a vertical 1×2 domino, and c_n denote the number of pavements of a $3 \times 2n$ rectangle such that each of the squares in the leftmost column of the $3 \times 2n$ rectangle is covered by a horizontal 1×2 domino. By (AP), we have $a_{2n} = b_n + b'_n + c_n$, and by symmetry, we have $b'_n = b_n$. Considering the tiling for the initial sections of the rectangle, we find that $b_{n+1} = 2b_n + b'_n + c_n = 3b_n + c_n$, $c_{n+1} = b_n + b'_n + c_n = 2b_n + c_n$ and $a_{2n} = c_{n+1}$ for all $n \geq 1$. This implies that we have $b_n = \frac{1}{2}(c_{n+1} - c_n)$, and we substitute this into the equation $b_{n+1} = 3b_n + c_n$ to get $\frac{1}{2}(c_{n+2} - c_{n+1}) = \frac{3}{2}(c_{n+1} - c_n) + c_n$, or equivalently, $c_{n+2} - 4c_{n+1} + c_n = 0$. The resulting characteristic equation is given by $x^2 - 4x + 1 = 0$, and its characteristic roots are $\alpha_1 = 2 + \sqrt{3}$ and $\alpha_2 = 2 - \sqrt{3}$. Thus, by result (I), the general solution of the recurrence relation is given by $c_n = A(2 + \sqrt{3})^n + B(2 - \sqrt{3})^n$, where A and B are constants to be determined. The initial conditions

$c_1 = 1$ and $c_2 = 2b_1 + c_1 = 3$ imply that

$$(2 + \sqrt{3})A + (2 - \sqrt{3})B = 1, \quad \text{and}$$
$$(2 + \sqrt{3})^2 A + (2 - \sqrt{3})^2 B = 3.$$

Solving the system for A and B gives us $A = \frac{\sqrt{3}-1}{2\sqrt{3}}$ and $B = \frac{\sqrt{3}+1}{2\sqrt{3}}$. So we have $c_n = \frac{1}{2\sqrt{3}}[(\sqrt{3}-1)(2+\sqrt{3})^n + (\sqrt{3}+1)(2-\sqrt{3})^n]$, and hence we have $a_{2n} = c_{n+1} = \frac{1}{2\sqrt{3}}[(\sqrt{3}+1)(2+\sqrt{3})^n + (\sqrt{3}-1)(2-\sqrt{3})^n]$.

Remark. The solution to the above problem is given by D. Ž. Djoković, University of Waterloo, Canada. We shall refer the reader to *Amer. Math. Monthly*, **81** (1974), 522–523 for further details.

37. From the first equation, we get $b_n = a_n - a_{n+1}$. We substitute this into the second equation to get $a_{n+1} - a_{n+2} = a_n + 3a_n - 3a_{n+1}$, or equivalently, $a_{n+2} = 4a_{n+1} - 4a_n$. The corresponding characteristic equation is given by $x^2 - 4x + 4 = 0$, and its characteristic root is $\alpha = 2$ of multiplicity 2. Thus, by result (II), the general solution of the recurrence relation is given by $a_n = (A + Bn) \cdot 2^n$, where A and B are constants to be determined. The initial conditions $a_0 = -1$ and $a_1 = a_0 - b_0 = -6$ imply that

$$A = -1, \quad \text{and}$$
$$2(A + B) = -6.$$

Solving the system for A and B gives us $A = -1$ and $B = -2$. So we have $a_n = -(1+2n)2^n$, and hence we have $b_n = a_n - a_{n+1} = (5+2n)2^n$.

38. From the first recurrence relation we have $b_{n-1} = -\frac{1}{2}(a_n + a_{n-1})$. We substitute this into the second relation to obtain to get $-\frac{1}{2}(a_{n+1}+a_n) - 2a_{n-1} + \frac{3}{2}(a_n + a_{n-1}) = 0$, or equivalently, $a_{n+1} - 2a_n + a_{n-1} = 0$. The resulting characteristic equation is given by $x^2 - 2x + 1 = 0$, and its characteristic root is $\alpha = 1$ of multiplicity 2. Thus, by result (II), the general solution of the recurrence relation is given by $a_n = A + Bn$, where A and B are constants to be determined. The initial conditions $a_0 = 1$ and $a_1 = -a_0 - 2b_0 = -1$, and hence we have

$$A = 1, \quad \text{and}$$
$$A + B = -1.$$

Solving the system for A and B gives us $A = 1$ and $B = -2$. So we have $a_n = 1 - 2n$ for all $n \geq 0$, and hence we have $b_n = -\frac{1}{2}(a_{n+1}+a_n) = 2n$ for all $n \geq 0$.

39. From the first equation, we get $b_{n-1} = -5a_n + \frac{9}{2}a_{n-1}$. We substitute this into the second equation to get $-25a_{n+1} + \frac{45}{2}a_n = -a_{n-1} - 15a_n + \frac{27}{2}a_{n-1}$, or equivalently, $2a_{n+1} = 3a_n - a_{n-1}$. The resulting characteristic equation is given by $2x^2 - 3x + 1 = 0$, and its characteristic roots are $\alpha_1 = 1$ and $\alpha_2 = \frac{1}{2}$. Thus, by result (I), the general solution of the recurrence relation is given by $a_n = A + B\left(\frac{1}{2}\right)^n$, where A and B are constants to be determined. The initial conditions $a_0 = 4$ and $a_1 = \frac{1}{10}(9a_0 - 2b_0) = 3$ imply that

$$A + B = 4, \quad \text{and}$$

$$A + \frac{1}{2}B = 3.$$

Solving the system for A and B gives us $A = B = 2$. So we have $a_n = 2\left(1 + \left(\frac{1}{2}\right)^n\right)$, and hence $b_n = -5a_{n+1} + \frac{9}{2}a_n = -1 + \left(\frac{1}{2}\right)^{n-2}$.

40. From the first equation, we get $b_{n-1} = 3a_n - 2a_{n-1}$. We substitute this into the second equation to get $9a_{n+1} - 6a_n - a_{n-1} - 6a_n + 4a_{n-1} = 0$, or equivalently, $3a_{n+1} = 4a_n - a_{n-1}$. The corresponding characteristic equation is given by $3x^2 - 4x + 1 = 0$, and its characteristic roots are $\alpha_1 = 1$ and $\alpha_2 = \frac{1}{3}$. Thus, by result (I), the general solution of the recurrence relation is given by $a_n = A + B\left(\frac{1}{3}\right)^n$, where A and B are constants to be determined. The initial conditions $a_0 = 2$ and $a_1 = \frac{1}{3}(2a_{n-1} + b_{n-1}) = 1$ imply that

$$A + B = 2, \quad \text{and}$$

$$A + \frac{1}{3}B = 1.$$

Solving the system for A and B gives us $A = \frac{1}{2}$ and $B = \frac{3}{2}$. So we have $a_n = \frac{1}{2}\left(1 + \left(\frac{1}{3}\right)^{n-1}\right)$, and hence $b_n = 3a_{n+1} - 2a_n = \frac{1}{2}\left(1 - \left(\frac{1}{3}\right)^{n-1}\right)$.

41. We let $c_n = \log_2 a_n$ and $d_n = \log_2 b_n$ for all $n \geq 0$. Then we have $2c_n = c_{n-1} + d_n$ and $2d_n = c_n + d_{n-1}$. From the first equation, we get $d_n = 2c_n - c_{n-1}$. We substitute this into the second equation to get $4c_n - 2c_{n-1} = c_n + 2c_{n-1} - c_{n-2}$, or equivalently, $3c_n - 4c_{n-1} + c_{n-2} = 0$. The corresponding characteristic equation is given by $3x^2 - 4x + 1 = 0$, and its characteristic roots are $\alpha_1 = 1$ and $\alpha_2 = \frac{1}{3}$. Thus, by result (I), the general solution of the recurrence relation is given by $c_n = A + B\left(\frac{1}{3}\right)^n$, where A and B are constants to be determined. The initial conditions $c_0 = \log_2 \frac{1}{8} = -3$ and $c_1 = \frac{1}{3}(2c_0 + d_0) = 0$ imply that

$$A + B = -3, \quad \text{and}$$

$$A + \frac{1}{3}B = 0.$$

Solving the system for A and B gives us $A = \frac{3}{2}$ and $B = -\frac{9}{2}$. So we have $c_n = \frac{3}{2}\left(1 - \left(\frac{1}{3}\right)^{n-1}\right)$, and hence $d_n = 2c_n - c_{n-1} = \frac{3}{2}\left(1 + \left(\frac{1}{3}\right)^{n-1}\right)$. This implies that $a_n = 2^{\frac{3}{2}\left(1-\left(\frac{1}{3}\right)^{n-1}\right)}$ and $b_n = 2^{\frac{3}{2}\left(1+\left(\frac{1}{3}\right)^{n-1}\right)}$. As $\lim\limits_{n\to\infty} \frac{3}{2}\left(\frac{1}{3}\right)^{n-1} = 0$, it follows that we have $\lim\limits_{n\to\infty} 2^{\frac{3}{2}\left(\frac{1}{3}\right)^{n-1}} = 1$, and therefore we have

$$\lim_{n\to\infty} a_n = \lim_{n\to\infty} 2^{\frac{3}{2}\left(1-\left(\frac{1}{3}\right)^{n-1}\right)} = 2^{\frac{3}{2}} = \lim_{n\to\infty} 2^{\frac{3}{2}\left(1+\left(\frac{1}{3}\right)^{n-1}\right)} = \lim_{n\to\infty} b_n.$$

42. (i) Let us pick any binary sequence $s_1 s_2 \cdots s_n$ of length n with an even number of 0's and an even number of 1's. If $s_1 = 0$, then $s_2 \cdots s_n$ is a binary sequence of length $n-1$ with an odd number of 0's and an even number of 1's, and if $s_1 = 1$, then $s_2 \cdots s_n$ is a binary sequence of length $n-1$ with an even number of 0's and an odd number of 1's. Consequently, it follows from (AP) that we have $a_n = b_{n-1} + c_{n-1}$. A similar reasoning as above would also show that $d_n = b_{n-1} + c_{n-1}$.

Next, let us pick any binary sequence $s_1 s_2 \cdots s_n$ of length n with an even number of 0's and an odd number of 1's. If $s_1 = 0$, then $s_2 \cdots s_n$ is a binary sequence of length $n-1$ with an odd number of 0's and an odd number of 1's, and if $s_1 = 1$, then $s_2 \cdots s_n$ is a binary sequence of length $n-1$ with an even number of 0's and an even number of 1's. Consequently, it follows from (AP) that we have $b_n = a_{n-1} + d_{n-1}$. A similar reasoning as above would also show that $c_n = a_{n-1} + d_{n-1}$.

(ii) As the initial conditions for the recurrence relations are $a_0 = 1$ and $b_0 = c_0 = d_0 = 0$, it follows from a similar argument as in Example 6.73 that we have the following system of equations:

$$A(x) - x(B(x) + C(x)) = 1,$$
$$B(x) - x(A(x) + D(x)) = 0,$$
$$C(x) - x(A(x) + D(x)) = 0, \quad \text{and}$$
$$D(x) - x(B(x) + C(x)) = 0.$$

Solving the system for $A(x), B(x), C(x)$ and $D(x)$ gives us $A(x) = \frac{1-2x^2}{1-4x^2}$, $B(x) = C(x) = \frac{x}{1-4x^2}$ and $D(x) = \frac{2x^2}{1-4x^2}$.

(iii) Firstly, we have

$$A(x) = \frac{1 - 2x^2}{1 - 4x^2}$$

$$= \frac{1}{2} + \frac{1}{2(1 - 2x)(1 + 2x)}$$

$$= \frac{1}{2} + \frac{1}{4} \left[\frac{1}{1 + 2x} + \frac{1}{1 - 2x} \right]$$

$$= \frac{1}{2} + \sum_{n=0}^{\infty} [(-2)^{n-2} + 2^{n-2}] x^n,$$

where the last equality follows from (5.1.4). This implies that we have $a_n = (-2)^{n-2} + 2^{n-2}$ for all $n \geq 1$.

Next, we have

$$B(x) = \frac{x}{1 - 4x^2}$$

$$= \frac{1}{4} \left[-\frac{1}{1 + 2x} + \frac{1}{1 - 2x} \right]$$

$$= \sum_{n=0}^{\infty} [-(-2)^{n-2} + 2^{n-2}] x^n,$$

where the last equality follows from (5.1.4). This implies that we have $b_n = -(-2)^{n-2} + 2^{n-2}$ for all $n \geq 0$. As $B(x) = C(x)$, it follows that we have $c_n = b_n = -(-2)^{n-2} + 2^{n-2}$ for all $n \geq 0$.

Finally, we have

$$D(x) = \frac{2x^2}{1 - 4x^2}$$

$$= -\frac{1}{2} + \frac{1}{2(1 - 2x)(1 + 2x)}$$

$$= -\frac{1}{2} + \sum_{n=0}^{\infty} [(-2)^{n-2} + 2^{n-2}] x^n,$$

so we have $d_n = (-2)^{n-2} + 2^{n-2}$ for all $n \geq 1$.

43. (i) Let $A(x), B(x)$ and $C(x)$ be the generating functions of the sequences $(a_n), (b_n)$ and (c_n) respectively. As the initial conditions for the recurrence relations are $a_0 = p$, $b_0 = q$ and $c_0 = r$, it follows from a similar argument as in Example 6.73 that we have the following system of equations:

$$A(x) - \frac{x}{2}(B(x) + C(x) - A(x)) = p,$$

$$B(x) - \frac{x}{2}(C(x) + A(x) - B(x)) = q, \quad \text{and}$$

$$C(x) - \frac{x}{2}(A(x) + B(x) - C(x)) = r.$$

Solving the system for $A(x)$ gives us $A(x) = \frac{-x(q+r)-2p}{(x-2)(x+1)}$. Since $\frac{x}{(x-2)(x+1)} = \frac{1}{3}\left[\frac{2}{x-2} + \frac{1}{x+1}\right]$ and $\frac{1}{(x-2)(x+1)} = \frac{1}{3}\left[\frac{1}{x-2} - \frac{1}{x+1}\right]$, it follows that we have

$$A(x) = \frac{1}{3}\left[-\frac{2(p+q+r)}{x-2} + \frac{2p-q-r}{x+1}\right]$$

$$= \frac{1}{3}\left[\frac{p+q+r}{1-\frac{x}{2}} + \frac{2p-q-r}{1+x}\right]$$

$$= \frac{1}{3}\sum_{n=0}^{\infty}\left[(p+q+r)\left(\frac{1}{2}\right)^n + (2p-q-r)(-1)^n\right]x^n,$$

where the last equality follows from (5.1.4). This implies that we have $a_n = \frac{1}{3}(p+q+r)\left(\frac{1}{2}\right)^n + (-1)^n\frac{1}{3}(2p-q-r)$ for all $n \geq 0$.

(ii) Suppose on the contrary that $2p - q - r \neq 0$. As

$$\lim_{n\to\infty}\frac{1}{3}(p+q+r)\left(\frac{1}{2}\right)^n = 0,$$

there exists some $N \in \mathbf{N}$, such that

$$\frac{1}{3}\left|(p+q+r)\left(\frac{1}{2}\right)^n\right| < \frac{1}{3}|2p-q-r|$$

for all $n \geq N$. Let us pick any $m \geq N$ for which

$$(-1)^m\frac{1}{3}(2p-q-r) = -\frac{1}{3}|2p-q-r|.$$

Then we have

$$a_m = \frac{1}{3}(p+q+r)\left(\frac{1}{2}\right)^m + (-1)^m\frac{1}{3}(2p-q-r)$$

$$\leq \frac{1}{3}\left|(p+q+r)\left(\frac{1}{2}\right)^m\right| - \frac{1}{3}|2p-q-r|$$

$$< \frac{1}{3}|2p-q-r| - \frac{1}{3}|2p-q-r|$$

$$= 0,$$

which contradicts the fact that $a_n > 0$ for all $n \geq 0$. So we must have $2p - q - r = 0$. By symmetry, we must have $2q - r - p = 0 = 2r - p - q$, and hence we must have $p = q = r$.

44. (i) We shall proceed to prove by induction on n that we have $\sum_{r=1}^{n} F_r = F_{n+2} - 1$ for all $n \geq 1$, with the base case $n = 1$ being trivial. Suppose that the statement holds for $n = k$, where $k \geq 1$. By

induction hypothesis, we have $\sum_{r=1}^{k} F_r = F_{k+2} - 1$. This implies that we have

$$\sum_{r=1}^{k+1} F_r = F_{k+1} + \sum_{r=1}^{k} F_r = F_{k+1} + F_{k+2} - 1 = F_{k+3} - 1,$$

which completes the induction step. We are done.

(ii) We shall proceed to prove by induction on n that we have $\sum_{r=1}^{n} F_{2r} = F_{2n+1} - 1$ for all $n \geq 1$, with the base case $n = 1$ being trivial. Suppose that the statement holds for $n = k$, where $k \geq 1$. By induction hypothesis, we have $\sum_{r=1}^{k} F_{2r} = F_{2k+1} - 1$. This implies that we have

$$\sum_{r=1}^{k+1} F_{2r} = F_{2k+2} + \sum_{r=1}^{k} F_{2r}$$
$$= F_{2k+2} + F_{2k+1} - 1$$
$$= F_{2k+3} - 1$$
$$= F_{2(k+1)+1} - 1,$$

which completes the induction step. We are done.

(iii) We shall proceed to prove by induction on n that we have $\sum_{r=1}^{n} F_{2r-1} = F_{2n}$ for all $n \geq 1$, with the base case $n = 1$ being trivial. Suppose that the statement holds for $n = k$, where $k \geq 1$. By induction hypothesis, we have $\sum_{r=1}^{k} F_{2r-1} = F_{2k}$. This implies that we have

$$\sum_{r=1}^{k+1} F_{2r-1} = F_{2k+1} + \sum_{r=1}^{k} F_{2r-1} = F_{2k+1} + F_{2k} = F_{2k+2} = F_{2(k+1)},$$

which completes the induction step. We are done.

(iv) We shall proceed to prove by induction on n that we have $\sum_{r=1}^{n} (-1)^{r+1} F_r = (-1)^{n+1} F_{n-1} + 1$ for all $n \geq 2$, with the base case $n = 2$ being trivial. Suppose that the statement holds for $n = k$, where $k \geq 2$. By induction hypothesis, we have $\sum_{r=1}^{k} (-1)^{r+1} F_r = (-1)^{k+1} F_{k-1} + 1$. This implies that we have

$$\sum_{r=1}^{k+1} (-1)^{r+1} F_r = (-1)^{k+2} F_{k+1} + \sum_{r=1}^{k} (-1)^{r+1} F_r$$

$$= (-1)^{k+2} F_{k+1} + (-1)^{k+1} F_{k-1} + 1$$
$$= (-1)^{k+2} F_k + (-1)^{k+2} F_{k-1} + (-1)^{k+1} F_{k-1} + 1$$
$$= (-1)^{k+2} F_k + 1,$$

which completes the induction step. We are done.

45. (i) We shall prove by induction on n that we have $F_{m+n} = F_m F_{n-1} + F_{m+1} F_n$ for all $m \geq 1$ and $n \geq 2$, with the base case $n = 2$ being trivial. Suppose that the statement holds for $n = k$, where $k \geq 2$. By induction hypothesis, we have $F_{m+k+1} = F_{m+1+k} = F_{m+1} F_{k-1} + F_{m+2} F_k$. This implies that we have

$$F_{m+k+1} = F_{m+1} F_{k-1} + F_{m+2} F_k$$
$$= F_{m+1}(F_{k+1} - F_k) + (F_{m+1} + F_m) F_k$$
$$= F_m F_k + F_{m+1} F_{k+1},$$

and this completes the induction step. We are done.

(ii) We shall prove by induction on n that we have

$$\begin{pmatrix} 1 & 1 \\ 1 & 0 \end{pmatrix}^n = \begin{pmatrix} F_{n+1} & F_n \\ F_n & F_{n-1} \end{pmatrix}$$

for all $n \geq 2$, with the base case $n = 2$ being trivial. Suppose that the statement holds for $n = k$, where $k \geq 2$. By induction hypothesis, we have

$$\begin{pmatrix} 1 & 1 \\ 1 & 0 \end{pmatrix}^k = \begin{pmatrix} F_{k+1} & F_k \\ F_k & F_{k-1} \end{pmatrix}.$$

This implies that we have

$$\begin{pmatrix} 1 & 1 \\ 1 & 0 \end{pmatrix}^{k+1} = \begin{pmatrix} 1 & 1 \\ 1 & 0 \end{pmatrix}^k \begin{pmatrix} 1 & 1 \\ 1 & 0 \end{pmatrix}$$
$$= \begin{pmatrix} F_{k+1} & F_k \\ F_k & F_{k-1} \end{pmatrix} \begin{pmatrix} 1 & 1 \\ 1 & 0 \end{pmatrix}$$
$$= \begin{pmatrix} F_{k+1} + F_k & F_{k+1} \\ F_k + F_{k+1} & F_k \end{pmatrix}$$
$$= \begin{pmatrix} F_{k+2} & F_{k+1} \\ F_{k+1} & F_k \end{pmatrix},$$

and this completes the induction step. We are done.

(iii) By part (ii), we have

$$F_{n+1} F_{n-1} - F_n^2 = \begin{vmatrix} F_{n+1} & F_n \\ F_n & F_{n-1} \end{vmatrix} = \begin{vmatrix} 1 & 1 \\ 1 & 0 \end{vmatrix}^n = (-1)^n.$$

(iv) We have

$$F_{n+1}^2 = (F_n + F_{n-1})^2$$
$$= (2F_{n-1} + F_{n-2})^2$$
$$= 4F_{n-1}^2 + 4F_{n-1}F_{n-2} + F_{n-2}^2$$
$$= 4F_{n-1}^2 + 4F_{n-1}(F_n - F_{n-1}) + F_{n-2}^2$$
$$= 4F_nF_{n-1} + F_{n-2}^2.$$

(v) We shall prove by induction on n that we have $(F_n, F_{n+1}) = 1$ for all $n \geq 1$, with the base case $n = 1$ being trivial. Suppose that the statement holds for $n = k$, where $k \geq 1$. By induction hypothesis, we have $(F_k, F_{k+1}) = 1$. Let d be a common divisor of F_{k+1} and F_{k+2}. Since $F_k = F_{k+2} - F_{k+1}$, it follows that d divides F_k, and hence d divides $(F_k, F_{k+1}) = 1$, which implies that $d = 1$. Consequently, we must have $(F_{k+1}, F_{k+2}) = 1$. This completes the induction step, and we are done.

46. (i) The identity follows by setting $m = n - 1$ in Problem 6.45(i).
    (ii) We have

$$F_{n+1}^2 - F_{n-1}^2 = (F_n + F_{n-1})F_{n+1} - F_{n-1}(F_{n+1} - F_n)$$
$$= F_{n+1}F_n + F_nF_{n-1}$$
$$= F_{2n},$$

where the last equality follows by setting $m = n$ in Problem 6.45(i).

(iii) We have

$$F_{n+1}^3 + F_n^3 - F_{n-1}^3$$
$$= F_{n+1}^2(F_n + F_{n-1}) + F_n^3 - F_{n-1}^3$$
$$= F_n(F_n^2 + F_{n+1}^2) + F_{n-1}(F_{n+1}^2 - F_{n-1}^2)$$
$$= F_nF_{2n+1} + F_{n-1}F_{2n} \quad \text{(by Problem 6.46(i) and (ii))}$$
$$= F_{3n},$$

where the last equality follows by setting $m = 2n$ in Problem 6.45(i).

47. (i) The identity evidently holds for $n = 1$, so let us assume that $n \geq 2$. We have

$$\sum_{r=1}^{n} F_r^2 = 1 + \sum_{r=2}^{n} F_r^2$$
$$= 1 + \sum_{r=2}^{n} F_r(F_{r+1} - F_{r-1})$$

$$= 1 + \sum_{r=2}^{n} F_r F_{r+1} - \sum_{r=2}^{n} F_r F_{r-1}$$

$$= 1 + \sum_{r=2}^{n} F_r F_{r+1} - \sum_{r=1}^{n-1} F_r F_{r+1}$$

$$= 1 + F_n F_{n+1} - F_1 F_2$$

$$= F_n F_{n+1}.$$

(ii) We have

$$\sum_{r=1}^{2n-1} F_r F_{r+1}$$

$$= \frac{1}{2} \sum_{r=1}^{2n-1} [(F_r + F_{r+1})^2 - F_{r+1}^2 - F_r^2]$$

$$= \frac{1}{2} \sum_{r=1}^{2n-1} F_{r+2}^2 - \frac{1}{2} \sum_{r=1}^{2n-1} F_{r+1}^2 - \frac{1}{2} \sum_{r=1}^{2n-1} F_r^2$$

$$= \frac{1}{2} \sum_{r=2}^{2n} F_{r+1}^2 - \frac{1}{2} \sum_{r=1}^{2n-1} F_{r+1}^2 - \frac{1}{2} F_{2n-1} F_{2n} \quad \text{(by part (i))}$$

$$= \frac{1}{2} [F_{2n+1}^2 - 1 - F_{2n-1} F_{2n}]$$

$$= \frac{1}{2} [(F_{2n} + F_{2n-1})^2 + F_{2n}^2 - F_{2n+1} F_{2n-1}]$$

$$\quad - \frac{1}{2} [F_{2n-1}(F_{2n+1} - F_{2n-1})]$$

(by Problem 6.45(iii))

$$= \frac{1}{2} [2F_{2n}^2 + 2F_{2n} F_{2n-1} + 2F_{2n-1}^2 - 2F_{2n-1} F_{2n+1}]$$

(by Problem 6.45(iii))

$$= F_{2n}^2 + F_{2n-1}(F_{2n} + F_{2n-1} - F_{2n+1})$$

$$= F_{2n}^2.$$

(iii) We have

$$\sum_{r=1}^{2n} F_r F_{r+1} = F_{2n} F_{2n+1} + \sum_{r=1}^{2n-1} F_r F_{r+1}$$

$$= F_{2n} F_{2n+1} + F_{2n}^2 \quad \text{(by part (ii))}$$

$$= F_{2n} F_{2n+2}$$

$$= F_{2n+1}^2 + (-1)^{2n+1} \quad \text{(by Problem 6.45(iii))}$$
$$= F_{2n+1}^2 - 1.$$

48. We shall prove by induction on n that we have $F_{2n+1} = \sum_{r=0}^{n} \binom{2n-r}{r}$ and $F_{2n+2} = \sum_{r=0}^{n} \binom{2n+1-r}{r}$ for all $n \geq 0$, with the base case $n = 0$ being trivial. Suppose that the statement holds for $n = k$, where $k \geq 0$. By induction hypothesis, we have $F_{2k+1} = \sum_{r=0}^{k} \binom{2k-r}{r}$ and $F_{2k+2} = \sum_{r=0}^{k} \binom{2k+1-r}{r}$. We have

$$
\begin{aligned}
F_{2k+3} &= F_{2k+2} + F_{2k+1} \\
&= \sum_{r=0}^{k} \binom{2k+1-r}{r} + \sum_{r=0}^{k} \binom{2k-r}{r} \\
&= \sum_{r=0}^{k} \binom{2k+1-r}{r} + \sum_{r=1}^{k+1} \binom{2k+1-r}{r-1} \\
&= \sum_{r=0}^{k+1} \binom{2k+1-r}{r} + \sum_{r=0}^{k+1} \binom{2k+1-r}{r-1} \\
&= \sum_{r=0}^{k+1} \binom{2k+2-r}{r} \quad \text{(by (2.14)).}
\end{aligned}
$$

Likewise, we have

$$
\begin{aligned}
F_{2k+4} &= F_{2k+3} + F_{2k+2} \\
&= \sum_{r=0}^{k+1} \binom{2k+2-r}{r} + \sum_{r=0}^{k} \binom{2k+1-r}{r} \\
&= \sum_{r=0}^{k+1} \binom{2k+2-r}{r} + \sum_{r=1}^{k+1} \binom{2k+2-r}{r-1} \\
&= \sum_{r=0}^{k+1} \binom{2k+2-r}{r} + \sum_{r=0}^{k+1} \binom{2k+2-r}{r-1} \\
&= \sum_{r=0}^{k+1} \binom{2k+3-r}{r} \quad \text{(by (2.14)).}
\end{aligned}
$$

This completes the induction step, and the desired identity follows.

49. We shall prove by induction on p that we have $F_{m+2n} = \sum_{r=0}^{p} \binom{p}{r} F_{m+2n-p-r}$ for all $p = 0, \ldots, n$, with the base case $p = 0$ being trivial. Suppose that the statement holds for $p = k$, where $0 \le k < n$. By induction hypothesis, we have $F_{m+2n} = \sum_{r=0}^{k} \binom{k}{r} F_{m+2n-k-r}$, which implies that we have

$$
\begin{aligned}
F_{m+2n} &= \sum_{r=0}^{k} \binom{k}{r} F_{m+2n-k-r} \\
&= \sum_{r=0}^{k} \binom{k}{r} F_{m+2n-k-1-r} + \sum_{r=0}^{k} \binom{k}{r} F_{m+2n-k-2-r} \\
&= \sum_{r=0}^{k+1} \binom{k}{r} F_{m+2n-k-1-r} + \sum_{r=1}^{k+1} \binom{k}{r-1} F_{m+2n-k-1-r} \\
&= \sum_{r=0}^{k+1} \binom{k}{r} F_{m+2n-k-1-r} + \sum_{r=0}^{k+1} \binom{k}{r-1} F_{m+2n-k-1-r} \\
&= \sum_{r=0}^{k+1} \binom{k+1}{r} F_{m+2n-k-1-r}.
\end{aligned}
$$

This completes the induction step, and thus we have $F_{m+2n} = \sum_{r=0}^{p} \binom{p}{r} F_{m+2n-p-r}$ for all $p = 0, \ldots, n$. In particular, we have

$$
F_{m+2n} = \sum_{r=0}^{n} \binom{n}{r} F_{m+n-r} = \sum_{j=0}^{n} \binom{n}{n-j} F_{m+j} = \sum_{j=0}^{n} \binom{n}{j} F_{m+j}.
$$

50. Let $a = \frac{\sqrt{5}+1}{2}$. Then it follows that we have $F_n = \frac{a^n}{\sqrt{5}} (1 + (-1)^{n+1} a^{-2n})$, which implies that we have

$$
\frac{F_n}{F_{n+1}} = \frac{1 + (-1)^{n+1} a^{-2n}}{a(1 + (-1)^{n+2} a^{-2n-2})}.
$$

As $0 < a^{-1} < 1$, it follows that we have $\lim_{n \to \infty} a^{-2n} = 0 = \lim_{n \to \infty} a^{-2n-2}$. Thus, we have

$$
\lim_{n \to \infty} \frac{F_n}{F_{n+1}} = \lim_{n \to \infty} \frac{1 + (-1)^{n+1} a^{-2n}}{a(1 + (-1)^{n+2} a^{-2n-2})} = \frac{1}{a} = \frac{\sqrt{5}-1}{2}.
$$

51. For each $n \in \mathbf{N}$, we let b_n and c_n denote the number of adult and baby rabbit pairs at the end of the n-th month respectively. Let $a_n = b_n + c_n$ denote the total number of rabbits at the end of the n-th month. Then

the number of adult rabbit pairs at the end of the n-th month is equal
to the total number of rabbit pairs at the end of the $(n-1)$-th month
and the number of baby rabbit pairs at the end of the n-th month is
equal to the number of adult rabbit pairs at the end of the $(n-1)$-th
month, so we have $b_n = a_{n-1}$ and $c_n = b_{n-1} = a_{n-2}$, which implies
that $a_n = b_n + c_n = a_{n-1} + a_{n-2}$. Since $a_1 = a_2 = 1$, we have $a_n = F_n$.

52. Let $a_0 = 1$ and

$$a_n = \left.\begin{vmatrix} 1 & -1 & 0 & 0 & \cdots & 0 & 0 \\ 1 & 1 & -1 & 0 & \cdots & 0 & 0 \\ 0 & 1 & 1 & -1 & \cdots & 0 & 0 \\ \vdots & \vdots & \vdots & \vdots & \ddots & \vdots & \vdots \\ 0 & 0 & 0 & 0 & \cdots & 1 & -1 \\ 0 & 0 & 0 & 0 & \cdots & 1 & 1 \end{vmatrix}\right\} n$$

$$\underbrace{}_{n}$$

for all $n \in \mathbf{N}$. We shall prove by induction on n that we have $F_{n+1} = a_n$
for all $n \geq 0$, with the base cases $n = 0$ and $n = 1$ being trivial.
Suppose that the statement holds for all $n \leq k$, where $k \geq 1$. By
induction hypothesis, we have $F_{k+1} = a_k$ and $F_k = a_{k-1}$. By applying
the cofactor expansion of a_{k+1} along the first row, it follows that we
have $a_{k+1} = a_k + b_k$, where

$$b_k = \left.\begin{vmatrix} 1 & -1 & 0 & 0 & \cdots & 0 & 0 \\ 0 & 1 & -1 & 0 & \cdots & 0 & 0 \\ 0 & 1 & 1 & -1 & \cdots & 0 & 0 \\ \vdots & \vdots & \vdots & \vdots & \ddots & \vdots & \vdots \\ 0 & 0 & 0 & 0 & \cdots & 1 & -1 \\ 0 & 0 & 0 & 0 & \cdots & 1 & 1 \end{vmatrix}\right\} k$$

$$\underbrace{}_{k}$$

Next, by applying the cofactor expansion of b_k along the first column,
we have $b_k = a_{k-1}$. It follows from the above equation that we have
$a_{k+1} = a_k + a_{k-1} = F_{k+1} + F_k = F_{k+2}$. This completes the induction
step, and we are done.

53. If the man's last step was to climb one step on the staircase, then there
were a_{n-1} ways in which he could have climbed the $(n-1)$-step staircase
where he covers one step or two steps at each step. Otherwise, if the
man's last step was to climb two steps on the staircase, then there were
a_{n-2} ways in which he could have climbed the $(n-2)$-step staircase

where he covers one step or two steps at each step. By (AP), we must have $a_n = a_{n-1} + a_{n-2}$, with $a_1 = 1, a_2 = 2$.

54. For each $n \in \mathbf{N}$, we let a_n denote the number of binary sequences of length n in which no two 0's are adjacent, and let $d = d_1 d_2 \cdots d_n$ be a binary sequence of length n in which no two 0's are adjacent. If $d_1 = 1$, then $d_2 d_3 \cdots d_n$ is a binary sequence of length $n - 1$ in which no two 0's are adjacent, and there are no restrictions on d_2. Since there are a_{n-1} binary sequences of length $n - 1$ in which no two 0's are adjacent, it follows that there are a_{n-1} binary sequences $d_1 d_2 \cdots d_n$ of length n in which no two 0's are adjacent and $d_1 = 1$.

Else, if $d_1 = 0$, then $d_2 d_3 \cdots d_n$ is a binary sequence of length $n - 1$ in which no two 0's are adjacent, and $d_2 = 1$. A similar argument as in Case 1 would now show that there are a_{n-2} binary sequences $d_1 d_2 \cdots d_n$ of length n in which no two 0's are adjacent and $d_1 = 0$. By (AP), we must have $a_n = a_{n-1} + a_{n-2}$. Since $a_1 = 2 = F_3$ and $a_2 = 3 = F_4$, we must have $a_n = F_{n+2}$ for all $n \geq 1$.

55. Let $n = m_1 + \cdots + m_k$ where $m_i > 1$ for all $i = 1, \ldots, k$. If $m_1 = 2$, then we have $n - 2 = m_2 + \cdots + m_k$ with $m_i > 1$ for all $i = 2, \ldots, k$, so there are a_{n-2} ways to express n as a sum of positive integers greater than 1 with the first number equal to 2. Else, if $m_1 > 2$, then we have $n - 1 = (m_1 - 1) + m_2 + \cdots + m_k$ with $m_1 - 1, m_i > 1$ for all $i = 2, \ldots, k$. Thus, there are a_{n-1} ways to express n as a sum of positive integers greater than 1 with the first number greater than 2. By (AP), we must have $a_n = a_{n-1} + a_{n-2}$. Since $a_2 = 1 = F_1$ and $a_3 = 1 = F_2$, we must have $a_n = F_{n-1}$.

56. For each $n \in \mathbf{N}$, we let a_n denote the number of subsets of $\{1, \ldots, n\}$ that contain no consecutive integers, and let A be a subset of $\{1, \ldots, n\}$ that contains no consecutive integers. If $n \in A$, then $n - 1 \notin A$, so $A \backslash \{n\}$ is a subset of $\{1, \ldots, n-2\}$ that contains no consecutive integers. Consequently, there are a_{n-2} such subsets A. Else, if $n \notin A$, then A is a subset of $\{1, \ldots, n-1\}$ that contains no consecutive integers, so there are a_{n-1} such subsets A. By (AP), we must have $a_n = a_{n-1} + a_{n-2}$. Since $a_1 = 2 = F_3$ and $a_2 = 3 = F_4$, we must have $a_n = F_{n+2}$.

57. We have

$$\sum_{j=0}^{n} \frac{\binom{n}{2j-n-1}}{5^j} = \frac{1}{5^n} \sum_{j=0}^{n} \binom{n}{2j-n-1} 5^{n-j}$$

$$= \frac{1}{5^n} \sum_{j=0}^{n} \binom{n}{2n - 2j + 1} 5^{n-j}$$

$$= \frac{1}{5^n} \sum_{i=0}^{n} \binom{n}{2i + 1} 5^i \quad \text{(by setting } i = n - j\text{)}$$

$$= \frac{1}{2 \cdot 5^n \sqrt{5}} \sum_{i=0}^{n} \binom{n}{2i + 1} 2(\sqrt{5})^{2i+1}$$

$$= \frac{1}{2 \cdot 5^n \sqrt{5}} \sum_{k=0}^{n} \binom{n}{k} (1 - (-1)^k)(\sqrt{5})^k$$

$$= \frac{1}{2 \cdot 5^n \sqrt{5}} [(1 + \sqrt{5})^n - (1 - \sqrt{5})^n]$$

$$= \frac{1}{2} \left(\frac{2}{5}\right)^n \frac{1}{\sqrt{5}} \left[\left(\frac{1 + \sqrt{5}}{2}\right)^n - \left(\frac{1 - \sqrt{5}}{2}\right)^n\right]$$

$$= \frac{1}{2}(0.4)^n F_n.$$

58. For each positive integer n, we let A_n denote the set of admissible ordered pairs of subsets of $\{1, \ldots, n\}$, and we let $a_n = |A_n|$. We seek to find a recurrence relation for a_n. To this end, we shall pick any $(S, T) \in A_n$, and consider the following cases:

Case 1: $n \notin S$ and $n \notin T$. Then (S, T) is an admissible ordered pair of subsets of $\{1, \ldots, n - 1\}$, and there are a_{n-1} such pairs by definition.

Case 2: $n \in S$. We let $T' = \{t - 1 \mid t \in T\}$. We claim that $(S \setminus \{n\}, T')$ is an admissible ordered pair of subsets of $\{1, \ldots, n - 1\}$. Indeed, since (S, T) is admissible and $n \in S$, it follows that we have $t > |S| \geq 1$ for all $t \in T$, so $t - 1 \in \{1, \ldots, n - 1\}$ for all $t \in T$. Now, we see that for all $s \in S \setminus \{n\}$, we have $s > |T| = |T'|$, and for all $t \in T$, we have $t - 1 > |S| - 1 = |S \setminus \{n\}|$, which completes the proof of the claim. Now, by letting L_n denote the set of admissible ordered pairs (X, Y) of subsets of $\{1, \ldots, n\}$ with $n \in X$, it is easy to see that the map $L_n \to A_{n-1}$, defined by $(X, Y) \mapsto (X \setminus \{n\}, Y')$, where $Y' = \{y - 1 \mid y \in Y\}$, is a bijection, so there are $|L_n| = |A_{n-1}| = a_{n-1}$ such pairs for this case.

Case 3: $n \in T$. A similar argument as in Case 2 shows that there are a_{n-1} such pairs in this case.

Case 4: $n \in S \cap T$. Let $S'' = \{s - 1 \mid s \in S \setminus \{n\}\}$ and $T'' = \{t - 1 \mid t \in T \setminus \{n\}\}$. We claim that (S'', T'') is an admissible ordered pair of subsets of $\{1, \ldots, n - 2\}$. Indeed, since (S, T) is admissible and $n \in S$, it follows that we have $t > |S| \geq 1$ for all $t \in T$, so $t - 1 \in \{1, \ldots, n - 2\}$ for all $t \in T \setminus \{n\} \subseteq \{1, \ldots, n - 1\}$. Likewise, we have $s - 1 \in \{1, \ldots, n - 2\}$ for all $s \in S \setminus \{n\}$. Now, we see that for all $s \in S \setminus \{n\}$, we have $s - 1 > |T| - 1 = |T \setminus \{n\}| = |T''|$, and for all $t \in T \setminus \{n\}$, we have $t - 1 > |S| - 1 = |S \setminus \{n\}| = |S''|$, which completes the proof of the claim. Now, by letting B_n denote the set of admissible ordered pairs (X, Y) of subsets of $\{1, \ldots, n\}$ with $n \in X \cap Y$, it is easy to see that the map $B_n \to A_{n-2}$, defined by $(X, Y) \mapsto (X'', Y'')$, where $X'' = \{x - 1 \mid x \in X \setminus \{n\}\}$ and $Y'' = \{y - 1 \mid y \in Y \setminus \{n\}\}$, is a bijection, so there are $|B_n| = |A_{n-2}| = a_{n-2}$ such pairs for this case.

By (4.1.1), the number of admissible ordered pairs (S, T) of subsets of $\{1, \ldots, n - 1\}$ such that $n \in S \cup T$ must be equal to $2a_{n-1} - a_{n-2}$, and hence by (AP), we must have $a_n = 3a_{n-1} - a_{n-2}$. The corresponding characteristic equation is $x^2 - 3x + 1 = 0$, and its characteristic roots are $\alpha_1 = \frac{3+\sqrt{5}}{2}$, and $\alpha_2 = \frac{3-\sqrt{5}}{2}$. Thus, by result (I), the general solution of the recurrence relation is given by $a_n = A\left(\frac{3+\sqrt{5}}{2}\right)^n + B\left(\frac{3-\sqrt{5}}{2}\right)^n$, where A and B are constants to be determined. For $n = 1$, the admissible ordered pairs are $(\emptyset, \emptyset), (\emptyset, \{1\})$ and $(\{1\}, \emptyset)$, so we have $a_1 = 3$. For $n = 2$, the admissible pairs are $(\emptyset, \emptyset), (\emptyset, \{1\}), (\emptyset, \{2\}), (\emptyset, \{1, 2\}), (\{1\}, \emptyset), (\{2\}, \emptyset), (\{1, 2\}, \emptyset)$ and $(\{2\}, \{2\})$, so we have $a_2 = 8$. The initial conditions $a_1 = 3$ and $a_2 = 8$ imply that

$$A\left(\frac{3+\sqrt{5}}{2}\right) + B\left(\frac{3-\sqrt{5}}{2}\right) = 3, \quad \text{and}$$

$$A\left(\frac{3+\sqrt{5}}{2}\right)^2 + B\left(\frac{3-\sqrt{5}}{2}\right)^2 = 8.$$

Solving the system for A and B gives us $A = \frac{3+\sqrt{5}}{2\sqrt{5}}$ $B = -\frac{3-\sqrt{5}}{2\sqrt{5}}$. So we have

$$a_n = \frac{1}{\sqrt{5}}\left[\left(\frac{3+\sqrt{5}}{2}\right)^{n+1} - \left(\frac{3-\sqrt{5}}{2}\right)^{n+1}\right]$$

for all $n \geq 1$. As $\frac{3 \pm \sqrt{5}}{2} = \left(\frac{1 \pm \sqrt{5}}{2} \right)^2$, it follows that we have

$$a_n = \frac{1}{\sqrt{5}} \left[\left(\frac{1 + \sqrt{5}}{2} \right)^{2n+2} - \left(\frac{1 - \sqrt{5}}{2} \right)^{2n+2} \right] = F_{2n+2}$$

for all $n \geq 1$. Therefore, the required answer is $a_{10} = F_{22} = 17711$.

59. Let us take a natural number N that satisfies the given conditions. Given any $j \in \{1, 3, 4\}$, we see that if the first digit of N is j, then there are a_{n-j} ways to form the remaining digits of N, so by (AP), we have $a_n = a_{n-1} + a_{n-3} + a_{n-4}$. The corresponding characteristic equation is given by $x^4 - x^3 - x - 1 = 0$. As $x^4 - x^3 - x - 1 = (x^2 + 1)(x^2 - x - 1)$, we see that the characteristic roots are $\alpha = \frac{1+\sqrt{5}}{2}$, $-\alpha^{-1} = \frac{1-\sqrt{5}}{2}$ and $\pm i$. Thus, by result (I), the general solution of the recurrence relation is given by $a_n = A\alpha^n + B(-1)^n \alpha^{-n} + C \cos \frac{n\pi}{2} + D \sin \frac{n\pi}{2}$, where A, B, C and D are constants to be determined. The initial conditions $a_1 = 1, a_2 = 1, a_3 = 2$ and $a_4 = 4$ imply that

$$A\alpha - B\alpha^{-1} + D = 1,$$
$$A\alpha^2 + B\alpha^{-2} - C = 1,$$
$$A\alpha^3 - B\alpha^{-3} - D = 2, \quad \text{and}$$
$$A\alpha^4 + B\alpha^{-4} + C = 4.$$

Solving the system first for A, B, C and D gives us $A = \frac{3\alpha+2}{(1+\alpha^2)^2}$ and $B = \frac{\alpha^3(2\alpha-3)}{(1+\alpha^2)^2}$. Now, we observe that $\alpha^2 = \alpha + 1$, which implies that

$$3\alpha + 2 = \alpha + 1 + 2\alpha + 1 = \alpha^2 + 2\alpha + 1 = (\alpha + 1)^2 = \alpha^4,$$
$$\alpha^3(2\alpha - 3) = 2\alpha^4 - 3\alpha^3 = 2(\alpha + 1)^2 - 3\alpha(\alpha + 1) = -\alpha^2 + \alpha + 2 = 1,$$
$$(1 + \alpha^2)^2 = (\alpha + 2)^2 = \alpha^2 + 4\alpha + 4 = 5\alpha^2.$$

Thus, we have $A = \frac{1}{5}\alpha^2$ and $B = \frac{1}{5}\alpha^{-2}$. Consequently, we have $C = \frac{2}{5}$ and $D = \frac{1}{5}$. Consequently, we have

$$a_n = \frac{1}{5} \left[\alpha^{n+2} + (-1)^n \alpha^{-n-2} + 2 \cos \frac{n\pi}{2} + \sin \frac{n\pi}{2} \right]$$

for all $n \geq 1$. In particular, we have

$$a_{2n} = \frac{1}{5}[\alpha^{2n+2} + \alpha^{-2n-2} + 2(-1)^n]$$
$$= \frac{1}{5}[(\alpha^{n+1})^2 + ((-1)^{n+2}\alpha^{-n-1})^2 + 2(-1)^{n+2}]$$
$$= \frac{1}{5}(\alpha^{n+1} + (-1)^{n+2}\alpha^{-n-1})^2$$

$$= \frac{1}{5}(\alpha^{n+1} - (-\alpha^{-1})^{n+1})^2$$

$$= \frac{1}{5}\left[\left(\frac{1+\sqrt{5}}{2}\right)^{n+1} - \left(\frac{1-\sqrt{5}}{2}\right)^{n+1}\right]^2$$

$$= F_{n+1}^2$$

for all $n \geq 1$, and the desired statement follows.

60. We note that for each $k = 1, \ldots, n-1$, we can write $x_1 \cdots x_n$ as the product of $x_1 \cdots x_k$ and $x_{k+1} \cdots x_n$. We see that there are a_k ways to place parentheses to indicate the order of multiplication of $x_1 x_2 \cdots x_k$, while there are a_{n-k} ways to place parentheses to indicate the order of multiplication of $x_{k+1} x_{k+2} \cdots x_n$. By (MP), there are $a_k a_{n-k}$ ways to place parentheses to indicate the multiplication of $x_1 x_2 \cdots x_n$, such that $x_1 x_2 \cdots x_n$ is written as the product of $x_1 \cdots x_k$ and $x_{k+1} \cdots x_n$. As $k \in \{1, \ldots, n-1\}$, it follows from (AP) that we have

$$a_n = a_1 a_{n-1} + a_2 a_{n-2} + \cdots + a_{n-1} a_1,$$

for all $n \geq 2$, where $a_1 = 1$.

61. (i) For $n = 1$, the only sequence that satisfies conditions (1) and (2) is $(1, -1)$, so we have $b_1 = 1$. For $n = 2$, the sequences that satisfy conditions (1) and (2) are $(1, 1, -1, -1)$ and $(1, -1, 1, -1)$, so we have $b_2 = 2$. For $n = 3$, the sequences that satisfy conditions (1) and (2) are $(1, 1, 1, -1, -1, -1)$, $(1, 1, -1, 1, -1, -1)$, $(1, 1, -1, -1, 1, -1)$, $(1, -1, 1, 1, -1, -1)$ and $(1, -1, 1, -1, 1, -1)$, so we have $b_3 = 5$.

(ii) Let A denote the set of all parenthesized expressions of $x_1 x_2 \cdots x_{n+1}$, and B denote the set of all sequences of $2n$ terms consisting of 1's and -1's that satisfies conditions (1) and (2). We shall seek to establish a bijection between A and B. To this end, we first observe that each parenthesized expression of $x_1 x_2 \cdots x_{n+1}$ contain exactly n pairs of parentheses, and the number of left parentheses to the right of x_i is at most $n - i$ for all $i = 1, 2, \ldots, n$. Henceforth, we will define $f: A \to B$ as follows: given a parenthesized expression E of $x_1 x_2 \cdots x_{n+1}$, we remove all right parentheses as well as x_{n+1} from E, and replace each left parenthesis by 1 and each x_i by -1 for each $i = 1, 2, \ldots, n$. Then it is clear from our above observation about the set B that f is a well-defined map.

It remains to show that f is surjective. To this end, we shall take any sequence S in B, and replace all 1's by left parentheses and the

i-th -1 by x_i, and we add x_{n+1} to the right of x_n. Subsequently, the right parentheses can be placed a unique way such that the resultant expression R is a parenthesized expression of $x_1 x_2 \cdots x_{n+1}$. It is clear from our above construction that $f(R) = S$. So f is a bijection, and we are done.

62. Let us label the $2n$ points on the circumference of the circle by $P_1, \ldots, P_{2n}$ in order. Let us pick any pairing of the $2n$ points by n nonintersecting chords. If P_1 is paired with the point P_k, where $2 \le k \le 2n$, then the chord connecting P_1 and P_k divides the circumference of the circle into two arcs consisting of $k - 2$ points, namely $P_2, P_3, \ldots, P_{k-1}$, and $2n - k$ points, namely $P_{k+1}, \ldots, P_{2n}$ respectively. As the chords must be nonintersecting, it follows that for any $i \in \{2, \ldots, k - 1\}$, the point P_i must be paired with the point P_j, where $j \in \{2, \ldots, k - 1\}$. This implies that k must be even, that is, $k = 2m$, where $m = 1, \ldots, n$. When $k = 2m$, the points $P_2, \ldots, P_{2m-1}$ can be paired off with $m - 1$ nonintersecting chords in a_{m-1} ways, and the points $P_{2m+1}, \ldots, P_{2n}$ can be paired off with $n - m$ nonintersecting chords in a_{n-m} ways. Thus, by (MP), there are $a_{m-1} a_{n-m}$ ways to pair of the $2n$ points by n nonintersecting chords, such that P_1 is paired with the point P_{2m}. Since $m \in \{1, \ldots, n\}$, it follows from (AP) that we have

$$a_n = a_0 a_{n-1} + a_1 a_{n-2} + \cdots + a_{n-1} a_0,$$

for all $n \ge 1$, where $a_0 = 1$.

63. We note that a term in q_n must be of the form $x_n^{a_1} x_{n-1}^{a_2} \cdots x_1^{a_n}$, where $a_i \ge 0$ for all $i = 1, \ldots, k$, $\sum_{i=1}^{k} a_i \le k$ for all $k = 1, \ldots, n - 1$ and $\sum_{i=1}^{n} a_i = n$. Let A_n denote the set of n-tuples $(a_1, \ldots, a_n)$ of nonnegative integers, such that $\sum_{i=1}^{k} a_i \le k$ for all $k = 1, \ldots, n - 1$ and $\sum_{i=1}^{n} a_i = n$. Let B_n denote the set of n-tuples $(b_1, \ldots, b_n)$ of nonnegative integers, such that $b_k \le b_{k+1}$, $b_k \le k$ for all $k = 1, \ldots, n - 1$ and $b_n = n$. Then it is easy to see that the map $A_n \mapsto B_n$, defined by $(a_1, \ldots, a_n) \mapsto (a_1, a_1 + a_2, \ldots, a_1 + \cdots + a_n)$, is a well-defined bijection. Thus, there exists a bijection between the set of terms in q_n and the elements of B_n.

Next, we let C_n denote the set of shortest routes from $(0,0)$ to (n, n) that do not meet the line $y = x + 1$. We seek to establish a bijection between B_n and C_n. To this end, we pick any $B = (b_1, \ldots, b_n) \in B_n$, and construct a shortest route $R(B) \in C_n$ as follows: we let $b_0 = 0$, and for $i = 1, \ldots, n$, we join $(i - 1, b_{i-1})$ to (i, b_{i-1}), and (i, b_{i-1}) to (i, b_i) if $b_{i-1} < b_i$. If $b_{i-1} = b_i$, then we join $(i - 1, b_{i-1})$ to (i, b_{i-1}) only. Then it is easy to see from the definitions of B_n and C_n that the map $B_n \to C_n$, defined by $B \mapsto R(B)$, is a well-defined bijection, so it follows that there exists a bijection between the set of terms in q_n and the set of shortest routes from $(0, 0)$ to (n, n) that do not meet the line $y = x + 1$.

Finally, we let s_n denote the number of shortest routes from $(0, 0)$ to (n, n), and t_n denote the number of shortest routes from $(0, 0)$ to (n, n) that meet the line $y = x + 1$. By (RP), t_n is the number of paths from $(-1, 1)$ to (n, n) which is equal to $\binom{2n}{n+1}$. We note that $s_n = \binom{2n}{n}$. By (CP), it follows that we have

$$
\begin{aligned}
\#(q_n) &= |C_n| \\
&= s_n - t_n \\
&= \binom{2n}{n} - \binom{2n}{n+1} \\
&= \binom{2n}{n} - \frac{n}{n+1}\binom{2n}{n} \quad \text{(by (2.1.3))} \\
&= \frac{1}{n+1}\binom{2n}{n} \\
&= \frac{1}{2n+1}\binom{2n+1}{n+1} \quad \text{(by (2.1.2))} \\
&= \frac{1}{2n+1}\binom{2n+1}{n} \quad \text{(by (2.1.1))}.
\end{aligned}
$$

64. Let A denote the set of arrangements x of the $2n$ integers $a_1, \ldots, a_n, b_1, \ldots, b_n$ that satisfy the given conditions, and let B denote the set of shortest paths from $(0, 0)$ to (n, n) that do not go above the line $y = x$, or equivalently, the set of shortest routes from $(0, 0)$ to (n, n) that do not meet the line $y = x + 1$. We seek to establish a bijection between B_n and C_n. To this end, we pick any arrangement $x = x_1 x_2 \ldots, x_{2n} \in A$, and construct a shortest route $R(x) \in C_n$ as follows: we let $(p_0, q_0) = (0, 0)$, and for $i = 1, \ldots, 2n$, we inductively define (p_i, q_i) as follows:

$$
(p_i, q_i) = \begin{cases} (p_{i-1} + 1, q_{i-1}) & \text{if } x_i = a_j \text{ for some } j \in \{1, 2, \ldots, n\}, \\ (p_{i-1}, q_{i-1} + 1) & \text{if } x_i = b_j \text{ for some } j \in \{1, 2, \ldots, n\}. \end{cases}
$$

We then define $R(x)$ to be the path obtained by joining (p_{i-1}, q_{i-1}) to (p_i, q_i) for all $i = 1, 2, \ldots, 2n$. It is easy to see from the definitions of A and B that the map $A \to B$, defined by $x \mapsto R(x)$, is a well-defined bijection, so it follows that there exists a bijection between the set of arrangements of the $2n$ integers $a_1, \ldots, a_n, b_1, \ldots, b_n$ that satisfy the given conditions and the set of shortest routes from $(0, 0)$ to (n, n) that do not meet the line $y = x + 1$. By our solution for Problem 6.63, we have

$$|A| = |B| = \frac{1}{2n + 1} \binom{2n + 1}{n} = \frac{1}{n + 1} \binom{2n}{n},$$

where the last equality follows from (2.1.1) and (2.1.2).

65. We note that any arrangement of the ballots that satisfies the conditions could be represented by a binary sequence $a_1 \cdots a_{m+n}$ consisting of m 1's and n 0's, such that the number of 1's in $a_1 \cdots a_k$ is more than the number of 0's in $a_1 \cdots a_k$ for all $k = 1, 2, \ldots, m + n$. In turn, it is easy to see that any such binary sequence can be represented by a shortest route from $(1, 0)$ to (m, n) that does not meet the line $y = x$. Let s_n denote the number of shortest routes from $(1, 0)$ to (m, n), and t_n denote the number of shortest routes from $(1, 0)$ to (m, n) that meets the line $y = x$. Then we have $s_n = \binom{m+n-1}{m-1}$. Next, by (RP), it follows that t_n is equal to the number of shortest routes from $(0, 1)$ to (m, n), which is then equal to $\binom{m+n-1}{m}$. By (CP), it follows that the number of arrangements of the ballots that satisfy the conditions is equal to

$$
\begin{aligned}
s_n - t_n &= \binom{m + n - 1}{m - 1} - \binom{m + n - 1}{m} \\
&= \binom{m + n - 1}{m - 1} - \frac{n}{m} \binom{m + n - 1}{m - 1} \quad \text{(by (2.1.3))} \\
&= \frac{m - n}{m} \binom{m + n - 1}{m - 1} \\
&= \frac{m - n}{m + n} \binom{m + n}{m} \quad \text{(by (2.1.2))}.
\end{aligned}
$$

66. (i) For convenience, we shall represent all mappings f that satisfy the given conditions by an n-tuple $(f(1), \ldots, f(n))$. For $n = 1$, the only such 1-tuple is (1), so we have $a_1 = 1$. For $n = 2$, the only such 2-tuples are $(1, 2), (2, 1)$ and $(1, 1)$, so we have $a_2 = 3$. For $n = 3$, the only such 3-tuples are $(1, 2, 3), (1, 3, 2), (2, 1, 3)$, $(2, 3, 1), (3, 1, 2), (3, 2, 1), (1, 2, 2), (2, 1, 2), (2, 2, 1), (1, 1, 2), (1, 2, 1)$, $(2, 1, 1)$ and $(1, 1, 1)$, so we have $a_3 = 13$.

(ii) Let us pick any mapping f that satisfies the given conditions, and let k denote the number of $j's$ for which $f(j) = 1$. There are $\binom{n}{k}$ ways to choose k distinct numbers $a_1, \ldots, a_k$ out of $\{1, \ldots, n\}$ so that $f(a_i) = 1$ for all $i = 1, \ldots, k$, and there a_{n-k} ways to assign a number in $\{2, \ldots, n - k + 1\}$ to each j for each $j \in \{1, \ldots, n\} \setminus \{a_1, \ldots, a_k\}$, so that the resultant mapping f satisfies the given conditions. By (MP), there are $\binom{n}{k} a_{n-k}$ such mappings f such that the number of $j's$ for which $f(j) = 1$ is equal to k. As $k \in \{1, 2, \ldots, n\}$, it follows from (AP) that we have $a_n = \sum\limits_{k=1}^{n} \binom{n}{k} a_{n-k}$ for all $n \geq 1$, where $a_0 = 1$.

(iii) We first observe from part (ii) that we have

$$2a_n = a_n + \sum_{k=1}^{n} \binom{n}{k} a_{n-k} = \sum_{k=0}^{n} \binom{n}{n-k} a_{n-k} = \sum_{j=0}^{n} \binom{n}{j} a_j$$

for all $n \geq 1$, where the last equality follows by setting $j = n - k$. Thus, we have

$$A(x) = \sum_{n=0}^{\infty} \frac{a_n}{n!} x^n$$

$$= a_0 + \frac{1}{2} \sum_{n=1}^{\infty} \sum_{j=0}^{n} \frac{a_j}{j!(n-j)!} x^n$$

$$= a_0 + \frac{1}{2} \sum_{n=0}^{\infty} \sum_{j=0}^{n} \frac{a_j}{j!(n-j)!} x^n - \frac{1}{2} a_0$$

$$= \frac{1}{2} a_0 + \frac{1}{2} \sum_{j=0}^{\infty} \frac{a_j}{j!} x^j \sum_{n=j}^{\infty} \frac{x^{n-j}}{(n-j)!}$$

$$= \frac{1}{2} + \frac{1}{2} \sum_{j=0}^{\infty} \frac{a_j}{j!} x^j \sum_{i=0}^{\infty} \frac{x^i}{i!} \quad \text{(by setting } i = n - j)$$

$$= \frac{1}{2}(1 + e^x A(x)),$$

or equivalently, $A(x) = \frac{1}{2-e^x}$ as desired.

(iv) We have

$$A(x) = \frac{1}{2 - e^x} \quad \text{(by part (iii))}$$

$$= \frac{1}{2\left(1 - \frac{e^x}{2}\right)}$$

$$= \frac{1}{2} \sum_{r=0}^{\infty} \left(\frac{e^x}{2} \right)^r \quad \text{(by (5.1.3))}$$

$$= \sum_{r=0}^{\infty} \frac{e^{rx}}{2^{r+1}}$$

$$= \sum_{r=0}^{\infty} \sum_{n=0}^{\infty} \frac{(rx)^n}{n!2^{r+1}} \quad \text{(by Example 5.4.1)}$$

$$= \sum_{n=0}^{\infty} \sum_{r=0}^{\infty} \frac{r^n}{2^{r+1}} \cdot \frac{x^n}{n!}.$$

By comparing the coefficients of x^n on both sides of the above equation, it follows that we have $a_n = \sum_{r=0}^{\infty} \frac{r^n}{2^{r+1}}$ for all $n \geq 0$ as desired.

67. (i) Let $n > 1$, and let us pick a $(n+1) \times (n+1)$ matrix A that satisfies the given conditions. The given conditions immediately imply that $a_{i,j} = 0$ or 1 for all $i, j = 1, \ldots, n$. Due to the symmetry and $\sum_{i=1}^{n+1} a_{i,j} = 1$ condition, each row and column has precisely one entry which is a 1 with the remaining entries being 0. Let us consider the following cases:

Case 1: $a_{n+1,n+1} = 1$. Then the matrix B obtained by removing the $(n+1)$-th rows and columns of A is a $n \times n$ matrix that satisfies the given conditions, and there are S_n such matrices B by definition. As there is a bijection between the set of $(n+1) \times (n+1)$ matrices A with $a_{n+1,n+1} = 1$ that satisfies the given conditions, and the set of $n \times n$ matrices B that satisfies the given conditions, it follows from (BP) that there are S_n such matrices A in this case.

Case 2: $a_{n+1,n+1} = 0$. Then there must exist a unique $i \in \{1, \ldots, n\}$ such that $a_{i,n+1} = a_{n+1,i} = 1$. Then the matrix B obtained by removing the i-th and $(n+1)$-th rows, and i-th and $(n+1)$-th columns of A is a $(n-1) \times (n-1)$ matrix that satisfies the given conditions, and there are S_{n-1} such matrices B by definition. As there is a bijection between the set of $(n+1) \times (n+1)$ matrices A with $a_{i,n+1} = a_{n+1,i} = 1$ that satisfies the given conditions, and the set of $(n-1) \times (n-1)$ matrices B that satisfies the given conditions for any fixed $i \in \{1, \ldots, n\}$, it follows from (BP) that there are S_{n-1} matrices A with $a_{i,n+1} = a_{n+1,i} = 1$ that satisfies the given conditions. As there are n choices for i, it follows from

(MP) that there are nS_{n-1} such matrices A in this case.

By (AP), it follows that we have $S_{n+1} = S_n + nS_{n-1}$ for all $n > 1$. As $S_2 = 2 = S_1 + 1 \cdot S_0$, it follows that we have $S_{n+1} = S_n + nS_{n-1}$ for all $n \geq 1$ as desired.

(ii) Let the exponential generating function for (S_n) be $S(x)$. By part (i), we have $\frac{S_{n+1}}{n!}x^{n-1} = \frac{S_n}{n!}x^{n-1} + \frac{S_{n-1}}{(n-1)!}x^{n-1}$ for all $n \geq 1$. Since $S_0 = S_1 = 1$, it follows that we have $x^{-1}(S'(x) - 1) = x^{-1}(S(x) - 1) + S(x)$, which would then give us $\frac{S'(x)}{S(x)} = 1 + x$. This implies that $\log_e S(x) = x + \frac{x^2}{2} + c$, or equivalently, $S(x) = e^{x + \frac{x^2}{2} + c}$, where c is a constant to be determined. As $S(0) = S_0 = 1$, it follows that we have $e^c = 1$ and $c = 0$. Hence, we have $\sum_{n=0}^{\infty} \frac{S_n}{n!}x^n = S(x) = e^{x + \frac{x^2}{2}}$ as desired.

68. By equation (2), we have $(n+1)^2 a_{n+1} - n^2 a_n = a_{n+1}$, which simplifies to the equation $a_{n+1} = \frac{n}{n+2}a_n$ for all $n \geq 1$. This implies that for all $n \geq 2$, we have

$$
\begin{aligned}
a_n &= \frac{n-1}{n+1}a_{n-1} \\
&= \frac{n-1}{n+1} \cdot \frac{n-2}{n} \cdot a_{n-2} \\
&= \cdots \\
&= \frac{n-1}{n+1} \cdot \frac{n-2}{n} \cdots \frac{1}{3} \cdot a_1 \\
&= \frac{n-1}{n+1} \cdot \frac{n-2}{n} \cdots \frac{1}{3} \cdot \frac{1}{2} \\
&= \frac{1}{n(n+1)}.
\end{aligned}
$$

As $a_1 = \frac{1}{2} = \frac{1}{1(1+1)}$, it follows that we have $a_n = \frac{1}{n(n+1)}$ for all $n \geq 1$.

69. Let $n \geq 1$. We first observe that we can write each positive integer $m < 2^n$ in the form $m = 2^\ell \cdot (2k - 1)$, where ℓ, k are integers satisfying $0 \leq \ell \leq n - 1$ and $1 \leq k \leq 2^{n-1-\ell}$. As the greatest odd divisor of 2^n is equal to 1, it follows that we have $a_n = 1 + \sum_{\ell=0}^{n-1} \sum_{k=1}^{2^{n-1-\ell}} (2k - 1)$. This implies that for all $n > 1$, we have

$$
\begin{aligned}
a_n &= 1 + \sum_{\ell=0}^{n-1} \sum_{k=1}^{2^{n-1-\ell}} (2k - 1) \\
&= 1 + \sum_{\ell=1}^{n-1} \sum_{k=1}^{2^{n-1-\ell}} (2k - 1) + \sum_{k=1}^{2^{n-1}} (2k - 1)
\end{aligned}
$$

$$= 1 + \sum_{j=0}^{n-2} \sum_{k=1}^{2^{n-2-j}} (2k-1) + (2^{n-1})^2$$

$$= a_{n-1} + 4^{n-1}.$$

We shall now proceed to solve for a_n with the initial condition $a_0 = 1$. It is easy to see that $a_n^{(h)} = A$, where A is a constant.

Next, let us find $a_n^{(p)}$. Since $f(n) = 4^{n-1}$ is an exponential function, and 4 is not a characteristic root, we let $a_n^{(p)} = B4^n$, where B is a constant. Since $a_n^{(p)}$ satisfies the equation $a_n - a_{n-1} = 4^{n-1}$, we must have

$$B4^n - B4^{n-1} = 4^{n-1}.$$

By equating the coefficients of 4^n, we get $B = \frac{1}{3}$, and hence we have $a_n^{(p)} = \frac{1}{3} \cdot 4^n$. By (6.4.2), the general solution of the equation $a_n = a_{n-1} + 4^{n-1}$ is given by $a_n = a_n^{(h)} + a_n^{(p)} = A + \frac{1}{3}4^n$. The initial condition $a_0 = 1$ imply that $1 = A + \frac{1}{3}$, or equivalently, $A = \frac{2}{3}$. Therefore, we have $a_n = \frac{1}{3}(4^n + 2)$ for all $n \geq 0$.

70. We shall prove by induction that we have $d_n = (-1)^{n-1}(n-1)2^{n-2}$ for all $n \geq 1$, with the base cases $n = 1$ and $n = 2$ being trivial. Suppose that the statement holds for $n = k$, where $k \geq 2$. By induction hypothesis, we have $d_k = (-1)^{k-1}(k-1)2^{k-2}$. Let A be the $(k+1) \times (k+1)$ matrix with entries $a_{i,j} = |i - j|$ for all $i, j = 1, \ldots, k+1$, and let us denote the i-th column of A by A_i for all $i = 1, \ldots, k+1$. Let B be the $(k+1) \times (k+1)$ matrix whose i-th column is A_i for $i = 1, \ldots, k$, and whose $(k+1)$-th column is $\frac{1}{k-1}(-A_1 - kA_k + (k-1)A_{k+1})$. Then we must have $d_{k+1} = \det A = \det B$. By denoting the entries of B by $b_{i,j}$, it follows that for all $i, j = 1, \ldots, k$, we have $b_{i,j} = a_{i,j}$, and $b_{k+1,j} = a_{k+1,j}$ for all $j = 1, 2, \ldots, k$. For $i = 1, 2, \ldots, k$, we have

$$b_{i,k+1} = \frac{1}{k-1}(-a_{i,1} - ka_{i,k} + (k-1)a_{i,k+1})$$

$$= \frac{1}{k-1}(-(i-1) - k(k-i) + (k-1)(k+1-i))$$

$$= 0.$$

Also, we have

$$b_{k+1,k+1} = \frac{1}{k-1}(-a_{k+1,1} - ka_{k+1,k} + (k-1)a_{k+1,k+1})$$

$$= \frac{1}{k-1}(-k - k)$$

$$= -\frac{2k}{k-1}.$$

By letting C be the matrix obtained by removing the $(k+1)$-th rows and columns of B, it follows from the definition of B that we have $\det C = d_k = (-1)^{k-1}(k-1)2^{k-2}$ by the induction hypothesis. Now, by applying the cofactor expansion of B along the $(k+1)$-th column, it follows that we have $d_{k+1} = \det B = -\frac{2k}{k-1}\det C = (-1)^k k2^{k-1}$. This completes the induction step, and we are done.

71. We first observe from the recurrence relation that we have $a_n - a_{n-1} = n(a_{n-1} - a_{n-2})$ for all $n \geq 3$. Henceforth, we let $b_n = a_{n+1} - a_n$ for all $n \geq 1$. Then it follows that we have $b_{n+1} = (n+2)b_n$ for all $n \geq 1$. This implies that for all $n \geq 2$, we have

$$
\begin{aligned}
b_n &= (n+1)b_{n-1} \\
&= (n+1)nb_{n-2} \\
&= \cdots \\
&= (n+1)\cdots 3 \cdot b_1 \\
&= (n+1)\cdots 3 \cdot 2 \\
&= (n+1)! \, .
\end{aligned}
$$

Since $b_1 = 2 = (1+1)!$, it follows that we have $b_n = (n+1)!$ for all $n \geq 1$. Thus, for all $n \geq 2$, we have

$$
\begin{aligned}
a_n &= a_1 + \sum_{i=1}^{n-1}(a_{i+1} - a_i) \\
&= 1 + \sum_{i=1}^{n-1} b_i \\
&= 1! + \sum_{i=1}^{n-1}(i+1)! \\
&= 1! + \sum_{i=2}^{n} i! \\
&= \sum_{i=1}^{n} i! \, .
\end{aligned}
$$

As $a_1 = 1 = 1!$, it follows that we have $a_n = \sum_{i=1}^{n} i!$ for all $n \geq 1$. Now, it is easy to verify that 11 divides a_4, a_8 and a_{10}. As $i!$ is a multiple of 11 for all $i \geq 11$, it follows that the values of n such that $11 | a_n$ are $n = 4, 8$ or $n \geq 10$.

72. Let $b_n = \sqrt{1 + 24a_n}$ for all $n \geq 0$. Then it follows that we have $a_n = \frac{b_n^2 - 1}{24}$ for all $n \geq 0$. By substituting $a_n = \frac{b_n^2 - 1}{24}$ into the recurrence relation, we have $\frac{b_n^2 - 1}{24} = \frac{1}{16}\left(1 + \frac{b_{n-1}^2 - 1}{6} + b_{n-1}\right)$, which simplifies to $(2b_n + b_{n-1} + 3)(2b_n - b_{n-1} - 3) = 0$ for all $n \geq 1$. As we have $b_n \geq 0$ for all $n \geq 0$, we must have $2b_n - b_{n-1} - 3 = 0$, or equivalently, $b_n = \frac{1}{2}b_{n-1} + \frac{3}{2}$ for all $n \geq 1$.

Now, we see that the characteristic equation of $b_n - \frac{1}{2}b_{n-1} = 0$ is $x - \frac{1}{2} = 0$, and its characteristic root is $\alpha = \frac{1}{2}$. Thus, we have $b_n^{(h)} = A \cdot 2^{-n}$. Next, let us find $b_n^{(p)}$. Since $f(n) = \frac{3}{2}$ is a polynomial in n of degree 0, and 1 is not a characteristic root, we let $b_n^{(p)} = B$, where B is a constant. Since $b_n^{(p)}$ satisfies the equation $2b_n - b_{n-1} - 3 = 0$, we must have $2B - B - 3 = 0$, or equivalently, $B = 3$, and hence we have $b_n^{(p)} = 3$. By (6.4.2), the general solution of the equation $b_n = \frac{1}{2}b_{n-1} + \frac{3}{2}$ is given by $b_n = b_n^{(h)} + b_n^{(p)} = A \cdot 2^{-n} + 3$. The initial condition $b_0 = \sqrt{1 + 24a_0} = 5$ imply that $A + 3 = 5$, or equivalently, $A = 2$. Therefore, we have $b_n = 3 + 2^{1-n}$ for all $n \geq 0$. Consequently, we have $a_n = \frac{1}{24}[(3 + 2^{1-n})^2 - 1]$ for all $n \geq 0$.

73. Let us first prove by induction on n that we have $a_n = F_{2n}$ for all $n \geq 1$. We first observe that we have

$$a_1 = \frac{3}{2}a_0 + \frac{1}{2}\sqrt{5a_0^2 + 4} = 1 = F_2,$$

so the base case $n = 1$ holds. Next, let us suppose that the case $n = k$ holds, where $k \geq 1$. By induction hypothesis, we have $a_k = F_{2k}$. We first observe that

$$5a_k^2 + 4$$
$$= 5F_{2k}^2 + 4$$
$$= (F_{2k+1} - F_{2k-1})^2 + 4(F_{2k+1}F_{2k-1} - (-1)^{2k}) + 4$$
$$\text{(by Problem 6.45(iii))}$$
$$= F_{2k+1}^2 - 2F_{2k+1}F_{2k-1} + F_{2k-1}^2 + 4F_{2k+1}F_{2k-1} - 4 + 4$$
$$= F_{2k+1}^2 + 2F_{2k+1}F_{2k-1} + F_{2k-1}^2$$
$$= (F_{2k+1} + F_{2k-1})^2.$$

This implies that we have

$$a_{k+1} = \frac{3}{2}a_k + \frac{1}{2}\sqrt{5a_k^2 + 4}$$
$$= \frac{3}{2}F_{2k} + \frac{1}{2}\sqrt{5F_{2k}^2 + 4}$$

$$= \frac{3}{2}F_{2k} + \frac{1}{2}(F_{2k+1} + F_{2k-1})$$

$$= F_{2k} + \frac{1}{2}F_{2k+1} + \frac{1}{2}(F_{2k} + F_{2k-1})$$

$$= F_{2k} + F_{2k+1}$$

$$= F_{2k+2},$$

which completes the induction step. So we have $a_n = F_{2n}$ for all $n \geq 1$ as claimed.

It remains to show that 1992 does not divide a_{2m+1} for all $m \geq 1$. We first note that $m^2 \equiv 0, 1 \pmod{3}$ for all integers m. By Problem 6.46(i), we have $F_{2n-1} = F_n^2 + F_{n-1}^2$, and by Problem 6.45(v), we have $(F_{n-1}, F_n) = 1$. This implies that we have 3 does not divide F_{2n-1} for all $n \geq 1$, and consequently, we have $F_{2n-1}^2 \equiv 1 \pmod{3}$ for all $n \geq 1$. By Problem 6.46(ii), we have $F_{4n} = F_{2n+1}^2 - F_{2n-1}^2 \equiv 0 \pmod{3}$ for all $n \geq 1$.

Next, we will show that 3 does not divide F_{4m+2} for all $m \geq 1$. Suppose on the contrary that $3|F_{4m+2}$ for some $m \geq 1$. Since $F_{4m+1} = F_{4m+2} - F_{4m}$ and $3|F_{4m}$, we see that $3|F_{4m+1}$, which is a contradiction. Hence 3 does not divide F_{4m+2} for all $m \geq 1$. As $a_{2m+1} = F_{4m+2}$, we see that 3 does not divide a_{2m+1} for all $m \geq 1$. As $3|1992$ and $a_{2m+1} = F_{4m+2}$, we must have 1992 to not divide a_{2m+1} for all $m \geq 1$.

74. Let $A(x)$ denote the generating function for the sequence (a_n). Then it follows that we have $na_n x^{n-3} = (n-2)a_{n-1}x^{n-3} - (n+1)x^{n-3}$ for all $n \geq 1$. In particular, since $(1-x)^{-2} = \sum_{n=0}^{\infty}(n+1)x^n$, it follows that we have

$$A' = \sum_{n=1}^{\infty} na_n x^{n-1}$$

$$= \sum_{n=1}^{\infty} \left[(n-2)a_{n-1} + (n+1)\right]x^{n-1}$$

$$= \sum_{n=1}^{\infty}(n-1)a_{n-1}x^{n-1} - \sum_{n=1}^{\infty}a_{n-1}x^{n-1} + \sum_{n=1}^{\infty}(n+1)x^{n-1}$$

$$= xA' - A + \frac{1}{x}\left[(1-x)^{-2} - 1\right]$$

$$= xA' - A + (2-x)(1-x)^{-2},$$

which would then give us

$$(1-x)^{-1}A' + (1-x)^{-2}A = (2-x)(1-x)^{-4}.$$

As we have

$$((1-x)^{-1}A)' = (1-x)^{-1}A' + (1-x)^{-2}A$$

and

$$(2-x)(1-x)^{-4} = (1-x)^{-3} + (1-x)^{-4},$$

it follows that we have

$$((1-x)^{-1}A)' = (1-x)^{-3} + (1-x)^{-4},$$

which would then give us

$$(1-x)^{-1}A = \frac{1}{2}(1-x)^{-2} + \frac{1}{3}(1-x)^{-3} + C,$$

or equivalently,

$$A = \frac{1}{2}(1-x)^{-1} + \frac{1}{3}(1-x)^{-2} + C(1-x),$$

where C is a constant to be determined. As $a_0 = 0$, we must have $0 = \frac{1}{2} + \frac{1}{3} + C$, or equivalently, $C = -\frac{5}{6}$. So we have

$$A = \frac{1}{2}(1-x)^{-1} + \frac{1}{3}(1-x)^{-2} - \frac{5}{6}(1-x).$$

By comparing the coefficients of x^n on both sides of the above equation, we must have $a_0 = 0$, $a_1 = 2$ and $a_2 = \frac{1}{2} + \frac{1}{3}(n+1) = \frac{1}{6}(2n+5)$ for all $n \geq 2$.

75. Let us first prove by induction on n that we have $a_{2n} = 0$ for all $n \geq 0$, with the base case $n = 0$ being trivial. Suppose that the statement holds for $n = k$, where $k \geq 0$. By induction hypothesis, we have $a_{2k} = 0$. This gives us $a_{2k+2} = \frac{4k^2}{(2k+2)(2k+1)}a_{2k} = 0$, which completes the induction step. Thus, we have $a_{2n} = 0$ for all $n \geq 0$ as claimed. Next, for all $n \geq 1$, we have $a_{2n+1} = \frac{(2n-1)^2}{2n(2n+1)}a_{2n-1}$. This implies that for all $n \geq 1$, we have

$$
\begin{aligned}
a_{2n+1} &= \frac{(2n-1)^2}{2n(2n+1)}a_{2n-1} \\
&= \frac{(2n-1)^2}{2n(2n+1)} \cdot \frac{(2n-3)^2}{(2n-2)(2n-1)}a_{2n-3} \\
&= \cdots \\
&= \frac{(2n-1)^2}{2n(2n+1)} \cdot \frac{(2n-3)^2}{(2n-2)(2n-1)} \cdots \frac{1^2}{2 \cdot 3}a_1 \\
&= \frac{(2n-1)^2(2n-3)^2 \cdots 1^2}{(2n+1)!}
\end{aligned}
$$

$$= \frac{1}{(2n+1)!} \cdot \left[\frac{2n(2n-1)(2n-2)\cdots 1}{2n(2n-2)(2n-4)\cdots 2}\right]^2$$

$$= \frac{1}{(2n+1)!} \cdot \left[\frac{(2n)!}{2^n \cdot n!}\right]^2$$

$$= \frac{1}{2n+1} \cdot \frac{(2n)!}{2^{2n}(n!)^2}$$

$$= \frac{1}{2^{2n}(2n+1)}\binom{2n}{n}.$$

As $a_1 = 1 = \frac{1}{2^{2\cdot 0}(2\cdot 0+1)}\binom{2\cdot 0}{0}$, it follows that we have $a_{2n+1} = \frac{1}{2^{2n}(2n+1)}\binom{2n}{n}$ for all $n \geq 0$.

76. By using the definitions of $f(x)$ and $g(x)$, we can write the recurrence relation as $3(a_n - a_{n-1})(a_{n-1}-1)^2 + (a_{n-1}-1)^3 = 0$, and the equation then simplifies to $(a_{n-1}-1)^2(3a_n - 2a_{n-1}-1) = 0$ for all $n \geq 1$. We shall proceed by induction on n that we have $a_n > 1$ for all $n \geq 0$, with the base case $n = 0$ being clear. Suppose that the assertion holds for some $n = k$, where $k \geq 0$. By induction hypothesis, we have $a_k > 1$. As we have $(a_k - 1)^2(3a_{k+1} - 2a_k - 1) = 0$ and $(a_k - 1)^2 \neq 0$, we must have $3a_{k+1} - 2a_k - 1 = 0$, or equivalently, $a_{k+1} = \frac{1}{3}(2a_k+1) > \frac{1}{3}(2\cdot 1+1) = 1$. This completes the induction step, and hence we have $a_n > 1$ for all $n \geq 0$. As an immediate consequence, we have $3a_n - 2a_{n-1} - 1 = 0$ for all $n \geq 0$.

Now, we see that the characteristic equation of $3a_n - 2a_{n-1} = 0$ is $3x - 2 = 0$, and its characteristic root is $\alpha = \frac{2}{3}$. Thus, we have $a_n^{(h)} = A\left(\frac{2}{3}\right)^n$. Next, let us find $a_n^{(p)}$. Since $f(n) = 1$ is a polynomial in n of degree 0, and 1 is not a characteristic root, we let $a_n^{(p)} = B$, where B is a constant. Since $a_n^{(p)}$ satisfies the equation $3a_n - 2a_{n-1} - 1 = 0$, we must have $3B - 2B - 1 = 0$, or equivalently, $B = 1$, and hence we have $a_n^{(p)} = 1$. By (6.4.2), the general solution of the equation $3a_n - 2a_{n-1} - 1 = 0$ is given by $a_n = a_n^{(h)} + a_n^{(p)} = A\left(\frac{2}{3}\right)^n + 1$. The initial condition $a_0 = 2$ imply that $A + 1 = 2$, or equivalently, $A = 1$. Therefore, we have $a_n = 1 + \left(\frac{2}{3}\right)^n$ for all $n \geq 0$.

77. Let $b_n = n(n-1)a_n$ for all $n \geq 0$. Then it follows from the given recurrence relation that we have $b_n = b_{n-1} - \frac{1}{n-2}b_{n-2}$, or equivalently, $(n-2)b_n = (n-2)b_{n-1} - b_{n-2}$ for all $n \geq 2$. Let $B(x)$ denote the generating function for the sequence (b_n). Then it follows that we have $(n-2)b_n x^{n-3} = (n-2)b_{n-1}x^{n-3} - b_{n-2}x^{n-3}$ for all $n \geq 2$. In particular, since $b_0 = b_1 = 0$, we deduce that $(x^{-2}B)' = (x^{-1}B)' - x^{-1}B$, which

would then give us $-2x^{-3}B + x^{-2}B' = -x^{-2}B + x^{-1}B' - x^{-1}B$, which subsequently simplifies to $\frac{B'}{B} = 1 + \frac{2}{x}$. This implies that $\log_e B = x + 2\log_e x + c$, or equivalently, $B = x^2 e^{x+c}$, where c is a constant to be determined. Now, the constant term of $x^{-2}B$ is both equal to e^c and $b_2 = 2a_2 = a_0 = 1$. So we must have $e^c = 1$, and hence $B = x^2 e^x$. As $e^x = \sum\limits_{n=0}^{\infty} \frac{x^n}{n!}$, it follows that we have $b_n = \frac{1}{(n-2)!}$ for all $n \geq 2$. Consequently, we have $a_n = \frac{1}{n(n-1)}b_n = \frac{1}{n!}$ for all $n \geq 2$. Thus, we have

$$\sum_{k=0}^{1992} \frac{a_k}{a_{k+1}} = \frac{a_0}{a_1} + \frac{a_1}{a_2} + \sum_{k=2}^{1992}(k+1)$$

$$= \frac{1}{2} + 4 + \sum_{k=1}^{1991}(k+2)$$

$$= \frac{9}{2} + \frac{1991 \cdot 1992}{2} + 1991 \cdot 2$$

$$= \frac{9}{2} + 1991 \cdot 998.$$

78. Let us take any representation of n as a sum of 1's and 2's, taking order into account. If the first summand is equal to 1, then the remaining summands sum up to $n - 1$, and there are $a(n - 1)$ ways to form the remaining summands. Else, if the first summand is equal to 2, then the remaining summands sum up to $n - 2$, and there are $a(n - 2)$ ways to form the remaining summands. By (AP), we have $a(n) = a(n-1)+a(n-2)$ for all $n \geq 3$. Since $a(1) = 1 = F_2$ and $a(2) = 2 = F_3$, we must have $a(n) = F_{n+1}$.

Next, for all $n \geq 2$, let us take any representation of n as a sum of integers greater than 1, taking order into account. Then either there is a single summand, namely n, or the first summand is equal to k, where $2 \leq k \leq n - 2$. In the latter case, there are $b(n - k)$ ways to form the remaining summands. By (AP), we must have $b(n) = 1 + \sum\limits_{k=2}^{n-2} b(n-k) = 1 + \sum\limits_{k=2}^{n-2} b(k)$. This implies that we have $b(n + 1) - b(n) = b(n - 1)$, or equivalently, $b(n + 1) = b(n) + b(n - 1)$ for all $n \geq 3$. As $b(2) = 1 = F_1$ and $b(3) = 1 = F_2$, we must have $b(n) = F_{n-1}$. Consequently, we must have $a(n) = F_{n+1} = b(n + 2)$ as desired.

79. We first observe from (5.1.6) that we have

$$(2-1)^{-n} = 2^{-n}\left(1 - \frac{1}{2}\right)^{-n} = 2^{-n}\sum_{r=0}^{\infty}\binom{r+n-1}{r}\left(\frac{1}{2}\right)^{r}$$

for all $n \geq 1$. Let $a_n = 2^{-n}\sum_{r=0}^{n-1}\binom{r+n-1}{r}\left(\frac{1}{2}\right)^{r}$ for all $n \geq 1$. Then for all $n > 1$, we have

$$a_n = 2^{-n}\sum_{r=0}^{n-1}\binom{r+n-1}{r}\left(\frac{1}{2}\right)^{r}$$

$$= 2^{-n}\sum_{r=0}^{n-1}\binom{r+n-2}{r}\left(\frac{1}{2}\right)^{r} + 2^{-n}\sum_{r=0}^{n-1}\binom{r+n-2}{r-1}\left(\frac{1}{2}\right)^{r}$$

(by (2.1.4))

$$= 2^{-n}\sum_{r=0}^{n-2}\binom{r+n-2}{r}\left(\frac{1}{2}\right)^{r} + \binom{2n-3}{n-1}2^{-2n+1}$$

$$\quad + 2^{-n}\sum_{r=1}^{n-1}\binom{r+n-2}{r-1}\left(\frac{1}{2}\right)^{r}$$

$$= \frac{1}{2}a_{n-1} + \binom{2n-3}{n-1}2^{-2n+1} + 2^{-n-1}\sum_{r=0}^{n-2}\binom{r+n-1}{r}\left(\frac{1}{2}\right)^{r}$$

$$= \frac{1}{2}a_{n-1} + \binom{2n-3}{n-2}2^{-2n+1}$$

$$\quad + 2^{-n-1}\sum_{r=0}^{n-1}\binom{r+n-1}{r}\left(\frac{1}{2}\right)^{r} - \binom{2n-2}{n-1}2^{-2n}$$

$$= \frac{1}{2}a_{n-1} + \binom{2n-3}{n-2}2^{-2n+1} + \frac{1}{2}a_n - \frac{2n-2}{n-1}\binom{2n-3}{n-2}2^{-2n}$$

(by (2.1.2))

$$= \frac{1}{2}a_n + \frac{1}{2}a_{n-1},$$

or equivalently, $a_n = a_{n-1}$. Thus, we have $a_n = a_1 = \frac{1}{2}$ for all $n \geq 1$ as desired.

80. It is easy to see that the function $f(x) = 2x$ satisfy the given conditions, so it only remains to show that $f(x) = 2x$ is the only function that satisfies the given conditions. Suppose on the contrary that we have $f(x_0) \neq 2x_0$ for some $x_0 > 0$. Without loss of generality, we may assume that $f(x_0) > 2x_0$, or equivalently, $2x_0 - f(x_0) < 0$. Let $a_n =$

$f^n(x_0)$ for all $n \geq 0$. Then it follows that for all $n \geq 2$, we have

$$a_n = f^2(f^{n-2}(x_0)) = 6f^{n-2}(x_0) - f(f^{n-2}(x_0)) = 6a_{n-2} - a_{n-1}.$$

The characteristic equation of the recurrence relation $a_n = 6a_{n-2} - a_{n-1}$ is $x^2 + x - 6 = 0$, and its characteristic roots are $\alpha_1 = 2$ and $\alpha_2 = -3$. Thus, by result (I), the general solution of the recurrence relation is given by $a_n = A \cdot 2^n + B(-3)^n$, where A and B are constants to be determined. The initial conditions $a_0 = x_0$ and $a_1 = f(x_0)$ imply that

$$A + B = x_0, \quad \text{and}$$
$$2A - 3B = f(x_0).$$

Solving the system for A and B gives us $A = \frac{3x_0 + f(x_0)}{5}$ and $B = \frac{2x_0 - f(x_0)}{5}$. So we have

$$f^n(x_0) = a_n = \frac{1}{5}[(3x_0 + f(x_0))2^n + (2x_0 - f(x_0))(-3)^n]$$

for all $n \geq 0$.

Now, we observe that we have $3x_0 + f(x_0) > 0$. As $\lim_{n \to \infty} \left(\frac{3}{2}\right)^n = \infty$, it follows that there exists some positive integer m, such that $\left(\frac{3}{2}\right)^{2m} > \frac{3x_0 + f(x_0)}{f(x_0) - 2x_0}$, or equivalently, $(f(x_0) - 2x_0)3^{2m} > (3x_0 + f(x_0))2^{2m}$. This implies that we have

$$f^{2m}(x_0) = \frac{1}{5}[(3x_0 + f(x_0))2^{2m} + (2x_0 - f(x_0))3^{2m}] < 0,$$

which is a contradiction. Likewise, a similar contradiction would occur if $f(x_0) < 2x_0$. So we must have $f(x) = 2x$ for all $x > 0$. This proves the uniqueness of the function f, and we are done.

81. Let us first rewrite the recurrence relation as

$$T_n - nT_{n-1} = 4(T_{n-1} - (n-1)T_{n-2}) - 4(T_{n-2} - (n-2)T_{n-3}).$$

We let $Q_n = T_{n+1} - (n+1)T_n$ for all $n \geq 0$. Then the above recurrence relation could be expressed in terms of the Q_n's as $Q_{n+2} = 4Q_{n+1} - 4Q_n$ for all $n \geq 0$. The corresponding characteristic equation is $x^2 - 4x + 4 = 0$, and its characteristic root is $\alpha = 2$ of multiplicity 2. Thus, by result (I), the general solution of the recurrence relation $Q_{n+2} = 4Q_{n+1} - 4Q_n$ is given by $Q_n = (A + Bn)2^n$, where A and B are constants to be determined. The initial conditions $Q_0 = T_1 - T_0 = 1$ and $Q_1 = T_2 - 2T_1 = 0$ imply that

$$A = 1, \quad \text{and}$$
$$2A + 2B = 0.$$

Solving the system for A and B gives us $A = 1$ and $B = -1$. Consequently, we have $T_{n+1} - (n+1)T_n = Q_n = (1-n)2^n$ for all $n \geq 0$.
Next, we let $T(x)$ denote the exponential generating function of T.
Then it follows that we have $\frac{T_{n+1}}{(n+1)!}x^{n+1} - x\frac{T_n}{n!}x^n = \frac{(2x)^{n+1}}{(n+1)!} - x \cdot \frac{(2x)^n}{n!}$
for all $n \geq 0$. As we have $T_0 = 2$ and $e^{2x} = \sum\limits_{n=0}^{\infty} \frac{(2x)^n}{n!}$, this implies that
we have $T(x) - 2 - xT(x) = e^{2x} - 1 - xe^{2x}$, which would then give us
$T(x) = \frac{1}{1-x} + e^{2x}$. As the coefficient of x^n in $\frac{1}{1-x} + e^{2x}$ is equal to
$1 + \frac{2^n}{n!}$ for all $n \geq 0$, it follows that we have $T_n = n! + 2^n$ for all $n \geq 0$.

82. Firstly, the characteristic equation for the recurrence relation $a_n = a_{n-1} + 2a_{n-2}$ is $x^2 - x - 2 = 0$, and its characteristic roots are $\alpha_1 = 2$ and $\alpha_2 = -1$. Thus, by result (I), the general solution of the recurrence relation is given by $a_n = A \cdot 2^n + B(-1)^n$, where A and B are constants to be determined. The initial conditions $a_0 = a_1 = 1$ gives

$$A + B = 1, \quad \text{and}$$
$$2A - B = 1.$$

Solving the system for A and B gives us $A = \frac{2}{3}$ and $B = \frac{1}{3}$. So we have
$a_n = \frac{1}{3}(2^{n+1} + (-1)^n)$ for all $n \geq 0$.
The characteristic equation for the recurrence relation $b_n = 2b_{n-1} + 3b_{n-2}$ is $x^2 - 2x - 3 = 0$, and its characteristic roots are $\alpha_1 = 3$ and $\alpha_2 = -1$. Thus, by result (I), the general solution of the recurrence relation is given by $b_n = C \cdot 3^n + D(-1)^n$, where C and D are constants to be determined. The initial conditions $b_0 = 1$ and $b_1 = 7$ imply that

$$C + D = 1, \quad \text{and}$$
$$3C - D = 7.$$

Solving the system for C and D gives us $C = 2$ and $D = -1$. So we have $b_n = 2 \cdot 3^n + (-1)^{n+1}$ for all $n \geq 0$.
Now, let us take a term k that appears in both sequences. Then we have $k = a_m = b_n$ for some $m, n \in \mathbf{N}$, that is, we have $\frac{1}{3}(2^m + (-1)^{m+1}) = 2 \cdot 3^{n-1} + (-1)^n$, or equivalently, $2^m = 2 \cdot 3^n + 3(-1)^n + (-1)^m$. If $m = 1, 2$, then we see that the only solution to the equation $2^m = 2 \cdot 3^n + 3(-1)^n + (-1)^m$ is $n = 1$, which would then correspond to $k = 1$. Suppose on the contrary that there exists a positive integer solution (m, n) to the equation $2^m = 2 \cdot 3^n + 3(-1)^n + (-1)^m$ with $m \geq 3$. Then it follows that we have $2^m \equiv 0 \pmod 8$. On the other hand, using the fact that we have $s^2 \equiv 1 \pmod 8$ for all odd s, it follows that we have $2 \cdot 3^n + 3(-1)^n \equiv 3, 5 \pmod 8$. As $(-1)^m \equiv$

$1, 7 \pmod 8$, we must have $2 \cdot 3^n + 3(-1)^n + (-1)^m \not\equiv 0 \pmod 8$, which is a contradiction. So the only positive integer solutions (m, n) to the equation $2^m = 2 \cdot 3^n + 3(-1)^n + (-1)^m$ are $(1, 1)$ and $(2, 1)$, and consequently we see that except for "1", there is no term which occurs in both sequences.

Remark. The solution is courtesy of John Scholes. We refer the reader to `https://mks.mff.cuni.cz/kalva/usa/usoln/usol732.html` for the original solution.

83. For each $m \geq 0$, we let $a_m = x_{2m+1}$ and $b_m = x_{2m+2}$. Then we have $a_0 = 2$, $b_0 = 3$, $a_m = a_{m-1} + b_{m-1}$ and $b_m = a_m + 2b_{m-1}$ for all $m \geq 1$. This gives us $b_{m-1} = a_m - a_{m-1}$, and we substitute $b_{m-1} = a_m - a_{m-1}$ into the equation $b_m = a_m + 2b_{m-1}$ to get $a_{m+1} - a_m = a_m + 2a_m - 2a_{m-1}$, or equivalently, $a_{m+1} - 4a_m + 2a_{m-1} = 0$ for all $m \geq 1$. The corresponding characteristic equation is given by $x^2 - 4x + 2 = 0$, and its characteristic roots are $\alpha_1 = 2 + \sqrt{2}$ and $\alpha_2 = 2 - \sqrt{2}$. Thus, by result (I), the general solution of the recurrence relation is given by $a_n = A(2 + \sqrt{2})^n + B(2 - \sqrt{2})^n$, where A and B are constants to be determined. The initial conditions $a_0 = 2$ and $a_1 = a_0 + b_0 = 5$ imply that

$$A + B = 2, \quad \text{and}$$
$$A(2 + \sqrt{2}) + B(2 - \sqrt{2}) = 5.$$

Solving the system for A and B gives us $A = \frac{4+\sqrt{2}}{4}$ and $B = \frac{4-\sqrt{2}}{4}$. So we have

$$x_{2n+1} = a_n = \frac{1}{4}\left[(4 + \sqrt{2})(2 + \sqrt{2})^n + (4 - \sqrt{2})(2 - \sqrt{2})^n\right],$$

and hence we have

$$x_{2n} = a_n - a_{n-1} = \frac{1}{4}\left[(1 + 2\sqrt{2})(2 + \sqrt{2})^n + (1 - 2\sqrt{2})(2 - \sqrt{2})^n\right].$$

We conclude that

$$x_n = \begin{cases} \frac{1}{4}\left[(4 + \sqrt{2})(2 + \sqrt{2})^{\frac{n-1}{2}} + (4 - \sqrt{2})(2 - \sqrt{2})^{\frac{n-1}{2}}\right] & \text{if } n \text{ is odd,} \\ \frac{1}{4}\left[(1 + 2\sqrt{2})(2 + \sqrt{2})^{\frac{n}{2}} + (1 - 2\sqrt{2})(2 - \sqrt{2})^{\frac{n}{2}}\right] & \text{if } n \text{ is even.} \end{cases}$$

84. For each $n \geq 1$, we let a_n be the number of sequences $(x_1, \ldots, x_n)$ with $x_i \in \{a, b, c\}$ for $i = 1, \ldots, n$, such that $x_1 = x_n = a$ and $x_i \neq x_{i+1}$ for $i = 1, \ldots, n - 1$, and let us pick such a sequence $(x_1, \ldots, x_n)$. Let us consider these cases:

Case 1: $x_3 = a$. Then $(x_3, \ldots, x_n)$ is a sequence of length $n - 2$ that satisfies the above conditions, and $x_2 \in \{b, c\}$. By (MP), there are $2a_{n-2}$ such sequences $(x_1, \ldots, x_n)$ for which $x_3 = a$.

Case 2: $x_3 \neq a$. Then $(a, x_3, \ldots, x_n)$ is a sequence of length $n - 1$ that satisfies the above conditions. Moreover, since $x_1 = a$ and $x_3 \neq a$, it follows that x_2 is uniquely determined by x_3. This implies that there is a bijection between the set of sequences $(x_1, \ldots, x_n)$ of length n with $x_3 \neq a$ that satisfy the above conditions, and the set of sequences $(x_2, \ldots, x_n)$ of length $n - 1$ that satisfy the above conditions. Hence, there are a_{n-1} such sequences $(x_1, \ldots, x_n)$ for which $x_3 \neq a$.

By (AP), we must have $a_n = a_{n-1} + 2a_{n-2}$ for all $n \geq 3$. The corresponding characteristic equation is given by $x^2 - x - 2 = 0$, and its characteristic roots are $\alpha_1 = 2$ and $\alpha_2 = -1$. Thus, by result (I), the general solution of the recurrence relation is given by $a_n = A \cdot 2^n + B(-1)^n$, where A and B are constants to be determined. The initial conditions $a_1 = 1$ and $a_2 = 0$ imply that

$$2A - B = 1, \quad \text{and}$$
$$4A + B = 0.$$

Solving the system for A and B gives us $A = \frac{1}{6}$ and $B = -\frac{2}{3}$. So we have $a_n = \frac{1}{6} \cdot 2^n - \frac{2}{3}(-1)^n = \frac{2}{3}[2^{n-2} + (-1)^{n+1}]$.

85. Let us first find $x_n^{(h)}$. The characteristic equation of $x_n - 14x_{n-1} + x_{n-2} = 0$ is $x^2 - 14x + 1 = 0$, and its characteristic roots are $\beta = 7 + 4\sqrt{3}$ and $\beta^{-1} = 7 - 4\sqrt{3}$. Thus, we have $x_n^{(h)} = A \cdot \beta^n + B \cdot \beta^{-n}$, where A and B are constants.

Next, let us find $x_n^{(p)}$. Since $f(n) = -4$ is a polynomial in n of degree 0, and 1 is not a characteristic root, we let $x_n^{(p)} = C$, where C is a constant. As $x_n^{(p)}$ satisfies the equation $x_n - 14x_{n-1} + x_{n-2} = -4$, we must have $C - 14C + C = -4$, which would then give us $C = \frac{1}{3}$, and hence we have $x_n^{(p)} = \frac{1}{3}$. By (6.4.2), the general solution of the equation $x_n - 14x_{n-1} + x_{n-2} = -4$ is given by $x_n = x_n^{(h)} + x_n^{(p)} = A \cdot \beta^n + B \cdot \beta^{-n} + \frac{1}{3}$. The initial conditions $x_1 = x_2 = 1$ imply that

$$A\beta + B\beta^{-1} + \frac{1}{3} = 1, \quad \text{and}$$
$$A\beta^2 + B\beta^{-2} + \frac{1}{3} = 1.$$

Solving the system for A and B gives us $A = \frac{26 - 15\sqrt{3}}{6}$ and $B = \frac{26 + 15\sqrt{3}}{6}$. Thus, we have $x_n = \frac{26 - 15\sqrt{3}}{6}\beta^n + \frac{26 + 15\sqrt{3}}{6}\beta^{-n} + \frac{1}{3}$ for all $n \geq 1$.

It remains to show that x_n is a perfect square for all $n \geq 1$. To this end, we first observe that we have $7 \pm 4\sqrt{3} = (2 \pm \sqrt{3})^2$ and $\frac{26 \pm 15\sqrt{3}}{6} = \left(\frac{9 \pm 5\sqrt{3}}{6}\right)^2$. As $(2 + \sqrt{3})(2 - \sqrt{3}) = 1$ and $2\left(\frac{9 - 5\sqrt{3}}{6}\right)\left(\frac{9 + 5\sqrt{3}}{6}\right) = \frac{1}{3}$, it follows that we have

$$
\begin{aligned}
x_n &= \frac{26 - 15\sqrt{3}}{6}\beta^n + \frac{26 + 15\sqrt{3}}{6}\beta^{-n} + \frac{1}{3} \\
&= \left[\left(\frac{9 - 5\sqrt{3}}{6}\right)(2 + \sqrt{3})^n\right]^2 + \left[\left(\frac{9 + 5\sqrt{3}}{6}\right)(2 - \sqrt{3})^n\right]^2 + \frac{1}{3} \\
&= \left[\left(\frac{9 - 5\sqrt{3}}{6}\right)(2 + \sqrt{3})^n + \left(\frac{9 + 5\sqrt{3}}{6}\right)(2 - \sqrt{3})^n\right]^2.
\end{aligned}
$$

We let $a_n = \left(\frac{9 - 5\sqrt{3}}{6}\right)(2 + \sqrt{3})^n + \left(\frac{9 + 5\sqrt{3}}{6}\right)(2 - \sqrt{3})^n$ for all $n \geq 1$. Then it follows that for all $n \geq 1$, we have $x_n = a_n^2$, and

$$
\begin{aligned}
&a_{n+2} - 4a_{n+1} + a_n \\
&= \left(\frac{9 - 5\sqrt{3}}{6}\right)(2 + \sqrt{3})^n[(2 + \sqrt{3})^2 - 4(2 + \sqrt{3}) + 1] \\
&\quad + \left(\frac{9 + 5\sqrt{3}}{6}\right)(2 - \sqrt{3})^n[(2 - \sqrt{3})^2 - 4(2 - \sqrt{3}) + 1] \\
&= 0.
\end{aligned}
$$

Moreover, we have $a_1 = \left(\frac{9 - 5\sqrt{3}}{6}\right)(2 + \sqrt{3}) + \left(\frac{9 + 5\sqrt{3}}{6}\right)(2 - \sqrt{3}) = 1$ and $a_2 = \left(\frac{9 - 5\sqrt{3}}{6}\right)(2 + \sqrt{3})^2 + \left(\frac{9 + 5\sqrt{3}}{6}\right)(2 - \sqrt{3})^2 = 1$. So we have $a_n \in \mathbf{Z}$ for all $n \geq 1$. Consequently, we have $x_n = a_n^2$ to be a perfect square for all $n \geq 1$ as desired.

86. For each $m = 0, 1, 2, 3, 4$ and $n \in \mathbf{N}$, we let $a(m, n)$ denote the number of words with n digits, such that each digit is either $0, 1, 2, 3$ or 4, the first digit is m, and adjacent digits must differ by exactly one, and we let a_n denote the number of words with n digits, such that each digit is either $0, 1, 2, 3$ or 4, and adjacent digits must differ by exactly one. Then it is clear that we have $a_n = \sum_{m=0}^{4} a(m, n)$, and for all $n \geq 2$ and $i = 1, 2, 3$, we have $a(0, n) = a(1, n - 1)$, $a(4, n) = a(3, n - 1)$ and $a(i, n) = a(i - 1, n) + a(i + 1, n)$. Moreover, by replacing each digit i in a word by $4 - i$, we see that $a(0, n) = a(4, n)$ and $a(1, n) = a(3, n)$.

Thus, for all $n \geq 3$, we have

$$a(0,n) = a(1, n-1) = a(0, n-2) + a(2, n-2), \qquad (6.2)$$

$$a(1,n) = a(0, n-1) + a(2, n-1)$$
$$= 2a(1, n-2) + a(3, n-2)$$
$$= 3a(1, n-2),$$

$$a(2,n) = a(1, n-1) + a(3, n-1)$$
$$= a(0, n-2) + 2a(2, n-2) + a(4, n-2)$$
$$= 2a(0, n-2) + 2a(2, n-2),$$

$$a(3,n) = a(1,n) = 3a(1, n-2), \quad \text{and}$$

$$a(4,n) = a(0,n) = a(0, n-2) + a(2, n-2).$$

This implies that for all $n \geq 3$, we have

$$a_n = \sum_{m=0}^{4} a(m,n) = 4a(0, n-2) + 6a(1, n-2) + 4a(2, n-2).$$

Next, we observe from equation (6.2) that we have

$$a(2,n) = a(1, n-1) + a(3, n-1) = 2a(1, n-1) = 2a(0,n)$$

for all $n \geq 3$. Moreover, if $n \geq 4$, then we have

$$a_n = 6a(0, n-2) + 6a(1, n-2) + 3a(2, n-2) = 3a_{n-2},$$

where we have used the identity $a(i, n-2) = a(4 - i, n-2)$ in the last equality.

Now, we let $b_n = a_{2n}$ and $c_n = a_{2n+1}$ for all $n \geq 1$. Then it follows that we have $b_{n+1} = 3b_n$ and $c_{n+1} = 3c_n$ for all $n \geq 1$, from which we deduce that we have $b_n = B \cdot 3^n$ and $c_n = C \cdot 3^n$ for all $n \geq 1$, where B and C are constants to be determined. The initial conditions $b_1 = a_2 = 8$ and $c_1 = a_3 = 14$ imply that $B = \frac{8}{3}$ and $C = \frac{14}{3}$. So we have $a_{2n} = b_n = 8 \cdot 3^{n-1}$ and $a_{2n+1} = c_n = 14 \cdot 3^{n-1}$ for all $n \geq 1$. We conclude that

$$a_n = \begin{cases} 5 & \text{if } n = 1, \\ 14 \cdot 3^{\frac{n-3}{2}} & \text{if } n \text{ is odd and } n > 1, \\ 8 \cdot 3^{\frac{n-2}{2}} & \text{if } n \text{ is even.} \end{cases}$$

87. Let us define the sequence (b_n) of integers by $b_1 = 2$, $b_2 = 7$ and $b_n = 3b_{n-1} + 2b_{n-2}$ for all $n \geq 3$. Then it is clear from the defined recurrence relation that we have $b_n > 0$ for all $n \geq 1$. We shall prove

by induction on n that we have $-b_n < 2(b_{n+2}b_n - b_{n+1}^2) \leq b_n$ for all $n \geq 1$, with the base cases $n = 1$ and $n = 2$ being trivial. Suppose that the inequality holds for all $1 \leq n \leq k$, where $k \geq 2$. By induction hypothesis, we have $-b_k < 2(b_{k+2}b_k - b_{k+1}^2) \leq b_k$. We first observe that $b_k = \frac{1}{2}(b_{k+2} - 3b_{k+1})$, $b_{k+1} = \frac{1}{2}(b_{k+3} - 3b_{k+2})$ and, which implies that we have $b_k = \frac{1}{4}(2b_{k+2} - 3(b_{k+3} - 3b_{k+2})) = \frac{1}{4}(11b_{k+2} - 3b_{k+3})$. Thus, we have

$$
\begin{aligned}
&-2(b_{k+2}b_{k+4} - b_{k+3}^2) \\
&= -2b_{k+2}(2b_{k+2} + 3b_{k+3}) + 2b_{k+3}^2 \\
&= -4b_{k+2}^2 - 6b_{k+2}b_{k+3} + 2b_{k+3}^2 \\
&= b_{k+2}(11b_{k+2} - 3b_{k+3}) - (b_{k+3} - 3b_{k+2})^2 \\
&\quad + 3b_{k+3}(b_{k+3} - 3b_{k+2}) - 6b_{k+2}^2 \\
&= 4(b_{k+2}b_k - b_{k+1}^2) + 6(b_{k+3}b_{k+1} - b_{k+2}^2)
\end{aligned}
$$

Therefore, we have

$$
\begin{aligned}
-2(b_{k+2}b_{k+4} - b_{k+3}^2) &= 4(b_{k+2}b_k - b_{k+1}^2) + 6(b_{k+3}b_{k+1} - b_{k+2}^2) \\
&> -2b_k - 3b_{k+1} \\
&= -b_{k+2}.
\end{aligned}
$$

Likewise, we have $-2(b_{k+2}b_{k+4} - b_{k+3}^2) \leq b_{k+2}$. This gives us

$$-b_{k+2} < -2(b_{k+2}b_{k+4} - b_{k+3}^2) \leq b_{k+2},$$

or equivalently, $-b_{k+2} < 2(b_{k+2}b_{k+4} - b_{k+3}^2) \leq b_{k+2}$, and this completes the induction step. Thus, we have $-b_n \leq 2(b_{n+2}b_n - b_{n+1}^2) \leq b_n$ for all $n \geq 1$ as claimed. As $b_n > 0$ for all $n \geq 1$, it follows that we have $-\frac{1}{2} \leq b_{n+2} - \frac{b_{n+1}^2}{b_n} \leq \frac{1}{2}$ for all $n \geq 1$.

Next, we will prove by induction on n that b_n is odd and $a_n = b_n$ for all $n \geq 2$, with the base cases $n = 2$ and $n = 3$ being trivial. Suppose that the statement holds for all $2 \leq n \leq k$, where $k \geq 3$. By induction hypothesis, we have $a_k = b_k$, $a_{k-1} = b_{k-1}$, and b_k and b_{k-1} are both odd. As $b_{k+1} = 3b_k + 2b_{k-1}$, $3b_k$ is odd and $2b_{k-1}$ is even, it follows that b_{k+1} is odd. Next, since b_{k-1} is odd, it follows that $\frac{b_k^2}{b_{k-1}} - \frac{1}{2}$ and $\frac{b_k^2}{b_{k-1}} + \frac{1}{2}$ are both non-integers, so that exists a unique integer m such that $\frac{b_k^2}{b_{k-1}} - \frac{1}{2} < m \leq \frac{b_k^2}{b_{k-1}} + \frac{1}{2}$. As $x = a_{k+1}$ and $x = b_{k+1}$ both satisfy the inequality $\frac{b_k^2}{b_{k-1}} - \frac{1}{2} < x \leq \frac{b_k^2}{b_{k-1}} + \frac{1}{2}$ (noting that $a_{k-1} = b_{k-1}$ and $a_k = b_k$) and a_{k+1} and b_{k+1} are both integers, it follows from the uniqueness of m that we must have $a_{k+1} = b_{k+1}$. This completes the induction step, and hence we have $a_n = b_n$ to be odd for all $n \geq 2$ as desired.

88. For all $i = 0, \ldots, n$, let us label the intersection points on the bottom (respectively top) horizontal line from left to right by $P_0, \ldots, P_n$ (respectively $Q_0, \ldots, Q_n$), so that we have $P_0 = A$ and $Q_n = B$. We first observe that P_i is connected to P_{i+1}, Q_i is connected to Q_{i+1} and P_i is connected to Q_{i+1} for all $i = 0, \ldots, n-1$, and P_j is connected to Q_j for all $j = 0, \ldots, n$.

For each $k \geq 0$, we let a_k denote the number of paths from P_0 to Q_k that does not pass through any intersection twice and does not go past the edge $P_k Q_k$. Also, for each $j \geq 1$, we let b_j denote the number of paths from P_0 to P_j that does not pass through any intersection twice and does not go past the edge $P_j Q_j$, and ends in either $Q_j \to P_j$ or $P_{j-1} \to P_j$, and the segment $Q_j \to P_{j-1}$ is not used. We let c_j denote the number of paths from P_0 to P_j that does not pass through any intersection twice and does not go past the edge $P_j Q_j$, and ends in $Q_j \to P_{j-1} \to P_j$. Then the number of paths from A to B that does not pass through any intersection twice is equal to a_n. It follows that the number of paths from P_0 to P_j that does not pass through any intersection twice and does not go past the edge $P_j Q_j$ is equal to $b_j + c_j$. We seek to establish a recurrence relation for a_n. To this end, we first observe that $a_0 = 1$, $a_1 = 3 = b_1$, $a_2 = 9$ and $c_2 = 2$.

First, let us fix $k \geq 2$, and take a path from P_0 to Q_k that does not pass through any intersection twice and does not go past the edge $P_k Q_k$. Then the path must end in either $Q_{k-1} \to Q_k$, $P_{k-1} \to Q_k$ or $P_{k-1} \to P_k \to Q_k$. By (AP), it follows that we have $a_k = a_{k-1} + 2b_{k-1} + 2c_{k-1}$. Next, let us take a path from P_0 to P_k that does not pass through any intersection twice and does not go past the edge $P_k Q_k$, and ends in either $Q_k \to P_k$ or $P_{k-1} \to P_k$, and the segment $Q_k \to P_{k-1}$ is not used. Then the path must end in either $Q_{k-1} \to Q_k \to P_k$, $P_{k-1} \to Q_k \to P_k$ or $P_{k-1} \to P_k$. By (AP), it follows that we have $b_k = a_{k-1} + 2b_{k-1} + 2c_{k-1} = a_k$, and hence we have

$$a_k = 3a_{k-1} + 2c_{k-1}. \tag{6.3}$$

Finally, let us fix $j \geq 3$, and take a path from P_0 to P_j that does not pass through any intersection twice and does not go past the edge $P_j Q_j$, and ends in $Q_j \to P_{j-1} \to P_j$. Then the path must end in either $Q_{j-2} \to Q_{j-1} \to Q_j \to P_{j-1} \to P_j$ or $P_{j-2} \to Q_{j-1} \to Q_j \to P_{j-1} \to P_j$. By (AP), it follows that we have $c_j = a_{j-2} + b_{j-2} + c_{j-2} = 2a_{j-2} + c_{j-2}$.

Now, we observe from equation (6.3) that we have $c_i = \frac{1}{2}(a_{i+1} - 3a_i)$ for all $i \geq 1$. By substituting this into the equation $c_n = 2a_{n-2} + c_{n-2}$ (where $n \geq 3$), it follows that we have

$$\frac{1}{2}(a_{n+1} - 3a_n) = 2a_{n-2} + \frac{1}{2}(a_{n-1} - 3a_{n-2}),$$

or equivalently, $a_{n+1} = 3a_n + a_{n-1} + a_{n-2}$ for all $n \geq 3$. Finally, since $a_3 = 3a_2 + 2c_2 = 3a_2 + 4 = 3a_2 + a_1 + a_0$, it follows that we have $a_{n+1} = 3a_n + a_{n-1} + a_{n-2}$ for all $n \geq 2$.

89. Let $n_k = 2^k + k - 2$ for all $k \geq 1$. Then we have $n_1 = 1$, $n_2 = 4$ and $n_{k+1} = n_k + 2^k + 1$ for all $k \geq 1$. We shall prove by induction on k that we have

$$a_{n_{k-1}+2i+1} = (2^{k-2} + i)^2 + 2^{k-2} - i, \quad \text{and}$$

$$a_{n_{k-1}+2j} = (2^{k-2} + j - 1)^2 + 2^{k-1}$$

for all $k \geq 2$, $0 \leq i \leq 2^{k-2}$ and $1 \leq j \leq 2^{k-2}$, with the base case $k = 2$ being trivial. Suppose that the statement holds for $k = m$, where $m \geq 2$. By induction hypothesis, we have $a_{n_{m-1}+2i+1} = (2^{m-2}+i)^2 + 2^{m-2} - i$ and $a_{n_{m-1}+2j} = (2^{m-2} + j - 1)^2 + 2^{m-1}$ for all $0 \leq i \leq 2^{m-2}$ and $1 \leq j \leq 2^{m-2}$. We will prove by induction on i and j that we have $a_{n_m+2i+1} = (2^{m-1}+i)^2 + 2^{m-1} - i$ and $a_{n_m+2j} = (2^{m-1}+j-1)^2 + 2^m$ for all $0 \leq i \leq 2^{m-1}$ and $1 \leq j \leq 2^{m-1}$. To this end, we first observe that

$$a_{n_m} = a_{n_{m-1}+2^{m-1}+1} = (2^{m-2} + 2^{m-2})^2 + 2^{m-2} - 2^{m-2} = (2^{m-1})^2.$$

This implies that

$$a_{n_m+1} = a_{n_m} + \lfloor\sqrt{a_{n_m}}\rfloor = (2^{m-1})^2 + 2^{m-1} = (2^{m-1} + 0)^2 + 2^{m-1} - 0,$$

which implies that the base case $i = 0$ holds. Next, we observe that

$$\begin{aligned}
a_{n_m+2} &= a_{n_m+1} + \lfloor\sqrt{a_{n_m+1}}\rfloor \\
&= (2^{2m-2} + 2^{m-1}) + \lfloor\sqrt{2^{2m-2} + 2^{m-1}}\rfloor \\
&= 2^{2m-2} + 2^{m-1} + 2^{m-1} \\
&= (2^{m-1} + j - 1)^2 + 2^m,
\end{aligned}$$

which implies that base case $j = 1$ holds.

Suppose that we have $a_{n_m+2i+1} = (2^{m-1}+i)^2 + 2^{m-1} - i$ and $a_{n_m+2j} = (2^{m-1}+j-1)^2 + 2^m$ for all $0 \leq i \leq \ell$ and $1 \leq j \leq \ell$, where $0 \leq \ell < 2^{m-1}$. By induction hypothesis, we have $a_{n_m+2\ell+1} = (2^{m-1} + \ell)^2 + 2^{m-1} - \ell$ and $a_{n_m+2\ell} = (2^{m-1} + \ell - 1)^2 + 2^m$ if $\ell > 0$. We will first compute

$a_{n_m+2\ell+2}$. To this end, we first observe that $a_{n_m+2\ell+1} > (2^{m-1} + \ell)^2$, since $\ell < 2^{m-1}$. On the other hand, we have

$$
\begin{aligned}
a_{n_m+2\ell+1} &= (2^{m-1} + \ell)^2 + 2^{m-1} - \ell \\
&< (2^{m-1} + \ell)^2 + 2(2^{m-1} + \ell) + 1 \\
&= (2^{m-1} + \ell + 1)^2.
\end{aligned}
$$

This implies that $\lfloor \sqrt{a_{n_m+2\ell+1}} \rfloor = 2^{m-1} + \ell$, and so we have

$$
\begin{aligned}
a_{n_m+2\ell+2} &= a_{n_m+2\ell+1} + \lfloor \sqrt{a_{n_m+2\ell+1}} \rfloor \\
&= (2^{m-1} + \ell)^2 + 2^{m-1} - \ell + 2^{m-1} + \ell \\
&= (2^{m-1} + (\ell + 1) - 1)^2 + 2^m.
\end{aligned}
$$

Next, we will compute $a_{n_m+2\ell+3}$. To this end, we first observe that $a_{n_m+2\ell+2} > (2^{m-1} + \ell)^2$, and

$$
\begin{aligned}
a_{n_m+2\ell+2} &= (2^{m-1} + \ell)^2 + 2^m \\
&< (2^{m-1} + \ell)^2 + 2(2^{m-1} + \ell) + 1 \\
&= (2^{m-1} + \ell + 1)^2.
\end{aligned}
$$

This implies that $\lfloor \sqrt{a_{n_m+2\ell+2}} \rfloor = 2^{m-1} + \ell$, and so we have

$$
\begin{aligned}
a_{n_m+2\ell+3} &= a_{n_m+2\ell+2} + \lfloor \sqrt{a_{n_m+2\ell+2}} \rfloor \\
&= (2^{m-1} + \ell)^2 + 2^m + 2^{m-1} + \ell \\
&= (2^{m-1} + \ell + 1)^2 - 2(2^{m-1} + \ell) - 1 + 2^m + 2^{m-1} + \ell \\
&= (2^{m-1} + \ell + 1)^2 + 2^{m-1} - \ell - 1.
\end{aligned}
$$

This completes the induction step on both i and j (and hence k), and therefore we have $a_{n_{k-1}+2i+1} = (2^{k-2} + i)^2 + 2^{k-2} - i$ and $a_{n_{k-1}+2j} = (2^{k-2} + j - 1)^2 + 2^{k-1}$ for all $k \geq 2$, $0 \leq i \leq 2^{k-2}$ and $1 \leq j \leq 2^{k-2}$ as claimed. It follows immediately from these equations that $a_{n_{k-1}+2i+1}$ and $a_{n_{k-1}+2j}$ are not perfect squares for all $0 \leq i \leq 2^{k-2}$ and $1 \leq j \leq 2^{k-2}$. Moreover, we have shown that $a_{n_m} = (2^{m-1})^2$ for all $m \geq 1$. Hence, a_n is a perfect square if and only if we have $n = n_m$ for some $m \geq 1$.

Remark. The solution to the above problem is given by R. Sherman. Lehman, University of California, Berkeley. We shall refer the reader to *Amer. Math. Monthly*, **85** (1978), 52–53 for further details.

90. For each positive integer h and s, we let the number of regions formed by the $h + s$ lines be $R(h, s)$. Then we see that conditions (i)-(iii) uniquely determines $R(h, s)$. Let us pick any non-horizontal line ℓ. By

conditions (ii) and (iii), we see that ℓ intersects the other $h + s - 1$ lines in exactly $h + s - 1$ (distinct) points. This implies that the number of new regions created by adding the line ℓ is equal to $h + s$, so we must have $R(h, s) = h + s + R(h, s - 1)$ where $R(h, 0) = h + 1$. As $R(h, 1) = 2h + 2$, it follows that we have for $n \geq 1$,

$$
\begin{aligned}
R(h, s) &= h + s + R(h, s - 1) \\
&= h + s + h + s - 1 + h + s + R(h, s - 2) \\
&= \cdots \\
&= h + s + h + s - 1 + \cdots + h + 2 + R(h, 1) \\
&= h + s + h + s - 1 + \cdots + h + 2 + 2h + 2 \\
&= \frac{(s + 2h)(s + 1) + 2}{2}.
\end{aligned}
$$

Now, if $R(h, s) = 1992$, then we must have $(s + 2h)(s + 1) = 3982$. As $3982 = 2 \cdot 11 \cdot 181$, $s + 2h$ and $s + 1$ have different parities and $s + 2h > s + 1 \geq 2$, we must have $(s + 2h, s + 1) = (1991, 2), (362, 11)$ or $(181, 22)$, which would then give us $(h, s) = (995, 1), (176, 10)$ or $(80, 21)$.

91. We first recall that $S(r, n)$ is equal to the number of ways to distribute r distinct objects into n identical boxes, such that no box is empty. Let us fix an object B, and let us suppose that there are $j + 1$ objects in the box that contains B, where $j = 0, \ldots, r - 1$. Since no box is empty, the largest possible value of j is $(r - 1) - (n - 1) = r - n$. This restriction does not matter in the subsequent discussion since $S(r - j - 1, n - 1) = 0$ if $j > r - n$. There are $\binom{r-1}{j}$ ways to choose j objects out of the remaining $r - 1$ distinct objects to be placed in the same box as B, and subsequently there are $S(r - j - 1, n - 1)$ ways to distribute the remaining $r - j - 1$ distinct objects into the remaining $n - 1$ distinct boxes, so that no box is empty. By (AP) and (MP), we must have

$$
\begin{aligned}
S(r, n) &= \sum_{j=0}^{r-1} \binom{r - 1}{j} S(r - j - 1, n - 1) \\
&= \sum_{j=0}^{r-n} \binom{r - 1}{r - j - 1} S(r - j - 1, n - 1) \\
&= \sum_{k=n-1}^{r-1} \binom{r - 1}{k} S(k, n - 1) \quad \text{(by setting } k = r - j - 1\text{)}.
\end{aligned}
$$

92. We have

$$
\begin{aligned}
B_r &= \sum_{n=1}^{r} S(r,n) \\
&= \sum_{n=1}^{r} \sum_{k=n-1}^{r-1} \binom{r-1}{k} S(k, n-1) \quad \text{(by Problem 6.91)} \\
&= \sum_{k=0}^{r-1} \sum_{n=1}^{k+1} \binom{r-1}{k} S(k, n-1) \\
&= \sum_{k=0}^{r-1} \binom{r-1}{k} \sum_{n=0}^{k} S(k, n) \\
&= \sum_{k=0}^{r-1} \binom{r-1}{k} \sum_{n=1}^{k} S(k, n) \\
&= \sum_{k=0}^{r-1} \binom{r-1}{k} B_k.
\end{aligned}
$$

93. We shall first proceed by induction on m that we have $P(m,n) = \binom{n+m-1}{m-1}$ for all $m \geq 1$ and $n \geq 0$. When $m = 1$, we have

$$
P(1, n) = \sum_{j=0}^{n} P(0, j) = P(0,0) + \sum_{j=1}^{n} P(0, j) = 1 = \binom{n+m-1}{m-1}
$$

for all $n \geq 0$. So this shows that the base case $m = 1$ holds. Suppose that the statement holds for $m = k$, where $k \geq 1$. By induction hypothesis, we have $P(k, j) = \binom{j+k-1}{k-1}$ for all $j \geq 0$. This implies that we have

$$
P(k+1, n) = \sum_{j=0}^{n} P(k, j) = \sum_{j=0}^{n} \binom{j+k-1}{k-1} = \binom{n+k}{k},
$$

where the last equality follows from (2.5.1). This completes the induction step, and hence we have $P(m,n) = \binom{n+m-1}{m-1}$ for all $m \geq 1$ and $n \geq 0$ as claimed. As $P(m,n) = 0 = \binom{n+m-1}{m-1}$ for all $m \geq 1$ and $n < 0$, it follows that we have $P(m,n) = \binom{n+m-1}{m-1}$ for all $m \geq 1$. Thus we

have

$$Q(m, n)$$

$$= P(m-1, n) + P(m-1, n-1) + P(m-1, n-2)$$

$$= \binom{n+m-2}{m-2} + \binom{n+m-3}{m-2} + \binom{n+m-4}{m-2}$$

$$= \binom{n+m-2}{m-2} + \binom{n+m-3}{m-2} + \binom{n+m-3}{m-1} - \binom{n+m-4}{m-1}$$

$$= \binom{n+m-2}{m-2} + \binom{n+m-2}{m-1} - \binom{n+m-4}{m-1}$$

$$= \binom{n+m-1}{m-1} - \binom{n+m-4}{m-1},$$

where the last three equalities follow from (2.1.4).

94. (i) We have

$$S_k(n) = \sum_{j=1}^{n} j^k$$

$$= 1 + \sum_{j=1}^{n-1} (j+1)^k$$

$$= n^{k+1} + \sum_{j=1}^{n-1} j^{k+1} - \sum_{j=1}^{n-1} (j+1)^{k+1} + \sum_{j=1}^{n-1} (j+1)^k$$

$$= n^{k+1} - \sum_{j=1}^{n-1} [(j+1)^{k+1} - (j+1)^k - j^{k+1}]$$

$$= n^{k+1} - \sum_{j=1}^{n-1} \left[j(j+1)^k - \binom{k}{k} j^{k+1} \right]$$

$$= n^{k+1} - \sum_{j=1}^{n-1} \left[\sum_{r=0}^{k} \binom{k}{r} j^{r+1} - \binom{k}{k} j^{k+1} \right]$$

$$= n^{k+1} - \sum_{j=1}^{n-1} \sum_{r=0}^{k-1} \binom{k}{r} j^{r+1}$$

$$= n^{k+1} - \sum_{r=0}^{k-1} \binom{k}{r} \sum_{j=1}^{n-1} j^{r+1}$$

$$= n^{k+1} - \sum_{r=0}^{k-1} \binom{k}{r} S_{r+1}(n-1).$$

(ii) By part (i), we have

$$S_k(n+1) = (n+1)^{k+1} - \sum_{r=0}^{k-1} \binom{k}{r} S_{r+1}(n)$$

$$= (n+1)^{k+1} - \sum_{r=0}^{k-2} \binom{k}{r} S_{r+1}(n) - kS_k(n). \qquad (6.4)$$

Next, it follows from the definition that we have

$$S_k(n+1) = S_k(n) + (n+1)^k.$$

Combining the previous equation with equation (6.4) gives us

$$(n+1)^{k+1} - \sum_{r=0}^{k-2} \binom{k}{r} S_{r+1}(n) - kS_k(n) = S_k(n) + (n+1)^k,$$

or equivalently,

$$(k+1)S_k(n) = (n+1)^{k+1} - (n+1)^k - \sum_{r=0}^{k-2} \binom{k}{r} S_{r+1}(n)$$

as required.

Printed in the United States
By Bookmasters